Operator Theory
Advances and Applications
Vol. 108

Editor:
I. Gohberg

Mathematical Results in Quantum Mechanics

QMath7 Conference, Prague, June 22–26, 1998

Jaroslav Dittrich
Pavel Exner
Miloš Tater
Editors

Springer Basel AG

Editors:
Jaroslav Dittrich
Pavel Exner
Miloš Tater
Department of Theoretical Physics
Nuclear Physics Institute
Academy of Science
25068 Rez near Prague
Czech Republic

1991 Mathematics Subject Classification 81-06

A CIP catalogue record for this book is available from the
Library of Congress, Washington D.C., USA

Deutsche Bibliothek Cataloging-in-Publication Data

Mathematical results in quantum mechanics / QMath7 Conference, Prague,
June 22–26, 1998. Jaroslav Dittrich ... ed. – Basel ; Boston ; Berlin : Birkhäuser, 1999
(Operator theory ; Vol. 108)
ISBN 978-3-0348-9754-9 ISBN 978-3-0348-8745-8 (eBook)
DOI 10.1007/978-3-0348-8745-8

Originally published by Birkhäuser Verlag in 1999
Softcover reprint of the hardcover 1st edition 1999
Printed on acid-free paper produced from chlorine-free pulp. TCF ∞
Cover design: Heinz Hiltbrunner, Basel

ISBN 978-3-0348-9754-9

9 8 7 6 5 4 3 2 1

Contents

Preface

At the age of almost three quarters of a century, quantum mechanics is by all accounts a mature theory. There were times when it seemed that it had borne its best fruit already and would give way to investigation of deeper levels of matter. Today this sounds like rash thinking. Modern experimental techniques have led to discoveries of numerous new quantum effects in solid state, optics and elsewhere. Quantum mechanics is thus gradually becoming a basis for many branches of applied physics, in this way entering our everyday life.

While the dynamic laws of quantum mechanics are well known, a proper theoretical understanding requires methods which would allow us to derive the abundance of observed quantum effects from the first principles. In many cases the rich structure hidden in the Schrödinger equation can be revealed only using sophisticated tools. This constitutes a motivation to investigate rigorous methods which yield mathematically well-founded properties of quantum systems.

Among the traditional meetings where such problems are discussed there was a conference series launched in 1987 at Dubna, Russia. The present volume collects some of the contributions presented at the seventh gathering of this series, so far the largest, held on June 22–26, 1998, at the Czech Technical University in Prague. The coverage is by far not exhaustive because of the conference size, the natural volume restriction, the scope of the series *Operator Theory: Advances and Applications*, and the decision of some participants not to write again about results they have published recently. These and other reasons have led to the fact that this book is just a selection representing less than one-third of the talks. It is clear that many interesting results had to be left out. To give the reader a better idea, we include at the end the titles of the other contributions together with the list of participants; it is easy in this wired world to ask anybody for a copy of a paper.

We want to thank all the participants who made the QMath7 conference interesting and full of inspiration indeed. We are no less grateful to the institutions which provided financial support: the Ministry of Education of the Czech Republic, the Physics Section of the Union of Czech Mathematicians and Physicists, and the Foundation for Support of Theoretical Physics at Slemeno. Last but not least we thank the referees who reviewed the contributions rapidly but attentively; their remarks and suggestions helped to improve the book.

Prague, November 1998

The editors

Part I

Plenary talks

Operator Theory:
Advances and Applications, Vol. 108

An Adiabatic Theorem without a Gap Condition

J. E. Avron and A. Elgart
Department of Physics, Technion, 32000 Haifa, Israel

Abstract

The basic adiabatic theorems of classical and quantum mechanics are overviewed and an adiabatic theorem in quantum mechanics without a gap condition is described.

1 Classical Adiabatic Invariants

Consider a (mathematical) pendulum whose period is slowly modulated, for example by shortening the length of the pendulum, fig 1, [1, 18]. The Hamiltonian

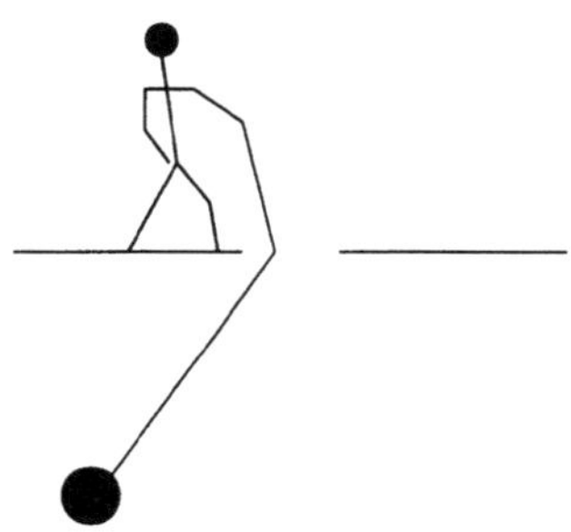

Figure 1: An adiabatic pendulum

describing the system is

$$H(s) = \frac{1}{2}\left(p^2 + \omega^2(s)\,x^2\right), \quad s = \frac{t}{\tau}. \tag{1}$$

t is the physical time, τ is the time scale. The adiabatic limit is $\omega\tau >> 1$. The period $\omega(s)$ is a smooth function which is time independent in the past, $s < 0$, and in the distant future, $s > 1$. A graph showing a possible variation of $\omega(s)$ is shown in fig. 2.

An adiabatic invariant is an approximately conserved quantity whose deviation from constant can be made arbitrarily small for large τ, uniformly in s and

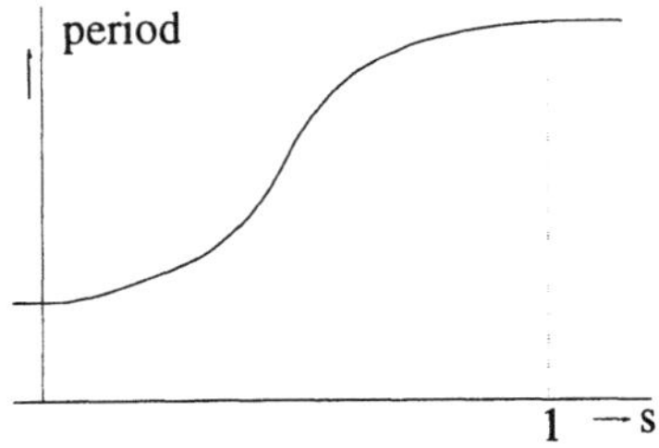

Figure 2: An adiabatic variation

for all times. For the Harmonic oscillator the adiabatic invariant is

$$S(s) = \frac{H(s)}{\omega(s)}. \tag{2}$$

The special properties of this particular combination of H and ω can be seen from its equations of motion:

$$\dot{S}(s) = \frac{\dot{\omega}(s)}{2\,\omega^2(s)}\,\left(\omega^2(s)\,x^2 - p^2\right). \tag{3}$$

$\dot{S}$ is compactly supported (because $\dot{\omega}$ is), and appears to be $O(1)$ in τ. But, for the (time independent) Harmonic oscillator the time average over one period of the kinetic energy equals the time average of the potential energy. So, for large τ, the change of S in one period is small: $\langle \Delta S\rangle = O\left(\frac{1}{\tau}\right)$. Because of this adiabatic invariants give precise information on the long time behavior even though the total variation in the Hamiltonian is finite.

A remarkable fact about adiabatic invariants is that *for large times* the error is essentially exponentially small with τ if $\omega(s)$ is smooth [18]:

$$|S(s) - S(0)| = O\left(\frac{1}{\tau^\infty}\right), \quad s > 1. \tag{4}$$

(The error is, in general, not exponentially small for $0 < s < 1$.) In certain circles an exponentially small error is sometimes taken to be the defining property of adiabatic invariant, so that proving an adiabatic theorem is taken to imply proving an exponentially small bound on the error. This, to our opinion, is not a satisfactory definition of the notion of adiabatic invariant, and instead we shall stick with the definition given above, namely, that adiabatic invariants are conserved up to an error that is uniformly bounded for all times, and can be made arbitrarily small with τ.

A link of classical adiabatic invariants with quantum mechanics that was emphasized by Ehrenfest [10] focused on the observation that adiabatic invariants are related to quantum numbers. For the (time independent) Harmonic oscillator the particular combination of H and ω in Eq. (2) is a function of quantum numbers:

$$\frac{E}{\omega} = \hbar\left(n + \frac{1}{2}\right). \tag{5}$$

2 The Quantum Adiabatic Theorem

Ehrenfest observation had much influence in the early days of quantum mechanics, and in particular motivated the work of Born and Fock [8] on the adiabatic theorem of quantum mechanics. In quantum theory one is interested is solving the initial value problem

$$i\,\partial_t \psi = H(s)\,\psi, \quad s = \frac{t}{\tau}, \tag{6}$$

with ψ a vector in Hilbert space and $H\left(\frac{t}{\tau}\right)$ a self adjoint operator. We shall assume, as we did in the previous section, that $H(s)$ is time independent in the past, $s < 0$, and distant future, $s > 1$, and is a smooth operator valued function of s. In the case that $H(s)$ is an unbounded operator, like the Schrödinger operator, the notion of smoothness needs some clarification. We shall not get into this here.

Changing variables one writes the initial value problem as

$$i\,\dot{\psi} = \tau\,H(s)\,\psi. \tag{7}$$

The adiabatic limit is $\tau \to \infty$. Adiabatic theorems in quantum mechanics relate the solutions of the initial value problem to spectral properties.

The oldest result of this kind is due to Born and Fock who studied Hamiltonians with *discrete and simple* spectrum, fig. 3.

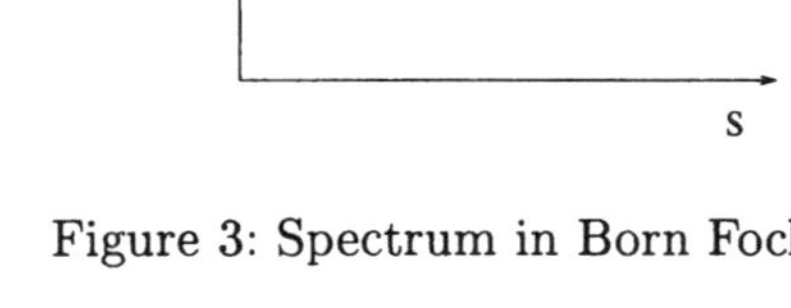

Figure 3: Spectrum in Born Fock Theory

Born and Fock showed that if the initial data are $\psi(0) = \phi(0)$, with $\phi(0)$ an eigenvector of $H(0)$, then $\psi(s)$ is close to an eigenvector $\phi(s)$ of $H(s)$ with particular choice of phase:

$$||\psi(s) - \phi(s)|| = O\left(\frac{1}{\tau}\right). \tag{8}$$

For large times, $s > 1$, outside the support of $\dot{H}(s)$, much stronger result hold: the error is essentially exponentially small in τ, see e.g. [6, 15, 17, 21].

3 The Adiabatic Theorem of Kato

Kato generalized the result of Born and Fock. He showed that the assumption of spectral simplicity of $H(s)$ can be removed, and so can the assumption that the spectrum is discrete, fig. 4. These generalizations are important for applications to atomic physics where some continuous spectrum is always present, and degeneracies are ubiquitous. But, perhaps more importantly, Kato introduced an essentially new method of proving the adiabatic theorem that we shall now describe. Kato's

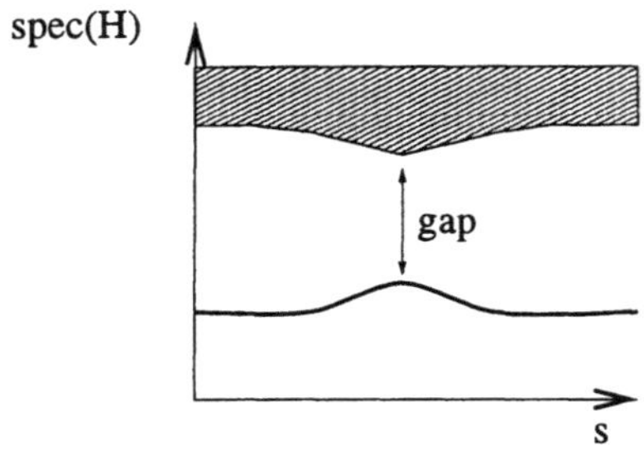

Figure 4: Spectrum in Kato's Theory

idea was to introduce a geometric evolution which satisfies the adiabatic theorem without an error. That is, a unitary $U_a(s)$, so that:

$$U_a(s)\, P(0) = P(s)\, U_a(s), \tag{9}$$

where $P(s)$ is a spectral projection for $H(s)$, and $U_a(0) = 1$. The second step is to compare the physical evolution, $U(s)$, generated by

$$i\, \dot{U}(s) = \tau\, H(s)\, U(s), \quad U(0) = 1. \tag{10}$$

with U_a and show that the two are close.

It turns out that both steps involve looking into commutator equations. If we let $H_a(s)$ denote the generator of the geometric evolution $U_a(s)$, it is not difficult to see that it must satisfy

$$\tau\, [H_a(s), P(s)] = i\, \dot{P}(s). \tag{11}$$

Using the fact that for any projection P

$$P(s)\, \dot{P}(s)\, P(s) = 0, \tag{12}$$

one checks that

$$H_a(s) = H(s) + \frac{i}{\tau}\,[\dot{P}(s), P(s)], \tag{13}$$

solves the commutator equation, with $H_a(s)$ which is manifestly close to $H(s)$.

To compare $U(s)$ and $U_a(s)$ let $\Omega(s) = U_a^*(s)\,U(s)$, $\Omega(0) = 1$. Using the equation of motion one finds

$$\dot{\Omega}(s) = i\,\tau\,U_a^*(s)\,(H_a(s) - H(s))\;U(s) = -U_a^*(s)[\dot{P}(s), P(s)]\,U(s), \tag{14}$$

which is compactly supported (since $\dot{P}(s)$ is) and $O(1)$ in τ. Now, like the situation for the classical adiabatic invariants, even though $\dot{\Omega}$ is not small, the change in Ω is small. This is where a second commutator equation enters. Suppose that the commutator equation

$$[H(s), X(s)] = [\dot{P}(s), P(s)] \tag{15}$$

has a smooth and bounded solution $X(s)$. Then,

$$\begin{aligned}
-\dot{\Omega}(s) &= U_a^*(s)[H(s), X(s)]\,U(s) = \\
&= U_a^*(s)\,(H_a(s)\,X(s) - X(s)\,H(s))\;U(s) + O\left(\frac{1}{\tau}\right) \\
&= \frac{i}{\tau}\left(\dot{U}_a^*(s)\,X(s)\,U(s) + U_a^*(s)\,X(s)\,\dot{U}(s)\right) + O\left(\frac{1}{\tau}\right) \\
&= \frac{i}{\tau}\left(\left(U_a^*(s)\,X(s)\,U(s)\right)^{\cdot} - U_a^*(s)\,\dot{X}(s)\,U(s)\right) + O\left(\frac{1}{\tau}\right).
\end{aligned} \tag{16}$$

From this it follows that $\Omega(s) - 1 = O\left(\frac{1}{\tau}\right)$.

The gap condition is a condition for the solvability of the commutator equation, Eq. (15). Indeed, suppose there is a gap in the spectrum so that the spectral projection P is associated with a contour Γ in the complex plane that lies entirely in the resolvent set, Fig. 5.

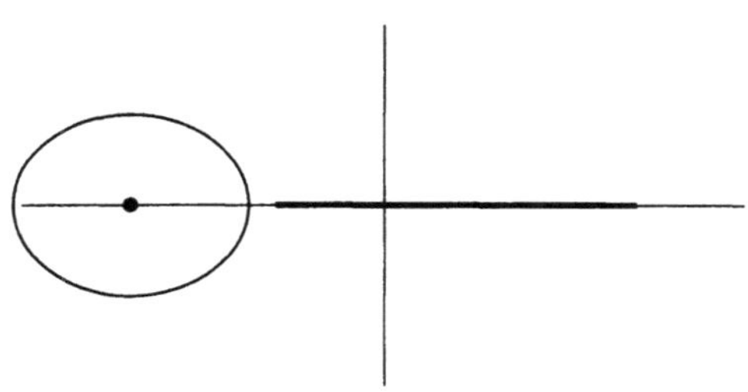

Figure 5: A contour Γ in the Complex Plane

A solution to the commutator equation is

$$X(s) = \frac{1}{2\pi i}\int_\Gamma R(s,z)\,\dot{P}(s)\,R(s,z)\,dz. \tag{17}$$

And, as usual $R(s,z) = (H(s) - z)^{-1}$ is bounded for $z \in \Gamma$. If the size of the gap is g then

$$\|X(s)\| = O\left(\frac{1}{g}\right). \tag{18}$$

Using Kato's method various adiabatic theorems have been proven see e.g. [3, 4, 17, 21].

4 The Role of the Gap Condition

The adiabatic theorem described in the previous section relied on a gap condition. How serious is this?

In the proof of Kato the gap condition guarantees the existence of a bounded solution to the commutator equation given by $X(s)$ of Eq. (17). The bound Eq.(18) blows up at the gap shrinks to zero and there is no *a-priori* bounded solution to the commutator equation. This suggests that the gap condition is essential.

A second argument leading to the same conclusion is a dimensional argument. The adiabatic limit needs a intrinsic time scale so that τ can be measured in dimensionless units. Otherwise the notion of large τ depends on a choice of a unit and is meaningless. In the case of the classical Harmonic oscillator the intrinsic time scale is set by ω. In the quantum case, a gap and Planck constant dictates an intrinsic time scale. In the absence of a gap, this time scale is lost. This suggests that the gap condition is essential and there should be no *general* adiabatic theorem in its absence.

Let us now describe two arguments that say the opposite. The first refers once again to the work of Born and Fock. Born and Fock (and also Kato) considered the more delicate adiabatic theorem for crossing energy levels, Fig. 6, and proved an adiabatic theorem in this case.

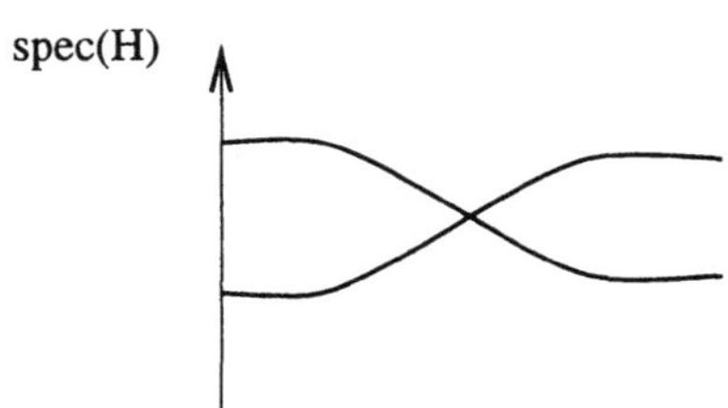

Figure 6: Crossing Eigenvalues in Born Fock Theory

Since the energy levels cross, the gap closes. For the case of linear crossing

Born and Fock showed that Eq. (8) is replaced by

$$\|\psi(s) - \phi(s)\| = O\left(\frac{1}{\sqrt{\tau}}\right). \tag{19}$$

The time scale for this problem is dictated by the slope of the energy curves at the crossing point. These results have since been considerably strengthened and extended [12]. This suggests that a gap conditions controls the rate at which the adiabatic limit is approached, but an adiabatic theorem does not really require a gap condition.

A second argument supporting the view that a gap condition is only technical is a physical argument. Gaps in the spectrum are indeed prevalent in quantum mechanical systems, but there are no gaps in quantum electrodynamics: The interaction with radiation eliminates the gaps. Suppose that a charged quantum mechanical system, initially at the ground state, is slowly rotated. The adiabatic theorem would fail if the number of photons generated by the slow rotation does not go to zero in the adiabatic limit. Let us estimate this number [1]. The power radiated by a charged system in classical electrodynamics is proportional to the acceleration squared, i.e. to τ^{-4}. Hence the total radiated energy is of the order τ^{-3}. Since a typical radiated photon will, presumably, have frequency of order $\frac{1}{\tau}$ the number of radiated photons is of order τ^{-2}. This goes to zero in the adiabatic limit. This argument, in spite of its shortcomings, suggests that the gap condition, at least in the context of QED, is not really essential.

5 Removing the Gap Condition

A general adiabatic theorem without a gap condition was given in [2]. The point is that all the adiabatic theorem really needs is a distinguished smooth family of finite dimensional spectral projections, so that the adiabatic evolution has a distinguished subspace to follow. The proof works for eigenvalues embedded in some essential spectrum, or for eigenvalues at the threshold of essential spectrum, as one would expect to find in QED, fig. 7. It is essential for this result that the distinguished spectral subspace is *finite dimensional*. Let us begin by stating the theorem:

Theorem: *Suppose that $P(s)$ is a finite rank spectral projection, which is at least twice differentiable (as a bounded operator), for the self-adjoint Hamiltonian $H(s)$, which is bounded and differentiable for all $s \in [0,1]$. Then, the evolution of the initial state $\psi(0) \in RangeP(0)$, according to Eq. (6), is such that in the adiabatic limit $\psi(s) \in RangeP(s)$ for all s.*

Remarks: The theorem is stated for bounded self adjoint operators $H(s)$. As it stands it does not apply to Schrödinger operator. The extension to unbounded

[1] We owe this argument to A. Ori.

operators is a technical problem which can be handled by known functional analytic methods. We choose not to phrase the result for the general case for several reasons. The first is that the technical issues will obscure the basic idea which is simple. The second is that the essence of the adiabatic theorem is an infrared problem. The unboundedness of Schrödinger operators is an ultraviolet problem. It is a conceptual advantage to keep the two issues separate.

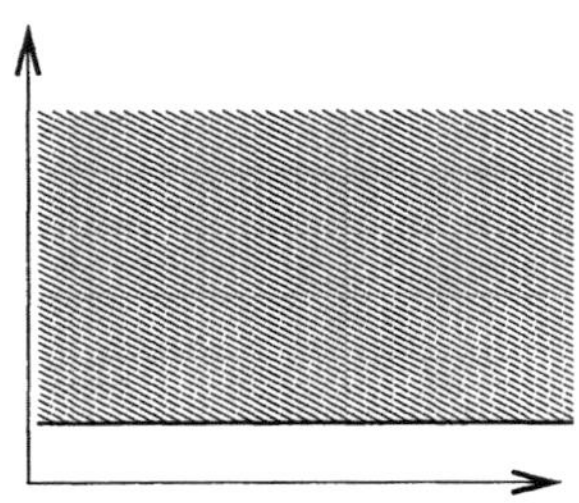

Figure 7: An Eigenvalue at Threshold

The basic idea is to replace Kato's commutator equation, Eq. (15), by a definition of a new quantity $Y(s)$:

$$[H(s), X(s)] = [\dot{P}(s), P(s)] + Y(s), \tag{20}$$

and take $X(s)$ to be

$$X_\Delta(s) = \frac{1}{2\pi i}\int_\Gamma dz\,(1 - F_\Delta(s))\,R(z,s)\,\dot{P}(s)R(z,s)\,(1 - F_\Delta(s)). \tag{21}$$

where $F_\Delta(s)$ is an approximate characteristic function of $H(s)$, which is Δ localized near the relevant eigenvalue, whose range is in *Range* $P_\perp(s)$. $X_\Delta(s)$ is bounded, by construction, for $\Delta > 0$ and its norm diverges as $\Delta \to 0$. At the same time, and this is the crucial point, $||Y_\Delta(s)|| \to 0$ provided $P(s)$ is finite dimensional. Chasing the argument of Kato one then finds that the adiabatic theorem holds, and the price one has to pay for the absence of a gap is the loss of control on the rate at which the adiabatic limit is approached. Instead of Eq. (8) on gets

$$||\psi(s) - \phi(s)|| = o(1)\,. \tag{22}$$

That is, the error can be made arbitrarily small with τ, but the rate is undetermined.

We conclude with an interpretation of the result. For an isolated eigenvalue the gap in the spectrum protects against tunneling out of the spectral subspace. In the case that the eigenvalue in question is embedded in essential spectrum there is no gap to protect against tunneling out. But, since the essential spectrum is associated with eigenfunctions supported near infinity, there is small overlap with the eigenfunction in question, and the protection against tunneling comes from this fact.

6 What Has Been Left Out

Adiabatic theorems of classical and quantum mechanics are a developed subject with rich and fertile history. In this short overview, based an a talk by one of us, we reviewed a small corner of this field, the one close to its foundations and characterized by elementary results. There are many beautiful and sophisticated results that we did not have the opportunity to review. These include: Classical adiabatic invariants for integrable systems to all orders [1, 18, 19]; Adiabatic invariants for chaotic systems [22, 7, 14]; Quantum adiabatic theorems to all orders [6, 17, 21, 15]; Landau-Zener formulas [13, 15]; Adiabatic invariants in scattering theory[20]; Adiabatic invariants in C^* algebras and models of quantum fields [9] and geometry and adiabatic curvature [5, 11].

Acknowledgments

This work was partially supported by a grant from the Israel Academy of Sciences, the Deutsche Forschungsgemeinschaft, and by the Fund for Promotion of Research at the Technion.

References

[1] V. Arnold, *Mathematical Methods of Classical Mechanics*, Springer (1980).

[2] J. E. Avron and A. Elgart, preprint, http://xxx.lanl.gov/abs/math-ph/9805022

[3] J. E. Avron, J. S. Howland and B. Simon, *Adiabatic theorems for dense point spectra, Comm. Math. Phys.* **128** (1990), 497–507.

[4] J. E. Avron, R. Seiler and L. G. Yaffe, *Adiabatic theorems and applications to the quantum Hall effect*, Comm. Math. Phys. **110** (1987), 33–49, (Erratum: Comm. Math. Phys. **153** (1993), 649-650).

[5] M.V. Berry, Proc. Roy. Soc. Lond. A **392**, 45, (1984); The quantum phase: Five years after, in *Geometric phases in physics* (A. Shapere and F. Wilczek, Eds., World Scientific, 1989).

[6] M.V. Berry, *Histories of adiabatic transition*, Proc. Roy. Soc. Lond. A **429**, 61-72, (1990).

[7] M.V. Berry and J.M. Robbins, *Chaotic classical and half classical adiabatic reactions: Geometric magnetism and deterministic friction*, Proc. Roy. Soc. Lond. A **442**, 659-672, (1993). Proc. Roy. Soc. A **392**, 45 (1984).

[8] M. Born and V. Fock, *Beweis des Adiabatensatzes*, Z. Phys. **51** (1928), 165–169.

[9] E. B. Davies and H. Spohn, *Open quantum systems with time-dependent Hamiltonians and their linear response*, J. Stat. Phys. **19**, 511, (1978).

[10] P. Ehrenfest, *Adiabatische Invarianten u. Quantentheorie*, Ann. d. Phys. **51**, 327 (1916)

[11] S. Golin, A. Knauf and S. Marmi, *The Hannay angles: Geometry, Adiabaticity and an example*, Comm. Math. Phys. **123**, 95-122, (1989).

[12] G. Hagedorn, *Adiabatic Expansions near Eigenvalue Crossings*, Ann.Phys. **196**, 278-295 ,(1989).

[13] V. Jakšić and J. Segert, *On the Landau Zener formula for two-level systems*, J. Math. Phys. **34**, 2807-2820, (1993).

[14] C. Jarzinski, *Multiple-time-scale approach to ergodic adiabatic systems: Another look*, Phys. Rev. Lett. **71**, 839, (1993).

[15] A. Joye and C.E. Pfister, *Exponential Estimates in Adiabatic Quantum Evolution*, Proceeding of the XII ICMP, Brisbane Australia (1997); *Quantum Adiabatic Evolution*, in *On Three Levels*, M. Fannes, C. Maes and A. Verbure Editors, Plenum, (1994).

[16] T. Kato, *On the adiabatic theorem of quantum mechanics*, Phys. Soc. Jap. **5** (1958), 435–439.

[17] M. Klein and R. Seiler, *Power law corrections to the Kubo formula vanish in quantum Hall systems*, Comm. Math. Phys. **128** (1990), 141.

[18] A. Lennard, *Adiabatic Invariance to All Orders*, Ann. Phys. **6**, (1959), 261-276.

[19] P. Lochak and C. Meunier, *Multiphase Averaging for Classical systems*, Springer, (1988).

[20] H. Narnhofer and W. Thirring, *Adiabatic theorem in quantum statistical mechanics*, Phys. Rev. A **26**, 3646, (1982).

[21] G. Nenciu, *Linear Adiabatic Theory: Exponential Estimates*, Comm. Math. Phys. **152**, 479-496, (1993)

[22] E. Ott, *Goodness of ergodic adiabatic invariants*, Phys. Rev. Lett. **42** (1979), 1628-1631.

Operator Theory:
Advances and Applications, Vol. 108

TWO-DIMENSIONAL PERIODIC PAULI OPERATOR. THE EFFECTIVE MASSES AT THE LOWER EDGE OF THE SPECTRUM

M.SH.BIRMAN, T.A.SUSLINA

St.Petersburg State University
Department of physics, Division of mathematical physics
St.Petersburg, 198904, Petrodvorets, Ul'yanovskaya 1

ABSTRACT. We calculate the tensor of effective masses for the two-dimensional periodic Pauli operator. The explicit representation for this tensor is given in terms of the magnetic field. It is proved that the tensor of effective masses is circular symmetric and always proportional to the unit matrix. We also consider the generalized Pauli operator with a variable metric. In the appendix we study the periodic elliptic operators of the second order and discuss the behaviour of the first band function near its minimum point.

§0. INTRODUCTION

1. Let $\mathbf{A} = \{A_1, A_2\}$ be a *periodic* magnetic potential, and let

$$B = \partial_1 A_2 - \partial_2 A_1 \tag{0.1}$$

be the corresponding magnetic field. We use the notation $\nabla = \{\partial_1, \partial_2\}$, $\mathbf{D} = \{D_1, D_2\} = -i\nabla$, $\nabla^* = -\mathrm{div}$. As usual,

$$\sigma_1 = \begin{pmatrix} 0 & 1 \\ 1 & 0 \end{pmatrix}, \quad \sigma_2 = \begin{pmatrix} 0 & -i \\ i & 0 \end{pmatrix}, \quad \sigma_3 = \begin{pmatrix} 1 & 0 \\ 0 & -1 \end{pmatrix}$$

are the Pauli matrices, $\mathbf{1}$ is the unit matrix. By Ω we denote the elementary cell of the period lattice Γ, and by $|\Omega|$ is denoted the area of Ω. It is convenient to consider simultaneously the pair of Pauli operators

$$P_\pm = (\mathbf{D} - \mathbf{A})^2 \pm B \tag{0.2}$$

Key words and phrases: Periodic operator, Pauli operator, Schrödinger operator, effective masses.

in $L_2(\mathbb{R}^2)$. The operators (0.2) are the blocks of the "vector" Pauli operator

$$\mathcal{P} := ((D_1 - A_1)\sigma_1 + (D_2 - A_2)\sigma_2)^2 = \begin{pmatrix} P_- & 0 \\ 0 & P_+ \end{pmatrix}. \tag{0.3}$$

By using standard arguments the periodic operators (0.2) are decomposed into direct integral of operators $P_\pm(\mathbf{k})$ acting in $L_2(\Omega)$. The operators

$$P_\pm(\mathbf{k}) = (\mathbf{D} + \mathbf{k} - \mathbf{A})^2 \pm B, \quad \mathbf{k} \in \mathbb{R}^2, \tag{0.4}$$

act in $L_2(\Omega)$ and are defined on periodic functions. The values of $\mathbf{k} \in \mathbb{R}^2$ (called *quasi-momentum*) are identified modulo the dual lattice. The operators (0.4) have discrete spectrum. Their eigenvalues

$$E_\pm(\mathbf{k}) = E_\pm^{(1)}(\mathbf{k}) \leqslant E_\pm^{(2)}(\mathbf{k}) \leqslant \dots \leqslant E_\pm^{(n)}(\mathbf{k}) \leqslant \dots$$

are called *band functions* for the operators (0.2). The spectrum of $P_\pm$ is the union of intervals (*bands*) equal to the ranges of the band functions. The intervals lying between the spectral bands (if they exist) are called gaps.

2. The behaviour of the band functions near the edges of the bands is of great interest for periodic problems. Almost nothing is known about this behaviour for multi-dimensional problems (dimension $d \geqslant 2$). The periodic Schrödinger operator

$$H = -\Delta + V(\mathbf{x}), \quad \mathbf{x} \in \mathbb{R}^d, \quad d \geqslant 2, \tag{0.5}$$

is an exception. For this operator the behaviour of the first band function $E(\mathbf{k})$ near the lower edge of the spectrum is known. In particular (see [KSi] and §4 below), *the function $E(\mathbf{k})$ has a non-degenerate minimum E_0, which is reached only at point $\mathbf{k} = 0$, and the next band function is strictly greater than E_0*. In such a situation we say that the lower edge of the spectrum of a periodic problem is *regular*. Without loss of generality we can assume that $E_0 = 0$. Thus, $E_0 = 0$ is a simple eigenvalue of the operator $H(0) = -\Delta + V(\mathbf{x})$, acting in $L_2(\Omega)$ on periodic functions. For the operators $H(\mathbf{k}) = (\mathbf{D} + \mathbf{k})^2 + V(\mathbf{x}), \ \mathbf{k} \neq 0$, the first eigenvalue $E(\mathbf{k})$ is positive. For the operator (0.5) we have

$$E(\mathbf{k}; H) = E(\mathbf{k}) = \gamma(\mathbf{k}) + O(|\mathbf{k}|^3), \quad \mathbf{k} \to 0,$$

where $\gamma(\mathbf{k})$ is *a positively definite quadratic form*. The value

$$\mu(\mathbf{k}_0) = (2\gamma(\mathbf{k}_0))^{-1}, \quad |\mathbf{k}_0| = 1,$$

is naturally called *the effective mass in the direction* $\mathbf{k}_0$ for the operator (0.5) at the point 0. When the operator (0.5) is perturbed by a decaying potential, there arises a number of questions which can be answered in terms of the form γ.

Already for the magnetic Schrödinger operator $M = (\mathbf{D} - \mathbf{A})^2 + V$ the behaviour of the first band function near its minimum point is not studied. An exception is the two-dimensional Pauli operator (0.2). For this operator we show that the lower edge of the spectrum (point 0) is *regular* and that the forms $\gamma(\mathbf{k}; P_\pm)$ admit convenient representations. Similar results will be obtained for some natural generalizations of the operators (0.2).

3. We use a well-known (cf., for example, [BSu1]) factorisation for the operators (0.2). By an appropriate gauge transformation, the potential $\mathbf{A}$ can be subject to the conditions

$$\partial_1 A_1 + \partial_2 A_2 = 0, \quad \int_\Omega \mathbf{A}\, d\mathbf{x} = 0. \tag{0.6}$$

Then there exists a real *periodic* function φ such that

$$\nabla\varphi = \{A_2, -A_1\}, \quad \int_\Omega \varphi\, d\mathbf{x} = 0. \tag{0.7}$$

For a given $\mathbf{A}$, φ can be found by quadratures or via Fourier series. Notice that, by (0.1), (0.7), we have

$$B = \Delta\varphi. \tag{0.8}$$

We introduce the notation

$$\omega_\pm = \exp(\pm\varphi), \tag{0.9}$$

$$Q_\pm(\mathbf{k}) = D_1 + k_1 \pm i(D_2 + k_2), \quad \mathbf{k} \in \mathbb{R}^2.$$

Then (0.3) easily implies that the operators (0.4) admit the representation

$$P_\pm(\mathbf{k}) = \omega_\pm^{-1} Q_\pm(\mathbf{k}) \omega_\pm^2 Q_\mp(\mathbf{k}) \omega_\pm^{-1}, \quad \mathbf{k} \in \mathbb{R}^2, \tag{0.10}$$

which can be rewritten in the form

$$P_\pm(\mathbf{k}) = \omega_\pm^{-1}(\nabla + i\mathbf{k})^*(\mathbf{1} \pm \sigma_2)\omega_\pm^2(\nabla + i\mathbf{k})\omega_\pm^{-1}. \tag{0.11}$$

Finally, using the notation $R_\pm(\mathbf{k}) = \omega_\pm Q_\mp(\mathbf{k})\omega_\pm^{-1}$, we have

$$P_\pm(\mathbf{k}) = R_\pm^*(\mathbf{k}) R_\pm(\mathbf{k}), \quad P_\mp(\mathbf{k}) = R_\pm(\mathbf{k}) R_\pm(\mathbf{k})^*. \tag{0.12}$$

Any of the relations (0.10)–(0.12) shows that

$$P_\pm(0)\omega_\pm = 0. \tag{0.13}$$

Moreover, the functions $\omega_\pm$ are unique solutions of the equations (0.13) up to constant multipliers. In addition,

$$\operatorname{Ker} P_\pm(\mathbf{k}) = \{0\}, \quad \mathbf{k} \neq 0. \tag{0.14}$$

All these facts show that the operators $P_+(\mathbf{k})$ and $P_-(\mathbf{k})$ are unitarily equivalent to each other for all $\mathbf{k} \in \mathbb{R}^2$. Therefore, the operators P_+ and P_- are unitarily equivalent and their band functions coincide. In particular, $E_+(\mathbf{k}) = E_-(\mathbf{k})$. Hence, the quadratic forms $\gamma(\mathbf{k}; P_\pm)$, defined by the relations

$$E(\mathbf{k}) := E(\mathbf{k}; P_\pm) = \gamma(\mathbf{k}; P_\pm) + O(|\mathbf{k}|^3), \quad \mathbf{k} \to 0,$$

coincide with each other. In order to prove that $E = 0$ is a regular lower edge of the spectrum of $P_\pm$, it is sufficient to show that the form $\gamma(\mathbf{k}; P_\pm)$ is positive definite. It can be done with the help of simple estimates. However, by direct calculations of the perturbation theory (see Subsection 1.4), we obtain

$$\begin{aligned}\gamma(\mathbf{k}_0; P_\pm) &= |\Omega|^2 \left(\int_\Omega \omega_\pm^2 \, d\mathbf{x}\right)^{-1} \left(\int_\Omega \omega_\pm^{-2} \, d\mathbf{x}\right)^{-1} \\ &= |\Omega|^2 \left(\int_\Omega \exp{(2\varphi)} \, d\mathbf{x}\right)^{-1} \left(\int_\Omega \exp{(-2\varphi)} \, d\mathbf{x}\right)^{-1}, \quad |\mathbf{k}_0| = 1.\end{aligned} \tag{0.15}$$

The formula (0.15) is rather efficient, since φ can be easily found for a given $\mathbf{A}$. Somewhat unexpectedly, it turns out that *the effective mass for the Pauli operator at the point* 0 *is independent of the direction.* We write $\gamma_0(P_\pm) := \gamma(\mathbf{k}_0; P_\pm)$.

4. Simultaneously with operators (0.2) we consider the Schrödinger operators

$$H_\pm = -\Delta + |\mathbf{A}|^2 \pm B. \tag{0.16}$$

They can be obtained from $P_\pm$ by opening the brackets in (0.2) and eliminating the terms of the first order. The operators

$$H_\pm(\mathbf{k}) = (\mathbf{D} + \mathbf{k})^2 + |\mathbf{A}|^2 \pm B, \quad \mathbf{k} \in \mathbb{R}^2, \tag{0.17}$$

are associated with the operators (0.16). They act in $L_2(\Omega)$ and are defined on periodic functions. The operators (0.16) will be called the *accompanying* operators for the Pauli operators (0.2). Similarly, the operators (0.17) are the accompanying operators for (0.4).

One can directly check that

$$H_\pm(\mathbf{k}) = \omega_\pm^{-1}(\nabla + i\mathbf{k})^* \omega_\pm^2 (\nabla + i\mathbf{k}) \omega_\pm^{-1}, \quad \mathbf{k} \in \mathbb{R}^2. \tag{0.18}$$

This implies that all operators (0.18) are positive and

$$\operatorname{Ker} H_\pm(0) = \{c\omega_\pm\}, \tag{0.19}$$

$$\operatorname{Ker} H_\pm(\mathbf{k}) = \{0\}, \quad \mathbf{k} \neq 0. \tag{0.20}$$

Thus, the lower point of the spectrum for operators (0.16) coincides with zero. It can be easily seen from (0.18) that the lower edge of the spectrum of $H_\pm$ is regular.

As in the case of the Pauli operators, we shall not check this with help of estimates. The positive definiteness of the forms $\gamma(\mathbf{k}; H_\pm)$ will follow (see Subsection 1.5) from their "explicit" representations.

The relations (0.19) can be used for an independent definition of the functions $\omega_\pm$. Obviously their product is automatically a constant. It is convenient to choose these functions so that $\omega_+\omega_- = 1$. Then (0.9) holds with $\varphi = \log\omega_+ = -\log\omega_-$.

It is useful to compare the representation (0.18) with (0.11). The matrix $\mathbf{1} \pm \sigma_2$ is nonnegative, but $\det(\mathbf{1} \pm \sigma_2) = 0$.

The expressions for $\gamma(\mathbf{k}_0; H_\pm)$ are not so simple as in (0.15). Further, the operators (0.16) are not unitarily equivalent to each other; $\gamma(\mathbf{k}_0; H_+)$ and $\gamma(\mathbf{k}_0; H_-)$, generally speaking, do not coincide. Since the coefficients in (0.18) are real, the expression (1.19) for $\gamma(\mathbf{k}_0; H_\pm)$ implies

$$\gamma(\mathbf{k}_0; H_\pm) = \gamma(-\mathbf{k}_0; H_\pm).$$

From (1.18), (1.19) it also follows that

$$\gamma_0(P_\pm) \leqslant \gamma(\mathbf{k}_0; H_\pm) < 1, \quad |\mathbf{k}_0| = 1. \tag{0.21}$$

The right inequality in (0.21) is strict. It means that the effective mass for $H_\pm$ at the point 0 is always greater than the physical mass. The left inequality in (0.21) shows that the effective mass for the Pauli operators at point 0 is not greater than the effective mass for each of the operators $H_\pm$ for any $\mathbf{k}_0$.

Remark. It can be shown that the left inequality in (0.21) is also strict with a single exception. The equality is reached if and only if the potential $\mathbf{A}$ depends only on one coordinate: $\mathbf{A}(\mathbf{x}) = \mathbf{A}(t\,\mathbf{x}_0)$, $|\mathbf{x}_0| = 1$, $t \in \mathbb{R}$, and $\mathbf{k}_0 = \pm\mathbf{x}_0$.

5. The paper is organized as follows. In §1, devoted to the Pauli operators, we give exact statements (for example, conditions on $\mathbf{A}$) and explain some of the facts presented above without proofs. In §2, 3 we consider some generalizations of the Pauli operators with a variable metric (the case of variable principal coefficients). Finally, regularity of the lower edge of the spectrum for elliptic divergence-type operators of the second order is discussed in the Appendix (see §4).

6. In what follows integrals without indication of the integration domain are understood as integrations over $\mathbb{R}^d$. The symbol $\langle\cdot,\cdot\rangle$ stands for the standard scalar product in $\mathbb{C}^d$. By c are denoted various constants which values are not important; symbol $\asymp$ means a two-sided estimate. By $W^1_\rho(\mathbb{R}^d)$, $W^1_\rho(\Omega)$ we denote the Sobolev spaces of functions which are integrable with the power ρ together with their first derivatives. The space $\widetilde{W}^1_\rho(\Omega)$ consists of Γ-periodic functions from $W^1_\rho(\Omega)$. By $(\cdot,\cdot)$ and $\|\cdot\|$ we denote the scalar product and the norm in $L_2(\Omega)$. We assume that $d = 2$ in §1–3 and $d \geqslant 2$ in §4.

7. Acknowledgement. We are grateful to Erwin Schrödinger Institute (ESI) in Vienna and especially T. Hoffmann-Ostenhof for hospitality. During our visit to the ESI in May–June 1998 we have completed this paper. We are also thankful to A. Laptev for helpful discussions concerning the English version of the text.

The paper has been discussed with Prof. D. Vassiliev and Prof. A. Sobolev during the authors visit to the University of Sussex in September 1998. Their remarks have been taken into account in the final version of the text.

§1. Two-dimensional periodic Pauli operators

1. We assume that the magnetic potential $\mathbf{A}$ satisfies the condition

$$\mathbf{A} \in L_\rho(\Omega), \quad \rho > 2. \tag{1.1}$$

It is subject to the gauge conditions (0.6), which can be satisfied (cf. [BSu2]) without violating (1.1). For the functions φ and $\omega_\pm$, defined by (0.7) and (0.9), (1.1) implies that

$$\varphi, \omega_\pm \in \widetilde{W}^1_\rho(\Omega) \subset \operatorname{Lip} \tau, \quad \tau = 1 - 2\rho^{-1}. \tag{1.2}$$

It is easy to see that under the condition (1.1)

$$\mathcal{D} = (D_1 - A_1)\sigma_1 + (D_2 - A_2)\sigma_2$$

is a selfadjoint operator in $L_2(\mathbb{R}^2; \mathbb{C}^2)$ defined on $W^1_2(\mathbb{R}^2; \mathbb{C}^2)$. It means (see (0.3)), that the Pauli operator $\mathcal{P} = \mathcal{D}^2$ is generated by the closed quadratic form

$$\int |\mathcal{D}u|^2 \, d\mathbf{x}, \quad u \in W^1_2(\mathbb{R}^2; \mathbb{C}^2).$$

Simultaneously, the blocks $P_\pm$ of the operator $\mathcal{P}$ are defined via quadratic forms. The same facts concern the operators $P_\pm(\mathbf{k})$ which admit the representations (0.10)–(0.12) understood in the sense of forms. The corresponding forms are given by

$$\begin{aligned} p_\pm(\mathbf{k})[u, u] &= \int_\Omega \omega_\pm^2 \langle (\mathbf{1} \pm \sigma_2)(\nabla + i\mathbf{k})\omega_\pm^{-1} u, (\nabla + i\mathbf{k})\omega_\pm^{-1} u\rangle \, d\mathbf{x} \\ &= \int_\Omega |R_\pm(\mathbf{k})u|^2 \, d\mathbf{x}, \quad u \in \widetilde{W}^1_2(\Omega), \quad \mathbf{k} \in \mathbb{R}^2. \end{aligned} \tag{1.3}$$

Both expressions in (1.3) are useful.

Representations (0.4) for the operators $P_\pm(\mathbf{k})$ have to be treated differently. From (0.8) it is clear that φ must be subject to additional (with respect to (1.2)) conditions. For example, we can understand (0.4) in the sense of forms under the condition

$$\Delta\varphi \in L_{\rho/2}(\Omega), \tag{1.4}$$

or, equivalently,

$$\Delta\omega_\pm \in L_{\rho/2}(\Omega). \tag{1.5}$$

In fact, we do not need to use representations (0.4). It is more convenient to take (0.10) or, equivalently, (0.11) for the *definition* of the operators $P_\pm(\mathbf{k})$. *It is this definition that we use in what follows.* Both expressions are understood in the sense of forms. It also concerns the operators $P_\pm$, which can be defined, for example, by the relations

$$P_\pm = \omega_\mp^{-1}\nabla^*(\mathbf{1} \pm \sigma_2)\omega_\pm^2\nabla\omega_\mp^{-1}, \tag{1.6}$$

understood in the sense of forms. The condition (1.4) allows us to use (0.2) instead of (1.6).

2. Let us now give a detailed explanation of (0.13) and (0.14). By (0.12), the equation $P_\pm(\mathbf{k})u = 0$ is equivalent to

$$R_\pm(\mathbf{k})u = 0, \quad u \in \widetilde{W}_2^1(\Omega), \quad \mathbf{k} \in \mathbb{R}^2. \tag{1.7}$$

We put $u = \omega_\pm v \exp(-i\mathbf{kx})$. Then (1.7) acquires the form $(D_1 \mp iD_2)v = 0$. Integrating by parts and using the quasi-periodicity of v, we obtain

$$0 = \int_\Omega |(D_1 \mp iD_2)v|^2\, d\mathbf{x} = \int_\Omega |\nabla v|^2\, d\mathbf{x}.$$

Consequently, $v = \text{const}$. For $\mathbf{k} \neq 0$ this necessarily implies that $v = 0$, which gives (0.14). For $\mathbf{k} = 0$ we have

$$\operatorname{Ker} P_\pm(0) = \{c\omega_\pm\}.$$

3. We can treat the operators $H_\pm$ and $H_\pm(\mathbf{k})$ similarly. The operators $H_\pm$ are defined by

$$H_\pm = \omega_\mp^{-1}\nabla^*\omega_\pm^2\nabla\omega_\mp^{-1}, \tag{1.8}$$

and the operators $H_\pm(\mathbf{k})$ are defined by (0.18). Both expressions are understood in the sense of quadratic forms. For example, the quadratic form $h_\pm(\mathbf{k})$ for the operator (0.18) is given by

$$h_\pm(\mathbf{k})[u, u] = \int_\Omega |\omega_\pm(\nabla + i\mathbf{k})\omega_\mp^{-1}u|^2\, d\mathbf{x}, \quad u \in \widetilde{W}_2^1(\Omega). \tag{1.9}$$

The representation (1.9) immediately implies (0.19), (0.20).

The conditions (1.4), (1.5) permit us to consider (0.16) and (0.17) instead of (1.8) and (0.18).

4. The effective masses for the Pauli operators. We shall prove (0.15) by using standard arguments from the spectral perturbation theory. For the sake of definiteness, we restrict ourselves to the case of the sign "+" and write here $\omega_+ = \omega$, $\gamma(\mathbf{k}_0; P_+) = \gamma$, $\varkappa = \|\omega\|^2$, $Q_\pm(0) = Q_\pm$. For $\mathbf{k}_0$ with $|\mathbf{k}_0| = 1$ we denote $\xi = k_{01} + ik_{02}$. Using (0.10) and putting $u = \omega v$, we rewrite the equation $P_+(t\mathbf{k}_0)u = Eu$ in the form

$$(Q_+ + t\xi)\omega^2(Q_- + t\bar{\xi})v = E\omega^2 v. \tag{1.10}$$

When $t = 0$ and $E = 0$ the equation (1.10) has a unique solution $v = 1$. Since the function $E_+(t\mathbf{k}_0)$ has a minimum at $t = 0$, then

$$E = \gamma t^2 + \dots, \quad v = 1 + tz + \dots, \tag{1.11}$$

for small values of t. Substituting (1.11) into (1.10) and matching the terms of order t and t^2, we obtain

$$Q_+\omega^2(Q_- z + \bar{\xi}) = 0, \tag{1.12}$$

$$\xi\omega^2(Q_- z + \bar{\xi}) + \bar{\xi} Q_+\omega^2 z = \gamma\omega^2. \tag{1.13}$$

The equation (1.12) is a (non-homogeneous) equation for z, and the solvability condition is automatically satisfied. From (1.12) and from the relation $\operatorname{Ker} Q_+ = \{c\}$, $c = \text{const}$, we find

$$\omega^2(Q_- z + \bar{\xi}) = c. \tag{1.14}$$

Substituting (1.14) into (1.13) and integrating over Ω, we obtain $\xi c|\Omega| = \varkappa\gamma$. On the other hand, multiplying (1.14) by ω^{-2} and integrating over Ω, we have: $\|\omega^{-1}\|^2(\xi c) = |\Omega|$. Excluding ξc from the last two equalities, we find that $\gamma = |\Omega|^2\|\omega\|^{-2}\|\omega^{-1}\|^{-2}$. This coincides with (0.15).

5. The effective masses for the accompanying Schrödinger operators. We will use notation $\varkappa_\pm = \|\omega_\pm\|^2$. The equation $H_\pm(t\mathbf{k}_0)u = Eu$ after the substitution $u = \omega_\pm v$ acquires the form

$$(\nabla + it\mathbf{k}_0)^*\omega_\pm^2(\nabla + it\mathbf{k}_0)v = \omega_\pm^2 Ev.$$

We put $E = \gamma_\pm(\mathbf{k}_0)t^2 + \dots, \quad v = 1 + tZ_\pm + \dots$. For $Z_\pm$ we obtain the equation

$$\nabla^*\omega_\pm^2(\nabla Z_\pm + i\mathbf{k}_0) = 0. \tag{1.15}$$

Furthermore, matching the terms of order t^2, we have

$$\omega_\pm^2 + i\nabla^*\omega_\pm^2 Z_\pm\mathbf{k}_0 - i\omega_\pm^2\langle\nabla Z_\pm, \mathbf{k}_0\rangle = \gamma_\pm(\mathbf{k}_0)\omega_\pm^2. \tag{1.16}$$

Let us multiply both sides of (1.15) by $Z_\pm$ and integrate over Ω. We also integrate (1.16) over Ω. Then

$$\int_\Omega \omega_\pm^2|\nabla Z_\pm|^2\,d\mathbf{x} + i\int_\Omega \omega_\pm^2\langle\mathbf{k}_0, \nabla Z_\pm\rangle\,d\mathbf{x} = 0, \tag{1.17}$$

$$\varkappa_\pm(\gamma_\pm(\mathbf{k}_0) - 1) = -i\int_\Omega \omega_\pm^2\langle\nabla Z_\pm, \mathbf{k}_0\rangle\,d\mathbf{x} = -\int_\Omega \omega_\pm^2|\nabla Z_\pm|^2\,d\mathbf{x}. \tag{1.18}$$

The latter directly yields the right inequality in (0.21). Relations (1.17), (1.18) also imply another representation for $\gamma_\pm(\mathbf{k}_0)$:

$$\varkappa_\pm\gamma_\pm(\mathbf{k}_0) = \varkappa_\pm\gamma(\mathbf{k}_0; H_\pm) = \int_\Omega \omega_\pm^2|\mathbf{k}_0 - i\nabla Z_\pm|^2\,d\mathbf{x}. \tag{1.19}$$

In contrast with Subsection 1.4, now we are not able to express $Z_\pm$ explicitly in terms of $\omega_\pm$. However, the relation (1.19) permits us to prove the left inequality in (0.21). Indeed, $(\mathbf{k}_0, \mathbf{k}_0 - i\nabla Z_\pm) = |\Omega|$, whence

$$|\Omega|^2 \leqslant \varkappa_\mp \|\omega_\pm(\mathbf{k}_0 - i\nabla Z_\pm)\|^2 = \varkappa_\mp\varkappa_\pm\gamma_\pm(\mathbf{k}_0). \tag{1.20}$$

By (0.15), this is the required inequality.

Analyzing whether equality is possible in (1.20), one can easily check what has been stated in Remark at the end of Subsection 0.4.

6. In this subsection we discuss one simple generalization. Let $f > 0$ be a Γ-periodic function such that $f + f^{-1} \in L_\infty$. Consider the operators

$$P_\pm(f) = f^{-1}P_\pm f^{-1}, \quad P_\pm(f;\mathbf{k}) = f^{-1}P_\pm(\mathbf{k})f^{-1}, \tag{1.21}$$

which are defined via their quadratic forms. For example, the form for $P_\pm(f;\mathbf{k})$ is given in terms of the form (1.3):

$$p_\pm(f;\mathbf{k})[u,u] = p_\pm(\mathbf{k})[f^{-1}u, f^{-1}u], \quad u \in f\widetilde{W}_2^1(\Omega).$$

The spectrum of $P_\pm(f;\mathbf{k})$ coincides with the spectrum of the quotient of forms

$$\frac{p_\pm(\mathbf{k})[v,v]}{\|fv\|^2}, \quad 0 \neq v \in \widetilde{W}_2^1(\Omega). \tag{1.22}$$

The last statement corresponds to the study of the problem in the weighted space $L_2(\Omega; f^2)$.

The accompanying Schrödinger operators are defined (in terms of quadratic forms) in a similar way:

$$H_\pm(f) = f^{-1}H_\pm f^{-1}, \quad H_\pm(f;\mathbf{k}) = f^{-1}H_\pm(\mathbf{k})f^{-1}.$$

The lower edge of the spectrum for each of the operators $P_\pm(f)$, $H_\pm(f)$ (it coincides with the point 0) is *regular*. Indeed, the quotient (1.22) admits two-sided estimates by the similar quotient with $f = 1$, for which the answer is already clear. A similar estimate can be done also for $H_\pm(f;\mathbf{k})$.

The following formulas for the effective masses for $P_\pm(f)$ are obtained in a similar way as in Subsection 1.4:

$$\gamma(\mathbf{k}_0; P_\pm(f)) = |\Omega|^2\|\omega_\mp\|^{-2}\|f\omega_\pm\|^{-2}.$$

The effective masses in this case are still independent of direction, but now $\gamma(P_+(f)) \neq \gamma(P_-(f))$. This is connected with the fact that the operators $P_+(f)$, $P_-(f)$ are not unitarily equivalent to each other. Following the considerations of Subsection 1.5, we obtain

$$\gamma(P_\pm(f)) \leqslant \gamma(\mathbf{k}_0; H_\pm(f)) < \|\omega_\pm\|^2\|\omega_\pm f\|^{-2}.$$

§2. The Pauli operators with a (normalized) metric

1. We describe the class of the periodic Pauli type operators with variable principal coefficients. The lower edge of the spectrum (the point 0) for these operators is also regular.

Let $g(\mathbf{x}) = \{g^{ij}(\mathbf{x})\}_{i,j=1,2}$ be a Γ-periodic matrix-valued function with real-valued entries such that

$$g(\mathbf{x}) > 0, \quad \mathbf{x} \in \Omega; \quad g + g^{-1} \in L_\infty,$$
$$\det g(\mathbf{x}) = 1. \tag{2.1}$$

Under the normalized condition (2.1) the formulas become much simpler. We will consider this case as the basic one. In §3 we will discuss the general case, although not with so many details as the one with condition (2.1).

Let $\omega_\pm$ be Γ-periodic functions such that

$$\omega_\pm(\mathbf{x}) > 0, \quad \omega_+\omega_- = 1, \quad \omega_+ + \omega_- \in L_\infty.$$

We consider these functions as initial objects and do not suppose them to be smooth. We could use this point of view already in §1. We introduce the operators (cf. (1.6))

$$P_\pm(g) = \omega_\mp^{-1}\mathbf{D}^*(g \pm \sigma_2)\omega_\pm^2\mathbf{D}\omega_\mp^{-1}. \tag{2.2}$$

Precisely, these are operators generated by the quadratic forms

$$p_\pm(g)[u,u] = \int \omega_\pm^2\langle(g \pm \sigma_2)\mathbf{D}\omega_\mp^{-1}u, \mathbf{D}\omega_\mp^{-1}u\rangle\, d\mathbf{x}, \quad u \in \omega_\pm W_2^1(\mathbb{R}^2).$$

By (2.1), $\det(g \pm \sigma_2) = 0$ and $\operatorname{tr}(g \pm \sigma_2) \geqslant 2$.

If $\omega_\pm$ satisfy conditions (1.2) and (cf. (1.5))

$$\nabla^* g \nabla \omega_\pm \in L_{\rho/2}(\Omega),$$

then, by direct calculations, the operators (2.2) acquire the form

$$P_\pm(g) = (\mathbf{D} - \mathbf{A})^* g(\mathbf{D} - \mathbf{A}) \pm B. \tag{2.3}$$

(It is more convenient to perform calculations with quadratic forms.) Here the periodic vector potential $\mathbf{A}$ and the magnetic field B are defined by the relations (cf. (0.7), (0.8))

$$\mathbf{A} = \sigma_2 g \mathbf{D}\varphi, \quad \varphi = \pm\log\omega_\pm,$$
$$B = \partial_1 A_2 - \partial_2 A_1 = -\nabla^* g \nabla\varphi.$$

The potential $\mathbf{A}$ automatically satisfies the gauge conditions

$$\operatorname{div} g\mathbf{A} = 0, \quad \int_\Omega g\mathbf{A}\, d\mathbf{x} = 0.$$

In order to check (2.3), one should use relation $g^{-1} = \sigma_2 g \sigma_2 = \sigma_2^{-1} g \sigma_2$, which is valid due to (2.1).

Representations (2.3) give a motivation to call the operators (2.2) *Pauli operators with a (normalized) metric.* We notice, that in contrast with the operators (1.6), the two operators (2.2), generally speaking, are not unitarily equivalent to each other.

2. Simultaneously with $P_\pm(g)$ we consider the corresponding operators $P_\pm(g;\mathbf{k})$ in $L_2(\Omega)$. They are generated by quadratic forms

$$p_\pm(g;\mathbf{k})[u,u] = \int_\Omega \omega_\pm^2 \langle (g \pm \sigma_2)(\mathbf{D}+\mathbf{k})v, (\mathbf{D}+\mathbf{k})v\rangle \, d\mathbf{x}, \quad u = \omega_\pm v, \quad v \in \widetilde{W}_2^1(\Omega). \tag{2.4}$$

Let us notice that $\int_\Omega \langle \sigma_2(\mathbf{D}+\mathbf{k})v, (\mathbf{D}+\mathbf{k})v\rangle \, d\mathbf{x} = 0$. Therefore,

$$p_\pm(g;\mathbf{k})[u,u] \asymp \int_\Omega \langle (g \pm \sigma_2)(\mathbf{D}+\mathbf{k})v, (\mathbf{D}+\mathbf{k})v\rangle \, d\mathbf{x} \asymp$$
$$\int_\Omega |(\mathbf{D}+\mathbf{k})v|^2 \, d\mathbf{x} \geqslant c|\mathbf{k}|^2 \|v\|^2 \asymp |\mathbf{k}|^2 \|u\|^2.$$

The third estimate in this series is valid for small values of $|\mathbf{k}|$, and it can be easily obtained in Fourier representation. This directly implies that

$$\operatorname{Ker} P_\pm(g;0) = \{c\omega_\pm\}; \quad \operatorname{Ker} P_\pm(g;\mathbf{k}) = \{0\}, \quad \mathbf{k} \neq 0,$$

and, moreover, that *the point* 0 *is a regular edge for* $P_\pm(g)$.

The spectral perturbation theory gives an opportunity to obtain rather convenient representations for the quadratic forms $\gamma(\mathbf{k}; P_\pm(g))$. Omitting the calculations, which can be done as in Subsection 1.4, we formulate the results.

3. We introduce the notation

$$J_\pm(w) = \int_\Omega \omega_\pm^2 \langle (g \pm \sigma_2)(\nabla w + i\mathbf{k}_0), \nabla w + i\mathbf{k}_0\rangle \, d\mathbf{x}, \quad w \in \widetilde{W}_2^1(\Omega), \quad |\mathbf{k}_0| = 1.$$

Let $z_\pm$ be a (unique up to an additive constant) weak solution of the equation

$$\operatorname{div} \omega_\pm^2 (g \pm \sigma_2)(\nabla z_\pm + i\mathbf{k}_0) = 0, \quad |\mathbf{k}_0| = 1. \tag{2.5}$$

The following two representations for $\gamma(\mathbf{k}_0; P_\pm(g))$ hold:

$$\gamma(\mathbf{k}_0; P_\pm(g)) = \|\omega_\pm\|^{-2} \int_\Omega \omega_\pm^2 (\langle g\mathbf{k}_0, \mathbf{k}_0\rangle - \langle (g \pm \sigma_2)\nabla z_\pm, \nabla z_\pm\rangle) \, d\mathbf{x}, \tag{2.6}$$

$$\gamma(\mathbf{k}_0; P_\pm(g)) = \|\omega_\pm\|^{-2} J_\pm(z_\pm). \tag{2.7}$$

The variational principle for the equation (2.5) yields

$$J_\pm(w) > J_\pm(z_\pm), \quad w \in \widetilde{W}_2^1(\Omega), \quad w \neq z_\pm. \tag{2.8}$$

It is convenient to use the last inequality in order to estimate $\gamma(\mathbf{k}_0; P_\pm(g))$ from above.

We are not able to obtain an explicit expression for $\gamma(\mathbf{k}_0; P_\pm(g))$ similar to (0.15). Also, generally speaking, the forms $\gamma(\mathbf{k}_0; P_\pm(g))$ depend on $\mathbf{k}_0$.

4. We introduce the accompanying Schrödinger operators $H_\pm(g)$, $H_\pm(g;\mathbf{k})$. The latter are generated by the quadratic forms

$$h_\pm(g;\mathbf{k})[u,u] = \int_\Omega \omega_\pm^2 \langle g(\mathbf{D}+\mathbf{k})v, (\mathbf{D}+\mathbf{k})v\rangle\, d\mathbf{x}, \quad u = \omega_\pm v, \quad v \in \widetilde{W}_2^1(\Omega).$$

Similarly to the estimate of the form (2.4), we find that

$$h_\pm(g;\mathbf{k})[u,u] \geqslant c\,|\mathbf{k}|^2 \|u\|^2.$$

Thus, the point 0 is a *regular* lower edge of the spectrum of the operator $H_\pm(g)$.

Now we shall describe formulas for $\gamma(\mathbf{k}_0; H_\pm(g))$. Let $Z_\pm$ be a weak solution of the equation

$$\operatorname{div} \omega_\pm^2 g(\nabla Z_\pm + i\mathbf{k}_0) = 0, \quad |\mathbf{k}_0| = 1. \tag{2.9}$$

We introduce the notation

$$J_\pm^S(w) = \int_\Omega \omega_\pm^2 \langle g(\nabla w + i\mathbf{k}_0), \nabla w + i\mathbf{k}_0\rangle\, d\mathbf{x}. \tag{2.10}$$

The following representations (cf. (2.6), (2.7)) hold

$$\gamma(\mathbf{k}_0; H_\pm(g)) = \|\omega_\pm\|^{-2} \int_\Omega \omega_\pm^2 (\langle g\mathbf{k}_0, \mathbf{k}_0\rangle - \langle g\nabla Z_\pm, \nabla Z_\pm\rangle)\, d\mathbf{x}, \tag{2.11}$$

$$\gamma(\mathbf{k}_0; H_\pm(g)) = \|\omega_\pm\|^{-2} J_\pm^S(Z_\pm). \tag{2.12}$$

The inequality

$$J_\pm^S(w) > J_\pm^S(Z_\pm), \quad w \in \widetilde{W}_2^1(\Omega), \quad w \neq Z_\pm, \tag{2.13}$$

is an analog of (2.8). From (2.11) (or from (2.10), (2.12) and (2.13) with $w = 1$) we have

$$\gamma(\mathbf{k}_0; H_\pm(g)) < \|\omega_\pm\|^{-2} \int_\Omega \omega_\pm^2 \langle g\mathbf{k}_0, \mathbf{k}_0\rangle\, d\mathbf{x}. \tag{2.14}$$

When $g = 1$, this estimate turns into the right inequality (0.21). The analog of the left inequality (0.21) also holds. Indeed, (2.9) implies that the vector $\nabla Z_\pm + i\mathbf{k}_0$ is purely imaginary, and consequently, $\langle \sigma_2(\nabla Z_\pm + i\mathbf{k}_0), \nabla Z_\pm + i\mathbf{k}_0\rangle = 0$. This and (2.8) yield

$$J_\pm^S(Z_\pm) = J_\pm(Z_\pm) \geqslant J_\pm(z_\pm).$$

Together with (2.7), (2.12), this means that (cf. (0.21))

$$\gamma(\mathbf{k}_0; P_\pm(g)) \leqslant \gamma(\mathbf{k}_0; H_\pm(g)). \tag{2.15}$$

§3. The Pauli operators with a metric (the general case)

1. Now we do not pose condition (2.1). Essentially, the results remain the same, but the formulas become somewhat more cumbersome. We restrict ourselves to a short description of the results.

Let us introduce the notation

$$\delta = \delta(\mathbf{x}) = (\det g(\mathbf{x}))^{1/2} \tag{3.1}$$

and let $\omega_\pm$ be Γ-periodic functions such that

$$\omega_\pm(\mathbf{x}) > 0, \quad \omega_+\omega_- = \delta^{-1}, \quad \omega_+ + \omega_- \in L_\infty.$$

Instead of the operators (2.2) we now consider more general operators

$$P_\pm(g) = \omega_\pm^{-1}\mathbf{D}^*(g \pm \delta\sigma_2)\mathbf{D}\omega_\pm^{-1}, \tag{3.2}$$

using the same notation. Precisely, these operators are generated by the quadratic forms

$$p_\pm(g)[u,u] = \int \omega_\pm^2 \langle (g \pm \delta\sigma_2)\mathbf{D}\omega_\pm^{-1}u, \mathbf{D}\omega_\pm^{-1}u\rangle \, d\mathbf{x}, \quad u \in \omega_\pm W_2^1(\mathbb{R}^2).$$

By virtue of (3.1), $\det(g \pm \delta\sigma_2) = 0$ and $\operatorname{tr}(g \pm \delta\sigma_2) \geqslant 2\delta$.

We introduce the notation

$$V_g = \langle g\nabla\Phi, \nabla\Phi\rangle - \operatorname{div}(g\nabla\Phi), \quad 2\Phi := \log\delta, \tag{3.3}$$

$$\mathbf{A} = \delta^{-1}\sigma_2 g\mathbf{D}\varphi, \quad \varphi := \pm(\log\omega_\pm + \Phi),$$

$$B = \partial_1 A_2 - \partial_2 A_1 = \nabla^*(\delta^{-1}g\nabla\varphi).$$

If the functions $\omega_\pm$, δ are subject to the conditions

$$\omega_\pm, \delta \in \widetilde{W}_\rho^1(\Omega), \quad \nabla^*(g\nabla\Phi) \in L_{\rho/2}(\Omega), \quad B \in L_{\rho/2}(\Omega), \quad \rho > 2,$$

then by direct, but somewhat tedious calculations, the operators (3.2) can be represented in the form

$$P_\pm(g) = (\mathbf{D} - \mathbf{A})^* g(\mathbf{D} - \mathbf{A}) + V_g(\mathbf{x}) \pm \delta B(\mathbf{x}).$$

In comparison with (2.3) there arises an additional electric potential V_g and an additional factor δ at the magnetic field B. The vector potential $\mathbf{A}$ satisfies the gauge conditions

$$\nabla^*(\delta^{-1}g\mathbf{A}) = 0, \quad \int_\Omega \delta^{-1}g\mathbf{A}\, d\mathbf{x} = 0.$$

We notice that the potential V_g has the form (4.19) (see Subsection 4.5 below). This is explained by the fact that the case of the operators (3.2) can be reduced to the case of the normalized metric $\delta^{-1}g$ and the weight function δ^{-1}. (In the simplest situation the case of the weighted space is discussed in Subsection 1.6. In accordance with (1.22), the operators (1.21) correspond to the space with weight function f^2).

Formally, the accompanying Schrödinger operators have the same form as under the condition (2.1):

$$H_{\pm}(g) = \omega_{\pm}^{-1}\mathbf{D}^*\omega_{\pm}^2 g\mathbf{D}\omega_{\pm}^{-1}.$$

However, g and the product $\omega_+\omega_-$ satisfy other conditions now.

2. As above, we introduce the operators $P_{\pm}(g;\mathbf{k})$, $H_{\pm}(g;\mathbf{k})$, acting in $L_2(\Omega)$. It can be shown that

$$\operatorname{Ker} P_{\pm}(g;0) = \operatorname{Ker} H_{\pm}(g;0) = \{c\omega_{\pm}\},$$

$$\operatorname{Ker} P_{\pm}(g;\mathbf{k}) = \operatorname{Ker} H_{\pm}(g;\mathbf{k}) = \{0\}, \quad \mathbf{k} \neq 0,$$

and that *the point* 0 *is a regular lower edge of the spectrum for the operators* $P_{\pm}(g)$, $H_{\pm}(g)$.

The following representation holds

$$\gamma(\mathbf{k}_0; P_{\pm}(g)) = \|\omega_{\pm}\|^{-2} \int_{\Omega} \omega_{\pm}^2 \langle (g \pm \delta\sigma_2)(\nabla z_{\pm} + i\mathbf{k}_0), \nabla z_{\pm} + i\mathbf{k}_0 \rangle \, d\mathbf{x},$$

where $z_{\pm}$ is a solution of the equation

$$\operatorname{div} \omega_{\pm}^2 (g \pm \delta\sigma_2)(\nabla z_{\pm} + i\mathbf{k}_0) = 0.$$

Similarly,

$$\gamma(\mathbf{k}_0; H_{\pm}(g)) = \|\omega_{\pm}\|^{-2} \int_{\Omega} \omega_{\pm}^2 \langle g(\nabla Z_{\pm} + i\mathbf{k}_0), \nabla Z_{\pm} + i\mathbf{k}_0 \rangle \, d\mathbf{x},$$

$$\operatorname{div} \omega_{\pm}^2 g(\nabla Z_{\pm} \pm i\mathbf{k}_0) = 0.$$

As in Subsection 2.4, these representations imply inequalities (2.14), (2.15).

§4. Appendix. The periodic elliptic divergence-type operators of the second order

We shall discuss here some transformations for multi-dimensional periodic operators. In terms of quadratic forms we deal with substitutions $u = \omega v$, where ω is a positive periodic solution of the corresponding homogeneous equation. We have used these substitutions systematically. They have proved to be very useful in applications to many problems and, in particular, to problems of spectral theory. For example, this approach has been applied in [B] and [BSo] when studying the discrete spectrum of periodic operators perturbed by decaying potentials. This method has been also used in [KSi] for the study of properties of the bottom edge of the spectrum of a periodic operator. We deal here with a similar problem, but go somewhat further. The results from [KSi] approximately correspond to the contents of our Subsection 4.2 applied to a Schrödinger operator. Clearly, the mentioned substitutions are useful not only in periodic cases (see, in particular, [KSi]).

1. Let Γ be a period lattice in $\mathbb{R}^d$ with an elementary cell Ω. By $|\Omega|$ we denote the volume of Ω and by $\|\cdot\|$ the norm in $L_2(\Omega)$. Let

$$g(\mathbf{x}) = \{g^{ij}(\mathbf{x})\}, \quad i,j = 1,\dots,d, \quad \mathbf{x} \in \mathbb{R}^d,$$

be a Γ-periodic matrix-valued function with real-valued entries such that

$$g(\mathbf{x}) > 0, \quad g + g^{-1} \in L_\infty, \tag{4.1}$$

and let V be a Γ-periodic real-valued function such that

$$V \in L_s(\Omega), \quad 2s > d. \tag{4.2}$$

In $L_2(\mathbb{R}^d)$ we consider the closed quadratic form

$$h[u,u] = \int (\langle g\mathbf{D}u, \mathbf{D}u\rangle + V|u|^2)\, d\mathbf{x}, \quad u \in W_2^1(\mathbb{R}^d). \tag{4.3}$$

Under the conditions (4.1), (4.2) the form (4.3) is semi-bounded from below. This form generates a selfadjoint operator H in $L_2(\mathbb{R}^d)$. The corresponding differential expression

$$H = \nabla^* g \nabla + V \tag{4.4}$$

is understood in the distributional sense. Simultaneously with the form (4.3) we consider the family of forms in $L_2(\Omega)$

$$h(\mathbf{k})[u,u] = \int_\Omega (\langle g(\mathbf{D}+\mathbf{k})u, (\mathbf{D}+\mathbf{k})u\rangle + V|u|^2)\, d\mathbf{x}, \quad u \in \widetilde{W}_2^1(\Omega), \quad \mathbf{k} \in \mathbb{R}^d. \tag{4.5}$$

The corresponding selfadjoint operators in $L_2(\Omega)$ are denoted by

$$H(\mathbf{k}) = (\mathbf{D}+\mathbf{k})^* g(\mathbf{D}+\mathbf{k}) + V, \quad \mathbf{k} \in \mathbb{R}^d. \tag{4.6}$$

Without loss of generality we assume that

$$0 = \inf_u \|u\|^{-2} h(0)[u,u], \quad 0 \neq u \in \widetilde{W}_2^1(\Omega). \tag{4.7}$$

(We can always satisfy (4.7) by adding a constant to V.) Since the embedding of $\widetilde{W}_2^1(\Omega)$ into $L_2(\Omega)$ is compact, then the infimum in (4.7) is reached on some (real-valued) function ψ and

$$\int_\Omega (\langle g\mathbf{D}\psi, \mathbf{D}\xi\rangle + V\psi\bar{\xi})\,d\mathbf{x} = 0, \quad \xi \in \widetilde{W}_2^1(\Omega). \tag{4.8}$$

Then the infimum in (4.7) is also reached at the function $\omega = |\psi|$ and this function satisfies the integral identity (4.8). Moreover, $\omega \in \operatorname{Lip}\tau$ with some (small) $\tau > 0$, and, by Harnak inequality, $\omega > 0$ (see [LU] concerning these facts under "minimal" conditions (4.1), (4.2)). It can be also shown that the functions ω, ω^{-1} are *multiplicators* in each of the spaces $W_2^1(\mathbb{R}^d)$, $\widetilde{W}_2^1(\Omega)$.

2. The relation (4.8) (with ψ replaced by ω) means that the equation $H(0)\psi = 0$ has a positive Γ-periodic solution ω of class $\operatorname{Lip}\tau$. By substitution $u = \omega v$, the forms (4.3), (4.5) turn into

$$h[u,u] = \int \omega^2\langle g\mathbf{D}v, \mathbf{D}v\rangle\,d\mathbf{x}, \quad u = \omega v, \quad v \in W_2^1(\mathbb{R}^d), \tag{4.9}$$

$$h(\mathbf{k})[u,u] = \int_\Omega \omega^2\langle g(\mathbf{D}+\mathbf{k})v, (\mathbf{D}+\mathbf{k})v\rangle\,d\mathbf{x}, \quad u = \omega v, \quad v \in \widetilde{W}_2^1(\Omega). \tag{4.10}$$

From (4.10) it directly follows that $\operatorname{Ker} H(0) = \{c\omega\}$. In addition, for small values of $|\mathbf{k}|$ we have

$$h(\mathbf{k})[u,u] \asymp \int_\Omega |(\mathbf{D}+\mathbf{k})v|^2\,d\mathbf{x} \geqslant c|\mathbf{k}|^2\|v\|^2 \asymp |\mathbf{k}|^2\|u\|^2.$$

Consequently, *the point* 0 *is a regular lower edge of the spectrum of* H.

3. The formulas (4.9), (4.10) for the forms h, $h(\mathbf{k})$ imply that the operators (4.4), (4.6) admit the representation

$$H = \omega^{-1}\mathbf{D}^*\omega^2 g\mathbf{D}\omega^{-1}, \tag{4.11}$$

$$H(\mathbf{k}) = \omega^{-1}(\mathbf{D}+\mathbf{k})^*\omega^2 g(\mathbf{D}+\mathbf{k})\omega^{-1}, \quad \mathbf{k} \in \mathbb{R}^d. \tag{4.12}$$

We can take (4.11), (4.12) as *the definition* of the operators H, $H(\mathbf{k})$, assuming that ω is an arbitrary Γ-periodic function such that

$$\omega(\mathbf{x}) > 0, \quad \omega + \omega^{-1} \in L_\infty. \tag{4.13}$$

Precisely, the relations (4.11), (4.12) should be understood in the sense of quadratic forms (4.9), (4.10). However, now the functions ω, ω^{-1} are no longer multiplicators in $W_2^1(\mathbb{R}^d)$ and $\widetilde{W}_2^1(\Omega)$.

Under conditions (4.13) *the lower edge of the spectrum of H is still regular.* Also, it is sufficient to use the definition (4.12) in order to obtain the formula for $\gamma(\mathbf{k}; H)$ by the perturbation theory. The arguments here are similar to that of Subsection 1.5. Let us describe our results.

4. We introduce the notation

$$J_g(w) = \int_\Omega \omega^2 \langle g(\nabla w + i\mathbf{k}_0), \nabla w + i\mathbf{k}_0\rangle \, d\mathbf{x}, \quad |\mathbf{k}_0| = 1, \quad w \in \widetilde{W}_2^1(\Omega). \tag{4.14}$$

Let $Z \in \widetilde{W}_2^1(\Omega)$ be a weak solution of the equation

$$\nabla^*(\omega^2 g(\nabla Z + i\mathbf{k}_0)) = 0. \tag{4.15}$$

The variational principle which is equivalent to the equation (4.15) gives

$$J_g(w) > J_g(Z), \quad w \in \widetilde{W}_2^1(\Omega), \quad w \neq Z. \tag{4.16}$$

The form $\gamma(\mathbf{k}_0; H)$ is expressed in terms of ω and Z by

$$\gamma(\mathbf{k}_0; H) = \|\omega\|^{-2} J_g(Z), \quad |\mathbf{k}_0| = 1. \tag{4.17}$$

The relations (4.14), (4.16) and (4.17) with $w = 1$ imply the upper estimate:

$$\gamma(\mathbf{k}_0; H) < \|\omega\|^{-2} \int_\Omega \omega^2 \langle g\mathbf{k}_0, \mathbf{k}_0\rangle \, d\mathbf{x},$$

or

$$\gamma(\mathbf{k}_0; H) < \langle \bar{g}\mathbf{k}_0, \mathbf{k}_0\rangle, \quad \bar{g} := \|\omega\|^{-2} \int_\Omega \omega^2 g \, d\mathbf{x}.$$

Now we estimate $\gamma(\mathbf{k}_0; H)$ from below. For any constant vector $\mathbf{l} \in \mathbb{R}^d$ we have

$$\int_\Omega \langle \mathbf{l}, \nabla Z + i\mathbf{k}_0\rangle \, d\mathbf{x} = -i\langle \mathbf{l}, \mathbf{k}_0\rangle |\Omega|,$$

whence,

$$|\Omega|^2 |\langle \mathbf{l}, \mathbf{k}_0\rangle|^2 \leqslant J_g(Z) \int_\Omega \omega^{-2} \langle g^{-1}\mathbf{l}, \mathbf{l}\rangle \, d\mathbf{x} = \langle a^{-1}\mathbf{l}, \mathbf{l}\rangle J_g(Z); \quad a^{-1} := \int_\Omega \omega^{-2} g^{-1} \, d\mathbf{x}.$$

Now, in accordance with (4.17),

$$\gamma(\mathbf{k}_0; H) \geqslant |\Omega|^2 \|\omega\|^{-2} |\langle \mathbf{l}, \mathbf{k}_0\rangle|^2 \langle a^{-1}\mathbf{l}, \mathbf{l}\rangle^{-1},$$

and, for $\mathbf{l} = a\mathbf{k}_0$, we have

$$\gamma(\mathbf{k}_0; H) \geqslant \langle \underline{g}\mathbf{k}_0, \mathbf{k}_0\rangle, \quad \underline{g} := |\Omega|^2 \|\omega\|^{-2} \left(\int_\Omega \omega^{-2} g^{-1} \, d\mathbf{x} \right)^{-1}.$$

In particular,

$$\gamma(\mathbf{k}_0; H) \geqslant |\Omega|^2 \|\omega\|^{-2} \|\omega^{-1}\|^{-2} \text{ when } g = \mathbf{1},$$

which is interesting to compare with (0.15), (0.21).

5. If ω satisfy an additional (with respect to (4.13)) smoothness condition, then we can turn from the representations (4.11), (4.12) back to (4.4), (4.6). Namely, let us assume that

$$\omega \in W^1_{2s}(\Omega), \quad \nabla^* g \nabla \omega \in L_s(\Omega), \quad 2s > d,$$

or, equivalently,

$$\log \omega =: \varphi \in W^1_{2s}(\Omega), \quad \nabla^* g \nabla \varphi \in L_s(\Omega), \quad 2s > d. \tag{4.18}$$

Then the operator H is reformed from (4.11) to (4.4). Here $V = \omega^{-1} \operatorname{div} g \nabla \omega$ or, equivalently,

$$V = \operatorname{div} g \nabla \varphi + \langle g \nabla \varphi, \nabla \varphi \rangle. \tag{4.19}$$

Conditions (4.18) yield (4.2). Thus, *the lower edge of the spectrum of the periodic operator* (4.4) *coincides with point* 0 *if and only if the potential V admits a representation* (4.19). *Moreover, $\varphi = \log \omega$, where ω is a positive Γ-periodic solution of the equation $\nabla^* g \nabla \omega + V\omega = 0$.*

What was mentioned before concerned also the operators $H_\pm$ in (0.16). The potential here has the form $|\mathbf{A}|^2 \pm B = |\nabla \varphi|^2 \pm \Delta \varphi$, where φ is defined by equations (0.7). Simultaneously $\omega_\pm = \exp(\pm\varphi)$ is a Γ-periodic solution of the equation $(-\Delta + |\mathbf{A}|^2 \pm B)\omega = 0$. We remind that the potential V_g in (3.3) also has the form (4.19).

Replacing in (4.19) φ by $t\varphi$, $t \in \mathbb{R}$, we have the *quadratic pencil* of potentials

$$V_t = t \operatorname{div} g \nabla \varphi + t^2 \langle g \nabla \varphi, \nabla \varphi \rangle, \quad t \in \mathbb{R}.$$

The lower edge of the spectrum for the corresponding operators is 0. Simultaneously, the function ω can be replaced by the function ω^t.

6. In conclusion, we discuss one generalization (only from a formal point of view). Let $\mathbf{A}$ be a vector (magnetic) potential. Simultaneously with (4.4) we consider the periodic operator in $L_2(\mathbb{R}^d)$

$$H(\mathbf{A}) = (\mathbf{D} - \mathbf{A})^* g (\mathbf{D} - \mathbf{A}) + V. \tag{4.20}$$

If V has the form (4.19), then, using the previous notation, we can rewrite the operator (4.20) in the form

$$H(A) = \omega^{-1} (\mathbf{D} - \mathbf{A})^* \omega^2 g (\mathbf{D} - \mathbf{A}) \omega^{-1}. \tag{4.21}$$

However, now this representation does not permit us to study the lower edge of the spectrum by the same method. If $\mathbf{A}$ is a gauge nontrivial periodic vector, then the lower point of the spectrum of the operator (4.21) is always positive. However, it is possible to rewrite the operator (4.20) in the form (4.21) also for a non-periodic $\mathbf{A}$. In the latter case this representation can be more useful.

References

[B] M.Sh.Birman, *On the spectrum of singular boundary-value problems*, Mat. Sb. **55** (1961), 125–174; English transl., Amer. Math. Soc. Trans. **53** (1966), 23–80.

[BSo] M.Sh.Birman and M.Z.Solomyak, *Estimates for the number of negative eigenvalues of the Schrödinger operator and its generalizations*, Estimates and Asymptotics for Discrete Spectra of Integral and Differential Equations, Adv. Soviet Math. **7** (1991), Amer. Math. Soc., Providence, RI, 1–55.

[BSu1] M.Sh.Birman and T.A.Suslina, *Two-dimensional periodic magnetic Hamiltonian is absolutely continuous*, Algebra i Analiz **9** (1997), no. 1, 32–48; English transl., St.Petersburg Math. J. **9** (1998), no. 1.

[BSu2] M.Sh.Birman and T.A.Suslina, *Absolute continuity of the two-dimensional periodic magnetic Hamiltonian with discontinuous vector-valued potential*, Algebra i Analiz **10** (1998), no. 4, 1–36; English transl., St.Petersburg Math. J. **10** (1999), no. 4.

[KSi] W.Kirsch and B.Simon, *Comparison theorems for the gap of Schrödinger operators*, J. Funct. Anal. **75** (1987), no. 2, 396–410.

[LU] O.A.Ladyzhenskaya and N.N.Ural'tseva, *Linear and quasilinear elliptic equations*, "Nauka", Moscow, 1964; English transl., Academic Press, New York, 1968.

Operator Theory:
Advances and Applications, Vol. 108

SPECTRAL PROBLEMS IN THE THEORY OF PHOTONIC CRYSTALS

Jean-Michel COMBES

Centre de Physique Théorique, CNRS Luminy, Case 907,
13288 Marseille – Cedex 9
and
PHYMAT, Université de Toulon et du Var,
B.P.132 - 83957 La Garde Cedex

Abstract

We present a short introduction to the rapidly growing field of photonic crystals and some of the related spectral problems. We review in particular the existence results for gaps in the frequency spectrum of the Maxwell operator in periodic dielectric medium and summarize the works of A. Figotin and P. Kuchment for periodic square structures. We also discuss the localization effect of random perturbations of periodic structures and the inhibition of radiative decay when gaps open.

1. Introduction

Photonic crystals are man-made periodic dielectric structures; their fascinating technological applications have stimulated a growing interest in the last few years on both experimental and theoretical sides. The aim of this non-specialist talk is to present some related spectral problems with the hope that they might look appealing to some members of our mathematical physics community. For an elementary introduction to the subject I recommend the recent book by J. Joannopoulos et al. [19] and for a more complete review the references [27],[26].

In contrast to solid state crystals the geometry of photonic crystals is not imposed on us by nature; thus the only constraint on building mathematical models is that they fit with present day technology in particular as far as length scales are concerned. For example, one of the most challenging question is to find geometries for which gaps open in the frequency spectrum of Maxwell operators; as a rule of thumb based on the Ioffe-Regel criterion gaps are expected to occur at wave lengths of the order of the lattice length scales restricting them presently to a range going from the microwave to nearly visible light regimes.

Existence of gaps implies that electromagnetic waves with given frequencies and polarizations cannot penetrate the dielectric medium; potential applications of this property are perfect mirrors, narrow band filters, polarizers, lossless curved wave-guides etc... Various structures which can open gaps have been investigated numerically in the last years (see [19],[27],[26] for the most significant references); on the mathematical side the main contribution is certainly the one of A. Figotin and P. Kuchment ([14],[15],[16]) and a large part of this review will be devoted to a description of their work.

Another important consequence of existence of gaps is the possibility to "localize" light, for example by drilling a cavity in a photonic crystal having frequency modes in the gap. This has promising applications, e.g. in designing high efficiency lasers.

On the other hand disorder is an unavoidable feature of man made devices; in analogy with electron localization in disordered semi-conductors one expects that localized electromagnetic waves occur with frequencies near the band edges of the spectrum. In fact disordered dielectric structures provide a nearly perfect testing ground for Anderson localization since in particular photons don't interact with each others in contrast to electrons. Also the wavelength of electromagnetic waves can be tuned very accurately thus offering the possibility of very reliable experimental tests of Anderson transition at mobility edges.

Finally one of the most intriguing aspect linked to the existence of photonic band gaps is the drastic modification of quantum electrodynamics phenomena in such media as compared to the homogeneous ones; in particular spontaneous emission of Rydberg atoms or electron-hole radiative recombination must be strongly inhibited if transition frequencies lie in a gap of the photonic crystal. This opens again many technological perspectives e.g. in the fields of semi-conductor lasers, solar cells or quantum computers.

In this talk some of the mathematical aspects of these features will be discussed. Concerning the gap problem I will give a condensed version of the works of A. Figotin and P. Kuchment [15] on 2-dimensional square structures which to my knowledge are the only existing exact mathematical results. Then I will turn to the random problem and present the recent results on band-edge localization. Finally I will briefly discuss the very challenging question of radiative decay in photonic crystals. This requires first a new canonical quantization procedure in non homogeneous media as described recently by A. Tip [28]; once this is done one is faced with the need for an extension of Resonance Theory to coupled matter-radiation systems.

2. The frequency spectrum in periodic dielectrics

We consider an isotropic, lossless, periodic dielectric medium with no free charges or currents and constant magnetic permeability which is assumed to be one. The Macroscopic Maxwell Equations for the electric (resp. magnetic) field $E_t(x) \equiv E(x,t)$ (resp. $H_t(x) \equiv H(x,t)$) read:

$$\frac{\partial H_t}{\partial t} = -\nabla \wedge E_t \tag{2.1}$$

$$\frac{\partial E_t}{\partial t} = \varepsilon^{-1} \nabla \wedge H_t \tag{2.2}$$

$$\nabla . \varepsilon E_t = 0 \tag{2.3a}$$

$$\nabla . H_t = 0 \tag{2.3b}$$

Here ε is the dielectric function which is assumed to be real (no absorption of photons) and frequency independent; these assumptions are meant to leave us with a nice linear self-adjoint problem but they correspond to situations which can be reached experimentally to a good approximation. Furthermore ε is assumed to have the periodicity of a Bravais lattice Γ in $\mathbb{R}^3$:

$$\varepsilon(x+\gamma) = \varepsilon(x) \qquad \forall x \in \mathbb{R}^3, \forall \gamma \in \Gamma .$$

In some cases of interest, ε can be homogeneous in some directions and thus depend on less than 3 variables. This is the case for example of layered dielectric or periodic arrays of dielectric cylinders as will be considered later. Notice that in cases of interest ε is not continuous but piecewise constant; the corresponding mathematical problem can be handled in a standard way (see e.g. [12]).

One can recast equations (2.1) and (2.2) as an evolution equation:

$$-i \frac{\partial F_t}{\partial t} = \mathbb{M} F_t$$

with

$$F_t \equiv \begin{pmatrix} H_t \\ E_t \end{pmatrix}$$

and

$$\mathbb{M} = \begin{bmatrix} 0 & i\nabla^{\times} \\ -i\varepsilon^{-1}\nabla^{\times} & 0 \end{bmatrix} \tag{2.4}$$

where $\nabla^{\times}$ stands for the curl.

The operator $\mathbb{M}$ is formally hermitian on

$$\mathcal{H} = L^2(\mathbb{R}^3, dx; \mathbb{C}^3) \oplus L^2(\mathbb{R}^3, \varepsilon dx; \mathbb{C}^3) . \tag{2.5}$$

One has

$$\mathbb{M}^2 = \begin{bmatrix} -\nabla^\times \varepsilon^{-1}\nabla^\times & 0 \\ 0 & -\varepsilon^{-1}\nabla^\times\nabla^\times \end{bmatrix} \tag{2.6}$$

which can be suitably defined as a self-adjoint operator by quadratic form techniques.

Simple calculations show that one has

$$\mathbb{M}^2 = \begin{bmatrix} -P_{\hbar}\nabla.\varepsilon^{-1}\nabla P^\perp & 0 \\ 0 & -\varepsilon^{-1}\nabla P^\perp \end{bmatrix} \tag{2.7}$$

where $P^\perp$ projects on "transverse fields" i.e. solutions of $\nabla.\psi = 0$. At this point it is worth mentioning that for non constant ε transversality of electric field is not preserved under time-evolution. It also follows from the form (2.7) that the spectrum of $\mathbb{M}^2$ has a richer structure than the spectrum of the scalar operators $-\varepsilon^{-1}\Delta$ or $\nabla.\varepsilon^{-1}\nabla$. since for non constant ε these operators don't commute with $P^\perp$. Define:

$$M_E \equiv -\varepsilon^{-1}\nabla^\times\nabla^\times \quad \text{acting on} \quad L^2(\mathbb{R}^3, \varepsilon dx; \mathbb{C}^3) \tag{2.8a}$$

and

$$M_H \equiv -\nabla^\times\varepsilon^{-1}\nabla^\times \quad \text{acting on} \quad L^2(\mathbb{R}^3, dx; \mathbb{C}^3); \tag{2.8b}$$

then ($\sigma(A)$ denoting the spectrum of A) one has:

$$\sigma(M_H) = \sigma(M_E) \tag{2.9}$$

In fact M_H and M_E are unitarily equivalent (see e.g. [13] for details on the construction of the corresponding unitary operator).

Notice however that in dimension ≥ 2 one has $\sigma(-\nabla.\varepsilon^{-1}\nabla) \neq \sigma(-\varepsilon^{-1}\Delta)$ (see e.g. [14],[15] in the case of two-dimensional square structures).

Assuming now that ε has the periodicity of a Bravais lattice $\Gamma \subset \mathbb{R}^3$ it is natural to perform a Floquet-Bloch reduction as usual. Let F (resp. $F^\star$) denote the primitive cell associated to a given basis of Γ (resp. to the corresponding dual basis of the dual lattice $\Gamma^\star$). Then the Floquet-Bloch transform is defined for $\psi \in C_0^\infty(\mathbb{R}^3; \mathbb{C}^3)$, by:

$$(U\psi)(x,k) = |F^\star|^{-1/2} \sum_{\gamma\in\Gamma} e^{-ik.\gamma}\psi(x+\gamma) \tag{2.10}$$

It is easy to see that U extends to a unitary operator from $L^2(\mathbb{R}^3, \varepsilon dx; \mathbb{C}^3)$ to $L^2(\mathbb{R}^3/\Gamma^\star, dk; \mathcal{H})$ where $\mathcal{H} \equiv L^2(F, \varepsilon dx; \mathbb{C}^3)$. Furthermore if M is an operator on $L^2(\mathbb{R}^d, \varepsilon dx; \mathbb{C}^3)$ having the periodicity of Γ then it admits an integral decomposition:

$$UM_X U^{-1} = \int_{\mathbb{R}^3/\Gamma^\star}^{\oplus} M_X(k)dk \tag{2.11}$$

with $X = E$ or H, as given by (2.8). One can easily identify $M_X(k)$ as the same differential operator as M_X but acting now on $L^2(F, \varepsilon dx; \mathbb{C}^3)$ for $X = E$ or $L^2(F, dx; \mathbb{C}^3)$ for $X = H$ with the usual Bloch boundary conditions for functions in the domains of $M_X(k)$ and their first order partial derivatives, namely:

$$\Psi(x+\gamma) = e^{ik.\gamma}\Psi(x) \quad \forall x \in \partial F, \forall \gamma \in \text{basis of } \Gamma \ . \tag{2.12}$$

The main consequence of (2.11) is:

$$\sigma(M_X) = \bigcup_{k \in \mathbb{R}^3/\Gamma^*} \sigma(M_X(k))$$

or equivalently:

$$\sigma(M_X) = \bigcup_{k \in B} \sigma(M_X(k)) \tag{2.13}$$

where B is the first Brillouin zone.

So finally (2.6), (2.9) and (2.13) imply:

$$\sigma(\mathbb{M}^2) = \bigcup_{k \in B} \sigma(M_X(k)), \ X = E \text{ or } H \ . \tag{2.14}$$

Thus, as for Schrödinger Hamiltonians investigation of the existence of gaps in the frequency spectrum of the Maxwell operator $\mathbb{M}$ given by (2.4) requires calculations of $\sigma(M_X(k))$ for all wave vectors k in the Brillouin zone B. It is commonly accepted that it is enough to investigate $\sigma(M_X(k))$ on the main symmetry lines of B since they correspond to those values of the wave vectors for which spectrum degeneracies occur in the homogeneous case; accordingly it is natural to expect that gap occurs if these degeneracies are removed by the periodic perturbation. The interested reader can find in [2] or [19] the Brillouin zone and these symmetry lines for the face centered cubic (fcc) structure which is one of the forms into which many solids crystallize as well as some photonic crystals obtained by colloidal suspensions of dielectric microspheres (see e.g. [30]).

Existence of gaps in more than one dimension is still a rather unexplored mathematical problem. For Schrödinger Hamiltonians there is generically an infinity of gaps in one dimension and the Bethe-Sommerfeld conjecture, now proved by M. Skriganov [25], asserts that there are only a finite number of them in dimension larger than 2. But existence results are mostly those due to B. Dahlberg and E. Trubowitz [10] and are restricted to dimension two. One can argue that gaps exist in the semi-classical regime from the smallness of the broadening of individual energy levels for each potential well due to the tunneling between identical wells. Such an argument does not apply to the Maxwell operators (2.8). At low frequency electromagnetic waves see an homogeneized medium and the frequencies are almost linear in the inverse wave-length; at high frequencies one is in the geometrical optics regime and interference effects are small.

So if gaps occur it must be at intermediate frequencies; in addition the whole thing is complicated by the vector nature of Maxwell equations.

Nevertheless there is now evidence that gaps exist; experimentally this was first observed by Yablonovitch et al [31]. There are also many numerical results for different structures (see e.g. [27],[26] for references). Numerical techniques range from usual truncated plane wave expansion methods to adaptation of the KKR method (see e.g. [2]) to Maxwell equations by A. Moroz [21], [22].

Concerning mathematical approaches the road has been opened by A. Figotin and P. Kuchment in a series of articles devoted to cylindrical square structures [14],[15],[16].

In addition to their intrinsic interest these works have led to some routes towards more general geometries which look rather promising [16]. Essentially it consists in considering some asymptotic regimes in which the problem can be reduced to the investigation of pseudo-differential operators of the Dirichlet-to-Neumann type on graphs, a rich field of investigation for amateurs in δ-perturbations [20]. The next section will be devoted to a condensed presentation of these results.

3. The cylindrical square structure

Cylindrical structures are obtained by drilling parallel cylinders in a piece of homogeneous dielectric. In the geometry considered in [14],[15] they have $\mathbb{Z}^2$ periodicity; their intersections with the plane $x_3 = 0$ consist of $\mathbb{Z}^2$ periodic translates of the squares $[d, 1] \times [d, 1]$ with $0 < d < 1$. Cylinders are filled e.g. with air having dielectric constant normalized to 1 for simplicity; the surrounding homogeneous medium has dielectric constant γ.

One is interested in the regime of large $r = \gamma d$, i.e. of high dielectric contrast, as specified by theorem 1 below. One considers only EM waves propagating perpendicularly to the cylinders hence depending only on two coordinates (x_1, x_2); roughly speaking one works in the improper invariant subspace $\mathcal{H}^\perp \equiv L^2(\mathbb{R}^2, dx_1 dx_2; \mathbb{C}^3) \oplus L^2(\mathbb{R}^2, \varepsilon dx_1 dx_2; \mathbb{C}^3)$ of $\mathbb{M}^2$ as defined by (2.6) and one looks for the spectrum of $\mathbb{M}$ and $\mathbb{M}^2$ "restricted" to $\mathcal{H}^\perp$; we will abuse notations by using the same symbols for these restrictions. Notice that clearly the gaps in their spectrum are not gaps for the original operators $\mathbb{M}$ and $\mathbb{M}^2$ which have spectrum respectively $\mathbb{R}$ and $\mathbb{R}^+$ in this geometry; nevertheless such gaps are obviously of physical interest.

We refer to [15] for the domain analysis and details on the well-known fact that the subspace of $\mathcal{H}^\perp$ satisfying the divergence condition splits into two other invariant subspaces corresponding to different polarizations namely

$$H\text{-modes} : E = (E_1, E_2, 0)\ ,\ H = (0, 0, H_3) \tag{3.1}$$

and

$$E\text{-modes} : E = (0, 0, E_3)\ ,\ H = (H_1, H_2, 0) \tag{3.2}$$

The corresponding spectra are different but the E-modes give the richer structure. Since the electric field is transverse for E-modes it is easy to see that $\omega \in \sigma(\mathbb{M}|_{\text{E-modes}})$ iff $\lambda = \omega^2 \in \sigma(A)$ where $A = -\varepsilon^{-1}\Delta$ acting on $L^2(\mathbb{R}^2, \varepsilon dx_1 x_2)$.

The main result of [15] can be rephrased as follows:

Theorem 1 *Let $r = \gamma d$; there exist $0 < c_1 < c_2 < \infty$ such that if*

i) $r > c_2$

ii) $rd^{1/3} < c_1$

iii) $M << r^2(\log r)^2$

then

$$\sigma(A) \cap [0, M] = \{\sigma_\ell \cup \sigma_a \cup \sigma_d\} \cap [0, M] \tag{3.3}$$

where

$$\sigma_\ell \subset [0, 4r^{-1}] + O(r^2 d^2) \qquad \textit{(Long Wavelength band)} \tag{3.4}$$

$$\sigma_a = \bigcup_{n,m \geq 1} B^a_{n,m} \tag{3.5}$$

with the "Air Bands" $B^a_{n,m}$ given by

$$B^a_{n,m} = \left[\pi^2(n^2 + m^2), \pi^2(n^2 + m^2) + O(r^{-1})(n + m)\right] + O(rd) \tag{3.6}$$

$$\sigma_d = \bigcup_{n \geq 1} B^d_n \tag{3.7}$$

with the "Dielectric Bands" given by

$$B^d_n = \left[2\pi n r^{-1}, 2\pi(n + \gamma_n) r^{-1}\right] + O(r^{-2}) \tag{3.8}$$

where

$$0 < \gamma_n < 1, \lim_{n \to \infty} \gamma_n = \frac{1}{2}\ . \tag{3.9}$$

Thus the spectrum of A has gaps for high dielectric contrast γ and small d which occur at intermediate energies. The spectrum is made up in this intermediate region of two types of bands. First the "Air Bands" corresponding to EM waves mostly localized in air regions, having size $O(r^{-1})$ and distance $O(1)$. Second the "Dielectric Bands" associated to EM waves having most of their energy in the high dielectric region; they also have size $O(r^{-1})$ thus very small. As shown in [14] this second type of spectrum does not show up for H-modes.

In the following we try to present a brief overview of the main ideas used in the proof and refer to [15] for more details sketch of the proof of theorem 1.

The Floquet-Bloch decomposition gives:

$$\sigma(A) = \bigcup_{k \in B} \sigma(A(k)) \tag{3.10}$$

where $B = [0,\pi]^2$ and $A(k)$ is $-\varepsilon^{-1}\Delta$ acting on $\mathcal{H} = L^2([0,1]^2, \varepsilon dx_1 dx_2)$ with the usual Bloch boundary conditions. The main ingredient in the analysis of $\sigma(A)$ is the separable approximation given by the new dielectric function:

$$\varepsilon_0(x_1,x_2) = \nu(x_1) + \nu(x_2) - 1 \tag{3.11}$$

defined as being $\mathbb{Z}^2$ periodic with

$$\nu(x) = \begin{cases} \gamma - 1 & , \quad 0 \le x < d \\ 0 & , \quad d \le x < 1 \end{cases} \tag{3.12}$$

The corresponding operator $A_0(k) = -\varepsilon_0^{-1}\Delta$ now acts on

$$\mathcal{H}_0 = L^2\left([0,1]^2, \varepsilon_0 dx_1 dx_2\right);$$

one looks for eigenfunctions $A_0(k)\varphi(k) = \lambda\varphi(k)$ in product form:

$$\varphi(k; x_1, x_2) = u_1(k_1; x_1) u_2(k_2; x_2) \tag{3.13}$$

with $k_j \in [0,\pi], j = 1,2...$ One obtains then two eigenvalue equations:

$$h(k_j)u_j(k_j) = E_j(\lambda, k_j)u_j(k_j) \ , \ j = 1,2 \ , \tag{3.14}$$

where $h(k) = -\frac{d^2}{dx^2} - \lambda\nu(x)$ with k-dependent Bloch boundary condition, and the *consistency condition*

$$\lambda = E_1(\lambda, k_1) + E_2(\lambda, k_2) \tag{3.15}$$

This separable approximation thus leads to a standard one dimensional problem with periodic square well-potential. As is well-known

$$\bigcup_{k\in[0,\pi]} \sigma(h(k)) = \bigcup_{n\geq 0} b_n(\lambda) \tag{3.16}$$

To obtain the bands $b_n(\lambda)$ is a simple (although lengthy) exercise (see e.g. [1], [24]); let $E(\lambda,k) = \mu^2(\lambda,k)$ and $M(E)$ be the Monodromy Matrix; then the discriminant equation $tr(M(E(\lambda,k)) = 2\cos k$ leads to

$$\left|\frac{\alpha}{2r}\sin(\mu(1-d)) - \cos(\mu(1-d)) - \cos k\right| \le rd \tag{3.17}$$

where α is related to $r = \gamma d$ by

$$|\alpha - \lambda r| = 0(r^2 d^2) \tag{3.18}$$

Taking the r.h.s. of (3.18) equal to zero one obtains, after rescaling μ and α by a factor $(1-d)$,

$$b_n^{KP}(\lambda) = \left[(n\pi)^2, \pi^2(n + \gamma_n(\lambda))^2\right] \text{ for } n \ge 1 \tag{3.19}$$

where

$$\gamma_n(\lambda) = \frac{2}{\pi} \arctan \left(\frac{2\pi(n + \gamma_n(\lambda))}{\lambda r} \right) \tag{3.20}$$

On the other hand if $b_0^{KP}(\lambda) = \left[E_0^-(\lambda), E_0^+(\lambda)\right]$ then (see [1]) one has

$$E_0^+(\lambda) \le 0 \ \text{ for } \alpha \le 4 \tag{3.21}$$

and from tunneling estimates

$$E_0^+(\lambda) \sim -\frac{\alpha^2}{4} \, (\alpha \to \infty), \ |E_0^+(\lambda) - E_0^-(\lambda)| = O(e^{-\frac{\alpha}{2}}) \tag{3.22}$$

Then it is easy to see that

$$b_n^{KP}(\lambda) = b_n(\lambda) + O(rd) \ \text{ if } \mu r d \ll \pi \tag{3.23}$$

which holds under the assumptions of theorem 1. The fact that the Kronig-Penney model is such a good approximation follows from the fact that the potential ν as given by (3.12) tends for r fixed and d going to zero towards $-\alpha\delta_0$ with $\alpha \sim \lambda r$.

Now the consistency condition (3.15) says

$$\lambda \in \sigma(A) \cap [0, M] \leftrightarrow \lambda \in b_n(\lambda) + b_m(\lambda) \tag{3.24}$$

$$\text{for some } (m, n) \in I\!N^2 \text{ and } 0 \le \lambda \le M.$$

The "Air Bands" correspond to $n, m \ge 1$; then $\lambda \sim \pi^2(n^2 + m^2)$, $\gamma_n(\lambda) = O(r^{-1})$ and (3.17) follows. Notice that the gaps open as long as $\lambda \ll r^2(\log r)^2$ from the statistics for sums $m^2 + n^2$ for m, n large.

The "Dielectric Bands" correspond to n or $m = 0$ and $n^2 + m^2 > 0$. Assume $\lambda \gg r^{-1}$ (so $\alpha \sim \lambda r \gg 1$); from (3.22) since the band $b_0(\lambda)$ is centered around $\frac{-\alpha^2}{4}$ and of size $O(e^{-\frac{\alpha}{2}})$ the consistency condition (3.15) gives

$$\lambda \in b_n(\lambda) - \frac{\lambda^2 r^2}{4} + O(rd) \ , \ n > 0.$$

In this regime $\lambda \ll \frac{\lambda^2 r^2}{4}$ so that from (3.19):

$$\lambda \in 2r^{-1} \left[n\pi, \pi(n + \gamma_n(\lambda))\right] + O(r^{-2}), \lim_{n\to\infty} \gamma_n(\lambda) = \frac{1}{2} \tag{3.25}$$

This corresponds to (3.8) for $1 \ll n < Mr$. Notice that the associated EM waves have the product form (3.13) and one of the two factors is strongly localized in the potential wells that is in the high dielectric region.

Finally the long wave-length band corresponds to $n = m = 0$; assume λ is $O(r^{-1})$ so that $\alpha = \lambda r = O(1)$; the consistency condition (3.15) says $\lambda \in 2b_0(\lambda) \cap I\!R^+$; it follows from (3.21) that this last set in non empty only for $\alpha \sim \lambda r \le 4$ leading to (3.4).

The question of how good is the separable approximation (3.11) can be answered as follows. First notice that $\varepsilon_0 \geq \varepsilon$ so that from the mini-max principle eigenvalues of $A_0(k)$ are lower bounds to those of $A(k)$. On the other hand if $A_0(k)\varphi_0(k) = \lambda_0\varphi_0(k)$ with $\varphi_0(k)$ as constructed above and $\|\varphi_0(k)\|_{\mathcal{H}_0} = 1$ an easy estimate gives $\|(A(\lambda) - \lambda_0)\varphi_0(k)\|_{\mathcal{H}} \leq cd\gamma^{1/2}$ and this is much smaller than r^{-1} if $rd^{1/3}$ is small which is assumption ii) of theorem 1. From this it immediately follows that gaps of size $O(r^{-1})$ survive when removing the separable approximation.

To conclude this chapter we remark that the above strategy certainly applies to 3 dimensional cubic structures with the result that dielectric gaps (if they exist) will only be $O(r^{-2})$. It can also be extended to more general structures by investigating the singular limit where the thickness d of the high dielectric region tends to zero while keeping the coefficient $r = d\gamma$ fixed (γ being the dielectric contrast). The physical interest of this regime is of course questionable since γ is of the order of a few units in the currently used dielectrics; nevertheless this approach can bring some insight into the geometries which favours opening of gaps (for example recent numerical work by Y. Ponomarev [23] shows that gaps survive for the square structure at physical values of the parameters d and γ). Denoting by Σ the support of the high dielectric region as d tends to zero, the problem boils down to investigate the kernel of $-\Delta - \lambda r\delta_\Sigma - \lambda\mathbb{1}$ [16]; this leads to problems of the Dirichlet-to-Neumann type for which reliable numerical implementation is available as shown in [20].

4. Localized electromagnetic waves

We now consider situations where the periodic dielectric function is randomly perturbed; so we assume that ε in (2.1-3) is a random function ε_ω on $\mathbb{R}^3$ of the form

$$\varepsilon_\omega(X) = \varepsilon_0(X)(1 + v_\omega(X)) \tag{4.1}$$

where ε_0 is Γ-periodic for some Bravais lattice $\Gamma \subset \mathbb{R}^3$ and v_ω denotes the random perturbation. In [13], [8] one considers the following Anderson type model:

$$v_\omega(X) = \sum_{j\in\Gamma} \lambda_j u(X - j) \tag{4.2}$$

where u is a non negative measurable function with compact support and the λ_j's are independent identically distributed random variables; they have density distribution g supported on a compact interval (in such away that ε_ω given by (4.1) remains bounded below).

Let us remark that this is certainly not the type of randomness which is met when designing photonic crystals. A more natural form would be for example $\varepsilon_\omega(X) = I + \sum_{j\in\Gamma} u(\lambda_j(X-j))$ with $\varepsilon_0(X) = I + \sum_{j\in\Gamma} u(X-j)$, u being supported in the primitive cell of Γ, the random variables λ_j describing e.g. the radius of individual dielectric spheres. Although the results presented below only concern the model (4.1-2) they certainly can be extended to this more realistic last one using techniques developed in [7]. However important models with Poisson or Bernoulli distributed dielectric spheres are to our knowledge completely unexplored.

Coming back to (4.1-2) the mathematical analysis boils down to the standard multiscale methods developed in [17], [11], [6], [3] for Schrödinger operators with Anderson potential (4.2) provided:

There exists a gap $[B_-, B_+]$ *in the spectrum of the unperturbed Maxwell operators (2.8a,b).* (4.3)

The spectrum Σ *of the randomly perturbed Maxwell operators (2.7) or (2.8) is deterministic i.e. almost surely independent of* ω*. Furthermore* Σ *has a strictly smaller gap* $[\widetilde{B_-}, \widetilde{B_+}] \subset [B_-, B_+]$ *(where* $\widetilde{B_-} = \widetilde{B_+}$ *is not excluded).* (4.4)

Provided some additional technical assumptions for which the interested reader can consult [8] or [13] one has:

Theorem 2 *If (4.3) and (4.4) hold there exist constants* $E_\pm$ *satisfying* $B_- < E_- < \widetilde{B_-}$ *and* $\widetilde{B_+} < E_+ < B_+$ *such that with probability one* $\Sigma \cap [E_-, E_+]$ *is pure point with exponentially decaying eigenfunctions.*

In other words strong enough disorder (on which $\widetilde{B_\pm}, E_\pm$ depend) produces localized electromagnetic modes; notice the corresponding squared frequencies are dense in the interval $[E_-, E_+]$.

To conclude this section it is worth mentioning that in [8] periodicity of ε_0 is not required but only (4.3), (4.4). This might be of interest in situations where there is a gap without having to assume an underlying periodic structure on the scale of the wavelength. In fact there exist crystalline materials that are transparent for light except for some frequencies in a gap (ordinary kitchen salt seems to be an example but not in the optical region; so you can't do this experiment at home!) Such a gap arises from avoided crossing of eigenvalues in a coupled phonon-photon system; however projecting out the electromagnetic part obtains a complex and frequency dependent dielectric function which does not fulfill the assumptions we made so far (see however [29] for a discussion of this situation).

5. Radiative decay in photonic crystals

As mentioned before it is expected that spontaneous emission of photons with frequencies in a gap of the Maxwell operator is expected to be strongly inhibited since energy conservation should prevent one-photon processes to occur and the multi-photon ones are expected to take place only on very large time scales (of the order of a day!) However description of matter-radiation couplings requires quantization of the electromagnetic field in a periodic dielectric medium. As compared to canonical quantization procedure in homogeneous media (see e.g. [9]) this requires in particular a redefinition of the usual gauge concept. As shown by A. Tip [28] this can be done in a systematic way. In particular in a suitably modified Coulomb gauge the EM field Hamiltonian has the standard form:

$$H_f = \sum_\alpha \int \omega a^\star(u_{\omega,\alpha}) a(u_{\omega,\alpha}) d\omega \tag{5.1}$$

where the eigenfunction expansion is in terms of solutions of

$$M_E u_{\lambda,\alpha} = \omega^2 u_{\lambda,\alpha} \tag{5.2}$$

with M_E given by (2.8a) (here α labels the degeneracies). Coupling to an hydrogen atom with infinite mass nucleus located at X_0 is described by:

$$H = \frac{1}{2m}\left[p - eA(X_0)\right]^2 - \frac{1}{4\pi\varepsilon(X_0)r} + H_f \tag{5.3}$$

where the vector potential A at point X_0 is

$$A(X_0) = \sum_\alpha \int \chi(\omega)\left[a^\star(u_{\omega,\alpha})\bar{u}_{\omega,\alpha}(X_0) + a(u_{\omega,\alpha})u_{\omega,\alpha}(X_0)\right]\frac{dw}{|w|^{1/2}}$$

(here χ is some ultra-violet cut-off).

Excited states of $H_{at} = \frac{1}{2m}P^2 - \frac{1}{4\pi\varepsilon(X_0)r}$ lie in the continuum of $H_0 = H_{at} + H_f$; interaction is expected to turn them into resonances. Without claim to mathematical rigor and within a two level, one-photon exchange approximation in Fermi's golden rule, one finds an approximate imaginary part of the resonance proportional to the local density of states:

$$N_f(X_\Theta) = \sum_\alpha \int \delta(\lambda_0 - \omega^2)|u_{\omega,\alpha}(X_0)|^2 \frac{dw}{|w|} \tag{5.4}$$

where λ_0 is the transition energy of the excited atomic state.

In particular the lifetime is infinite (in this approximation) if the density of states vanishes at such frequencies hence in particular in a gap. This result strongly supports the conjecture that radiative decay is inhibited in such a situation. It calls for a more rigorous investigation of the spectral properties and resonances for operators of type (5.2); the mathematical technology for this is rapidly growing with in particular the recent works of V. Jakšić and C.A. Pillet ([18]) and V. Bach, J. Fröhlich, I.M. Sigal ([4],[5]). In case randomness is added the spectrum of H is now a mixture of dense point spectrum and continuous spectrum which is of course the worst situation one can imagine in perturbation theory; however one also expects that the density of states (5.3) is very small in the localization region and that the qualitative picture described above remains essentially correct.

Acknowledgments

I want to thank warmly the organizers of this conference for giving me the opportunity to present this promising field of research. Most of what I learned on this subject comes from enjoyable discussions with my collaborators and friends P. Hislop and A. Tip.

References

[1] S. Albeverio, F. Gesztesy, R. Hoegh-Krohn and H. Holden; Solvable models in Quantum Mechanics. *Springer Verlag* 1988.

[2] N.W. Ashcroft and N.D. Mermin; *Solid State Physics* (Saunders College 1976).

[3] J.M. Barbaroux, J.M. Combes, P. Hislop; *Helv. Phys. Acta* **70**, 16 (1997).

[4] V. Bach, J. Fröhlich, I.M. Sigal; *Lett. Math. Phys.* **34**, 183 (1995).

[5] V. Bach, J. Fröhlich, I.M. Sigal; Preprint SFB 288 n° 275, Berlin (1997).

[6] J.M. Combes, P. Hislop; *J. Funct. Anal.* **124**, 149 (1994).

[7] J.M. Combes, P. Hislop and E. Mourre; *Trans. Amer. Math. Soc.* **348**, 4883 (1996).

[8] J.M. Combes, P. Hislop and A. Tip; Preprint Univ. Kentucky; to appear in *Ann. Inst. H. Poincaré.*

[9] C. Cohen-Tannoudji, J. Dupont-Roc and G. Grynberg; Photons and Atoms (Wiley, New-York 1989).

[10] B. Dahlberg, E. Trubowitz; *Comment. Meth. Helv.* **57**, 130-134 (1982).

[11] H. Dreyfus, A. Klein; *Commun. Math. Phys.* **124**, 285 (1989).

[12] H.S. Dorren, A. Tip; *J. Math. Phys.* **32**, 3060 (1991).

[13] A. Figotin, A. Klein; *Commun. Math. Phys.* **184**, 411 (1997).

[14] A. Figotin, P. Kuchment; *SIAM J. Appl. Math* **56**, 68 (1996).

[15] A. Figotin, P. Kuchment; *SIAM J. Appl. Math* **56**, 1561 (1996).

[16] A. Figotin, P. Kuchment; To appear in *SIAM J. Appl. Math.*

[17] J. Fröhlich, T. Spencer; *Commun. Math. Phys.* **88**, 151 (1983).

[18] V. Jakšić, C.A. Pillet; *Commun. Math. Phys.* **178**, 627 (1996).

[19] J.D. Joannopoulos, R.D. Meade and J.N. Winn; *Photonic Crystals* (Princeton University Press 1995).

[20] P. Kuchment, L. Kunyansky; Preprint, Wichita Univ. (1997).

[21] A. Moroz; *Phys. Rev.* **B51**, 2068 (1995).

[22] A. Moroz, C. Sommers; To appear in *Phys. Lett. A*.

[23] I. Ponomarev; Preprint, Wichita Univ. June 1997.

[24] M. Reed, B. Simon; *Methods of Modern Mathematical Physics* Vol IV, ch XIII. Academic Press, New-York 1978.

[25] M.M. Skriganov; *Invent. Math.* **80**, 107 (1985).

[26] Photonic Band Gap Materials; Ed. C.M. Soukoulis (Kluwer, Dordrecht 1996).

[27] Photonic Band Gap and Localization; Ed. C.M. Soukoulis (Plenum, New-York 1993).

[28] A. Tip; *Phys. Rev. A* **56**, 5022 (1997).

[29] A. Tip; *Phys. Rev. A* **57**, 4818 (1998).

[30] I.I. Tahran and G.H. Watson; *Phys. Rev. Lett.* **76**, 315 (1997).

[31] E. Yablonovitch, T.J. Gmitter and K.M. Leung; *Phys. Rev. Lett.* **67**, 2295 (1991).

Operator Theory:
Advances and Applications, Vol. 108

OPTIMAL EIGENVALUES FOR SOME LAPLACIANS AND SCHRÖDINGER OPERATORS DEPENDING ON CURVATURE

Pavel Exner [a,c] *, **Evans M. Harrell** [b] †
and
Michael Loss [b] ‡

a) Nuclear Physics Institute, Academy of Sciences, CZ–25068 Řež
b) School of Mathematics, Georgia Tech, Atlanta, GA 30332-0160, USA
c) Doppler Institute, Czech Technical Univ., Břehová 7, CZ-11519 Prague
exner@ujf.cas.cz, harrell@math.gatech.edu, loss@math.gatech.edu

Abstract

We consider Laplace operators and Schrödinger operators with potentials containing curvature on certain regions of nontrivial topology, especially closed curves, annular domains, and shells. Dirichlet boundary conditions are imposed on any boundaries. Under suitable assumptions we prove that the fundamental eigenvalue is maximized when the geometry is round.

We also comment on the use of coordinate transformations for these operators and mention some open problems.

I. Introduction

In this article we present some theorems in optimal spectral geometry which were suggested by situations where the physics is strongly controlled by the geometry of an interface, i.e., a lower-dimensional structure. Two specific interests are electrical properties of nanoscale structures (quantum wires, waveguides, and resonators) [DuEx], and the slow evolution of interfaces in reaction-diffusion systems such as Allen-Cahn [AlFu].

By "optimal spectral geometry" we mean the determination of the geometry which maximizes or minimizes a certain eigenvalue of a differential operator. The archetypal result of this genre is the Faber–Krahn theorem [Fab], [Kra]:

*Work supported by GA AS No.1048801
†Work supported by N.S.F. grant DMS-9622730
‡Work supported by N.S.F. grant DMS–9500840

1. If one considers the Dirichlet problem for the Laplacian on a region of specified volume (or area, in two dimensions), then it is the ball (disk) which uniquely minimizes the fundamental eigenvalue.

Other well-known theorems of optimal spectral geometry include:

2. (Weinberger [Wei]). If one considers the Neumann problem for the Laplacian on a region of specified volume, then the fundamental eigenvalue is trivially 0. The first positive eigenvalue is uniquely *maximized* by the ball.

3. (Hersch [Her1]). If one considers the Laplacian on closed, simply connected surfaces of specified area embedded in R^3, then, as in the previous situation, the fundamental eigenvalue is trivially 0. The first positive eigenvalue is then uniquely maximized by the sphere.

These theorems are sometimes termed "isoperimetric" in analogy with the classical isoperimetric theorem whereby the ball minimizes the surface area given a fixed volume. The simplicity of the optimizers in these situations may convey a misleading idea of the subtlety of the analysis involved. They involve more than simple rearrangement of the energy functional, and indeed the optimizers of higher eigenvalues are not so easy to characterize or even discover. Moreover, the theorem of Hersch is not true in all dimensions. For reviews of this subject, consult [AsBe], [Ban], and [Her2].

An exception to the statement about higher eigenvalues, however is the recent result in [HaLo], that the *second* eigenvalue of the Laplace operator penalized by the square of the mean curvature is uniquely maximized by the sphere. This result holds in any dimension. Interestingly, in two dimensions analogous facts can be proved, using Hersch's technique of conformal transplantation, for the second eigenvalue of the Laplace operator penalized by a substantially larger family of potentials quadratic in curvature (see [Har]).

In this article we shall present some new theorems where the fundamental eigenvalue is optimized by round geometry, and we attempt to shed light on the role of curvature in the spectra of Laplace and Schrödinger operators. In Section II, we consider the Laplacian on certain non–simply–connected domains and show that, under some circumstances, the optimization of the fundamental eigenvalue contrasts with the Faber–Krahn theorem. In Section III we review a transformation which has long been used to understand connections between curvature and spectra for quantum wires and waveguides, from the point of view of quadratic forms. This has consequences for further conjectures on spectral optimization and for the study of Schrödinger operators depending on curvature. In the final section we prove a new theorem on spectral optimization for some one-dimensional Schrödinger operators.

II. Isoperimetric spectral theorems for annuli and spherical shells

In sections II–III we consider annular domains $D \subset R^{\nu+1}$, consisting of the points on one side of a closed, sufficiently smooth non–self–intersecting subset Ω of dimension ν, and within a distance d of Ω. Our theorems will apply when $\nu = 1$ or 2, and where d is sufficiently small, thus corresponding to physical structures such as quantum wires which form closed loops, or thin capacitors or resonating cavities with special geometries. We shall refer to Ω as the *inner edge* or the *outer edge* of the domain. The area (or length) of Ω will be written as $|\Omega|$.

The edge will be assumed sufficiently smooth that its principal curvatures are defined and bounded at all points (i.e., $\partial\Omega \in C^2$), and it is restricted so that all principal curvatures are bounded in magnitude by $1/d$. (We choose the convention for plane curves which allows the curvature to have either sign. Later, when we treat surfaces, they will be assumed convex, so the principal curvatures will be positive.) This is an important constraint, which will be assumed throughout the article. In the language of differential geometry, it allows the existence of a Fermi coordinate system for D, consisting of a globally defined coordinate $r :=$ distance from Ω, which is orthogonal to the coordinates on the smooth "level surfaces" $\Omega_r := \{x \in D : dist(x, \Omega) = r\}$. (We shall not make special assumptions about the coordinates on the level surfaces.) A set D satisfying these assumptions will be called a *smooth annular domain*, and d will be its *thickness*.

Theorem 1: *a) With $dim(\Omega) = 1$, fix the length $|\Omega|$ and the thickness d,and consider the Dirichlet problem for the Laplacian on all smooth annular domains with Ω as one of the edges, either inner or outer. Then the fundamental eigenvalue λ_1 is uniquely maximized when Ω is a circle.*

b) With $dim(\Omega) = 2$, fix the surface area $|\Omega|$ and the volume, and consider the Dirichlet problem for the Laplacian on all smooth annular domains D with convex outer edges Ω. Then the fundamental eigenvalue λ_1 is uniquely maximized when Ω is a sphere.

Remark. The assumptions in part a) are tantamount to fixing the area of the domain D, and thus when Ω is the outer edge statements a) and b) are analogous. The claim when Ω is the inner edge (or, by nearly the same proof, a central level curve) is an additional fact which appears to be valid only when Ω is one–dimensional.

Proof. a) We first consider the case $\nu = 1$, and normalize so that $|\Omega| = 2\pi$. By the Rayleigh principle,

$$\lambda_1 = \inf \iint |\nabla\zeta|^2 \, d^2x,$$

where the infimum is taken over smooth functions ζ on the closure of D which vanish on its boundary, normalized in L^2. We write this in the orthogonal coordinate system defined by r and $s :=$ arclength of Ω of the point nearest to $\mathbf{x}$, as measured counterclockwise from some reference position:

$$\lambda_1 = \inf \int_0^d \int_0^{2\pi} \left(\frac{1}{(1 \pm \kappa(s)\, r)} \zeta_s^2 + (1 \pm \kappa(s)\, r)\, \zeta_r^2 \right) ds\, dr$$

(e.g., see [Ban], p. 143). Here, κ is the curvature of Ω at s, and we have chosen the orientation whereby the plus sign corresponds to Ω being the inner edge, and the minus sign to it being the outer edge. Suppose now that a smooth test function ζ which vanishes on the boundary of D is independent of s. Then

$$\lambda_1 \leq \int_0^d \int_0^{2\pi} \left((1 \pm \kappa(s)\, r)\, \zeta_r^2 \right) \, ds \, dr = \int_0^d \left(2\pi \, (1 \pm r)\, \zeta_r^2 \right) \, dr.$$

The inequality would be strict if the true ground state were to depend on s. The final expression, however, is equivalent to the one for the corresponding annulus, restricted to the set of test functions independent of s. Since the fundamental eigenfunction for the circular annulus is attained in the set of functions independent of s, we conclude

$$\lambda_1 \leq \lambda_1 \,(\text{annulus})\,.$$

Because equality requires that the fundamental eigenfunction be independent of s, one of the two terms in the eigenvalue equation which includes $\kappa(s)$ is zero due to the presence of ζ_s. This forces the only remaining term to be independent of s for any curve Ω which maximizes λ_1. The only possibility is a circle.

b) Next we turn to the case $\nu = 2$, and normalize so that $|\Omega| = 4\pi$, as for the unit sphere. We begin as before, by using the coordinate r, supplemented by coordinates orthogonal to r on the level surfaces Ω_r. It is convenient to denote $A(r) := |\Omega_r|$. Assuming that the test functions depend only on r, the Rayleigh principle states (for normalized test functions):

$$\lambda_1 \leq \int_0^d \zeta_r^2 \, A(r) dr.$$

We recall here that now the thickness d depends on the domain D; specifically, they are connected by

$$\int_0^d A(r) dr = Vol(D).$$

Let us now change variables to r' defined so that

$$A_0(r') dr' = A(r) dr,$$

where $A_0(r') := 4\pi(1 - r')^2 =$ the area of the sphere of radius $1 - r'$. We find:

$$\lambda_1 \leq \int_0^{d'} \zeta_{r'}^2 \left(\frac{A(r)}{A_0(r')} \right)^2 A_0(r') dr',$$

where d' is the thickness of the spherical shell with the same volume as D. (For brevity we use informal notation for functions of transformed variables, etc.)

We next claim that the expression in parentheses is strictly smaller than 1 for $r > 0$, unless Ω is a sphere. This is because the growth rate of the volume (here, area) element of the level surfaces is the sum of the principal curvatures (e.g., [Kar], eq. (1.5.4) or [Spi], p. 418), and hence

$$A'(r) = - \int_{\Omega_r} (\kappa_1 + \kappa_2)\, \mathrm{d}\sigma \tag{2.1}$$

(cf. [Spi], p. 426). The negative sign arises here because Ω is the outer edge of the domain. At this stage we remark that if Ω is convex, then so are all the level surfaces in a smooth annular domain bounded by Ω, and equation (2.1) can be integrated in a closed form discovered by Steiner ([San], p. 325):

$$A(r) = A(0) - r \int_{\Omega_0} (\kappa_1 + \kappa_2)\; d\sigma + 4\pi r^2$$

Now recall a classical geometric lemma (e.g., [HiCo], p. 225): *Among all closed surfaces of fixed area, bounding a convex set, the sphere is the unique minimizer of the total mean curvature,* $\int_{\Omega_r} (\kappa_1 + \kappa_2)\, \mathrm{d}\sigma$. We see that for a convex surface of a given area, the sphere is the unique maximizer of the negative quantity $A'(r)$. for $r < d$. It then also follows that on this interval, $r' \leq r$ and $0 < A(r) \leq A_0(r) \leq A_0(r')$ with equality only in the case of the sphere.

The argument then proceeds as in part a): The upper bound obtained is equal to the Rayleigh quotient for the spherical shell, with radial test functions. Since the lowest eigenfunction for the spherical shell is radial, $\lambda_1 \leq \lambda_1(\text{shell})$, as claimed. ■

Theorem 1 contrasts with the Faber–Krahn theorem, which might lead one to expect that round annular domains were minimizers rather than maximizers. There are indeed other situations where λ_1 is maximized for the Laplace operator when the volume is fixed. One of these is addressed in a theorem of Payne and Weinberger [PaWe], on annular domains of a given area, where the outer edge is subject to Dirichlet conditions and the inner edge to Neumann conditions. (For some extensions of this theorem, see [Ban], section 4.3.)

Moreover, for the pure Dirichlet problem, the following is a special case of a theorem in [HaKrKu]: *Let the ball B_1 be contained inside the ball B_2. Fix the radii of $B_{1,2}$ but not the position of B_1 in B_2. Among all domains of the form $B_2 \setminus B_1$, the one with the highest fundamental Dirichlet eigenvalue for the Laplacian is the one where they are concentric.*

In these cases as well as in Theorem 1, the domain is allowed to vary within a class of non-simply connected domains. The contrast with Faber–Krahn thus seems to be a topological effect.

For reasons to be explored in the final section, we conjecture that the conclusion of the theorem is false if the dimension is increased to $\nu > 2$.

III. The natural coordinate transformation for the Dirichlet quadratic form

In this section we begin with a coordinate transformation that is old and quite standard (e.g., [daC], [DuEx]), using the Fermi coordinate system of the previous section. Our excuse for thus risking the reader's boredom is that we intend to use the transformation to illuminate an essential geometric feature of the effective potential–energy term which emerges. Also, we shall do the transformation in the context of quadratic forms rather than operators, which appears to us more direct and offers the possibility of relaxing some conditions of smoothness.

We continue to assume that our domain is a smooth annular domain with inner or outer edge Ω. We construct a Fermi coordinate system using r, the distance from Ω, as one coordinate, orthogonal to the rest. We denote the components of the gradient which are parallel to the level surfaces of constant r by $\nabla_{||} f$.

We split the components of the Dirichlet form for the Laplace operator as:

$$\int_D |\nabla_{||} \zeta|^2 \, \mathrm{d}^{\nu+1}x + \int_D |\zeta_r|^2 \, \mathrm{d}^{\nu+1}x$$

and transform the second term (only) as follows. Fix a coordinate system on the edge Ω, and for any point $\mathbf{x}$ in D, choose as its coordinates other than r the coordinates of the closest point on the edge. Let dV^ν denote the volume element on Ω. Then

$$\int_D |\zeta_r|^2 \, \mathrm{d}^{\nu+1}x = \int_\Omega \int_0^d |\zeta_r|^2 \rho(\mathbf{x}) \mathrm{dr} \, \mathrm{dV}^\nu,$$

where ρ is the *volume growth factor*, a familiar quantity in differential geometry [Kar], [Spi]. We write the test function as

$$\zeta = \frac{1}{\sqrt{\rho}} \cdot (\sqrt{\rho}\, \zeta)$$

and use the product rule in the form

$$((fg)')^2 = f^2 (g')^2 + g^2 (f')^2 + \frac{1}{2} (f^2)' (g^2)'$$

to find

$$\int_0^d |\zeta_r|^2 \rho \mathrm{dr} = \int_0^\mathrm{d} \left(|(\sqrt{\rho}\zeta)_\mathrm{r}|^2 + \frac{1}{4} \left(\frac{\rho_\mathrm{r}}{\rho}\right)^2 \zeta^2 \rho + \frac{\rho}{2} \left(\frac{1}{\rho}\right)_\mathrm{r} (\rho\zeta^2)_\mathrm{r} \right) \mathrm{dr}.$$

When the final term is integrated by parts, the full Dirichlet form takes on the appearance

$$\int_A |\nabla_{||} \zeta|^2 \, \mathrm{d}^{\nu+1}x + \int_A q(\mathbf{x}) \, \zeta^2 \, \mathrm{d}^{\nu+1}\mathrm{x} + \int_\Omega \int_0^\mathrm{d} |(\sqrt{\rho}\zeta)_\mathrm{r}|^2 \, \mathrm{dr} \, \mathrm{dV}^\nu. \quad (3.1)$$

The effective potential in the middle contribution is

$$q(\mathbf{x}) := -\frac{1}{4}\left(\frac{\rho_{\mathrm{r}}}{\rho}\right)^2 + \frac{1}{2}\frac{\rho_{\mathrm{rr}}}{\rho}.$$

In order to elucidate the geometric meaning of the effective potential, we recall two elementary facts about the geometry of level surfaces:

$$\begin{aligned} (i) \qquad \frac{\partial}{\partial r}\rho &= \pm\left(\sum_{j=1}^{\nu} \kappa_j\right)\rho, \\ (ii) \quad \frac{\partial}{\partial r}\sum_{j=1}^{\nu} \kappa_j &= \mp \sum_{j=1}^{\nu} \kappa_j^2. \end{aligned}$$

Here, κ_j are the principal curvatures *of the level surface at coordinate* r, and the choice of sign corresponds as usual to the choice of orientation. The signs of (i) and (ii) anticorrelate. (The two-dimensional version of (i) made an appearance in the previous section.)

A short calculation reveals an identity with unambiguous sign regardless of orientation:

Proposition 2: *The Dirichlet form for the Laplacian is given by (3.1) with*

$$q(\mathbf{x}) = \frac{1}{4}\left(\sum_{\mathrm{j}=1}^{\nu} \kappa_{\mathrm{j}}\right)^2 - \frac{1}{2}\sum_{\mathrm{j}=1}^{\nu} \kappa_{\mathrm{j}}^2. \tag{3.2}$$

Effective potentials of the form (3.2) have long been familiar in the study of microstructures with $\nu = 1$ or 2 [daC], [DuEx], [ExŠe]. For the most part they have heretofore appeared as asymptotic calculations with the principal curvatures calculated on the edge, rather than on the level surfaces. In [DuEx] and [ExŠe] a variant effective potential throughout the domain is obtained by a "straightening" transformation. Roughly speaking, the latter articles replace the first term in (3.1) by an integral on a flat manifold, thus avoiding metric tensors, at the price of some derivatives of curvature in the potential energy.

We content ourselves here with a few simple observations about the uses of Proposition 2, which we hope to explore further in a future article.

First, Proposition 2 is a tool for obtaining spectral bounds. A representative such bound is:

Corollary 3:

$$\lambda_1 \geq \frac{\pi^2}{\mathrm{d}^2} + \inf(\mathrm{q}(\mathbf{x})).$$

Proof. This follows immediately once it is realized that the final term in (3.1) can be considered as the Rayleigh quotient for the simple operator $-d^2/dx^2$ with

Dirichlet boundary conditions at 0 and d, because the normalization is unchanged: $\|\sqrt{\rho}\zeta\|_{L^2(D,\mathrm{d}r\mathrm{d}V^\nu)} = \|\zeta\|_{L^2(D,\ \mathrm{d}^{\nu+1}x)}$.

The first term of (3.1) is dropped and in the second the potential is replaced by its infimum. ■

Actually, sharper bounds than Corollary 3 are obtainable using analogues of the Boggio and Hardy inequalities [Bog], [Dav].

Next we observe that although the effective potential is nonpositive if $\nu = 1$ ($q(\mathbf{x}) = -\frac{\kappa^2}{4}$) or 2 ($q(\mathbf{x}) = -\frac{(\kappa_1-\kappa_2)^2}{4}$), as soon as $\nu > 2$ the effective potential becomes positive for the sphere and many other hypersurfaces. While we have neither an isoperimetric theorem nor a counterexample to offer when $\nu > 2$, the simplified model in the next section indicates that the situation can change dramatically when the effective potential may be positive.

IV. Spectral optimization for some Schrödinger operators depending on curvature

In order to begin the analysis of the case where the effective potential is quadratic in the curvature but potentially positive, we consider the lowest eigenvalue of a family of one-dimensional operators parametrized by a real coupling constant g,

$$H(g) := -\frac{d^2}{ds^2} + g\,\kappa^2.$$

We have thus simplified the situation of the previous section by reducing the dimension, but have allowed the effective potential to be positive when $g > 0$. This model is also a natural generalization of one examined in [HaLo], corresponding to the case $g = -1$.

The operator $H(g)$ is defined on a closed planar curve normalized to have length 1. If $g < 0$, it is straightforward to see that of all curves, the circle maximizes the fundamental eigenvalue, as one might conjecture from Theorem 1 by considering infinitesimally thin annular domains approximating a closed curve. This would correspond only to $g = -1/4$, of course, but for all $g < 0$ the statement is easy to obtain from the Rayleigh–Ritz inequality by considering the trial function $\zeta = 1$ and making a simple estimate with the Cauchy–Schwarz inequality.

The situation is more subtle when $g > 0$.

Theorem 4: *a) Suppose that* $0 < g < 1/4$. *Then the circle is the unique curve which minimizes the fundamental eigenvalue* λ_1.

b) Suppose that $1 < g$. *Then the circle does not minimize the fundamental eigenvalue* λ_1.

Remarks. The questions of the critical value of g and the nature of the transition remain open. We conjecture that there is no legitimate minimizing curve when $g > 1$.

Proof. a) Assume first that $0 < g < 1/4$. The minimal value of λ_1, which we denote λ_*, is

$$\inf_{\kappa} \inf_{\zeta} \int \left(\left(\frac{d\zeta}{ds} \right)^2 + g\, \kappa^2 \zeta^2 \right) ds,$$

where the normalized ζ varies over the class of smooth periodic positive functions, while κ is the curvature function defining a planar curve. The assumption of positivity may be imposed because of the positivity property of ground states of Schrödinger operators. We relax the conditions on κ so that we require only that it be a function with integral 2π, i.e., we do not explicitly require that the curve be closed. We shall see that the minimizing curve is still a circle and therefore closed, so this causes no harm.

Because the quantity in question is an iterated infimum, it may be calculated in the other order. By Cauchy-Schwarz's inequality

$$2\pi = \int \frac{\kappa}{\zeta} \zeta ds \leq \left(\int \frac{1}{\zeta^2} ds \right)^{1/2} \left(\int \kappa^2 \zeta^2 ds \right)^{1/2},$$

with equality only if

$$\kappa = \left(2\pi / \int \frac{1}{\zeta^2} ds \right) \frac{1}{\zeta^2}.$$

Hence λ_* is identical to the infimum of the unusual functional

$$E(\zeta) := \int \left(\frac{d\zeta}{ds} \right)^2 ds + \frac{4\pi^2 g}{\int \left(\frac{1}{\zeta^2} \right) ds}.$$

By choosing the trial function which becomes exact for the circle, i.e., $\zeta \equiv 1$, it follows that $\lambda_* \leq E(\zeta) = 4g\pi^2 < \pi^2$ for $g < 1/4$.

In order to establish the existence of a minimizer for $E(\zeta)$, we need a pointwise estimate:

Lemma 5: *If $E(\zeta) \leq \pi^2$ for a positive test function ζ normalized in L^2, then*

$$\inf_s (\zeta(s)) > 1 - \frac{\sqrt{E(\zeta)}}{\pi}.$$

Proof of Lemma 5.

$$E(\zeta) > \int_0^1 (\zeta')^2 ds = \int_0^1 (\zeta - \zeta_{\min})'^2 ds \geq \pi^2 \int_0^1 (\zeta - \zeta_{\min})^2 ds,$$

because $\zeta - \zeta_{min}$ is an admissible test function for the operator $-d^2/ds^2$ with an additional Dirichlet boundary condition at the position of ζ_{min}, and that operator is bounded below by its lowest eigenvalue π^2, in the sense of quadratic forms. Hence

$$E(\zeta) > \pi^2 \left(\int_0^1 \zeta^2 ds - 2\,\zeta_{\min} \int_0^1 \zeta ds + \zeta_{\min}^2 \right),$$

which implies the claimed bound when we use the Cauchy-Schwarz inequality to replace $\int_0^1 \zeta \, ds \le \sqrt{\int_0^1 \zeta^2 ds} = 1$ and then solve for ζ_{min}. ■

We now continue with the proof of Theorem 4. Because of the lemma, if $\lambda_* < \pi^2$, then any minimizing sequence for $E(\zeta)$ is bounded in the Sobolev space H^1_{per}. By a standard compactness theorem, a subsequence converges uniformly to a limit ζ_* which, by the lemma, is strictly positive. The same function is also a weak limit in the H^1_{per} sense of a subsequence, from which it follows that $E(\zeta_*) = \lambda_*$ as claimed.

We have thus established the existence of the minimizer ζ_*, which is a non-negative periodic function on the interval $[0,1]$. Next we observe that ζ_* satisfies the Euler equation for the functional E, which is found to be

$$-\zeta_*'' + M\,\frac{1}{\zeta_*^3} = C\,\zeta_*, \tag{4.1}$$

where

$$M = \frac{4\pi^2 g}{\left(\int_0^1 \frac{1}{\zeta_*^2}\, ds\right)^2}$$

and C is a Lagrange multiplier. Since $E(\zeta_*) = \lambda_*$, we find $C = \lambda^*$ by multiplying (4.1) by ζ_* and integrating by parts.

If the minimizer is a constant function, $\zeta_*(s) = 1$, we have $\lambda_* = M = 4\pi^2 g$. It remains to be seen whether ζ_* may be nonconstant. We multiply equation (4.1) by ζ_*' and integrate; this yields

$$\zeta_*'^2 + \frac{M}{\zeta_*^2} + \lambda_*\zeta_*^2 = C' \tag{4.2}$$

for some C'. The minimizer is normalized by assumption, so integrating (4.2) we find $C' = 2\lambda_*$. Hence the last equation can be rewritten as

$$\zeta_*^2\zeta_*'^2 = \lambda_* - M - \lambda_*\left(\zeta_*^2 - 1\right)^2.$$

It follows that ζ_* oscillates between its (positive) minimal value $\sqrt{1-\sqrt{1-M/\lambda_*}}$ and the maximum $\sqrt{1+\sqrt{1-M/\lambda_*}}$, being strictly monotonic between them. The corresponding solution of (4.1) is given by

$$\zeta_*^2 = 1 + \sqrt{1-M/\lambda_*}\,\cos\left(2\sqrt{\lambda_*}(s-s_0)\right);$$

it is unique if we fix the point s_0 where the maximum is reached. However, a nonconstant function of this type cannot be periodic on $[0,1]$, because $\lambda_* < \pi^2$.

b) Suppose now that $g > 1$. The lowest eigenvalue for the circle is $4\pi^2 g$, so we need to show that lower eigenvalues are attainable, which we proceed to do with an explicit example. (While our example will have a discontinuous curvature, it can be approximated arbitrarily well by curves for which $\kappa(s)$ is perturbed arbitrarily slightly in the L^2 sense, which implies that the eigenvalues are shifted by arbitrarily small amounts [ReSi].) Consider the thin stadium–shaped curve defined by $\kappa(s) = \frac{\pi}{\varepsilon}$ for $1/2 - \epsilon < s < 1/2$ and $1 - \epsilon < s < 1$, and 0 otherwise, for some $\epsilon << 1/2$. Now estimate the fundamental eigenvalue using the Rayleigh-Ritz inequality and the trial function $\sin\left(\frac{\pi s}{\frac{1}{2} - \varepsilon}\right)$ for $0 < s < 1/2 - \epsilon$ and 0 otherwise; the result is $\lambda_1 \leq \left(\frac{\pi}{\frac{1}{2} - \varepsilon}\right)^2$, which can be made arbitrarily close to $4\pi^2$. ■

Remark: The proof of a) shows that the circle is also a minimizer if $g = 1/4$, however we do not know whether it is the unique minimizer.

Acknowledgments

The authors are grateful to Mark Ashbaugh, Michiel van den Berg, Rick Laugesen, and the referee for comments and references. E.H. wishes to thank the Erwin Schrödinger Institut and the Nuclear Physics Institute at Řež u Prahy for hospitality while some of this work was done.

References

[AlFu] Nicholas D. Alikakos and Giorgio Fusco, The spectrum of the Cahn-Hilliard operator for generic interface in higher space dimensions, *Indiana U. Math. J.* **4**, 1993, pp. 637–674.

[AsBe] Mark S. Ashbaugh and Rafael D. Benguria, Isoperimetric inequalities for eigenvalue ratios, pp. 1–36 in: A. Alvino, E. Fabes, and G. Talenti, eds., Partial Differential Equations of Elliptic Type, Cortona, 1992. Cambridge: Cambridge University Press, 1994.

[Ban] Catherine Bandle, Isoperimetric inequalities and applications, Pitman Monographs and Studies in Mathematics **7**. Boston: Pitman, 1980.

[Bog] Tommaso Boggio, Sull'equazione del moto vibratorio delle membrane elastiche, *Rend. Accad. Lincei*, sci. fis., ser. 5 **16**(1907)386-393.

[daC] R.C.T. da Costa, Quantum mechanics of a constrained particle, Phys Rev. **A23**(1981) 1982–1987 (and later articles).

[Dav] E.B. Davies, A review of Hardy inequalities, preprint 1998 available electronically as http://xxx.lanl.gov/abs/math.SP/9809159.

[DuEx] Pierre Duclos and Pavel Exner, Curvature–induced bound states in quantum waveguides in two and three dimensions, *Rev. Math. Phys.* **7**(1995)73–102.

[ExŠe] Pavel Exner and Pavel Šeba, Bound states in curved quantum waveguides, *J. Math. Phys.* **30**(1989)2574–2580.

[Fab] G. Faber, Beweis, dass unter allen homogenen Membranen von gleicher Fläche und gleicher Spannung die kreisförmige den tiefsten Grundton gibt, *Sitzungsber. der mathematisch-physikalischen Klasse der Bayer. Akad. der Wiss. zu München* (1923)169–172.

[Har] Evans M. Harrell II, On the second eigenvalue of the Laplace operator penalized by curvature, *Journal of Differential Geometry and Applications* **6**(1996)397–400.

[HaKrKu] Evans M. Harrell II, Pawel Kröger, and Kazuhiro Kurata, work in progress.

[HaLo] Evans M. Harrell II and Michael Loss, On the Laplace operator penalized by mean curvature, *Commun. Math. Physics* **195**(1998)643-650.

[Her1] Joseph Hersch, Quatre propriétés isopérimetriques de membranes sphériques homogènes, *C.R. Acad. Sci. Paris*, sér A-B **270**, 1970, pp. A1645–1648.

[Her2] Joseph Hersch, Isoperimetric monotonicity: Some properties and conjectures (connections between isoperimetric inequalities), *SIAM Review* **30** (1988)551–577.

[HiCo] D. Hilbert & S. Cohn–Vossen. Geometry and the Imagination. New York: Chelsea, 1952.

[Kar] Hermann Karcher, Riemannian comparison constructions, pp. 170-222 in: S. Chern, editor, Global Differential Geometry, Studies in Mathematics **27**. Washington: Math. Assoc. Amer., 1989.

[Kra] E. Krahn, Über Minimaleigenschaften der Kugel in drei und mehr Dimensionen, *Acta Comm. Univ. Tartu* (Dorpat) **A9**(1926)1–44.

[PaWe] L.E. Payne and H.F. Weinberger, Some isoperimetric inequalities for membrane frequencies and torsional rigidity, *J. Math. Anal. Appl.* **2**(1961) 210–216.

[ReSi] M. Reed and B.Simon, Methods of Modern Mathematical Physics, **IV**: Analysis of Operators. New York: Academic Press, 1978.

[Spi] Michael Spivak, A Comprehensive Introduction to Differential Geometry, **IV**, second edition. Houston: Publish or Perish, 1979.

[San] L.A. Santalo, Integral Geometry, pp.303–350, in: S. Chern, ed., *Global Differential Geometry*, Studies in Mathematics **27**. Washington: Math. Assoc. Amer., 1989.

[Wei] H.F. Weinberger, An isoperimetric inequality for the n–dimensional free membrane problem, *J. Rat. Mech. Anal.* **5**(1956)633–636.

Operator Theory:
Advances and Applications, Vol. 108

The Spectral Shift Operator

Fritz Gesztesy,[a] Konstantin A. Makarov,[a] and Serguei N. Naboko[b]

a) Department of Mathematics, University of Missouri, Columbia, MO 65211, USA
b) Department of Mathematical Physics, St. Petersburg State University, 198904 St. Petersburg, Russia
fritz@math.missouri.edu
http://www.math.missouri.edu/people/faculty/fgesztesypt.html
makarov@azure.math.missouri.edu
naboko@snoopy.phys.spbu.ru

Abstract

We introduce the concept of a spectral shift operator and use it to derive Krein's spectral shift function for pairs of self-adjoint operators. Our principal tools are operator-valued Herglotz functions and their logarithms. Applications to Krein's trace formula and to the Birman-Solomyak spectral averaging formula are discussed.

1 Introduction

Krein's spectral shift function [51], [52], [54] has received enormous attention in the past due to its widespread applications in a variety of fields including scattering theory ([5], Ch. 19, [9], [10], [12], [13], [44], [46], [52], [84], [88], Ch. 8), relative index theory ([3], [14], [15], [38], [59], Ch. X), spectral averaging ([2], [6], [11], [27], [42], [43], [66], [79]–[81]) and its application to localization properties of random Hamiltonians ([17], [20], [21], [22], Ch. VIII, [24], [25], [47]–[49], [64], Ch. V, [78], [79]), eigenvalue counting functions and spectral asymptotics ([10], [23], [29], [30], [67]–[69]), semi-classical approximation ([70]–[72], [73]) and trace formulas for one-dimensional Schrödinger and Jacobi operators ([31]–[33], [35], [39]). Detailed reviews on the spectral shift function and its applications up to 1993 were published by Birman and Yafaev [12], [13]. These two papers contain extensive bibliographies and the interested reader will find many additional references therein. Historically, the concept of a spectral shift function was introduced by I. M. Lifshits [56], [57].

Our main contribution to this circle of ideas is the introduction of a spectral shift operator $\Xi(\lambda, H_0, H)$ for a.e. $\lambda \in \mathbb{R}$, associated with a pair of self-adjoint operators H_0, $H = H_0 + V$ with $V \in \mathcal{B}_1(\mathcal{H})$ ($\mathcal{H}$ a complex separable Hilbert space). In the special cases of sign-definite perturbations $V \geq 0$ and $V \leq 0$, $\Xi(\lambda, H_0, H)$ turns out to be a trace class operator in $\mathcal{H}$, whose trace coincides with Krein's spectral shift function $\xi(\lambda, H_0, H)$ for the pair (H_0, H). While the special case $V \geq 0$ has previously been studied by Carey [18], our aim here is to treat the case of general interactions V by separately introducing the positive and negative parts $V_\pm = (|V| \pm V)/2$ of V. In general, if V is not sign-definite, then $\Xi(\lambda, H_0, H)$ is not necessarily of trace class. However, we will introduce trace class operators $\Xi_\pm(\lambda)$ naturally associated with $V_\pm$, acting in distinct Hilbert spaces $\mathcal{H}_\pm$, such that

$$\xi(\lambda, H_0, H) = \mathrm{Tr}_{\mathcal{H}_+}(\Xi_+(\lambda)) - \mathrm{Tr}_{\mathcal{H}_-}(\Xi_-(\lambda)) \text{ for a.e. } \lambda \in \mathbb{R}. \tag{1.1}$$

(An alternative approach to $\xi(\lambda, H_0, H)$, which does not rely on separately introducing V_+ and V_-, will be discussed elsewhere [36].)

Our main techniques are based on operator-valued Herglotz functions (continuing some of our recent investigations in this area [34], [37], [40]) and especially, on a detailed study of logarithms of Herglotz operators in Section 2. In Section 3 we introduce the spectral shift operator $\Xi(\lambda, H_0, H)$ associated with the pair (H_0, H) and relate it to Krein's spectral shift function $\xi(\lambda, H_0, H)$ and his celebrated trace formula [51]. Finally, Section 4 provides an application to spectral averaging originally due to Birman and Solomyak [11] and hints at operator-valued generalizations thereof. A number of additional applications of this formalism will appear elsewhere [36].

2 Logarithms of Operator-Valued Herglotz Functions

The principal purpose of this section is to study operator-valued Herglotz functions and particularly their logarithms and associated representation theorems. In this manner we shall obtain operator-valued generalizations of some of the classical results on exponential Herglotz representations studied by Aronszajn and Donoghue [4].

In the following $\mathcal{H}$ denotes a complex separable Hilbert space with scalar product $(\cdot, \cdot)_\mathcal{H}$ (linear in the second factor) and norm $\|\cdot\|_\mathcal{H}$, $I_\mathcal{H}$ the identity operator in $\mathcal{H}$, $\mathcal{B}(\mathcal{H})$ the Banach space of bounded linear operators defined on $\mathcal{H}$, $\mathcal{B}_p(\mathcal{H})$, $p \geq 1$ the standard Schatten-von Neumann ideals of $\mathcal{B}(\mathcal{H})$ (cf., e.g., [41], [77]) and $\mathbb{C}_+$ (resp., $\mathbb{C}_-$) the open complex upper (resp., lower) half-plane. Moreover, real and imaginary parts of a bounded operator $T \in \mathcal{B}(\mathcal{H})$ are defined as usual by $\mathrm{Re}(T) = (T + T^*)/2$, $\mathrm{Im}(T) = (T - T^*)/(2i)$.

Definition 2.1. $M : \mathbb{C}_+ \to \mathcal{B}(\mathcal{H})$ is called an *operator-valued Herglotz function* (in short, a *Herglotz operator*) if M is analytic on $\mathbb{C}_+$ and $\mathrm{Im}(M(z)) \geq 0$ for all $z \in \mathbb{C}_+$.

Theorem 2.2. (Birman and Entina [8], de Branges [26], Naboko [60]–[62].) *Let $M : \mathbb{C}_+ \to \mathcal{B}(\mathcal{H})$ be a Herglotz operator.*
(i) Then there exist bounded self-adjoint operators $A = A^* \in \mathcal{B}(\mathcal{H})$, $0 \leq B \in \mathcal{B}(\mathcal{H})$, a Hilbert space $\mathcal{K} \supseteq \mathcal{H}$, a self-adjoint operator $L = L^*$ in $\mathcal{K}$, a bounded nonnegative operator $0 \leq R \in \mathcal{B}(\mathcal{K})$ with $R|_{\mathcal{K}\ominus\mathcal{H}} = 0$ such that

$$M(z) = A + Bz + R^{1/2}(I_{\mathcal{K}} + zL)(L - z)^{-1}R^{1/2}|_{\mathcal{H}} \tag{2.1a}$$
$$= A + (B + R|_{\mathcal{H}})z + (1 + z^2)R^{1/2}(L - z)^{-1}R^{1/2}|_{\mathcal{H}}. \tag{2.1b}$$

(ii) Let $p \geq 1$. Then $M(z) \in \mathcal{B}_p(\mathcal{H})$ for all $z \in \mathbb{C}_+$ if and only if $M(z_0) \in \mathcal{B}_p(\mathcal{H})$ for some $z_0 \in \mathbb{C}_+$. In this case necessarily $A, B, R \in \mathcal{B}_p(\mathcal{H})$.
(iii) Let $M(z) \in \mathcal{B}_1(\mathcal{H})$ for some (and hence for all) $z \in \mathbb{C}_+$. Then $M(z)$ has normal boundary values $M(\lambda+i0)$ for (Lebesgue) a.e. $\lambda \in \mathbb{R}$ in every $\mathcal{B}_p(\mathcal{H})$-norm, $p > 1$. Moreover, let $\{E_L(\lambda)\}_{\lambda\in\mathbb{R}}$ be the family of orthogonal spectral projections of L in $\mathcal{K}$. Then $R^{1/2}E_L(\lambda)R^{1/2}$ is $\mathcal{B}_1(\mathcal{H})$-differentiable for a.e. $\lambda \in \mathbb{R}$ and denoting the derivative by $d(R^{1/2}E_L(\lambda)R^{1/2})/d\lambda$, $\mathrm{Im}(M(z))$ has normal boundary values $\mathrm{Im}(M(\lambda + i0))$ for a.e. $\lambda \in \mathbb{R}$ in $\mathcal{B}_1(\mathcal{H})$-norm given by

$$\lim_{\varepsilon\downarrow 0} \|\pi^{-1}\mathrm{Im}(M(\lambda + i\varepsilon)) - d(R^{1/2}E_L(\lambda)R^{1/2}|_{\mathcal{H}})/d\lambda\|_{\mathcal{B}_1(\mathcal{H})} = 0 \ \text{a.e.} \tag{2.2}$$

Actually, the normal boundary values $M(\lambda + i0)$ in Theorem 2.2 (iii) can be replaced by nontangential ones. Since this distinction will play no role in the remainder of this paper we omit the corresponding details.

Theorem 2.2 (i) follows by considering the quadratic form $(\varphi, M(z)\varphi)_{\mathcal{H}}$, resulting in a scalar Herglotz function, in combination with Naimark's dilation theorem (cf., e.g., Theorem 1 in Appendix I of [16]). Details can be found in [60]. For Theorem 2.2 (ii), (iii) we refer to [61], [62]. In particular, p cannot be chosen equal to 1 in Theorem 2.2 (iii) (cf. [61]). Moreover, if $M(z_0) \in \mathcal{B}_p(\mathcal{H})$ for some $z_0 \in \mathbb{C}_+$ and some $p > 1$, then $M(z)$ need not even have boundary values $M(\lambda + i0)$ in the weak topology of $\mathcal{H}$ for a.e. $\lambda \in \mathbb{R}$. In fact, the quadratic form $(f, M(\lambda + i\varepsilon)f)_{\mathcal{H}}$ may converge for f in a fixed set $\mathcal{D} \subset \mathcal{H}$ (independent of λ) to $(f, M(\lambda + i0)f)_{\mathcal{H}}$ for a.e. $\lambda \in \mathbb{R}$ as $\varepsilon \downarrow 0$, however, $M(\lambda + i0)$ may be a densely defined unbounded operator in $\mathcal{H}$ for a.e. $\lambda \in \mathbb{R}$ (cf. [61]).

Originally, the existence of normal limits $M(\lambda + i0)$ for a.e. $\lambda \in \mathbb{R}$ in $\mathcal{B}_2(\mathcal{H})$-norm, in the special case $A = 0$, $B = -R|_{\mathcal{H}}$, assuming $M(z) \in \mathcal{B}_1(\mathcal{H})$, was proved by de Branges [26] in 1962. (The more general case in (2.1) can easily be reduced to this special case.) In his paper [26], de Branges also proved the existence of normal limits $\mathrm{Im}(M(\lambda+i0))$ for a.e. $\lambda \in \mathbb{R}$ in $\mathcal{B}_1(\mathcal{H})$-norm and obtained (2.2). These results and their implications on stationary scattering theory were subsequently studied in detail by Birman and Entina [7], [8]. (Textbook representations of this material can also be found in [5], Ch. 3.)

Invoking the family of orthogonal spectral projections $\{E_L(\lambda)\}_{\lambda\in\mathbb{R}}$ of L, (2.1) then yields the familiar representation

$$M(z) = A + Bz + \int_{\mathbb{R}} (1+\lambda^2) d(R^{1/2} E_L(\lambda) R^{1/2}|_{\mathcal{H}})((\lambda - z)^{-1} - \lambda(1+\lambda^2)^{-1}), \tag{2.3}$$

where for our purpose it suffices to interpret the integral in (2.3) in the weak sense. Further results on representations of the type (2.3) can be found in [16], Sect. I.4 and [76].

Before we continue our investigations on Herglotz operators a few comments concerning our terminology are perhaps in order. The representation (2.3) in the special case of scalar Herglotz functions is due to Riesz and Herglotz, respectively, Nevanlinna. The former authors discussed the analog of (2.3) in the open unit disk $D \subset \mathbb{C}$, whereas the latter studied (2.3) in $\mathbb{C}_+$. As a consequence, functions of the type (2.3) are frequently called Herglotz or Nevanlinna functions. Moreover, researchers in the former Soviet Union coined the term R-functions. Here we follow the traditional terminology in mathematical physics which seems to prefer the notion of Herglotz functions.

Since we are interested in logarithms of Herglotz operators, questions of their invertibility naturally arise. The following result clarifies the situation.

Lemma 2.3. *Suppose M is a Herglotz operator with values in $\mathcal{B}(\mathcal{H})$. If $M(z_0)^{-1} \in \mathcal{B}(\mathcal{H})$ for some $z_0 \in \mathbb{C}_+$ then $M(z)^{-1} \in \mathcal{B}(\mathcal{H})$ for all $z \in \mathbb{C}_+$.*

Proof. Suppose there is a $z_1 \in \mathbb{C}_+$ and a sequence $\{e_n\}_{n\in\mathbb{N}} \subset \mathcal{H}$ such that $\|e_n\|_{\mathcal{H}} = 1$, $n \in \mathbb{N}$ but $\lim_{n\to\infty} \|M(z_1)e_n\|_{\mathcal{H}} = 0$. Then $\lim_{n\to\infty}(e_n, \operatorname{Im}(M(z_1))e_n)_{\mathcal{H}} = \lim_{n\to\infty} \|(\operatorname{Im}(M(z_1)))^{1/2} e_n\|_{\mathcal{H}} = 0$ and hence

$$\lim_{n\to\infty} \|\operatorname{Im}(M(z_1))e_n\|_{\mathcal{H}} = 0. \tag{2.4}$$

By (2.1a),

$$\begin{aligned} \operatorname{Im}(M(z)) &= \operatorname{Im}(z)(B + R^{1/2}(I_{\mathcal{K}} + L^2)((L - \operatorname{Re}(z))^2 + (\operatorname{Im}(z))^2)^{-1} R^{1/2}|_{\mathcal{H}}) \\ &= \operatorname{Im}(z)(B + R^{1/2} C(z) R^{1/2}|_{\mathcal{H}}), \end{aligned} \tag{2.5}$$

where $C(z) = (I_{\mathcal{K}} + L^2)((L - \operatorname{Re}(z))^2 + (\operatorname{Im}(z))^2)^{-1} > 0$ is invertible in $\mathcal{K}$, $C(z)^{-1} \in \mathcal{B}(\mathcal{K})$, $z \in \mathbb{C}_+$. Thus, (2.4) implies $\lim_{n\to\infty}((e_n, Be_n)_{\mathcal{H}} + (R^{1/2}|_{\mathcal{H}} e_n, C(z_1) R^{1/2}|_{\mathcal{H}} e_n)_{\mathcal{H}}) = 0$ and hence $\lim_{n\to\infty} \|B^{1/2} e_n\|_{\mathcal{H}} = 0$, and $\lim_{n\to\infty} \|Be_n\|_{\mathcal{H}} = 0$, $\lim_{n\to\infty} \|R^{1/2}|_{\mathcal{H}} e_n\|_{\mathcal{H}} = 0$. Consequently,

$$M(z_1)e_n = Ae_n + z_1 Be_n + (R^{1/2}(I_{\mathcal{K}} + z_1 L)(L - z_1)^{-1}) R^{1/2}|_{\mathcal{H}} e_n \tag{2.6}$$

with $R^{1/2}(I_{\mathcal{K}} + z_1 L)(L - z_1)^{-1} \in \mathcal{B}(\mathcal{K})$ yields $\lim_{n\to\infty} \|Ae_n\|_{\mathcal{H}} = 0$. Applying (2.1a) again we infer $\lim_{n\to\infty} \|M(z)e_n\|_{\mathcal{H}} = 0$ for all $z \in \mathbb{C}_+$ contradicting our hypothesis $M(z_0)^{-1} \in \mathcal{B}(\mathcal{H})$. This argument shows two facts. First, by choosing

$e_n = e \in \ker(M(z_1))$ it yields $\ker(M(z)) = \{0\}$ for all $z \in \mathbb{C}_+$, and second, it proves the boundedness of $M(z)^{-1} : \operatorname{ran}(M(z)) \to \mathcal{H}$, $z \in \mathbb{C}_+$. In particular, one infers that $\operatorname{ran}(M(z)) = \operatorname{dom}(M(z)^{-1})$ is a closed subspace of $\mathcal{H}$. By exactly the same argument one derives $\{0\} = \ker(M(z)^*) = (\operatorname{ran}(M(z))^\perp$, $z \in \mathbb{C}_+$ and thus, $M(z)^{-1} \in \mathcal{B}(\mathcal{H})$ for all $z \in \mathbb{C}_+$. □

Lemma 2.3 admits the following generalization in connection with parametrices, that is, generalized inverses, familiar from treatments of the Calkin algebra. In the following we denote by $\mathcal{F}(\mathcal{H})$ the algebra of finite-rank operators on $\mathcal{H}$ and recall that T is called a left (resp., right) parametrix of $M(z_0)$ if $(TM(z_0) - I_\mathcal{H}) \in \mathcal{F}(\mathcal{H})$ (resp., $(M(z_0)T - I_\mathcal{H}) \in \mathcal{F}(\mathcal{H})$).

Lemma 2.4. *Suppose M is a Herglotz operator with values in $\mathcal{B}(\mathcal{H})$. If $M(z_0)$ has a left (resp., right) parametrix for some $z_0 \in \mathbb{C}_+$, then $M(z)$ has a left and right parametrix for all $z \in \mathbb{C}_+$.*

Proof. From the proof of Lemma 2.3 one infers

$$\ker(M(z_1)) = \ker(M(z_2)) \text{ for all } z_1, z_2 \in \mathbb{C}_+, \tag{2.7a}$$

$$\ker(M(z)^*) = \ker(M(z)) = (\operatorname{ran}(M(z)))^\perp, \quad z \in \mathbb{C}_+ \tag{2.7b}$$

and hence $M(z)$ can be decomposed as

$$M(z) = \begin{pmatrix} \widetilde{M}(z) & 0 \\ 0 & 0 \end{pmatrix}, \quad \mathcal{H} = \widetilde{\mathcal{H}} \oplus \ker(M(i)), \tag{2.8}$$

where

$$\ker(\widetilde{M}(z)) = \{0\}, \quad \overline{\operatorname{ran}(\widetilde{M}(z))} = \widetilde{\mathcal{H}}, \quad z \in \mathbb{C}_+. \tag{2.9}$$

From $TM(z_0) - I_\mathcal{H} = F \in \mathcal{F}(\mathcal{H})$ one concludes $\ker(M(z_0)) \subseteq \ker(I_\mathcal{H} + F)$ and hence

$$\dim(\ker(M(z))) < \infty \text{ for all } z \in \mathbb{C}_+. \tag{2.10}$$

Denote by $\widetilde{P}$ the orthogonal projection onto $\widetilde{\mathcal{H}}$, $\widetilde{\mathcal{H}} = \widetilde{P}\mathcal{H}$ then $\widetilde{P}TM(z_0)\widetilde{P} = \widetilde{P}T\widetilde{P}\widetilde{M}(z_0) = \widetilde{P}(I_\mathcal{H} + F)\widetilde{P} = I_{\widetilde{\mathcal{H}}} + \widetilde{F}$. In order to prove that $\widetilde{M}(z_0)^{-1} \in \mathcal{B}(\widetilde{\mathcal{H}})$ we argue as follows: Suppose there is a sequence $\{e_n\}_{n\in\mathbb{N}} \subset \widetilde{\mathcal{H}}$, $\|e_n\|_{\widetilde{\mathcal{H}}} = 1$ such that

$$\lim_{n\to\infty} \|\widetilde{M}(z_0)e_n\|_{\widetilde{\mathcal{H}}} = 0. \tag{2.11}$$

Hence $\lim_{n\to\infty} \|\widetilde{P}T\widetilde{P}\widetilde{M}(z_0)e_n\|_{\widetilde{\mathcal{H}}} = \lim_{n\to\infty} \|(I_{\widetilde{\mathcal{H}}} + \widetilde{F})e_n\|_{\widetilde{\mathcal{H}}} = 0$. Next, let $\widetilde{E}_{-1}$ be the spectral projection onto the geometric eigenspace of $\widetilde{F}$ corresponding to the eigenvalue -1, and

$$\widetilde{E}_{-1} \dot{+} \widetilde{E}_{-1}^\perp = I_{\widetilde{\mathcal{H}}} \tag{2.12}$$

(here $\dot{+}$ denotes the direct sum but not necessarily the orthogonal direct sum $\oplus$). Then

$$\lim_{n\to\infty} \|(I_{\tilde{\mathcal{H}}} + \widetilde{F})\widetilde{E}_{-1}^{\perp} e_n\|_{\tilde{\mathcal{H}}} = 0. \tag{2.13}$$

Since $I_{\tilde{\mathcal{H}}} + \widetilde{F}$ is boundedly invertible on $\widetilde{E}_{-1}^{\perp}\widetilde{\mathcal{H}}$, one infers

$$\lim_{n\to\infty} \|\widetilde{E}_{-1}^{\perp} e_n\|_{\tilde{\mathcal{H}}} = 0. \tag{2.14}$$

Introducing $\tilde{e}_n = \widetilde{E}_{-1}e_n$, $n \in \mathbb{N}$, one concludes from (2.11), (2.12)–(2.14) that $\lim_{n\to\infty} \|\tilde{e}_n\|_{\tilde{\mathcal{H}}} = 1$, $\lim_{n\to\infty} \|\widetilde{M}(z_0)\tilde{e}_n\|_{\tilde{\mathcal{H}}} = 0$. Since $\{\tilde{e}_n\}_{n\in\mathbb{N}}$ is compact, there exists a subsequence $\{\tilde{e}_{n_k}\}_{k\in\mathbb{N}} \subseteq \{\tilde{e}_n\}_{n\in\mathbb{N}}$ such that $\lim_{k\to\infty} \|\tilde{e}_{n_k} - \tilde{e}_0\|_{\tilde{\mathcal{H}}} = 0$, $\tilde{e}_0 \in \operatorname{ran}(\widetilde{E}_{-1})$, $\|\tilde{e}_0\|_{\tilde{\mathcal{H}}} = 1$. Hence $\lim_{k\to\infty} \|\widetilde{M}(z_0)\tilde{e}_{n_k}\|_{\tilde{\mathcal{H}}} = 0$ implying the contradiction $\widetilde{M}(z_0)\tilde{e}_0 = 0$. Thus, $\widetilde{M}(z_0)^{-1} \in \mathcal{B}(\widetilde{\mathcal{H}})$ and hence $\widetilde{M}(z)^{-1} \in \mathcal{B}(\widetilde{\mathcal{H}})$, $z \in \mathbb{C}_+$ by Lemma 2.3. In particular, $\operatorname{ran}(\widetilde{M}(z)) = \widetilde{\mathcal{H}}$, $z \in \mathbb{C}_+$ and hence $M(z) = \begin{pmatrix} \tilde{M}(z) & 0 \\ 0 & 0 \end{pmatrix}$ has left and right parametrices taking into account (2.10). □

Concerning boundary values at the real axis we also recall

Lemma 2.5. (Naboko [63].) *Suppose $(M - I_{\mathcal{H}})$ is a Herglotz operator with values in $\mathcal{B}_1(\mathcal{H})$. Then the boundary values $M(\lambda+i0)$ exist for a.e. $\lambda \in \mathbb{R}$ in $\mathcal{B}_p(\mathcal{H})$-norm, $p > 1$ and $M(\lambda + i0)$ is a Fredholm operator for a.e. $\lambda \in \mathbb{R}$ with index zero a.e.,*

$$\operatorname{ind}(M(\lambda + i0)) = 0 \ \textit{for a.e.}\ \lambda \in \mathbb{R}. \tag{2.15}$$

Moreover,

$$\ker(M(\lambda + i0)) = \ker(M(i)) = (\operatorname{ran}(M(\lambda + i0)))^{\perp} \ \textit{for a.e.}\ \lambda \in \mathbb{R}. \tag{2.16}$$

In addition, if $M(z_0)^{-1} \in \mathcal{B}(\mathcal{H})$ for some (and hence for all) $z_0 \in \mathbb{C}_+$, then

$$M(\lambda + i0)^{-1} \in \mathcal{B}(\mathcal{H}) \ \textit{for a.e.}\ \lambda \in \mathbb{R}. \tag{2.17}$$

Generalizations and a counter example to (2.17) if $(M - I_{\mathcal{H}}) \in \mathcal{B}_1(\mathcal{H})$ is replaced by $(M - I_{\mathcal{H}}) \in \mathcal{B}_p(\mathcal{H})$, $p > 1$ can be found in [63].

Next, let T be a bounded *dissipative* operator, that is,

$$T \in \mathcal{B}(\mathcal{H}), \quad \operatorname{Im}(T) \geq 0. \tag{2.18}$$

In order to define the logarithm of T we use the integral representation

$$\log(z) = -i \int_0^{\infty} d\lambda\, ((z + i\lambda)^{-1} - (1 + i\lambda)^{-1}), \quad z \neq -iy,\, y \geq 0, \tag{2.19}$$

with a cut along the negative imaginary z-axis. We use the symbol $\log(\cdot)$ in (2.19) in order to distinguish it from the integral representation

$$\ln(z) = \int_{-\infty}^{0} d\lambda\, ((\lambda - z)^{-1} - \lambda(1+\lambda^2)^{-1}), \quad z \in \mathbb{C}\backslash(-\infty, 0] \tag{2.20}$$

with a cut along the negative real axis. Both representations will be used later and it is easily verified that $\log(\cdot)$ and $\ln(\cdot)$ coincide for $z \in \mathbb{C}_+$. In particular, one verifies that (2.19) and (2.20) are Herglotz functions, that is, they are analytic in $\mathbb{C}_+$ and

$$0 < \operatorname{Im}(\log(z)),\ \operatorname{Im}(\ln(z)) < \pi, \quad z \in \mathbb{C}_+. \tag{2.21}$$

Lemma 2.6. *Suppose $T \in \mathcal{B}(\mathcal{H})$ is dissipative and $T^{-1} \in \mathcal{B}(\mathcal{H})$. Define*

$$\log(T) = -i \int_0^{\infty} d\lambda\, ((T + i\lambda)^{-1} - (1 + i\lambda)^{-1} I_{\mathcal{H}}) \tag{2.22}$$

in the sense of a $\mathcal{B}(\mathcal{H})$-norm convergent Riemann integral. Then
(i) $\log(T) \in \mathcal{B}(\mathcal{H})$.
(ii) If $T = zI_{\mathcal{H}}$, $z \in \mathbb{C}_+$, then $\log(T) = \log(z) I_{\mathcal{H}}$.
(iii) Suppose $\{P_n\}_{n\in\mathbb{N}} \subset \mathcal{B}(\mathcal{H})$ is a family of orthogonal finite-rank projections in $\mathcal{H}$ with $\text{s-lim}_{n\to\infty} P_n = I_{\mathcal{H}}$. *Then*

$$\operatorname*{s-lim}_{n\to\infty}((I_{\mathcal{H}} - P_n) + P_n T P_n) = T$$

and

$$\begin{aligned}&\operatorname*{s-lim}_{n\to\infty} \log((I_{\mathcal{H}} - P_n) + P_n(T + i\varepsilon)P_n)\\ &= \operatorname*{s-lim}_{n\to\infty} P_n(\log(P_n(T + i\varepsilon)P_n|_{P_n\mathcal{H}})P_n = \log(T + i\varepsilon I_H), \quad \varepsilon > 0.\end{aligned}$$

(iv) $\text{n-lim}_{\varepsilon\downarrow 0} \log(T + i\varepsilon I_{\mathcal{H}}) = \log(T)$.
(v) $e^{\log(T)} = T$.

Proof. (i) Clearly $\log(T) \in \mathcal{B}(\mathcal{H})$ since

$$\|\log(T)\| \le \int_0^{\delta} d\lambda\, (\|(T + i\lambda)^{-1}\| + 1) + \int_{\delta}^{\infty} d\lambda\, (\|T\| + 1)\lambda^{-2} \tag{2.23}$$

using $\|(T + i\lambda)^{-1}\| \le \lambda^{-1}$, $\lambda > 0$. Moreover, by $\|(T + i\lambda)^{-1}\| \le \|T^{-1}\|(1 - \|T^{-1}\|\lambda)^{-1}$ for $0 < \lambda < \|T^{-1}\|^{-1}$, choosing $\delta = 2^{-1}\|T^{-1}\|^{-1}$ is sufficient to bound the first integral in (2.23).

(ii) is obvious from (2.19).

(iii) For any $f \in \mathcal{H}$, $\varepsilon > 0$,

$$\left\| \int_0^{\infty} d\lambda\, (((I_{\mathcal{H}} - P_n) + P_n(T + i\varepsilon)P_n + i\lambda)^{-1} - (T + i\varepsilon + i\lambda)^{-1}) f \right\|$$

$$\leq 2\int_0^\delta d\lambda\,((\varepsilon+\lambda)^{-1})\|f\|$$
$$+\int_\delta^N d\lambda\,\|((I_{\mathcal{H}}-P_n)+P_nTP_n-T+i\varepsilon(P_n-I_{\mathcal{H}}))(T+i\varepsilon+i\lambda)^{-1}f\|\delta^{-1}$$
$$+\int_N^\infty d\lambda\,\lambda^{-2}\|((I_{\mathcal{H}}-P_n)+P_n(T+i\varepsilon)P_n-T-i\varepsilon I_{\mathcal{H}})\|\|f\|. \tag{2.24}$$

Taking $\delta>0$ sufficiently small and N sufficiently large such that the first and third integrals in (2.24) are sufficiently small uniformly with respect to $n\in\mathbb{N}$ for fixed $f\in\mathcal{H}$, it suffices to let $n\to\infty$ in the second integral in (2.24) for fixed δ and N.

(iv) One estimates,

$$\|\log(T+i\varepsilon)-\log(T)\| = \left\|\int_0^\infty d\lambda\,((T+i\varepsilon+i\lambda)^{-1}-(T+i\lambda)^{-1})\right\|$$
$$\leq\int_0^\delta d\lambda\,(\|T^{-1}\|(1-(\varepsilon+\lambda)\|T^{-1}\|)^{-1}+\|T^{-1}\|(1-\lambda\|T^{-1}\|)^{-1})$$
$$+\varepsilon\int_\delta^\infty d\lambda\,\lambda^{-2}\leq 2\delta(\|T^{-1}\|^{-1}-(\varepsilon+\delta))^{-1}+\varepsilon\delta^{-1} \tag{2.25}$$

for $\varepsilon+\delta<\|T^{-1}\|^{-1}$. Taking $\delta\leq 2^{-1}\|T^{-1}\|^{-1}$, $\varepsilon<\delta/2$ and $\varepsilon\downarrow 0$ then yields the result since $\delta>0$ can be chosen arbitrarily small.

(v) clearly holds for dissipative $n\times n$ matrices and hence

$$e^{\log((I_{\mathcal{H}}-P_n)+P_n(T+i\varepsilon)P_n)}=(I_{\mathcal{H}}-P_n)+P_n(T+i\varepsilon)P_n,\quad \varepsilon>0 \tag{2.26}$$

upon decomposing $\mathcal{H}=P_n\mathcal{H}\oplus(I_{\mathcal{H}}-P_n)\mathcal{H}$, where P_n are orthogonal rank-n projections. Since $\text{s-lim}_{n\to\infty}e^{A_n}=e^A$ whenever $A_n,A\in\mathcal{B}(\mathcal{H})$ and $\text{s-lim}_{n\to\infty}A_n=A$ (simply use $e^B=\sum_{m=0}^\infty B^m/(m!)$, $B\in\mathcal{B}(\mathcal{H})$ and $A_n^m-A^m=\sum_{\ell=0}^{m-1}A_n^\ell(A_n-A)A^{m-1-\ell}$), one infers from (iii) and (2.26) $e^{\log(T+i\varepsilon)}=T+i\varepsilon$, $\varepsilon>0$. Together with (iv) this yields (v). □

Lemma 2.7. *Suppose $T\in\mathcal{B}(\mathcal{H})$ is dissipative and $T^{-1}\in\mathcal{B}(\mathcal{H})$. Let L be the minimal self-adjoint dilation of T in the Hilbert space $\mathcal{K}\supseteq\mathcal{H}$. Then*

$$\text{Im}(\log(T))=\pi P_{\mathcal{H}}E_L((-\infty,0))|_{\mathcal{H}}, \tag{2.27}$$

where $P_{\mathcal{H}}$ is the orthogonal projection in $\mathcal{K}$ onto $\mathcal{H}$ and $\{E_L(\lambda)\}_{\lambda\in\mathbb{R}}$ is the family of orthogonal spectral projections of L in $\mathcal{K}$. In particular,

$$0\leq\text{Im}(\log(T))\leq\pi I_{\mathcal{H}}. \tag{2.28}$$

Proof. By Sz.-Nagy's dilation theorem (see the corresponding result in [86], Ch. III, Sect. 2, Theorem 2.1 for contractions), one infers

$$(T+i\lambda)^{-1}=P_{\mathcal{H}}(L+i\lambda)^{-1}|_{\mathcal{H}},\quad \lambda>0, \tag{2.29}$$

where L is the minimal self-adjoint dilation of T in $\mathcal{K}$. Then the existence of $T^{-1} \in \mathcal{B}(\mathcal{H})$ and (2.29) yield

$$E_L(\{0\}) = 0. \tag{2.30}$$

In order to prove (2.30) one can argue as follows. Consider the contraction $S = (T-i)(T+i)^{-1}$. According to Theorem 3.2, Ch. I, Sect. 3 in [86], every contraction on the Hilbert space $\mathcal{H}$ corresponds to a decomposition of $\mathcal{H}$ into an orthogonal sum $\mathcal{H} = \mathcal{H}_0 \oplus \mathcal{H}_1$ of two reducing subspaces of S such that the part of S on $\mathcal{H}_0$ is unitary, and the part of S on $\mathcal{H}_1$ is completely non-unitary. Moreover, this decomposition is unique. (We recall that a contraction is called completely non-unitary if there are no non-zero subspaces reducing this contraction to a unitary operator.) Since T is invertible, the unitary part of the contraction S does not have the eigenvalue -1. Since the minimal unitary dilation of a completely non-unitary contraction has absolutely continuous spectrum (see [85]), we conclude that the minimal unitary dilation of S does not have the eigenvalue -1 and hence the kernel of the minimal self-adjoint dilation L of the dissipative operator T is trivial, proving (2.30). Next, (2.22) implies

$$\begin{aligned}
\operatorname{Im}(\log(T)) &= -\int_0^\infty d\lambda \operatorname{Re}((T+i\lambda)^{-1} - (1+i\lambda)^{-1} I_{\mathcal{H}}) \\
&= -\int_0^\varepsilon d\lambda \operatorname{Re}((T+i\lambda)^{-1} - (1+i\lambda)^{-1} I_{\mathcal{H}}) \\
&\quad - P_{\mathcal{H}} \int_\varepsilon^\infty d\lambda \operatorname{Re}((L+i\lambda)^{-1} - (1+i\lambda)^{-1} I_{\mathcal{K}})|_{\mathcal{H}} \\
&= -\int_0^\varepsilon d\lambda \operatorname{Re}((T+i\lambda)^{-1} - (1+i\lambda)^{-1} I_{\mathcal{H}}) \\
&\quad - P_{\mathcal{H}} \int_\varepsilon^\infty d\lambda \,(L(L^2+\lambda^2)^{-1} - (1+\lambda^2)^{-1} I_{\mathcal{K}})|_{\mathcal{H}}, \quad \varepsilon > 0.
\end{aligned} \tag{2.31}$$

Let us prove that

$$\operatorname*{s-lim}_{\varepsilon \downarrow 0} \int_\varepsilon^\infty d\lambda \, L(L^2+\lambda^2)^{-1} = (\pi/2)(P_{L,+} - P_{L,-}), \tag{2.32}$$

where $P_{L,+} = E_L((0,\infty))$ and $P_{L,-} = E_L((-\infty,0))$ are the spectral projections of L corresponding to the half-lines $(0,\infty)$ and $(-\infty,0)$. Then (2.27) and hence (2.28) follow from (2.31), (2.32) and the fact that

$$\lim_{\varepsilon \downarrow 0} \left\| \int_0^\varepsilon d\lambda \operatorname{Re}((T+i\lambda)^{-1} - (1+i\lambda)^{-1} I_{\mathcal{H}}) \right\| = 0. \tag{2.33}$$

Using the estimate $|\int_\varepsilon^\infty d\lambda \, \mu(\mu^2+\lambda^2)^{-1}| \leq (\pi/2)$, $\mu \in \mathbb{R}$, $\varepsilon > 0$ by the spectral theorem for the self-adjoint operator L, we infer that the family of operators $\int_\varepsilon^\infty d\lambda \, L(L^2+\lambda^2)^{-1}$ is uniformly bounded in $\varepsilon > 0$ and therefore, it suffices to check the convergence (2.32) on a dense set in $\mathcal{K}$. A natural candidate for this set is $\mathcal{D} = \{\bigcup_{\delta>0} E_L(\mathbb{R}\backslash(-\delta,\delta))f \mid f \in \mathcal{K}\}$, which is dense in $\mathcal{K}$ since by (2.30) the kernel of L is trivial. For $f \in \mathcal{D}$ we have

$$\begin{aligned}
\int_\varepsilon^\infty d\lambda\, L(L^2+\lambda^2)^{-1}f &= \int_\varepsilon^\infty d\lambda \int_{\mathbb{R}\backslash(-\delta,\delta)} \mu(\mu^2+\lambda^2)^{-1} dE_L(\mu) f \\
&= \int_\delta^\infty ((\pi/2) - \arctan(\varepsilon/\mu)) dE_L(\mu) f - \int_{-\infty}^{-\delta} ((\pi/2) - \arctan(\varepsilon/|\mu|)) dE_L(\mu) f \\
&= (\pi/2)(E_L((\delta,\infty)) - E_L((-\infty,-\delta)))f \\
&\quad - \int_\delta^\infty \arctan(\varepsilon/\mu) dE_L(\mu) f + \int_{-\infty}^{-\delta} \arctan(\varepsilon/|\mu|) dE_L(\mu) f \qquad (2.34)
\end{aligned}$$

for all $\delta > 0$, $\delta = \delta(f)$ small enough. For fixed $f \in \mathcal{D}$, going to the limit $\varepsilon \to 0$ in (2.34), we get

$$\begin{aligned}
\lim_{\varepsilon \downarrow 0} \int_\varepsilon^\infty d\lambda\, L(L^2+\lambda^2)^{-1} f &= (\pi/2)(E_L((\delta,\infty)) - E_L((-\infty,\delta)))f \\
&= (\pi/2)(P_{L,+} - P_{L,-})f, \qquad (2.35)
\end{aligned}$$

proving (2.32). □

Combining Lemmas 2.3 and 2.7 one can prove the following result.

Lemma 2.8. *Suppose $M : \mathbb{C}_+ \longrightarrow \mathcal{B}(\mathcal{H})$ is a Herglotz operator and $M(z_0)^{-1} \in \mathcal{B}(\mathcal{H})$ for some (and hence for all) $z_0 \in \mathbb{C}_+$. Then $\log(M) : \mathbb{C}_+ \longrightarrow \mathcal{B}(\mathcal{H})$ is a Herglotz operator and*

$$0 \le \operatorname{Im}(\log(M(z))) \le \pi I_{\mathcal{H}}, \quad z \in \mathbb{C}_+. \qquad (2.36)$$

Proof. Clearly

$$\log(M(z)) = -i \int_0^\infty d\lambda\, ((M(z)+i\lambda)^{-1} - (1+i\lambda)^{-1} I_{\mathcal{H}}), \quad z \in \mathbb{C}_+ \qquad (2.37)$$

is analytic for $z \in \mathbb{C}_+$ since $M(z)^{-1} \in \mathcal{B}(\mathcal{H})$ for all $z \in \mathbb{C}_+$ by Lemma 2.3. An application of Lemma 2.7 then yields (2.36). □

Thus applying (2.1a) to $\log(M(z))$ one infers

$$\log(M(z)) = C + Dz + \widetilde{R}^{1/2}(I_{\widetilde{\mathcal{K}}} + z\widetilde{L})(\widetilde{L} - z)^{-1}\widetilde{R}^{1/2}|_{\mathcal{H}}, \quad z \in \mathbb{C}_+ \qquad (2.38)$$

for some Hilbert space $\widetilde{\mathcal{K}} \supseteq \mathcal{H}$, some bounded self-adjoint operators C, $0 \le D \in \mathcal{B}(\mathcal{H})$, a bounded nonnegative operator $0 \le \widetilde{R} \in \mathcal{B}(\mathcal{H})$ with $\widetilde{R}|_{\widetilde{\mathcal{K}}\ominus\mathcal{H}} = 0$, and a self-adjoint operator $\widetilde{L} = \widetilde{L}^*$ in $\widetilde{\mathcal{K}}$. By comparison with scalar Herglotz functions one would expect that $D = 0$. That this is indeed the case is proved next.

Lemma 2.9. $D = 0$ *in the representation* (2.38) *for* $\log(M(z))$, $z \in \mathbb{C}_+$.

Proof. Consider $z = iy$, $y \uparrow \infty$. Then

$$||y^{-1}\log(M(iy)) - Di|| = O(y^{-1}) \tag{2.39}$$

and by (2.36),

$$\begin{aligned}\pi y^{-1} I_{\mathcal{H}} &\geq y^{-1}\mathrm{Im}(\log(M(iy)))\\ &= D + y^{-1}\mathrm{Im}(\widetilde{R}^{1/2}(I_{\widetilde{\mathcal{K}}} + iy\widetilde{L})(\widetilde{L} - iy)^{-1}\widetilde{R}^{1/2}|_{\mathcal{H}})\\ &= D + y^{-1}(... \geq 0...) \geq D \geq 0\end{aligned} \tag{2.40}$$

and hence $D = 0$ since y^{-1} can be chosen arbitrarily small. □

Introducing the family of orthogonal spectral projections $\{E_{\widetilde{L}}(\lambda)\}_{\lambda\in\mathbb{R}}$ of $\widetilde{L}$ in $\widetilde{\mathcal{K}}$ one can thus rewrite (2.38) as

$$\begin{aligned}&\log(M(z)) = C\\ &+ \int_{\mathbb{R}} (1+\lambda^2) d(\widetilde{R}^{1/2} E_{\widetilde{L}}(\lambda)\widetilde{R}^{1/2}|_{\mathcal{H}})((\lambda - z)^{-1} - \lambda(1+\lambda^2)^{-1}), \quad z \in \mathbb{C}_+\end{aligned} \tag{2.41}$$

and hence

$$\mathrm{Im}(\log(M(z))) = \mathrm{Im}(z)\int_{\mathbb{R}} d(\widetilde{R}^{1/2}E_{\widetilde{L}}(\lambda)\widetilde{R}^{1/2}|_{\mathcal{H}})(1+\lambda^2)|\lambda - z|^{-2} \tag{2.42}$$

for $z \in \mathbb{C}_+$, interpreting both integrals in (2.41) and (2.42) in the weak sense for simplicity. Again by comparison with scalar Herglotz functions one expects that $d(\widetilde{R}^{1/2}E_{\widetilde{L}}(\lambda)\ \widetilde{R}^{1/2}|_{\mathcal{H}})$ is a purely absolutely continuous operator-valued measure on $\mathbb{R}$. This is confirmed by the following result ($\mathrm{Tr}_{\mathcal{H}}(\cdot)$ denotes the trace of trace class operators in $\mathcal{H}$).

Theorem 2.10. *Suppose* $M : \mathbb{C}_+ \longrightarrow \mathcal{B}(\mathcal{H})$ *is a Herglotz operator and* $M(z_0)^{-1} \in \mathcal{B}(\mathcal{H})$ *for some (and hence for all)* $z_0 \in \mathbb{C}_+$. *Then there exists a family of bounded self-adjoint weakly (Lebesgue) measurable operators* $\{\Xi(\lambda)\}_{\lambda\in\mathbb{R}} \subset \mathcal{B}(\mathcal{H})$,

$$0 \leq \Xi(\lambda) \leq I_{\mathcal{H}} \text{ for a.e. } \lambda \in \mathbb{R} \tag{2.43}$$

such that

$$\log(M(z)) = C + \int_{\mathbb{R}} d\lambda\, \Xi(\lambda)((\lambda - z)^{-1} - \lambda(1+\lambda^2)^{-1}), \quad z \in \mathbb{C}_+ \tag{2.44}$$

the integral taken in the weak sense, where $C = C^* \in \mathcal{B}(\mathcal{H})$. *Moreover, if* $\mathrm{Im}(\log(M(z_0))) \in \mathcal{B}_1(\mathcal{H})$ *for some (and hence for all)* $z_0 \in \mathbb{C}_+$, *then*

$$0 \leq \Xi(\lambda) \in \mathcal{B}_1(\mathcal{H}) \text{ for a.e. } \lambda \in \mathbb{R}, \tag{2.45}$$

$$0 \le \mathrm{Tr}_{\mathcal{H}}(\Xi(\cdot)) \in L^1_{\mathrm{loc}}(\mathbb{R}; d\lambda), \quad \int_{\mathbb{R}} d\lambda\,(1+\lambda^2)^{-1}\,\mathrm{Tr}_{\mathcal{H}}(\Xi(\lambda)) < \infty, \tag{2.46}$$

and

$$\mathrm{Tr}_{\mathcal{H}}(\mathrm{Im}(\log(M(z)))) = \mathrm{Im}(z) \int_{\mathbb{R}} d\lambda\ \mathrm{Tr}_{\mathcal{H}}(\Xi(\lambda))|\lambda - z|^{-2}, \quad z \in \mathbb{C}_+. \tag{2.47}$$

Proof. Let $f \in \mathcal{H}$ and denote

$$d\omega_f(\lambda) = (1+\lambda^2)d(\widetilde{R}^{1/2}f, E_{\widetilde{L}}(\lambda)\widetilde{R}^{1/2}f)_{\mathcal{H}} = (1+\lambda^2)d\|E_{\widetilde{L}}(\lambda)\widetilde{R}^{1/2}f\|^2_{\mathcal{H}}. \tag{2.48}$$

Then

$$0 \le (f, \mathrm{Im}(\log(M(z)))f)_{\mathcal{H}} \le \pi\|f\|^2_{\mathcal{H}}, \quad z \in \mathbb{C}_+, \quad f \in \mathcal{H} \tag{2.49}$$

proves that $d\omega_f$ is purely absolutely continuous, $d\omega_f = d\omega_{f,ac}$ by standard arguments (see, e.g., [4]). Thus,

$$d\omega_f(\lambda) = \xi_f(\lambda)d\lambda \text{ with } 0 \le \xi_f(\lambda) \le \|f\|^2_{\mathcal{H}} \quad \text{for all } f \in \mathcal{H} \text{ and a.e. } \lambda \in \mathbb{R}. \tag{2.50}$$

By (2.48), $\xi_f(\lambda)$ defines a quadratic form

$$\xi_f(\lambda) = (f, \Xi(\lambda)f)_{\mathcal{H}} \text{ for some } 0 \le \Xi(\lambda) \le I_{\mathcal{H}} \text{ and a.e. } \lambda \in \mathbb{R} \tag{2.51}$$

proving (2.43) and (2.44). The representation (2.38) (with $D = 0$) implies

$$\mathrm{Im}(\log(M(z)) = \widetilde{R}^{1/2}(I_{\mathcal{K}} + \widetilde{L}^2)((\widetilde{L} - \mathrm{Re}(z))^2 + (\mathrm{Im}(z))^2)^{-1}\widetilde{R}^{1/2}|_{\mathcal{H}} \tag{2.52}$$

and hence $\mathrm{Im}(\log(M(z_0))) \in \mathcal{B}_1(\mathcal{H})$ for some $z_0 \in \mathbb{C}_+$ implies $\mathrm{Im}(\log(M(z))) \in \mathcal{B}_1(\mathcal{H})$ for all $z \in \mathbb{C}_+$. In particular,

$$\mathrm{Im}(\log(M(i))) = \int_{\mathbb{R}} d\lambda\,(1+\lambda^2)^{-1}\Xi(\lambda) \tag{2.53}$$

then proves

$$\mathrm{Tr}_{\mathcal{H}}(\mathrm{Im}(\log(M(i)))) = \int_{\mathbb{R}} d\lambda\,(1+\lambda^2)^{-1}\,\mathrm{Tr}_{\mathcal{H}}(\Xi(\lambda)) < \infty \tag{2.54}$$

by the monotone convergence theorem. Hence $0 \le \mathrm{Tr}_{\mathcal{H}}(\Xi(\lambda)) = \|\Xi(\lambda)\|_1 < \infty$ for a.e. $\lambda \in \mathbb{R}$ completes the proof. □

Remark 2.11. For simplicity we focused on dissipative operators thus far. Later we will also encounter operators $S \in \mathcal{B}(\mathcal{H})$ with $-S$ dissipative, that is, $\mathrm{Im}(S) \le 0$. In this case S^* is dissipative and one can simply define $\log(S)$ by

$$\log(S) = (\log(S^*))^*, \tag{2.55}$$

with $\log(S^*)$ defined as in (2.22). Moreover,

$$\log(\widehat{M}(z)) = \widehat{C} - \int_{\mathbb{R}} d\lambda\, \widehat{\Xi}(\lambda)((\lambda - z)^{-1} - \lambda(1+\lambda^2)^{-1}), \quad z \in \mathbb{C}_+, \tag{2.56}$$

$$\widehat{C} = \widehat{C}^* \in \mathcal{B}(\mathcal{H}) \text{ and } 0 \le \widehat{\Xi}(\lambda) \le I_{\mathcal{H}} \text{ for a.e. } \lambda \in \mathbb{R}, \tag{2.57}$$

whenever $\widehat{M}$ is analytic in $\mathbb{C}_+$ and $\operatorname{Im}(\widehat{M}(z)) \le 0$, $z \in \mathbb{C}_+$.

Remark 2.12. Theorem 2.10 represents the operator-valued generalization of the exponential Herglotz representation for scalar Herglotz functions studied in detail by Aronszajn and Donoghue [4] (see also Carey and Pepe [19]). Theorem 2.10 is not the first attempt in this direction. Carey [18], in 1976, considered the case $M(z) = I_{\mathcal{H}} + K^*(H_0 - z)^{-1}K$ (i.e., $A - R^{1/2}LR^{1/2}|_{\mathcal{H}} = I_{\mathcal{H}}$, $B = 0$, $(1+L^2)^{1/2}R^{1/2}|_{\mathcal{H}} = K$ when compared to (2.1a)) and established

$$M(z) = \exp\left(\int_{\mathbb{R}} d\lambda\, B(\lambda)(\lambda - z)^{-1}\right) \tag{2.58}$$

for a summable operator function $B(\lambda)$, $0 \le B(\lambda) \le I_{\mathcal{H}}$ (i.e., the analog of $\Xi(\lambda)$ in (2.44)). Although Carey's proof also uses Naimark's dilation theorem as described in Theorem 1 of Appendix I of [16], it is different from ours and does not utilize the integral representation (2.22) for logarithms.

Remark 2.13. At first glance it may seem that we have been a bit pedantic in introducing various branches of the logarithm in (2.19) and (2.20). Actually, these branches are just a special case of the following family of branches

$$\log_\alpha(z) = \int_{\gamma_\alpha} d\zeta\, ((z - \zeta)^{-1} - (1 - \zeta)^{-1}), \quad z \in \mathbb{C}\backslash\gamma_\alpha, \tag{2.59}$$

where γ_α denotes the ray, $\gamma_\alpha = \{\zeta \in \mathbb{C}_+ \,|\, \zeta = re^{i\alpha}, 0 \le r < \infty, \alpha \in [\pi, 2\pi]\}$. In particular, $\log(\cdot) = \log_{3\pi/2}(\cdot)$ and since $\int_{-\infty}^0 d\lambda\, ((1-\lambda)^{-1} - \lambda(1+\lambda^2)^{-1}) = 0$, one infers $\ln(\cdot) = \log_\pi(\cdot)$. That some care has to be taken in connection with a consistent choice of branches especially for operator-valued branches of the logarithm is illustrated in the following Remark 2.14 and in Remark 3.3.

Remark 2.14. In view of our applications in Sections 3 and 4 it seems worthwhile to recall in connection with our hypothesis $\operatorname{Im}(\log(M(z_0))) \in \mathcal{B}_1(\mathcal{H})$ in Theorem 2.10 that if $A \in \mathcal{B}_1(\mathcal{H})$ and $\det_{\mathcal{H}}(I_{\mathcal{H}} + A) \neq 0$, then $\log(I_{\mathcal{H}} + A) \in \mathcal{B}_1(\mathcal{H})$ by (2.22) (using $(I_{\mathcal{H}} + A + i\lambda)^{-1} - (1 + i\lambda)^{-1}I_{\mathcal{H}} = (1+\lambda)^{-1}(I_{\mathcal{H}} + A + i\lambda)^{-1}A$.) Moreover, $\det_{\mathcal{H}}(I_{\mathcal{H}} + A) = \prod_{n\in\mathbb{N}}(1 + \lambda_n(A))$, where $\lambda_n(A)$ denote the eigenvalues of A repeated according to their algebraic multiplicity, then shows

$$\operatorname{Tr}_{\mathcal{H}}(\log(I_{\mathcal{H}} + A)) = \log(\det_{\mathcal{H}}(I_{\mathcal{H}} + A)). \tag{2.60}$$

In fact, (2.60) holds for any branch $\log_\alpha(\cdot)$ and hence, in particular, for the branch $\ln(\cdot)$ on either side of (2.60). (Here $\det_{\mathcal{H}}(\cdot)$ denotes the Fredholm determinant for operators in $\mathcal{H}$.)

3 The Spectral Shift Operator

The main purpose of this section is to introduce the concept of a spectral shift operator (cf. Definition 3.5) and a new approach to Krein's basic trace formula [51].

Let $\mathcal{H}$ be a complex separable Hilbert space and assume the following hypothesis for the remainder of this section.

Hypothesis 3.1. *Let H_0 be a self-adjoint operator in $\mathcal{H}$ with domain* $\operatorname{dom}(H_0)$, *J a bounded self-adjoint operator with $J^2 = I_{\mathcal{H}}$, and $K \in \mathcal{B}_2(\mathcal{H})$ a Hilbert-Schmidt operator.*

Introducing

$$V = KJK^* \tag{3.1}$$

we define the self-adjoint operator

$$H = H_0 + V, \quad \operatorname{dom}(H) = \operatorname{dom}(H_0) \tag{3.2}$$

in $\mathcal{H}$.

Given Hypothesis 3.1 we decompose $\mathcal{H}$ and J according to

$$J = \begin{pmatrix} I_+ & 0 \\ 0 & -I_- \end{pmatrix}, \quad \mathcal{H} = \mathcal{H}_+ \oplus \mathcal{H}_-, \tag{3.3}$$

$$J_+ = \begin{pmatrix} I_+ & 0 \\ 0 & 0 \end{pmatrix}, \quad J_- = \begin{pmatrix} 0 & 0 \\ 0 & I_- \end{pmatrix}, \quad J = J_+ - J_-, \tag{3.4}$$

with $I_\pm$ the identity operator in $\mathcal{H}_\pm$. Moreover, we introduce the following bounded operators

$$\Phi(z) = J + K^*(H_0 - z)^{-1}K : \mathcal{H} \to \mathcal{H}, \tag{3.5}$$

$$\Phi_+(z) = I_+ + J_+K^*(H_0 - z)^{-1}K|_{\mathcal{H}_+} : \mathcal{H}_+ \to \mathcal{H}_+, \tag{3.6}$$

$$\widetilde{\Phi}_-(z) = I_- - J_-K^*(H_+ - z)^{-1}K|_{\mathcal{H}_-} : \mathcal{H}_- \to \mathcal{H}_-, \tag{3.7}$$

for $z \in \mathbb{C}\backslash\mathbb{R}$, where

$$V_+ = KJ_+K^*, \tag{3.8}$$

$$H_+ = H_0 + V_+, \quad \operatorname{dom}(H_+) = \operatorname{dom}(H_0). \tag{3.9}$$

Lemma 3.2. *Assume Hypothesis* 3.1. *Then Φ, Φ_+, and $-\widetilde{\Phi}_-$ are Herglotz operators in $\mathcal{H}$, $\mathcal{H}_+$, and $\mathcal{H}_-$, respectively. In addition ($z \in \mathbb{C}\backslash\mathbb{R}$),*

$$\Phi(z)^{-1} = J - JK^*(H - z)^{-1}KJ, \tag{3.10}$$

$$\Phi_+(z)^{-1} = I_+ - J_+K^*(H_+ - z)^{-1}K|_{\mathcal{H}_+}, \tag{3.11}$$

$$\widetilde{\Phi}_-(z)^{-1} = I_- + J_-K^*(H - z)^{-1}K|_{\mathcal{H}_-}. \tag{3.12}$$

Proof. It suffices to consider Φ_+. Since I_+ and H_0 are self-adjoint, $\Phi_+(z) = (J_+ + J_+K^*(H_0 - z)^{-1}KJ_+)|_{\mathcal{H}_+}$ is clearly analytic in $\mathbb{C}\backslash\mathbb{R}$ and satisfies $\operatorname{Im}(\Phi_+(z)) \geq 0$ for $z \in \mathbb{C}_+$. Relation (3.11) is then an elementary consequence of the second resolvent equation,

$$(H_+ - z)^{-1} = (H_0 - z)^{-1} - (H_0 - z)^{-1}V_+(H_+ - z)^{-1} \tag{3.13a}$$

$$= (H_0 - z)^{-1} - (H_+ - z)^{-1}V_+(H_0 - z)^{-1}, \tag{3.13b}$$

the fact $J_+^2 = J_+$, and simply follows by multiplying the right-hand sides of (3.6) and (3.11) in either order. □

Remark 3.3. In the following we need to make use of the formula

$$d\operatorname{Tr}_{\mathcal{K}}(\log(I_{\mathcal{K}} + F(z)))/dz = \operatorname{Tr}_{\mathcal{K}}(F'(z)(I_{\mathcal{K}} + F(z))^{-1}). \tag{3.14}$$

This result is proven, for instance, in [28], Ch. I, Sect. 6.11 or [41], Sect. IV.1 for analytic $F(\cdot) \in \mathcal{B}_1(\mathcal{K})$ in some region $\Omega \subset \mathbb{C}$ such that $(I_{\mathcal{K}} + F(\cdot))^{-1} \in \mathcal{B}(\mathcal{K})$ in Ω using the standard branch $\ln(\cdot)$. In this case equation (3.14) is first proven in the finite-dimensional case, followed by taking the limit $n \to \infty$ upon replacing $F(z)$ by $P_nF(z)P_n$, with P_n a family of orthogonal projections in $\mathcal{K}$ strongly converging to $I_{\mathcal{K}}$ as $n \to \infty$ (cf., [41], p. 163). This strategy of proof extends to all branches $\log_\alpha(\cdot)$ introduced in Remark 2.13. More generally, we have the following result,

$$d\operatorname{Tr}_{\mathcal{K}}(\varphi(F(s)))/ds = \operatorname{Tr}_{\mathcal{K}}(\varphi'(F(s))F'(s)), \tag{3.15}$$

with $F(s) \in \mathcal{B}_1(\mathcal{K})$ defined on an interval $s_1 \leq s \leq s_2$, continuously differentiable in $\mathcal{B}_1(\mathcal{K})$-norm, and $\varphi(z)$ any scalar function, holomorphic in some domain $\mathcal{D} \subset \mathbb{C}$ with a Jordan boundary and $\operatorname{spec}(F(s)) \subset \mathcal{D}$ for all $s \in [s_1, s_2]$.

Lemma 3.4. *Assume Hypothesis* 3.1 *and* $\mathbb{C}\backslash\mathbb{R}$. *Then*

$$\operatorname{Tr}_{\mathcal{H}}((H_0 - z)^{-1} - (H_+ - z)^{-1}) = d\operatorname{Tr}_{\mathcal{H}_+}(\log(\Phi_+(z)))/dz, \tag{3.16a}$$

$$\operatorname{Tr}_{\mathcal{H}}((H_+ - z)^{-1} - (H - z)^{-1}) = d\operatorname{Tr}_{\mathcal{H}_-}(\log(\widetilde{\Phi}_-(z)))/dz. \tag{3.16b}$$

Proof. It suffices to prove (3.16a). Applying (3.13) repeatedly one infers

$$\begin{aligned}
&\operatorname{Tr}_{\mathcal{H}}((H_0 - z)^{-1} - (H_+ - z)^{-1}) \\
&= \operatorname{Tr}_{\mathcal{H}}((H_0 - z)^{-1}V_+[(H_0 - z)^{-1} - (H_+ - z)^{-1}V_+(H_0 - z)^{-1}]) \\
&= \operatorname{Tr}_{\mathcal{H}}(J_+K^*(H_0 - z)^{-2}KJ_+[J_+ - J_+K^*(H_+ - z)^{-1}KJ_+]) \\
&= \operatorname{Tr}_{\mathcal{H}}(J_+\Phi_+'(z)J_+^2\Phi_+(z)^{-1}J_+) = \operatorname{Tr}_{\mathcal{H}_+}(\Phi_+'(z)\Phi_+(z)^{-1}) \\
&= d\operatorname{Tr}_{\mathcal{H}_+}(\log(\Phi_+(z)))/dz, \quad z \in \mathbb{C}\backslash\mathbb{R},
\end{aligned} \tag{3.17}$$

where we used $J_+^2 = J_+$,

$$\operatorname{Tr}_{\mathcal{H}}(AB) = \operatorname{Tr}_{\mathcal{H}}(BA) \tag{3.18}$$

for $A, B \in \mathcal{B}(\mathcal{H})$ with AB, $BA \in \mathcal{B}_1(\mathcal{H})$ (cf. Corollary 3.8 in [77]), and (3.14). □

Next, applying Theorem 2.10 and Remark 2.11 to $\Phi_+(z)$ and $\widetilde{\Phi}_-(z)$ one infers the existence of two families of bounded operators $\{\Xi_\pm(\lambda)\}_{\lambda\in\mathbb{R}}$ defined for (Lebesgue) a.e. $\lambda \in \mathbb{R}$ and satisfying

$$\begin{aligned} &0 \leq \Xi_\pm(\lambda) \leq I_\pm, \quad \Xi_\pm(\lambda) \in \mathcal{B}_1(\mathcal{H}_\pm) \text{ for a.e. } \lambda \in \mathbb{R}, \\ &||\Xi_\pm(\cdot)||_1 \in L^1(\mathbb{R}; (1+\lambda^2)^{-1}d\lambda) \end{aligned} \tag{3.19}$$

and

$$\begin{aligned} \log(\Phi_+(z)) &= \log(I_+ + J_+K^*(H_0 - z)^{-1}K|_{\mathcal{H}_+}) \\ &= C_+ + \int_{\mathbb{R}} d\lambda\, \Xi_+(\lambda)((\lambda - z)^{-1} - \lambda(1+\lambda^2)^{-1}), \end{aligned} \tag{3.20a}$$

$$\begin{aligned} \log(\widetilde{\Phi}_-(z)) &= \log(I_- - J_-K^*(H_+ - z)^{-1}K|_{\mathcal{H}_-}) \\ &= C_- - \int_{\mathbb{R}} d\lambda\, \Xi_-(\lambda)((\lambda - z)^{-1} - \lambda(1+\lambda^2)^{-1}) \end{aligned} \tag{3.20b}$$

for $z \in \mathbb{C}\backslash\mathbb{R}$, with $C_\pm = C_\pm^* \in \mathcal{B}_1(\mathcal{H})$.

Equations (3.20) motivate the following

Definition 3.5. $\Xi_+(\lambda)$ (resp., $\Xi_-(\lambda)$) is called the *spectral shift operator* associated with $\Phi_+(z)$ (resp., $\widetilde{\Phi}_-(z)$). Alternatively, we will refer to $\Xi_+(\lambda)$ as the spectral shift operator associated with the pair (H_0, H_+) and occasionally use the notation $\Xi_+(\lambda, H_0, H_+)$ to stress the dependence on (H_0, H_+), etc.

Moreover, we introduce

$$\xi_\pm(\lambda) = \mathrm{Tr}_{\mathcal{H}_\pm}(\Xi_+(\lambda)), \quad 0 \leq \xi_\pm \in L^1(\mathbb{R}; (1+\lambda^2)^{-1}d\lambda) \text{ for a.e. } \lambda \in \mathbb{R}. \tag{3.21}$$

Actually, taking into account the simple behavior of $\Phi_+(iy)$ and $\widetilde{\Phi}_-(iy)$ as $|y| \to \infty$, one can improve (3.20a) and (3.20b) as follows.

Lemma 3.6. *Assume Hypothesis* 3.1 *and define* $\xi_\pm$ *as in* (3.21). *Then*

$$0 \leq \xi_\pm \in L^1(\mathbb{R}; d\lambda), \tag{3.22}$$

and (3.20a) *and* (3.20b) *simplify to*

$$\log(\Phi_+(z)) = \int_{\mathbb{R}} d\lambda\, \Xi_+(\lambda)(\lambda - z)^{-1}, \tag{3.23a}$$

$$\log(\widetilde{\Phi}_-(z)) = -\int_{\mathbb{R}} d\lambda\, \Xi_-(\lambda)(\lambda - z)^{-1}. \tag{3.23b}$$

Moreover, for a.e. $\lambda \in \mathbb{R}$,

$$\lim_{\varepsilon\downarrow 0} ||\Xi_+(\lambda) - \pi^{-1}\mathrm{Im}(\log(\Phi_+(\lambda + i\varepsilon)))||_{\mathcal{B}_1(\mathcal{H}_+)} = 0, \tag{3.24a}$$

$$\lim_{\varepsilon\downarrow 0} ||\Xi_-(\lambda) + \pi^{-1}\mathrm{Im}(\log(\widetilde{\Phi}_-(\lambda + i\varepsilon)))||_{\mathcal{B}_1(\mathcal{H}_-)} = 0. \tag{3.24b}$$

Proof. It suffices to consider $\xi_+(\lambda)$ and $\Phi_+(z)$. Since

$$\|\log(\Phi_+(y))\|_1 = O(|y|^{-1}) \text{ as } |y| \to \infty \tag{3.25}$$

by the Hilbert-Schmidt hypothesis on K and the fact $\|(H_0 - iy)^{-1}\| = O(|y|^{-1})$ as $|y| \to \infty$, the scalar Herglotz function $\mathrm{Tr}_{\mathcal{H}_+}(\log(\Phi_+(z)))$ satisfies

$$|\,\mathrm{Tr}_{\mathcal{H}_+}(\log(\Phi_+(z)))| = O(|y|^{-1}) \text{ as } |y| \to \infty. \tag{3.26}$$

By standard results (see, e.g., [4], [45]), (3.26) yields

$$\mathrm{Tr}_{\mathcal{H}_+}(\log(\Phi_+(z))) = \int_{\mathbb{R}} d\omega_+(\lambda)(\lambda - z)^{-1}, \quad z \in \mathbb{C}\backslash\mathbb{R}, \tag{3.27}$$

where ω_+ is a finite measure,

$$\int_{\mathbb{R}} d\omega_+(\lambda) = -i \lim_{y\uparrow\infty} (y\,\mathrm{Tr}_{\mathcal{H}_+}(\log(\Phi_+(z)))) < \infty. \tag{3.28}$$

Moreover, the fact that $\mathrm{Im}(\log(\Phi_+(z)))$ is uniformly bounded with respect to $z \in \mathbb{C}_+$ yields that ω_+ is purely absolutely continuous,

$$d\omega_+(\lambda) = \xi_+(\lambda)d\lambda, \quad \xi_+ \in L^1(\mathbb{R}; d\lambda), \tag{3.29}$$

where

$$\begin{aligned}\xi_+(\lambda) &= \pi^{-1} \lim_{\varepsilon\downarrow 0}(\mathrm{Im}(\mathrm{Tr}_{\mathcal{H}_+}(\log(\Phi_+(\lambda + i\varepsilon))))) = \mathrm{Tr}_{\mathcal{H}_+}(\Xi_+(\lambda)) \\ &= \pi^{-1} \lim_{\varepsilon\downarrow 0}(\mathrm{Im}(\log(\det{}_{\mathcal{H}_+}(\Phi_+(\lambda + i\varepsilon))))) \text{ for a.e. } \lambda \in \mathbb{R}.\end{aligned} \tag{3.30}$$

In order to prove (3.24a) we first observe that $\mathrm{Im}(\log(\Phi_+(\lambda+i\varepsilon)))$ takes on boundary values $\mathrm{Im}(\log(\Phi_+(\lambda + i0)))$ for a.e. $\lambda \in \mathbb{R}$ in $\mathcal{B}_1(\mathcal{H}_+)$-norm by (2.2). Next, choosing an orthonormal system $\{e_n\}_{n\in\mathbb{N}} \subset \mathcal{H}_+$, we recall that the quadratic form $(e_n, \mathrm{Im}(\log(\Phi(\lambda + i0)))e_n)_{\mathcal{H}_+}$ exists for all $\lambda \in \mathbb{R}\backslash\mathcal{E}_n$, where $\mathcal{E}_n$ has Lebesgue measure zero. Thus one observes,

$$\begin{aligned}&\lim_{\varepsilon\downarrow 0}(e_m, \mathrm{Im}(\log(\Phi_+(\lambda + i\varepsilon)))e_n)_{\mathcal{H}_+} = (e_m, \mathrm{Im}(\log(\Phi_+(\lambda + i0)))e_n)_{\mathcal{H}_+} \\ &= \pi(e_m, \Xi_+(\lambda)e_n)_{\mathcal{H}_+} \text{ for } \lambda \in \mathbb{R}\backslash\{\mathcal{E}_m \cup \mathcal{E}_n\}.\end{aligned} \tag{3.31}$$

Let $\mathcal{E} = \cup_{n\in\mathbb{N}}\mathcal{E}_n$, then $|\mathcal{E}| = 0$ ($|\cdot|$ denoting the Lebesgue measure on $\mathbb{R}$) and hence

$$\begin{aligned}&(f, \mathrm{Im}(\log(\Phi_+(\lambda + i0)))g)_{\mathcal{H}_+} = \pi(f, \Xi_+(\lambda)g)_{\mathcal{H}_+} \\ &\text{for } \lambda \in \mathbb{R}\backslash\mathcal{E} \text{ and } f, g \in \mathcal{D} = \text{lin.span}\,\{e_n \in \mathcal{H}_+ \,|\, n \in \mathbb{N}\}.\end{aligned} \tag{3.32}$$

Since $\mathcal{D}$ is dense in $\mathcal{H}_+$ and $\Xi_+(\lambda) \in \mathcal{B}(\mathcal{H}_+)$ one infers $\mathrm{Im}(\log(\Phi_+(\lambda + i0))) = \pi\Xi_+(\lambda)$ for a.e. $\lambda \in \mathbb{R}$, completing the proof. □

Of course (3.24a), (3.24b) (and the method of proof) immediately extend to a.e. nontangential limits to the real axis.

Assuming Hypothesis 3.1 we define

$$\xi(\lambda) = \xi_+(\lambda) - \xi_-(\lambda) \text{ for a.e. } \lambda \in \mathbb{R} \tag{3.33}$$

and call $\xi(\lambda)$ (resp., $\xi_+(\lambda)$, $\xi_-(\lambda)$) the spectral shift function associated with the pair (H_0, H) (resp., (H_0, H_+), (H_+, H)), sometimes also denoted by $\xi(\lambda, H_0, H)$, etc., to underscore the dependence on the pair involved.

M. Krein's basic trace formula [51] is now obtained as follows.

Theorem 3.7. *Assume Hypothesis* 3.1. *Then* ($z \in \mathbb{C}\backslash\{\operatorname{spec}(H_0) \cup \operatorname{spec}(H)\}$)

$$\operatorname{Tr}_{\mathcal{H}}((H-z)^{-1} - (H_0 - z)^{-1}) = -\int_{\mathbb{R}} d\lambda\, \xi(\lambda)(\lambda - z)^{-2}. \tag{3.34}$$

Proof. Let $z \in \mathbb{C}\backslash\mathbb{R}$. By (3.27) and (3.29) we infer

$$\operatorname{Tr}_{\mathcal{H}_+}(\log(\Phi_+(z))) = \int_{\mathbb{R}} d\lambda\, \xi_+(\lambda)(\lambda - z)^{-1}, \tag{3.35}$$

$$\operatorname{Tr}_{\mathcal{H}_-}(\log(\widetilde{\Phi}_-(z))) = -\int_{\mathbb{R}} d\lambda\, \xi_-(\lambda)(\lambda - z)^{-1}. \tag{3.36}$$

Adding (3.16a) and (3.16b), differentiating (3.35) and (3.36) with respect to z proves (3.34) for $z \in \mathbb{C}\backslash\mathbb{R}$. The result extends to all $z \in \mathbb{C}\backslash\{\operatorname{spec}(H_0) \cup \operatorname{spec}(H)\}$ by continuity of $((H-z)^{-1} - (H_0 - z)^{-1})$ in $\mathcal{B}_1(\mathcal{H})$-norm. □

In particular, $\xi(\lambda)$ introduced in (3.33) is Krein's original spectral shift function (up to normalization). As noted in Section 2, the spectral shift operator $\Xi_+(\lambda)$ in the particular case $V = V_+$, and its relation to Krein's spectral shift function $\xi_+(\lambda)$, was first studied by Carey [18] in 1976.

Remark 3.8. (i) As shown originally by M. Krein [51], the trace formula (3.34) extends to

$$\operatorname{Tr}(f(H) - f(H_0)) = \int_{\mathbb{R}} d\lambda\, \xi(\lambda) f'(\lambda) \tag{3.37}$$

for appropriate functions f. This fact has been studied by numerous authors and we refer, for instance, to [5], Ch. 19, [10]–[12], [52], [53], [65], [82], [83], [87], [88], Ch. 8 and the references therein.

(ii) While we focus here on pairs of self-adjoint operators (H_0, H), M. Krein in his original paper [51] also considered pairs of unitary operators. Given the conformal equivalence of $\mathbb{C}_+$ and the open interior of the unit disk D in $\mathbb{C}$, this corresponds precisely to the study of Nevanlinna, respectively, Riesz-Herglotz functions in $\mathbb{C}_+$, respectively, D. Pairs of unitary operators are also studied, for instance, in [11], [12], [58], [82], [83]. The trace formula in the case of non-self-adjoint and non-unitary pairs of operators is a rather challenging problem. The interested reader

can get some insight into this matter by consulting [1], [53], [55], [74] and the extensive literature cited therein. Similarly, the case of non-trace class perturbations and associated trace formulas has been studied by various authors. We refer, for instance, to [75], where Hilbert-Schmidt perturbations of unitary operators are treated in depth.
(iii) While scattering theory for the pair (H_0, H) is not discussed in this paper, we remark that $\xi(\lambda)$, for a.e. $\lambda \in \operatorname{spec}_{ac}(H_0)$ (the absolutely continuous spectrum of H_0), is related to the scattering operator at fixed energy λ by the Birman-Krein formula [9],

$$\det_{\mathcal{H}_\lambda}(S(\lambda, H_0, H)) = e^{-2\pi i \xi(\lambda)} \text{ for a.e. } \lambda \in \operatorname{spec}_{ac}(H_0). \tag{3.38}$$

Here $S(\lambda, H_0, H)$ denote the fibers in the direct integral representation of the scattering operator

$$S(H_0, H) = \int_{\operatorname{spec}_{ac}(H_0)}^{\oplus} d\lambda\, S(\lambda, H_0, H) \text{ in } \mathcal{H} = \int_{\operatorname{spec}_{ac}(H_0)}^{\oplus} d\lambda\, \mathcal{H}_\lambda$$

with respect to the absolutely continuous part $H_{0,ac}$ of H_0. This fundamental connection, originally due to Birman and Krein [9], is further discussed in [5], Ch. 19, [12], [13], [18], [46], [52], [84], [88], Ch. 8 and the literature cited therein.
(iv) The standard identity ([41], Sect. IV.3)

$$\operatorname{Tr}_{\mathcal{H}}((H-z)^{-1} - (H_0 - z)^{-1}) = -d\log(\det_{\mathcal{H}}(I_{\mathcal{H}} + V(H_0 - z)^{-1}))/dz \tag{3.39}$$

together with the trace formula (3.34) yields the well-known connection between perturbation determinants and $\xi(\lambda)$, also due to M. Krein [51]

$$\log(\det_{\mathcal{H}}(I_{\mathcal{H}} + V(H_0 - z)^{-1})) = \int_{\mathbb{R}} d\lambda\, \xi(\lambda)(\lambda - z)^{-1}, \tag{3.40}$$

$$\xi(\lambda) = \lim_{\varepsilon \downarrow 0} \pi^{-1} \operatorname{Im}(\log(\det_{\mathcal{H}}(I_{\mathcal{H}} + V(H_0 - (\lambda + i0))^{-1}))) \text{ for a.e. } \lambda \in \mathbb{R}, \tag{3.41}$$

$$\operatorname{Tr}_{\mathcal{H}}(V) = \int_{\mathbb{R}} d\lambda\, \xi(\lambda), \qquad \int_{\mathbb{R}} d\lambda\, |\xi(\lambda)| \leq ||V||_1. \tag{3.42}$$

This is discussed in more detail, for instance, in [5], Ch. 19, [12], [18], [50], [52], [54], [82]–[84], [88]. Relation (3.41) and the analog of (2.2) for $d(AE_H(\lambda)B)/d\lambda$, where $H = H_0 + V$, $V = B^*A$, $A, B \in \mathcal{B}_2(\mathcal{H})$, $V = V^*$, leads to the expression

$$\xi(\lambda) = (-2\pi i)^{-1} \operatorname{Tr}_{\mathcal{H}}(\log(I_{\mathcal{H}} - 2\pi i(I_{\mathcal{H}} - A(H - \lambda - i0)^{-1}B^*)(d(AE_H(\lambda)B^*)/d\lambda)))$$

for a.e. $\lambda \in \mathbb{R}$ (cf., e.g., [5], Sects. 3.4.4 and 19.1.4).
(v) The invariance principle for wave operators of the pair (H_0, H) implies

$$\xi(\lambda) = \xi(\lambda, H_0, H) = \operatorname{sgn}(\Psi'(\lambda))\xi(\Psi(\lambda), \Psi(H_0), \Psi(H)) \tag{3.43}$$

for a certain class of admissible functions Ψ. In certain cases (e.g., if H_0 and H are bounded from below) this can be used to define ξ even though $H - H_0 = V$ is not of trace class as long as $(\Psi(H) - \Psi(H_0)) \in \mathcal{B}_1(\mathcal{H})$. Prime candidates for Ψ in such cases are semigroup ($\Psi(\lambda) = e^{-t\lambda}, t > 0$) and resolvent ($\Psi(\lambda) = (\lambda - z)^{-1}$, $z \in \mathbb{C}\backslash\mathbb{R}$ or $z < E_0$ for some $E_0 \in \mathbb{R}$) functions. Pertinent facts in this connection can be found in [44], [83], and [88], Sect. 8.11.
(vi) For simplicity we chose a single Hilbert space formulation throughout this section. However, every result immediately extends to the case where $K \in \mathcal{B}_2(\mathcal{K}, \mathcal{H})$, $J = J^* \in \mathcal{B}(\mathcal{K})$, $J^2 = I_{\mathcal{K}}$ and $\mathcal{K}$ is another complex separable Hilbert space.

Remark 3.9. Suppose H_0, H_1, and H_2 are self-adjoint operators in $\mathcal{H}$ with $(H_j - H_k) \in \mathcal{B}_1(\mathcal{H})$ for all $j, k \in \{0, 1, 2\}$. Denoting by $\xi(\lambda, H_j, H_k)$ the Krein spectral shift function of the pair (H_j, H_k) such that

$$\mathrm{Tr}_{\mathcal{H}}((H_k - z)^{-1} - (H_j - z)^{-1}) = -\int_{\mathbb{R}} d\lambda\, \xi(\lambda, H_j, H_k)(\lambda - z)^{-2}, \tag{3.44}$$

the chain rule,

$$\xi(\lambda, H_0, H_2) = \xi(\lambda, H_0, H_1) + \xi(\lambda, H_1, H_2) \text{ a.e.}, \tag{3.45}$$

together with

$$\xi(\lambda, H_j, H_k) = -\xi(\lambda, H_k, H_j) \text{ a.e.}, \tag{3.46}$$

and

$$\xi(\lambda, H_j, H_k) \geq 0 \text{ a.e. if } (H_k - H_j) \geq 0, \tag{3.47}$$

imply the monotonicity property

$$\xi(\lambda, H_0, H_2) \geq \xi(\lambda, H_0, H_1) \text{ a.e. if } H_2 \geq H_1. \tag{3.48}$$

Here (3.47) is clear from (3.19), (3.21). Equation (3.46) follows from (3.44), and (3.45) is a consequence of (3.41) observing the facts

$$\begin{gathered} I_{\mathcal{H}} + V(H_0 - z)^{-1} = (H_0 + V - z)(H_0 - z)^{-1}, \quad V \in \mathcal{B}(\mathcal{H}), \\ \det_{\mathcal{H}}((I_{\mathcal{H}} + A)(I_{\mathcal{H}} + B)) = \det_{\mathcal{H}}(I_{\mathcal{H}} + A)\det_{\mathcal{H}}(I_{\mathcal{H}} + B), \quad A, B \in \mathcal{B}_1(\mathcal{H}). \end{gathered}$$

Given the monotonicity property (3.48) of Krein's spectral shift function, it is natural to inquire whether or not this property is shared by the spectral shift operator. More precisely, one might ask whether or not

$$\Xi(\lambda, H_0, H_2) \overset{?}{\geq} \Xi(\lambda, H_0, H_1) \text{ a.e. if } H_2 \geq H_1 \geq H_0. \tag{3.49}$$

The following simple counter example destroys such hopes.

Example 3.10. Let $H_0 = 0$, $K \in \mathcal{B}_2(\mathcal{H})$, $J = I_{\mathcal{H}}$, and hence $\Phi(z) = I_{\mathcal{H}} - K^*Kz^{-1}$, $z \in \mathbb{C}\backslash\{0\}$. Then

$$\lim_{\varepsilon\downarrow 0} \| \log(I_{\mathcal{H}} - K^*K(\lambda + i\varepsilon)^{-1}) - \log(I_{\mathcal{H}} - K^*K\lambda^{-1})\|_{\mathcal{B}_1(\mathcal{H})} = 0 \tag{3.50}$$

for $\lambda \in \mathbb{R}\backslash\{\text{spec}(K^*K) \cup \{0\}\}$ and

$$\pi^{-1}\text{Im}(\log(I_{\mathcal{H}} - K^*K\lambda^{-1})) = \Xi(\lambda), \quad \lambda \in \mathbb{R}\backslash\{\text{spec}(K^*K) \cup \{0\}\}. \tag{3.51}$$

Decomposing the self-adjoint operator $I_{\mathcal{H}} - K^*K\lambda^{-1}$, $\lambda \in \mathbb{R}\backslash\{\text{spec}(K^*K) \cup \{0\}\}$ into its positive and negative spectral parts then yields $\Xi(\lambda) = \theta(K^*K - \lambda I_{\mathcal{H}})$, $\lambda \in \mathbb{R}\backslash\{\text{spec}(K^*K)\cup\{0\}\}$, where $\theta(\cdot)$ denotes the usual step function (i.e., $\theta(x) = 1$ for $x > 0$ and $\theta(x) = 0$ for $x < 0$).

Next we choose $\mathcal{H} = \mathbb{C}^2$, $0 < a < b < c < 1$, $ac - b^2 \geq 0$, and

$$K_1^*K_1 = \begin{pmatrix} 1 & b \\ b & 1 \end{pmatrix}, \quad \text{spec}(K_1^*K_1) = \{1 - b, 1 + b\}, \tag{3.52}$$

with eigenvectors $2^{-1/2}(1, \pm 1)^t$ associated to the eigenvalues $1 \pm b$ and

$$K_2^*K_2 = \begin{pmatrix} 1+a & 0 \\ 0 & 1+c \end{pmatrix}, \quad \text{spec}(K_2^*K_2) = \{1 + a, 1 + c\}, \tag{3.53}$$

with eigenvectors $(1, 0)^t$ (resp., $(0, 1)^t$) associated to the eigenvalue $1 + a$ (resp., $1 + c$). Finally, choosing $\lambda \in (1 + a, 1 + b)$ then yields

$$\Xi_1(\lambda) = \theta(K_1^*K_1 - \lambda I_{\mathbb{C}^2}) = E_{K_1^*K_1}(\{1 + b\}), \tag{3.54}$$

$$\Xi_2(\lambda) = \theta(K_2^*K_2 - \lambda I_{\mathbb{C}^2}) = E_{K_2^*K_2}(\{1 + c\}), \tag{3.55}$$

where $\{E_{K^*K}(\lambda)\}_{\lambda\in\mathbb{R}}$ denotes the family of orthogonal spectral projections of K^*K. Clearly $H_2 = K_2^*K_2 \geq K_1^*K_1 = H_1 \geq H_0 = 0$ but $\Xi_2(\lambda) \not\geq \Xi_1(\lambda)$ for $\lambda \in (1 + a, 1 + b)$ since one-dimensional self-adjoint projections cannot satisfy an order relation unless one is a real multiple of the other. (Note, however, that $\text{Tr}(\Xi_2(\lambda)) = 1 + c > 1 + b = \text{Tr}(\Xi_1(\lambda))$ in accordance with (3.48).)

This example shows, in particular, that the chain rule (3.45) for $\xi(\lambda)$ does not extend to $\Xi(\lambda)$.

4 Spectral Averaging: An Operator-Theoretic Approach

In this section we apply the formalism developed in Sections 2 and 3 to provide an effortless proof of spectral averaging and its relation to Krein's spectral shift function as originally proven by Birman and Solomyak [11].

For the basic setup of this section we assume the following hypothesis.

Hypothesis 4.1. *Let H_0 be a self-adjoint operator in $\mathcal{H}$ with* $\mathrm{dom}(H_0)$, *and assume $\{V(s)\}_{s\in\Omega} \subset \mathcal{B}_1(\mathcal{H})$ to be a family of self-adjoint trace class operators in $\mathcal{H}$, where $\Omega \subseteq \mathbb{R}$ denotes an open interval. Moreover, suppose that $V(s)$ is continuously differentiable with respect to $s \in \Omega$ in trace norm.*

To begin our discussions we temporarily assume that $V(s) \geq 0$, that is, we suppose

$$V(s) = K(s)K(s)^*, \quad s \in \Omega \tag{4.1}$$

for some $K(s) \in \mathcal{B}_2(\mathcal{H})$, $s \in \Omega$. Given Hypothesis 4.1 we define the self-adjoint operator $H(s)$ in $\mathcal{H}$ by

$$H(s) = H_0 + V(s), \quad \mathrm{dom}(H(s)) = \mathrm{dom}(H_0), \quad s \in \Omega. \tag{4.2}$$

In analogy to (3.5) and (3.6) we introduce in $\mathcal{H}$ ($s \in \Omega$, $z \in \mathbb{C}\backslash\mathbb{R}$),

$$\Phi(z,s) = I_{\mathcal{H}} + K(s)^*(H_0 - z)^{-1}K(s) \tag{4.3}$$

and hence infer from Lemma 3.2 that

$$\Phi(z,s)^{-1} = I_{\mathcal{H}} - K(s)^*(H(s) - z)^{-1}K(s). \tag{4.4}$$

The following is an elementary but useful result needed in the context of Theorem 4.3.

Lemma 4.2. *Assume Hypothesis* 4.1 *and* (4.1). *Then* ($s \in \Omega$, $z \in \mathbb{C}\backslash\mathbb{R}$),

$$d\,\mathrm{Tr}_{\mathcal{H}}(\log(\Phi(z,s)))/ds = \mathrm{Tr}_{\mathcal{H}}(V'(s)(H(s) - z)^{-1}). \tag{4.5}$$

Proof. By (3.15), (4.3), and (4.4) one infers for $z = iy$, $|y| > 0$ sufficiently large,

$$\begin{aligned}
d\,\mathrm{Tr}_{\mathcal{H}}(\log(\Phi(z,s)))/ds &= d\,\mathrm{Tr}_{\mathcal{H}}(\log(I_{\mathcal{H}} - V(s)(H_0 - z)^{-1}))/ds \\
&= (d/ds)\sum_{j=0}^{\infty}(-1)^j(j+1)^{-1}\,\mathrm{Tr}_{\mathcal{H}}((V(s)(H_0 - z)^{-1})^{j+1}) \\
&= \mathrm{Tr}_{\mathcal{H}}(V'(s)(H_0 - z)^{-1}\sum_{j=0}^{\infty}(-1)^j(V(s)(H_0 - z)^{-1})^j) \\
&= \mathrm{Tr}_{\mathcal{H}}(V'(s)(H(s) - z)^{-1})
\end{aligned} \tag{4.6}$$

by repeated use of (3.18) and (3.13a). Analytic continuation of (4.6) with respect to $z \in \mathbb{C}\backslash\mathbb{R}$ then proves (4.5). □

Next, applying Lemma 3.6 to $\Phi(z,s)$ in (4.3) one infers ($s \in \Omega$),

$$\log(\Phi(z,s)) = \int d\lambda\, \Xi(\lambda,s)(\lambda - z)^{-1}, \tag{4.7}$$

$$0 \leq \Xi(\lambda, s) \leq I_{\mathcal{H}}, \quad \Xi(\lambda, s) \in \mathcal{B}_1(\mathcal{H}) \text{ for a.e. } \lambda \in \mathbb{R}, \tag{4.8}$$
$$\|\Xi(\cdot, s)\|_1 \in L^1(\mathbb{R}; d\lambda),$$

where $\Xi(\lambda, s)$ is associated with the pair $(H_0, H(s))$, assuming $H(s) \geq H_0$, $s \in \Omega$.

Our principle result on averaging the spectral measure of $\{E_{H(s)}(\lambda)\}_{\lambda\in\mathbb{R}}$ of $H(s)$ then reads as follows.

Theorem 4.3. *Assume Hypothesis* 4.1 *and* $[s_1, s_2] \subset \Omega$. *Let* $\xi(\lambda, s)$ *be the spectral shift function associated with the pair* $(H_0, H(s))$ (*cf.* (3.33)), *where* $H(s)$ *is defined by* (4.2) (*and we no longer suppose* $H(s) \geq H_0$). *Then*

$$\int_{s1}^{s_2} ds\,(d(\mathrm{Tr}_{\mathcal{H}}(V'(s)E_{H(s)}(\lambda)))) = (\xi(\lambda, s_2) - \xi(\lambda, s_1))d\lambda. \tag{4.9}$$

Proof. First we prove (4.9) in the case $V(s) \geq 0$. The monotone convergence theorem, (4.7), and Lemma 4.2 then yield ($z \in \mathbb{C}\backslash\mathbb{R}$),

$$\begin{aligned}
&\mathrm{Tr}_{\mathcal{H}}\left(\int_{\mathbb{R}} d\lambda\,((\lambda - \mathrm{Re}(z))^2 + (\mathrm{Im}(z))^2)^{-1}\mathrm{Im}(z)(\Xi(\lambda, s_2) - \Xi(\lambda, s_1))\right)\\
&= \int_{\mathbb{R}} d\lambda\,((\lambda - \mathrm{Re}(z))^2 + (\mathrm{Im}(z))^2)^{-1}\mathrm{Im}(z)(\xi(\lambda, s_2) - \xi(\lambda, s_1))\\
&= \mathrm{Tr}_{\mathcal{H}}(\mathrm{Im}(\log(\Phi(z, s_2)))) - \mathrm{Tr}_{\mathcal{H}}(\mathrm{Im}(\log(\Phi(z, s_1))))\\
&= \int_{s_1}^{s_2} ds\left(\frac{d}{ds}\mathrm{Tr}_{\mathcal{H}}(\mathrm{Im}(\log(\Phi(z, s))))\right) = \int_{s_1}^{s_2} ds\,\mathrm{Tr}_{\mathcal{H}}(V'(s)\mathrm{Im}((H(s) - z)^{-1})).
\end{aligned}$$

By the spectral theorem applied to $H(s)$ one obtains

$$\mathrm{Im}((H(s) - z)^{-1}) = \mathrm{Im}(z)\int_{\mathbb{R}} dE_{H(s)}(\lambda)((\lambda - \mathrm{Re}(z))^2 + (\mathrm{Im}(z))^2)^{-1} \tag{4.10}$$

and hence

$$\begin{aligned}
&\int_{\mathbb{R}} d\lambda\,((\lambda - \mathrm{Re}(z))^2 + (\mathrm{Im}(z))^2)^{-1}(\xi(\lambda, s_2) - \xi(\lambda, s_1))\\
&= \int_{s_1}^{s_2} ds\,\mathrm{Tr}_{\mathcal{H}}\left(\int_{\mathbb{R}} dE_{H(s)}(\lambda)((\lambda - \mathrm{Re}(z))^2 + (\mathrm{Im}(z))^2)^{-1}V'(s)\right).
\end{aligned} \tag{4.11}$$

Decomposing the self-adjoint trace class operator $V'(s)$ into its positive and negative parts,

$$V'(s) = (V'(s))_+ - (V'(s))_-, \quad 0 \leq (V'(s))_\pm \in \mathcal{B}_1(\mathcal{H}), \tag{4.12}$$

the monotone convergence theorem yields

$$\int_{\mathbb{R}} d\lambda\,((\lambda - \mathrm{Re}(z))^2 + (\mathrm{Im}(z))^2)^{-1}(\xi(\lambda, s_2) - \xi(\lambda, s_1))$$

$$
\begin{aligned}
&= \int_{s_1}^{s_2} ds \int_{\mathbb{R}} ((\lambda - \mathrm{Re}(z))^2 + (\mathrm{Im}(z))^2)^{-1} (d(\mathrm{Tr}_{\mathcal{H}}((V'(s))_+^{1/2} E_{H(s)}(\lambda)(V'(s))_+^{1/2}) \\
&\quad - d(\mathrm{Tr}_{\mathcal{H}}((V'(s))_-^{1/2} E_{H(s)}(\lambda)(V'(s))_-^{1/2}))) \\
&= \int_{s_1}^{s_2} ds \int_{\mathbb{R}} ((\lambda - \mathrm{Re}(z))^2 + (\mathrm{Im}(z))^2)^{-1} d(\mathrm{Tr}_{\mathcal{H}}(V'(s) E_{H(s)}(\lambda))) \\
&= \int_{\mathbb{R}} ((\lambda - \mathrm{Re}(z))^2 + (\mathrm{Im}(z))^2)^{-1} \int_{s_1}^{s_2} ds (d(\mathrm{Tr}_{\mathcal{H}}(V'(s) E_{H(s)}(\lambda)))) \qquad (4.13)
\end{aligned}
$$

using Fubini's theorem in the last step. Thus, by the uniqueness property of Poisson transforms,

$$
(\xi(\lambda, s_2) - \xi(\lambda, s_1)) d\lambda = \int_{s_1}^{s_2} ds \, (d(\mathrm{Tr}_{\mathcal{H}}(V'(s) E_{H(s)}(\lambda)))). \qquad (4.14)
$$

In the case of arbitrary $V(s) \in \mathcal{B}_1(\mathcal{H})$ (not necessarily nonnegative), we argue as follows. Define (cf. (4.12))

$$
W = \left(\int_{s_1}^{s_2} ds \, (V'(s))_- \right) - V(s_1), \qquad (4.15)
$$

then $W \in \mathcal{B}_1(\mathcal{H})$ and $V(s) + W \geq 0$ for all $s \in [s_1, s_2]$. Equation (4.9) now follows from the chain rule (3.45) for spectral shift functions

$$
\xi(\lambda, H_0, H(s)) = \xi(\lambda, H_0, H_0 - W) + \xi(\lambda, H_0 - W, H(s)). \qquad (4.16)
$$

Indeed, $(V(s) + W)' = V'(s)$, and $\xi(\lambda, H_0, H_0 - W)$ is independent of s and hence drops out on the right-hand side of (4.9). Moreover, the pair $(H_0 - W, H(s))$ only involves the nonnegative perturbation $V(s) + W$ of $H_0 - W$ in $H(s) = H_0 - W + (V(s) + W)$ so that Lemma 4.2 becomes applicable as in (4.14) in the first part of our proof. □

Remark 4.4. (i) In the special case of averaging over the boundary condition parameter for half-line Sturm-Liouville operators (effectively a rank-one resolvent perturbation problem), Theorem 4.3 has first been derived by Javrjan [42], [43]. The case of rank-one perturbations was recently treated in detail by Simon [79]. The general case of trace class perturbations is due to Birman and Solomyak [11] using an approach of Stieltjes' double operator integrals. Birman and Solomyak treat the case $V(s) = sV$, $V \in \mathcal{B}_1(\mathcal{H})$, $s \in [0, 1]$. As explained in [6], [11], and [12], the authors were interested in a real analysis approach to the spectral shift function in contrast to M. Krein's complex analytic treatment. In this way the local integrability of ξ could not be obtained directly (only its property of being a generalized function was obtained) although it follows of course from the uniqueness ξ (up to additive constants). While our operator theoretic approach is intrinsically complex-analytic, and hence very much in M. Krein's spirit, it leads to a natural proof of the absolute continuity of the (signed) measure on the right-hand side

of (4.9). A short proof of (4.9) (assuming $V'(s) \geq 0$) has recently been given by Simon [80].
(ii) We note that variants of (4.9) in the context of one-dimensional Sturm-Liouville operators (i.e., variants of Javrjan's results in [42], [43]) have been repeatedly rediscovered by several authors. In particular, the absolute continuity of averaged spectral measures (with respect to boundary condition parameters or coupling constants of rank-one perturbations) has been used to prove localization properties of one-dimensional random Schrödinger operators (see, e.g., [17], [20], [21], [22], Ch. VIII, [24], [25], [47]–[49], [64], Ch. V, [78], [79]).
(iii) We emphasize that Theorem 4.3 applies to unbounded operators (and hence to random Schrödinger operators bounded from below) as long as appropriate relative trace class conditions (either with respect to resolvent or semigroup perturbations) are satisfied.
(iv) In the special case $V'(s) \geq 0$, the measure

$$d(\mathrm{Tr}_{\mathcal{H}}(V'(s)E_{H(s)}(\lambda))) = d(\mathrm{Tr}_{\mathcal{H}}(V'(s)^{1/2}E_{H(s)}(\lambda)V'(s)^{1/2}))$$

in (4.9) represents a positive measure.
(v) The result (4.9) is not restricted to a one-dimensional parameter space $s \in [s_1, s_2]$. In fact, if $\gamma(\mathbf{s_1}, \mathbf{s_2})$ denotes an oriented piecewise C^1-path connecting $\mathbf{s_1} \in \mathbb{R}^n$ and $\mathbf{s_2} \in \mathbb{R}^n$, one obtains analogously,

$$\int_{\gamma(\mathbf{s_1},\mathbf{s_2})} d\mathbf{s} \cdot (d(\mathrm{Tr}_{\mathcal{H}}((\nabla V)(\mathbf{s})E_{H(\mathbf{s})}(\lambda)))) = (\xi(\lambda, \mathbf{s_2}) - \xi(\lambda, \mathbf{s_1}))d\lambda. \tag{4.17}$$

We omit further details.

In the special case of a sign-definite perturbation of H_0 of the form sKK^*, one can in fact prove an operator-valued averaging formula as follows.

Theorem 4.5. *Assume Hypothesis* 3.1 *and* $J = I_{\mathcal{H}}$. *Then*

$$\int_0^1 ds\, d(K^*E_{H_0+sKK^*}(\lambda)K) = \Xi(\lambda)d\lambda, \tag{4.18}$$

where $\Xi(\cdot)$ *is the spectral shift operator associated with*

$$\Phi(z) = I_{\mathcal{H}} + K^*(H_0 - z)^{-1}K, \quad z \in \mathbb{C}\backslash\mathbb{R}, \tag{4.19}$$

that is,

$$\log(\Phi(z)) = \int_{\mathbb{R}} d\lambda\, \Xi(\lambda)(\lambda - z)^{-1}, \quad z \in \mathbb{C}\backslash\mathbb{R}, \tag{4.20}$$

$$0 \leq \Xi(\lambda) \in \mathcal{B}_1(\mathcal{H}) \text{ for a.e. } \lambda \in \mathbb{R}, \quad \|\Xi(\cdot)\|_1 \in L^1(\mathbb{R}; d\lambda). \tag{4.21}$$

Proof. An explicit computation shows

$$(\lambda - \Phi(z))^{-1} = -(1 - \lambda)^{-1}$$

$$\times (I_{\mathcal{H}} - (1-\lambda)^{-1}K^*(H_0 + (1-\lambda)^{-1}KK^* - z)^{-1}K) \in \mathcal{B}(\mathcal{H}) \tag{4.22}$$

for all $\lambda < 0$. Since $\log(\Phi(z)) = \ln(\Phi(z))$ for $z \in \mathbb{C}_+$ as a result of analytic continuation, one obtains

$$\begin{aligned} \log(\Phi(z)) &= \int_{\mathbb{R}} d\lambda\, \Xi(\lambda)(\lambda - z)^{-1} \\ = \ln(\Phi(z)) &= \int_{-\infty}^{0} d\lambda\, ((\lambda - \Phi(z))^{-1} - \lambda(1+\lambda^2)^{-1} I_{\mathcal{H}}) \\ &= \int_{-\infty}^{0} d\lambda\, (1-\lambda)^{-2} K^*(H_0 + (1-\lambda)^{-1}KK^* - z)^{-1}K \\ &= \int_0^1 ds \int_{\mathbb{R}} d(K^* E_{H_0 + sKK^*}(\lambda)K)(\lambda - z)^{-1} \\ &= \int_{\mathbb{R}} (\lambda - z)^{-1} \int_0^1 ds\, d(K^* E_{H_0+sKK^*}(\lambda)K) \end{aligned} \tag{4.23}$$

proving (4.18). (Here the interchange of the λ and s integrals follows from Fubini's theorem considering (4.23) in the weak sense.) □

As a consequence of Theorem 4.5 one obtains

$$\int_{s_1}^{s_2} ds\, d(K^* E_{H_0+sKK^*}(\lambda)K) = \Xi(\lambda, s_2) - \Xi(\lambda, s_1), \tag{4.24}$$

where $\Xi(\lambda, s)$ is the spectral shift operator associated with $\Phi(z,s) = I_{\mathcal{H}} + sK^*(H_0 - z)^{-1}K$, $s \in [s_1, s_2]$.

Acknowledgments. F. G. would like to thank P. Exner for his kind invitation to take part in this conference, and all organizers for providing a most stimulating meeting. He also thanks H. Holden for the generous invitation to spend seven weeks in the summer of 1998 at the Norwegian University of Science and Technology, NTNU, Trondheim, during the final stages of this work. The extraordinary hospitality at the Department of Mathematical Sciences at NTNU and financial support by the Research Council Norway, Grant No. 107510/410 are gratefully acknowledged.

K. A. M. gratefully acknowledges financial support by the University of Missouri Research Board, Grant No. 98–119.

S. N. N. thanks H. Holden for making it possible to spend a week in July of 1998 at the Norwegian University of Science and Technology, NTNU, Trondheim.

References

[1] V. M. Adamjan and B. S. Pavlov, *A trace formula for dissipative operators,* Vestnik Leningrad. Univ. Math. **12**, 85–91 (1980).

[2] A. B. Aleksandrov, *The multiplicity of the boundary values of inner functions,* Sov. J. Contemp. Math. Anal. **22**, No. 5, 74–87 (1987).

[3] W. O. Amrein and K. B. Sinha, *On pairs of projections in a Hilbert space,* Lin. Algebra Appl. **208/209**, 425–435 (1994).

[4] N. Aronszajn and W. F. Donoghue, *On exponential representations of analytic functions in the upper half-plane with positive imaginary part,* J. d'Anal. Math. **5**, 321–388 (1956-57).

[5] H. Baumgärtel and M. Wollenberg, *Mathematical Scattering Theory,* Birkhäuser, Basel, 1983.

[6] M. Sh. Birman, *Spectral shift function and double operator integrals,* in Linear and Complex Analysis Problem Book 3, V. P. Havin and N. K. Nikolski (eds.), Lecture Notes in Mathematics **1573**, Springer, Berlin, 1994, pp. 272–273.

[7] M. Sh. Birman and S. B. Entina, *A stationary approach in the abstract theory of scattering,* Sov. Math. Dokl. **5**, 432–435 (1964).

[8] M. Sh. Birman and S. B. Entina, *The stationary method in the abstract theory of scattering,* Math. USSR Izv. **1**, 391–420 (1967).

[9] M. Sh. Birman and M. G. Krein, *On the theory of wave operators and scattering operators,* Sov. Math. Dokl. **3**, 740–744 (1962).

[10] M. Sh. Birman and A. B. Pushnitski, *Spectral shift function, amazing and multifaceted,* Integr. Eqs. Operator Th. **30**, 191–199 (1998).

[11] M. Sh. Birman and M. Z. Solomyak, *Remarks on the spectral shift function,* J. Sov. Math. **3**, 408–419 (1975).

[12] M. Sh. Birman and D. R. Yafaev, *The spectral shift function. The work of M. G. Krein and its further development,* St. Petersburg Math. J. **4**, 833–870 (1993).

[13] M. Sh. Birman and D. R. Yafaev, *Spectral properties of the scattering matrix,* St. Petersburg Math. J. **4**, 1055–1079 (1993).

[14] D. Bollé, F. Gesztesy, H. Grosse, W. Schweiger, and B. Simon, *Witten index, axial anomaly, and Krein's spectral shift function in supersymmetric quantum mechanics,* J. Math. Phys. **28**, 1512–1525 (1987).

[15] N. Borisov, W. Müller, and R. Schrader, *Relative index theorems and supersymmetric scattering theory,* Commun. Math. Phys. **114**, 475–513 (1988).

[16] M. S. Brodskii, *Triangular and Jordan Representations of Linear Operators,* Amer. Math. Soc., Providence, RI, 1971.

[17] D. Buschmann and G. Stolz, *Two-parameter spectral averaging and localization for non-monotoneous random Schrödnder operators,* preprint, 1998.

[18] R. W. Carey, *A unitary invariant for pairs of self-adjoint operators,* J. reine angewandte Math. **283**, 294–312 (1976).

[19] R. W. Carey and W. D. Pepe, *The phase shift and singular measures,* Indiana Univ. Math. J. **22**, 1049–1064 (1973).

[20] R. Carmona, *One-dimensional Schrödinger operators with random or deterministic potentials: new spectral types,* J. Funct. Anal. **51**, 229–258 (1983).

[21] R. Carmona, *Absolute continuous spectrum of one-dimensional Schrödinger operators,* in Differential Equations, I. W. Knowles and R. T. Lewis (eds.), North-Holland, Amsterdam, 1984, pp. 77–86.

[22] R. Carmona and J. Lacroix, *Spectral Theory of Random Schrödinger Operators,* Birkhäuser, Boston, 1990.

[23] T. Christiansen, *Spectral asymptotics for compactly supported perturbations of the Laplacian on* $\mathbb{R}^n$, Commun. Part. Diff. Eq. **23**, 933–948 (1998).

[24] J.-M. Combes and P. D. Hislop, *Localization for continuous random Hamiltonians in d-dimensions,* J. Funct. Anal. **124**, 149–180 (1994).

[25] J. M. Combes, P. D. Hislop, and E. Mourre, *Spectral averaging, perturbation of singular spectrum, and localization,* Trans. Amer. Math. Soc. **348**, 4883–4894 (1996).

[26] L. de Branges, *Perturbations of self-adjoint transformations,* Amer. J. Math. **84**, 543–560 (1962).

[27] R. del Rio, B. Simon, and G. Stolz, *Stability of spectral types for Sturm-Liouville operators,* Math. Res. Lett. **1**, 437–450 (1994).

[28] J. Dixmier, *Les Algèbres d'Opérateurs dans l'Espace Hilbertien,* 2nd ed., Gauthier-Villars, Paris, 1969.

[29] J.-P. Eckmann and C.-A. Pillet, *Zeta functions with Dirichlet and Neumann boundary conditions for exterior domains,* Helv. Phys. Acta **70**, 44–65 (1997).

[30] R. Geisler, V. Kostrykin, and R. Schrader, *Concavity properties of Krein's spectral shift function,* Rev. Math. Phys. **7**, 161–181 (1995).

[31] F. Gesztesy and H. Holden, *On new trace formulae for Schrödinger operators,* Acta Appl. Math. **39**, 315–333 (1995).

[32] F. Gesztesy, H. Holden, and B. Simon, *Absolute summability of the trace relation for certain Schrödinger operators,* Commun. Math. Phys. **168**, 137–161 (1995).

[33] F. Gesztesy, H. Holden, B. Simon, and Z. Zhao, *Higher order trace relations for Schrödinger operators,* Rev. Math. Phys. **7**, 893–922 (1995).

[34] F. Gesztesy, N. J. Kalton, K. A. Makarov, and E. Tsekanovskii, *Some Applications of Operator-Valued Herglotz Functions*, preprint, 1998.

[35] F. Gesztesy, M. Krishna, and G. Teschl, *On isospectral sets of Jacobi operators,* Commun. Math. Phys. **181**, 631–645 (1996).

[36] F. Gesztesy and K. A. Makarov, in preparation.

[37] F. Gesztesy, K. A. Makarov, and E. Tsekanovskii, *An Addendum to Krein's Formula,* J. Math. Anal. Appl. **222**, 594–606 (1998).

[38] F. Gesztesy and B. Simon, *Topological invariance of the Witten index,* J. Funct. Anal. **79**, 91–102 (1988).

[39] F. Gesztesy and B. Simon, *The ξ function,* Acta Math. **176**, 49–71 (1996).

[40] F. Gesztesy and E. Tsekanovskii, *On matrix-valued Herglotz functions,* preprint, 1997.

[41] I. C. Gohberg and M. G. Krein, *Introduction to the Theory of Linear Nonselfadjoint Operators,* Amer. Math. Soc., Providence, RI, 1969.

[42] V. A. Javrjan, *On the regularized trace of the difference between two singular Sturm-Liouville operators,* Sov. Math. Dokl. **7**, 888–891 (1966).

[43] V. A. Javrjan, *A certain inverse problem for Sturm-Liouville operators,* Izv. Akad. Nauk Armjan. SSR Ser. Math. **6**, 246–251 (1971). (Russian.)

[44] A. Jensen and T. Kato, *Asymptotic behavior of the scattering phase for exterior domains,* Commun. Part. Diff. Eqs. **3**, 1165–1195 (1978).

[45] I. S. Kac and M. G. Krein, *R-functions-analytic functions mapping the upper halfplane into itself,* Amer. Math. Soc. Transl. (2) **103**, 1–18 (1974).

[46] T. Kato, *Monotonicity theorems in scattering theory,* Hadronic J. **1**, 134–154 (1978).

[47] V. Kostrykin and R. Schrader, *Scattering theory approach to random Schrödinger operators in one dimension,* Rev. Math. Phys., to appear.

[48] S. Kotani, *Lyapunov exponents and spectra for one-dimensional random Schrödinger operators,* Contemp. Math. **50**, 277–286 (1986).

[49] S. Kotani and B. Simon, *Localization in general one-dimensional random systems,* Commun. Math. Phys. **112**, 103–119 (1987).

[50] M. G. Krein, *A contribution to the theory of linear non-selfadjoint operators,* Sov. Math. Dokl. **1**, 38–40 (1960).

[51] M. G. Krein, *Perturbation determinants and a formula for the traces of unitary and self-adjoint operators,* Sov. Math. Dokl. **3**, 707–710 (1962).

[52] M. G. Krein, *On certain new studies in the perturbation theory for self-adjoint operators,* in *M. G. Krein, Topics in Differential and Integral Equations and Operator Theory,* I. Gohberg (ed.), Birkhäuser, Basel, 1983.

[53] M. G. Krein, *On perturbation determinants and a trace formula for certain classes of pairs of operators,* Amer. Math. Soc. Trans. (2) **145**, 39–84 (1989).

[54] M. G. Krein and V. A. Javrjan, *On spectral shift functions arising in perturbations of a positive operator,* J. Operator Th. **6**, 155–191 (1981). (Russian.)

[55] H. Langer, *Eine Erweiterung der Spurformel der Störungstheorie,* Math. Nachr. **30**, 123–135 (1965).

[56] I. M. Lifshits, *On a problem of perturbation theory,* Uspekhi Mat. Nauk **7**, No. 1, 171–180 (1952).

[57] I. M. Lifshits, *Some problems of the dynamic theory of nonideal crystal lattices,* Nuovo Cimento Suppl. **3**, (Ser. X), 716–734 (1956).

[58] A. Mohapatra and K. B. Sinha, *Spectral shift function and trace formula for unitaries - a new proof,* Integr. Eqs. Operator Th. **24**, 285–297 (1996).

[59] W. Müller, *Relative zeta functions, relative determinants and scattering theory,* Commun. Math. Phys. **192**, 309–347 (1998).

[60] S. N. Naboko, *Uniqueness theorems for operator-valued functions with positive imaginary part, and the singular spectrum in the Friedrichs model,* Ark. Mat. **25**, 115–140 (1987).

[61] S. N. Naboko, *Boundary values of analytic operator functions with a positive imaginary part,* J. Sov. Math. **44**, 786–795 (1989).

[62] S. N. Naboko, *Nontangential boundary values of operator-valued R-functions in a half-plane,* Leningrad Math. J. **1**, 1255–1278 (1990).

[63] S. N. Naboko, *On the conditions for existence of wave operators in the nonselfadjoint case,* Amer. Math. Soc. Transl. (2) **157**, 127–149 (1993).

[64] L. Pastur and A. Figotin, *Spectra of Random and Almost-Periodic Operators,* Springer, Berlin, 1992.

[65] V. V. Peller, *Hankel operators in the perturbation theory of unitary and self-adjoint operators,* Funct. Anal. Appl. **19**, 111-123 (1985).

[66] A. Poltoratski, *The Krein spectral shift and rank one perturbations of spectra,* St. Petersburg Math. J., to appear.

[67] A. B. Pushnitski, *Representation for the spectral shift function for perturbations of a definite sign,* preprint, 1997.

[68] A. B. Pushnitski, *Spectral shift function of the Schrödinger operator in the large coupling constant limit,* preprint, 1998.

[69] A. B. Pushnitski, *Integral estimates for the spectral shift function,* preprint, 1998.

[70] D. Robert, *Semi-classical and high energy asymptotics of the scattering phase for perturbations of elliptic operators,* in Mathematical Physics X, K. Schmüdgen (ed.), Springer, Berlin, 1992, pp.442–446.

[71] D. Robert, *Semi-classical approximation in quantum mechanics. A survey of old and recent mathematical results,* Helv. Phys. Acta **71**, 44–116 (1998).

[72] D. Robert, *Semi-classical asymptotics for the spectral shift function,* preprint.

[73] D. Robert and V. Sordoni, *Generalized determinants for Sturm-Liouville problems on the real line,* CRM Proceedings and Lecture Notes **12**, 251–259 (1997).

[74] A. V. Rybkin, *The spectral shift function, the characteristic function of a contraction, and a generalized integral,* Russian Acad. Sci. Sb. Math. **83**, 237–281 (1995).

[75] A. V. Rybkin, *On A-integrability of the spectral shift function of unitary operators arising in the Lax-Phillips scattering theory,* Duke Math. J. **83**, 683–699 (1996).

[76] Yu. L. Shmul'yan, *On operator R-functions,* Siberian Math. J. **12**, 315–322 (1971).

[77] B. Simon, *Trace Ideals and their Applications,* Cambridge University Press, Cambridge, 1979.

[78] B. Simon, *Localization in general one dimensional random systems, I. Jacobi matrices,* Commun. Math. Phys. **102**, 327–336 (1985).

[79] B. Simon, *Spectral analysis of rank one perturbations and applications,* CRM Proceedings and Lecture Notes **8**, 109–149 (1995).

[80] B. Simon, *Spectral averaging and the Krein spectral shift,* Proc. Amer. Math. Soc. **126**, 1409–1413 (1998).

[81] B. Simon and T. Wolff, *Singular continuous spectrum under rank one perturbations and localization for random Hamiltonians,* Commun. Pure Appl. Math. **39**, 75–90 (1986).

[82] K. B. Sinha, *On the theorem of M. G. Krein,* preprint, 1975, unpublished.

[83] K. B. Sinha and A. N. Mohapatra, *Spectral shift function and trace formula,* Proc. Indian Acad. Sci. (Math. Sci) **104**, 819–853 (1994).

[84] A. V. Sobolev, *Efficient bounds for the spectral shift function,* Ann. Inst. H. Poincaré **A58**, 55–83 (1993).

[85] B. Sz.-Nagy and C. Foias, *Sur les contractions de l'espace de Hilbert. IV,* Acta Sci. Math. **21**, 251–259 (1960).

[86] B. Sz.-Nagy and C. Foias, *Harmonic Analysis of Operators on Hilbert Space,* North-Holland, Amsterdam, 1970.

[87] D. Voiculescu, *On a trace formula of M. G. Krein,* in Operator Theory: Advances and Applications, Vol. 24, Birkhäuser, Basel, 1987, pp. 329–332.

[88] D. R. Yafaev, *Mathematical Scattering Theory,* Amer. Math. Soc., Providence, RI, 1992.

Operator Theory:
Advances and Applications, Vol. 108

On the scattering operator for the Schrödinger equation with a time-dependent potential

Arne Jensen
Department of Mathematics and MaPhySto*
Institute for Electronic Systems
Aalborg University, Fredrik Bajers Vej 7E
DK-9220 Aalborg Ø, Denmark
matarne@math.auc.dk

1 Introduction

In this paper we give some results on the scattering operator for the Schrödinger equation with a time-dependent potential. We consider the free Schrödinger equation

$$i\partial_t u(t,x) = -\Delta_x u(t,x), \qquad u(s,x) = u_0(x), \tag{1}$$

and the full Schrödinger equation

$$i\partial_t v(t,x) = -\Delta_x v(t,x) + V(t,x)v(t,x), \qquad v(s,x) = v_0(x). \tag{2}$$

Here V is a potential depending explicitly on time. The solution to (1) is given by $u(t) = e^{-i(t-s)H_0}u_0$, where $H_0 = -\Delta_x$ with the domain being the usual Sobolev space of order 2, $\mathcal{D}(H_0) = H^2(\mathbf{R}^d)$. If we assume $V(t,x)$ a real-valued function, such that $V \in L^1(\mathbf{R}; L^\infty(\mathbf{R}^d))$, then associated with (2) is a unitary propagator on $L^2(\mathbf{R}^d)$, denoted by $U(t,s)$, such that the solution to (2) is given by $v(t) = U(t,s)v_0$, see for example [8, 9] and references therein. More precisely, $v(t)$ solves the equation in the sense that v satisfies the integral equation

$$v(t) = U_0(t-s)v_0 - i\int_s^t U_0(t-\tau)V(\tau)v(\tau)\,d\tau, \tag{3}$$

i.e. v is a *mild* solution to the Cauchy problem (2). The propagator satisfies $U(t,t) = 1$ and $U(t,s)U(s,r) = U(t,r)$ for all $t,s,r \in \mathbf{R}$. Furthermore, $(t,s) \mapsto U(t,s)$ is strongly continuous.

---

*Centre for Mathematical Physics and Stochastics, funded by a grant from the Danish National Research Foundation.

For this class of V's one has a scattering theory associated with the solutions to the equations (1), (2). The wave operators are given by

$$W_{\pm}(s) = \operatorname*{s-lim}_{t \to \pm\infty} U(s,t)U_0(t-s). \tag{4}$$

The limits exist on all of $L^2(\mathbf{R}^d)$ and are unitary. The scattering operator is given by

$$S(s) = W_+(s)^{-1}W_-(s). \tag{5}$$

In the paper [3] we studied the scattering problem in the space-time framework, i.e. we considered the problem in the spaces $L^r(\mathbf{R}; L^q(\mathbf{R}^d))$ for a certain range of q, r. This approach was first used by Kato in [4] to study a class of nonlinear Schrödinger equations. In [3, Theorem 4.10] we obtained a representation formula for the scattering operator, using a purely time-dependent method of proof. In this paper we give a different derivation of this formula. We use the stationary scattering theory, in the formulation due to Kuroda [5, 6], combined with the stationary formulation of scattering theory for explicitly time-dependent potentials, in the form given by Howland [2]. For the case of potentials periodic in time the stationary scattering theory has been applied in [7] to derive a representation formula for the associated scattering matrix. The difference with the case considered here is that without a periodicity assumption there is no scattering matrix associated with the original problem, since $U_0(t)$ and $S(s)$ do not commute for any $t \neq 0$. There is a large literature on scattering theory for Schrödinger operators with time-dependent potentials. See for example [1] and references therein.

2 Preliminaries

We start by defining various spaces and operators needed to formulate the problem. We write $\mathcal{H} = L^2(\mathbf{R}^d)$ and introduce

$$\mathcal{K} = L^2(\mathbf{R}) \otimes \mathcal{H} \cong L^2(\mathbf{R}; \mathcal{H}) \cong L^2(\mathbf{R}^{d+1}). \tag{6}$$

We use the identifications of the three spaces without comment in the sequel. Let

$$\widetilde{K}_0 = -i\frac{d}{dx} \otimes I + I \otimes H_0 \tag{7}$$

with domain $H^1(\mathbf{R}) \otimes_{\text{alg}} H^2(\mathbf{R}^d)$. Then $\widetilde{K}_0$ is essentially self-adjoint on this domain. The closure is denoted by K_0.

We introduce the unitary operator

$$(\Upsilon f)(t,x) = (U_0(t)f(t,\cdot))(x) \tag{8}$$

on $\mathcal{K}$. Furthermore, we introduce the partial Fourier transform in the t-variable

$$(\Phi f)(\tau, x) = \frac{1}{\sqrt{2\pi}} \int_{-\infty}^{\infty} e^{-it\tau} f(t, x)\, dt. \tag{9}$$

These operators are combined to yield $F_0 = \Phi\Upsilon^*$. Viewing this operator as a map from $\mathcal{K}$ to $L^2(\mathbf{R}_\tau; \mathcal{H})$, we see that it defines a spectral representation for K_0. This means that we have

$$(F_0 K_0 f)(\tau) = \tau \cdot (F_0 f)(\tau), \tag{10}$$

initially for $f \in H^1(\mathbf{R}) \otimes_{\text{alg}} H^2(\mathbf{R}^d)$. Note that this result implies $\mathcal{D}(K_0) = \{f \in \mathcal{K} \,|\, F_0 f \in L^{2,1}(\mathbf{R}; \mathcal{H})\}$. Here $L^{2,s}$ denotes the usual weighted space in the t-variable.

One further ingredient in Kuroda's formulation of stationary scattering theory is the trace operator. We define

$$(\gamma(\tau)f)(x) = (F_0 f)(\tau, x). \tag{11}$$

This operator is well-defined on the space

$$\mathcal{K}^s = \Upsilon(L^{2,s}(\mathbf{R}; \mathcal{H})). \tag{12}$$

for any $s > 1/2$. It maps $\mathcal{K}^s$ boundedly into $\mathcal{H}$. For $s < 0$ we let $\mathcal{K}^s = (\mathcal{K}^{-s})^*$ and use the natural duality induced by the scalar product on $\mathcal{K}$ to get a scale ($s > 0$) $\mathcal{K}^s \hookrightarrow \mathcal{K} \hookrightarrow \mathcal{K}^{-s}$. For the sake of consistency we write $\mathcal{K} = \mathcal{K}^0$.

The limiting absorption principle holds for K_0. Given the comments above, the proof is well-known, and is omitted.

Proposition 1. *Assume $s > 1/2$. Then the boundary values*

$$(K_0 - \tau \mp i0)^{-1} = \lim_{\varepsilon \downarrow 0} (K_0 - \tau \mp i\varepsilon)^{-1} \tag{13}$$

exist in operatornorm on $\mathcal{B}(\mathcal{K}^s, \mathcal{K}^{-s})$. The boundary values are Hölder-continuous in τ.

Some of the results from [3] will be needed. We recall the necessary definitions, using the same notation for ease of reference. We introduce the spaces

$$L(B) = L^\infty(\mathbf{R}; \mathcal{H}), \qquad L(B') = L^1(\mathbf{R}; \mathcal{H}). \tag{14}$$

There is a natural duality between these two spaces, obtained from the inner product on $L^2(\mathbf{R}; \mathcal{H})$.

We note the following result. The proof is a simple consequence of the definitions and is omitted.

Lemma 2. *Let $s > 1/2$. We then have the continuous embeddings $\mathcal{K}^s \hookrightarrow L(B')$ and $L(B) \hookrightarrow \mathcal{K}^{-s}$.*

For each $s \in \mathbf{R}$ define an operator $\Gamma_0(s) \in \mathcal{B}(\mathcal{H}, L(B))$ by

$$\Gamma_0(s)\varphi = U_0(t-s)\varphi. \tag{15}$$

The adjoint relative to the duality mentioned above is

$$\Gamma_0(s)^* f = \int_{-\infty}^{\infty} U_0(s-t) f(t)\, dt, \tag{16}$$

and $\Gamma_0(s)^* \in \mathcal{B}(L(B'), \mathcal{H})$.

We impose the following

Assumption 3. *Let $V \in L^1(\mathbf{R}; L^\infty(\mathbf{R}^d))$ be a real-valued function.*

Then, as mentioned in the introduction, we have a propagator $U(t,s)$ associated with (2). We define four operators $G_\pm^0, G_\pm \in \mathbf{B}(L(B'), L(B))$ by

$$(G_\pm^0 f)(t) = \int_{\pm\infty}^{t} U_0(t-s) f(s)\, ds, \tag{17}$$

$$(G_\pm f)(t) = \int_{\pm\infty}^{t} U(t,s) f(s)\, ds. \tag{18}$$

Lemma 4. *Let V satisfy Assumption 3. Then $1 + iG_-^0 V$ is invertible in $\mathcal{B}(L(B))$ with inverse given by $1 - G_- V$. Similarly, $1 + iG_+^0 V$ is invertible with inverse given by $1 - G_+ V$.*

Proof. First we note that Assumption 3 implies $V \in \mathcal{B}(L(B), L(B'))$. Then we use that the following identities hold in $\mathcal{B}(L(B'), L(B))$:

$$G_-^0 - G_- = iG_-^0 V G_- = iG_- V G_-^0, \tag{19}$$

$$G_+^0 - G_+ = iG_+^0 V G_+ = iG_+ V G_+^0, \tag{20}$$

see [3, Lemma 3.7], whose proof is valid also under Assumption 3. The remainder of the proof is now a straightforward computation. □

3 Stationary scattering theory

We now briefly outline the stationary scattering theory applied to our problem. We need the following lemma, cf. [2, equation (1.8)].

Lemma 5. *Let $f \in \mathcal{K}^s$, $s > 1/2$, $\sigma \in \mathbf{R}$, and $\varepsilon > 0$. Then we have*

$$((K_0 - \sigma - i\varepsilon)^{-1} f)(t) = i \int_{-\infty}^{t} e^{i\sigma(t-t')} e^{-\varepsilon(t-t')} U_0(t-t') f(t')\, dt'. \tag{21}$$

Proof. We note $(K_0 - \sigma - i\varepsilon)^{-1} f = \Upsilon\Phi^*(\tau - \sigma - i\varepsilon)^{-1}\Phi\Upsilon^* f$. Since $f \in L(B')$ by Lemma 2, the result then follows using well-known results on the Fourier transform and convolutions. □

We will introduce a slightly stronger assumption on V, in order to be able to apply the stationary scattering theory.

Assumption 6. *Let $V(t,x)$ be a real-valued function such that for some $\beta > 1$ we have $(1+|t|)^\beta V(t,x) \in L^\infty(\mathbf{R}; L^\infty(\mathbf{R}^d))$.*

Let V satisfy Assumption 6. We define $K = K_0 + V$ on $\mathcal{K}$, with domain $\mathcal{D}(K) = \mathcal{D}(K_0)$. Since V is a bounded self-adjoint operator on $\mathcal{K}$, K is self-adjoint on this domain. Briefly stated, Howland's method [2] consists in applying the scattering theory to the pair K_0, K.

To establish the connection with the results in [3], we need the following lemma.

Lemma 7. *Let V satisfy Assumption 6. Assume $1/2 < s < \beta - 1/2$. Then we have the following results.*
(i) *The operators $1 + iVG^0_\pm$ are invertible on $\mathcal{K}^s$.*
(ii) *Let $\sigma \in \mathbf{R}$. Then $1 + V(K_0 - \sigma \mp i0)^{-1}$ are invertible on $\mathcal{K}^s$.*

Proof. It suffices to consider one of the cases. Assumption 6 and the restriction $1/2 < s < \beta - 1/2$ imply that $1 + iVG^0_-$ is bounded on $\mathcal{K}^s$. Assume $f \in \mathcal{K}^s$ and $(1 + iVG^0_-)f = 0$. Then Lemmas 2 and 5 imply $f = 0$. Let $g \in \mathcal{K}^s$. By Lemma 5 there exists $f \in L(B')$ such that $(1 + iVG^0_-)f = g$. But then $f = -iVG^0_- f + g$ shows that $f \in \mathcal{K}^s$. To prove (ii) we start by taking limits in (21) to get

$$(K_0 - i0)^{-1} f = iG^0_- f \tag{22}$$

for $f \in \mathcal{K}^s$, $s > 1/2$. Let $\mathcal{M}_\sigma$ denote the unitary operator of multiplication by $e^{-it\sigma}$ on $\mathcal{K}^s$, $s \in \mathbf{R}$. We note that

$$K_0 - \sigma = \mathcal{M}^*_\sigma K_0 \mathcal{M}_\sigma. \tag{23}$$

Using this result, part (i), and a limiting argument, part (ii) follows. □

The modified trace operators are defined by

$$\gamma_\pm(\tau) f = \gamma(\tau)\left(1 + V(K_0 - \tau \mp i0)^{-1}\right)^{-1} f \tag{24}$$

on $\mathcal{K}^s$, $1/2 < s < \beta - 1/2$, and the modified spectral representations by

$$(F_\pm f)(\tau) = \gamma_\pm(\tau) f, \tag{25}$$

initially on the same space. A standard argument then shows that $F_\pm$ extend to unitary operators on $\mathcal{K}$.

In Howland's theory the connection between the wave operators defined in (4) and the wave operators $\mathsf{W}_{\pm} = s\text{-}\lim_{\sigma\to\pm\infty} e^{i\sigma K}e^{-i\sigma K_0}$ is given by $(\mathsf{W}_{\pm}f)(t) = W_{\pm}(t)f(t)$. The connection with the stationary theory presented here is summarized in the relation $\mathsf{W}_{\pm} = F_{\pm}^{*}F_0$. A careful examination of the proof in [6] shows that it applies to the present case. We will omit the details.

The connection with the scattering operator defined in (5) is then given by $(\mathsf{S}f)(t) = (\mathsf{W}_{+}^{-1}\mathsf{W}_{-}f)(t) = S(t)f(t)$. On the other hand, the stationary scattering theory yields a representation for the decomposition of S in the spectral representation for K_0 given by F_0. Using this connection we get the following result, which is the main result connecting Howland's theory with the space-time scattering theory from [3]. We have retained the formulation given in that paper.

Theorem 8. *Let V satisfy Assumption 6. Then the scattering operator $S(s)$ from (5) has a representation*

$$S(s) = 1 - i\Gamma_0(s)^{*}V(1+iG_{-}^{0}V)^{-1}\Gamma_0(s). \tag{26}$$

Proof. We have from the stationary scattering theory (see [5, 6]) that the scattering matrix given by $(F_0\mathsf{S}f)(\tau) = \mathsf{S}(\tau)(F_0f)(\tau)$ is represented as

$$\mathsf{S}(\tau) = 1 - 2\pi i\gamma(\tau)\left(1+V(K_0-\tau-i0)^{-1}\right)^{-1}V\gamma(\tau)^{*}. \tag{27}$$

We now translate this representation into the terms used in [3]. Recalling the definitions of F_0 and $\gamma(\tau)$, we find that $\gamma(\tau) = \gamma(0)\mathcal{M}_{\tau}$. Combining this relation with (23) we find $\mathsf{S}(\tau) = \mathsf{S}(0)$, such that the scattering matrix is independent of the spectral parameter τ. As already observed by Howland [2, Remark (3), p. 325], the scattering matrix in our spectral representation is also given by multiplication by the constant operator $S(0)$. Thus we have the relation $\mathsf{S}(0) = S(0)$. Now for any $f \in \mathcal{K}^{s}$, $s > 1/2$,

$$\gamma(0)f = (F_0f)(0) = \frac{1}{\sqrt{2\pi}}\int_{-\infty}^{\infty} U_0(-t)f(t)\,dt = \frac{1}{\sqrt{2\pi}}\Gamma_0(0)^{*}f. \tag{28}$$

Using this relation together with (22), we find

$$S(0) = 1 - i\Gamma_0(0)^{*}(1+iVG_{-}^{0})^{-1}V\Gamma_0(0). \tag{29}$$

From (15) follows $\Gamma_0(s) = \Gamma_0(0)U_0(-s)$. Furthermore, $S(s) = U_0(s)S(0)U_0(-s)$. Finally, we have $V(1+iG_{-}^{0}V)^{-1} = (1+iVG_{-}^{0})^{-1}V$. Combining these results equation (26) follows. □

Some applications of the formula (26) are given in [3]. We need Assumption 6 to use the stationary scattering theory in our proof. The results in [3] show that the formula is valid also under Assumption 3.

References

[1] J. Dereziński and C. Gérard, *Scattering Theory of Classical and Quantum N-Particle Systems*, Texts and Monographs in Physics, Springer-Verlag, Berlin, Heidelberg, New York, 1997.

[2] J. Howland, *Stationary scattering theory for time-dependent Hamiltonians*, Math. Ann. **207** (1974), 315–335.

[3] A. Jensen, *Space-time scattering for the Schrödinger equation*, Ark. Mat., in press.

[4] T. Kato, *An $L^{q,r}$-theory for nonlinear Schrödinger equations*, Spectral and scattering theory and applications (Tokyo, Japan) (K. Yajima, ed.), Advanced Studies in Pure Mathematics, vol. 23, Kinokuniya, Tokyo, 1994, pp. 223–238.

[5] S. T. Kuroda, *Scattering theory for differential operators, I, II*, J. Math. Soc. Japan **25** (1973), 75–104, 222–234.

[6] ———, *An introduction to scattering theory*, Aarhus University, Mathematics Institute, Lecture Notes Series No. 51, 1980.

[7] S. Nakamura, *Integral kernels of the scattering matrices for time-periodic Schrödinger equations*, J. Funct. Anal. **76** (1988), 176–192.

[8] K. Yajima, *Existence of solutions for Schrödinger evolution equations*, Commun. Math. Phys. **110** (1987), 415–426.

[9] ———, *Schrödinger evolution equations with magnetic fields*, J. d'Analyse Math. **56** (1991), 29–76.

Operator Theory:
Advances and Applications, Vol. 108

$\mathcal{H}_{-2}$–construction and some applications *

S. T. Kuroda and Hiroshi Nagatani †

Department of Mathematics,
Gakushuin University,
1-5-1 Mejiro Toshima-ku, Tokyo 171-8588 Japan
e-mail: kuroda@math.gakushuin.ac.jp

1 Introduction

The present work is motivated by the works of Gesztesy and Simon ([2]) and Kiselev and Simon ([3]). In these works the authors considered rank one perturbations which are "singular" with respect to the unperturbed operator H_0 and proposed a model of construction which we want to call the "$\mathcal{H}_{-2}$ resolvent construction". The main purpose of the present work is to show that a kind of the iterated second resolvent equation, which is at the heart of the construction, is of rather general nature and can serve to parameterize all closed or selfadjoint operators by bounded operators from the space $\mathcal{H}_2 = \mathcal{D}(|\mathcal{H}_{,}|)$ to the space $\mathcal{H}_{-2} = \mathcal{D}(|H_0|)^*$. In this way the construction will enable us to classify all operators by the level of "singularity or regularity" with respect to H_0.

In Section 2 a sketch of the main results of the construction is given. A more detailed account with proof will be given elsewhere ([5]). In this report we confine ourselves to the case of selfadjoint perturbed operators. In [5] a systematic application of the general theory to the problem of one-center point interactions in one-dimension is also given. Instead of repeating that result we shall make a preliminary report on the applicability of our approach to approximation problems (Section 3). We shall show that at least we can recapture the known result of approximation of δ interaction by scaled short-range potentials ([1]). The approximation problem for δ' interaction seems subtler and will be left to future study.

More interesting is the applicability of the theory to higher-dimensional singular interactions such as surface interactions in three dimension. We shall not touch on it in this report. (For some recent results we refer to [4], [6], [7].)

---

*Work supported by Science Research Grant, Ministry of Education, Gr. No. 07640243
†Partly supported by the ITOH Scholarship Foundation

The bibliography of this report will be limited to references relevant to this presentation. More references will be found in [5].

The authors wish to thank the organizers of the conference and the editors of these proceedings, especially Professor Pavel Exner, for their interest in this work and their patience as editors.

2 A sketch of the general theory

2.1 Basic construction

We shall work in a Hilbert space $\mathcal{H}$. The norm and the inner product in $\mathcal{H}$ will be denoted by $\|\cdot\|$ and $(\cdot,\cdot)$, respectively. We use the following notations. $\mathcal{L}(X,Y)$ denotes the set of all bounded linear operators from a Hilbert space X to a Hilbert space Y. As usual, $\mathcal{L}(X)=\mathcal{L}(X,X)$. For a closed operator A in $\mathcal{H}$, the domain, the range, and the null space are denoted by $\mathcal{D}(A)$, $\mathcal{R}(A)$, and $\mathcal{N}(A)$, respectively. The resolvent set of A is denoted by $\rho(A)$ and the resolvent by $R(z;A)=(A-z)^{-1}$.

Fix a selfadjoint operator H_0 in $\mathcal{H}$ which we regard as an unperturbed operator and introduce a scale of Hilbert spaces $\mathcal{H}_s$ associated with H_0; that is, for $s\geq 0$, $\mathcal{H}_s$ is $\mathcal{D}(H_0)$ with $\|u\|_s=\|(|H_0|+1)^{s/2}u\|$ and for $s<0$, $\mathcal{H}_s$ is the completion of $\mathcal{H}$ by $\|\cdot\|_s$. We regard $\mathcal{H}_{-s}=\mathcal{H}_s^*$ and $\mathcal{H}_s=\mathcal{H}_{-s}^*$, $s\geq 0$, through the relation $(u,\varphi)_{s,-s}=(u,\varphi)$, $\forall u\in\mathcal{H}_s$, $\forall\varphi\in\mathcal{H}\subset\mathcal{H}_{-s}$. In this scale $\mathcal{H}_0=\mathcal{H}$. Hereafter, we use $\mathcal{H}_0$ instead of $\mathcal{H}$.

We use the notation $R_0(z)=R(z;H_0)$ to denote the resolvent of H_0. By means of the extension by continuity or by means of the restriction, $R_0(z)$ and H_0-z induce operators in $\mathcal{L}(\mathcal{H}_s,\mathcal{H}_{s+2})$ and $\mathcal{L}(\mathcal{H}_{s+2},\mathcal{H}_s)$, respectively. As operators in such spaces they are inverse to each other. We also use the following operator $W(z,w)$ which goes the scale up by four.

$$\begin{aligned} W(z,w) &= (z-w)R_0(z)R_0(w) \\ &= (z-w)R_0(w)R_0(z)\in\mathcal{L}(\mathcal{H}_s,\mathcal{H}_{s+4}), \quad z,\ w\in\rho(H_0). \end{aligned} \tag{2.1}$$

Hereafter, we fix a number $c\in\rho(H_0)$. Various quantities introduced below, such as ρ_T, depend also on c, but this dependence on c will not be shown explicitly in the notation. Let $T\in\mathcal{L}(\mathcal{H}_2,\mathcal{H}_{-2})$. We define ρ_T and $R_T(z)$ by

$$\rho_T=\{z\in\rho(H_0)\,|\,\exists\,(1+TW(z,c))^{-1}\in\mathcal{L}(\mathcal{H}_{-2})\}, \tag{2.2}$$

$$\begin{aligned} R_T(z) &= R_0(z)-R_0(z)(1+TW(z,c))^{-1}TR_0(z) \\ &= R_0(z)-R_0(z)T(1+W(z,c)T)^{-1}R_0(z), \qquad z\in\rho_T. \end{aligned} \tag{2.3}$$

Theorem 2.1 ([5], Theorem 2.4) *Let $T\in\mathcal{L}(\mathcal{H}_2,\mathcal{H}_{-2})$. Suppose that T satisfies*

$$u-TR_0(c)u=0, \quad u\in\mathcal{H}_0 \Longrightarrow u=0, \tag{2.4}$$

$$T - T^* = TW(\bar{c}, c)T^* = T^*W(\bar{c}, c)T. \tag{2.5}$$

Then, there exists a uniquely determined selfadjoint operator $H(T)$ in $\mathcal{H}$ such that

$$\rho_T = \rho(H(T)) \cap \rho(H_0), \tag{2.6}$$

$$R(z; H(T)) = R_T(z), \quad \forall z \in \rho(H(T)) \cap \rho(H_0). \tag{2.7}$$

The mapping $T \longmapsto H(T)$ gives a one-to-one correspondence between the set of all $T \in \mathcal{L}(\mathcal{H}_2, \mathcal{H}_{-2})$ satisfying (2.6) and (2.7) and the set of all selfadjoint operators H satisfying $c \in \rho(H_0)$. In particular, if c is non-real, then all selfadjoint operators are produced as $H(T)$.

Since $W(c, c) = 0$, we see by putting $z = c$ in (2.3) that

$$R(c; H(T)) = R_0(c) - R_0(c)TR_0(c), \tag{2.8}$$

$$\mathcal{D}(H(T)) = (I - R_0(c)T)\mathcal{D}(H_0). \tag{2.9}$$

Equation (2.9) is particularly useful for determining the domain of $H(T)$.

2.2 Relations between T and $H(T)$

The correspondence between T and $H(T)$ enables us to relate some useful properties of $H(T)$ to simple properties of T. The following two theorems are examples. The first theorem characterizes the case of a so-called additive perturbation and the second is concerned with the case that $H(T)$ is a selfadjoint extension of the restrictions of H_0 to a linear subspace of $\mathcal{D}(H_0)$

Theorem 2.2 ([5], Theorem 2.10) *Suppose that $T \in \mathcal{L}(\mathcal{H}_2, \mathcal{H}_{-2})$ and T satisfies (2.4) and (2.5). Then the following relations, (2.10) and (2.11), are equivalent.*

$$T \in \mathcal{L}(\mathcal{H}_2, \mathcal{H}_0) \text{ and } 1 \in \rho(TR_0(c)), \tag{2.10}$$

$$\mathcal{D}(H(T)) = \mathcal{D}(H_0). \tag{2.11}$$

In this case, $V \equiv H(T) - H_0$ is in $\mathcal{L}(\mathcal{H}_2, \mathcal{H}_0)$ and T and V are related as follows:

$$V = (1 - TR_0(c))^{-1}T = T(1 - R_0(c)T)^{-1}, \tag{2.12}$$

$$T = (1 + VR_0(c))^{-1}V = V(1 + R_0(c)V)^{-1}, \tag{2.13}$$

where all the inverses appearing in (2.12) and (2.13) exist either in $\mathcal{L}(\mathcal{H}_2)$ or in $\mathcal{L}(\mathcal{H}_{-2})$.

Theorem 2.3 ([5], Theorem 2.13) *Let $\mathcal{N}$ be a linear subspace of $\mathcal{H}_2$. Then $H(T)$ is a selfadjoint extension of $H_0|_{\mathcal{N}}$, the restriction of H_0 to $\mathcal{N}$, if and only if $\mathcal{N}(T) \supset \mathcal{N}$. In particular, if c is non-real, all selfadjoint extensions of $H_0|_{\mathcal{N}}$ are produced in this way.*

2.3 A convergence theorem

The following convergence theorem is a natural and immediate consequence of the construction.

Theorem 2.4 *Suppose that $T_n \in \mathcal{L}(\mathcal{H}_2, \mathcal{H}_{-2})$ satisfies conditions (2.4) and (2.5) and suppose that T_n converges to $T \in \mathcal{L}(\mathcal{H}_2, \mathcal{H}_{-2})$ in the operator norm. Suppose further that T satisfies (2.4). Then, T also satisfies (2.5) and we have $H(T)$. For $H(T_n)$ and $H(T)$ the convergence relation*

$$\lim_{n\to\infty} R(z; H(T_n)) = R(z; H(T)), \quad \forall z \in \rho(H(T)) \quad \text{(operator norm convergence).} \tag{2.14}$$

holds.

Proof. The fact that T satisfies (2.5) follows by taking the limit. (2.14) is an immediate consequence of (2.7), (2.3), and the norm continuity of the operation of taking the inverse. //

3 δ interaction and approximation by potentials

In this section, which is partly a review, we shall try to recapture a convergence theorem for one-dimensional point interaction by our approach. The main concern is not to obtain new results, but to test the applicability of our approach to these problems.

Let $\mathcal{H} = L^2(\mathbf{R})$ and let

$$H_0 = -\frac{d^2}{dx^2} \quad \text{with the domain } \mathcal{D}(H_0) = H^2(\mathbf{R}), \tag{3.1}$$

where $H^s(\mathbf{R})$, $s \in \mathbf{R}$, is the Sobolev space. Let $\mathcal{N}_0 = \{f \in H^2(\mathbf{R}) \mid f(0) = f'(0) = 0\}$. By Theorem 2.3 all selfadjoint extensions of $H_0|_{\mathcal{N}_0}$ are given by $H(T)$ with T satisfying $\mathcal{N}(T) \supset \mathcal{N}_0$. T with $\mathcal{N}(T) \supset \mathcal{N}_0$. must have the form

$$\begin{aligned} Tf &= (c_{11}(f,\delta) + c_{12}(f,\delta'))\delta + (c_{21}(f,\delta) + c_{22}(f,\delta'))\delta' \\ &= (c_{11}f(0) - c_{12}f'(0))\delta + (c_{21}f(0) - c_{22}f'(0))\delta', \end{aligned} \tag{3.2}$$

where δ is the δ function. (Note that $\delta \in H^{-1}(\mathbf{R})$ and $\delta' \in H^{-2}(\mathbf{R})$.) Since $\mathcal{R}(T_\lambda) \cap \mathcal{H}_0 = \{0\}$, condition (2.4) is automatically satisfied.

Hereafter we take $c = i$. Let $C_T = (c_{jk})_{j,k=1,2}$. In [5] it is shown that T given by (3.2) satisfies (2.5) if and only if $U_T \equiv \frac{1}{\sqrt{2}} C_T + i$ is a two by two unitary matrix. In [5] a systematic investigation is also carried out on relations between U_T and boundary conditions at $x = 0$ which determine the domain of $H(T)$. In the present report we shall not reproduce it.

3.1 δ interaction in one dimension

In this subsection we shall deal with the case that only c_{11} is non-zero and put for simplicity $c_{11} = \lambda$. Then, the corresponding T is given as follows:

$$T = T_\lambda f = \lambda(f, \delta)\delta = \lambda f(0)\delta. \tag{3.3}$$

It is shown in [5, §4] that the unitarity of U_T is equivalent to $\lambda = \sqrt{2}(-i + e^{i\theta})$, $0 \leq \theta < 2\pi$ and the domain of $H_\lambda \equiv H(T_\lambda)$ is given as follows ([5, Example 4.9]): for $e^{i\theta} \neq 1$, i.e. $\lambda \neq \sqrt{2}(1 - i)$

$$\mathcal{D}(H_\lambda) = \{f \in H^2(\mathbf{R}_+ \oplus \mathbf{R}_-) \mid f(0+) = f(0-),\ f'(0+) - f'(0-) = tf(0)\}, \tag{3.4}$$

where t is related to λ or to θ as

$$t = \frac{\sqrt{2}\lambda}{e^{i\pi/4}(1 - i - \lambda/\sqrt{2})} = \sqrt{2}\,\frac{\cos\theta + \sin\theta - 1}{1 - \cos\theta}; \tag{3.5}$$

$$\mathcal{D}(H_{\sqrt{2}(1-i)}) = \{f \in H^2(\mathbf{R}_+ \oplus \mathbf{R}_-) \mid f(0+) = f(0-) = 0\} \quad \text{(Dirichlet decoupling)}. \tag{3.6}$$

We now proceed to the problem of approximating δ interaction by potential interactions. The problem is handled in [1, I.3.2] under the title "approximations by means of local scaled short-range interactions." As a guide to more complicated problems we shall illustrate how the result and the reasoning given in [1] is put in our framework. To this end we take up the simplest case. We take a real $V \in L^1(\mathbf{R}) \cap L^\infty(\mathbf{R})$ and put

$$H^{(n)} = H_0 + V_n, \quad V_n(x) = nV(nx). \tag{3.7}$$

Here and hereafter, we use the notation V_n to denote the function $V_n(x)$ as well as the operator of multiplication by V_n.

$H^{(n)}$ is a bounded perturbation of H_0 and hence by Theorem 2.2, $H^{(n)}$ is expressed as $H^{(n)} = H(T^{(n)})$ with $T^{(n)}$ given by

$$T^{(n)} = (1 + V_n R_0(i))^{-1} V_n. \tag{3.8}$$

For simplicity of notation we put

$$D = (\cdot, \delta)\delta \in \mathcal{L}(\mathcal{H}_1, \mathcal{H}_{-1}), \quad C_V = \int_{-\infty}^{\infty} V(x)dx. \tag{3.9}$$

We also regard that $V_n \in \mathcal{L}(\mathcal{H}_1, \mathcal{H}_{-1})$ and obtain

$$\lim_{n\to\infty} V_n = C_V D, \quad \text{(norm convergence in } \mathcal{L}(\mathcal{H}_1, \mathcal{H}_{-1})\text{)}. \tag{3.10}$$

As a matter of fact the convergence takes place in the Hilbert-Schmidt norm ([1]). An easy proof is to consider the kernel of V_n and D in the Fourier space and observe that the square of the Hilbert-Schmidt norm of the kernel $k(\xi, \eta)$ in $\mathcal{L}(\mathcal{H}_1, \mathcal{H}_{-1})$ is $\displaystyle\int_{\mathbf{R}^2} \frac{|k(\xi,\eta)|^2}{(1+\xi^2)(1+\eta^2)}$, and to use the bounded convergence theorem.

Since the convergence in $\mathcal{L}(\mathcal{H}_1, \mathcal{H}_{-1})$ implies the convergence in $\mathcal{L}(\mathcal{H}_2, \mathcal{H}_{-2})$, it follows from (3.8) that

$$\lim_{n\to\infty} T^{(n)} = C_V(1 + C_V D R_0(i))^{-1} D \equiv T_V. \tag{3.11}$$

Here, we used the fact that $(1 + C_V D R_0(i))^{-1}$ exists as bounded operators, which is easily verified. An explicit computation of the middle member of (3.8) shows that, with T_λ defined by (3.3) we have

$$T_V = T_\lambda, \quad \lambda = \frac{2C_V}{2 + C_V e^{\pi i/4}}. \tag{3.12}$$

Thus, we have seen that $H^{(n)}$ converges in the norm resolvent sense to $H(T_V)$ and using (3.5) we find that the domain of $H(T_V)$ is given by (3.4) with t exactly equal to C_V. We note that the above discussion will be restricted neither to scaled approximation nor to bounded potentials. But we do not go into further details. In particular, to approximate the Dirichlet decoupling we must arrange so that $\int_{-\infty}^{\infty} V_n(x)dx \to \infty$.

The above discussion illustrates how one can recapture an approximation result given in [1, I.3.2] within our approach. We admit that for each concrete problem a more direct computation of the resolvent kernel will provide a quicker way. However, it is of some interest to put the problem into a general perspective.

3.2 δ' interaction in one dimension

δ' interaction arises when only c_{22} is non-zero in (3.2). The first part of the argument is parallel to the case of δ interaction. We put $c_{22} = \mu$. Then, the corresponding T is given as follows:

$$T = T'_\mu f = \mu(f, \delta')\delta' = -\mu f'(0)\delta. \tag{3.13}$$

μ takes the value $\mu = \sqrt{2}(-i + e^{i\theta})$, $0 \le \theta < 2\pi$ and the domain of $H'_\mu \equiv H(T'_\mu)$ is given as follows: for $e^{i\theta} \ne -1$, i.e. $\mu \ne -\sqrt{2}(1+i)$

$$\mathcal{D}(H'_\mu) = \{f \in H^2(\mathbf{R}_+ \oplus \mathbf{R}_-) \mid f'(0+) = f'(0-),\ f(0+) - f(0-) = tf'(0)\}, \tag{3.14}$$

where t is related to μ or to θ as

$$t = \frac{\sqrt{2}\mu}{e^{i\pi/4}(1 + i + \mu/\sqrt{2})} = \sqrt{2}\frac{\sin\theta - \cos\theta - 1}{1 + \cos\theta}. \tag{3.15}$$

$$\mathcal{D}(H'_{-\sqrt{2}(1+i)}) = \{f \in H^2(\mathbf{R}_+ \oplus \mathbf{R}_-) \mid f'(0+) = f'(0-) = 0\}$$
$$\text{(Neumann decoupling).} \tag{3.16}$$

The approximation problem seems subtler. One sensible choice of approximating interaction would be $A_n = -n\frac{d}{dx}a(nx)\frac{d}{dx}$. However, not only this is not bounded but also $H_0 + A_n$ may not make sense as a reasonable operator. A tentative remedy could be to introduce

$$\phi_n(\xi) = \begin{cases} \xi, & |\xi| < n, \\ n, & |\xi| \geq n, \end{cases} \tag{3.17}$$

and to consider

$$A_n = n\phi_n(\xi) \cdot a(nx) \cdot \phi_n(\xi), \quad H'^{(n)} = H_0 + A_n \tag{3.18}$$

with $a \in L^1(\mathbf{R}) \cap L^\infty(\mathbf{R})$. We have not yet obtained any sensible conclusion.

References

[1] S. Albeverio, F. Gesztesy, R. Høegh-Krohn, and H. Holden, Solvable Models in Quantum Mechanics, Springer-Verlag, Berlin-Heidelberg-New York, 1988.

[2] F. Gesztesy and B. Simon, Rank one perturbation at infinite coupling, J. Funct. Anal. **128**(1995), 245–252.

[3] A. Kiselev and B. Simon, Rank one perturbations with infinitesimal coupling, J. Functional Anal. **130**(1995), 345–356.

[4] H. Nagatani, Scattering theory for non-local surface interactions of rank one, Preprint, 1998.

[5] S. T. Kuroda and H. Nagatani, $\mathcal{H}_{-2}$-construction of general type and its application to point interactions, Preprint, 1998.

[6] K. Watanabe, On the self-adjoint operators defined by the H_{-2} construction: Finite rank and trace class perturbations, Preprint.

[7] K. Watanabe, Smooth perturbations of the self-adjoint operators defined by the H_{-2} con struction, Preprint.

Operator Theory:
Advances and Applications, Vol. 108

Scattering with time periodic potentials and cyclic states

Philippe A. Martin
Institut de physique théorique, Ecole Polytechnique Fédérale
CH-1015, Lausanne, Switzerland

Abstract

We consider a scattering system $H(t) = H_0 + V(\omega t)$ with free Hamiltonian H_0 having an absolutely continuous spectrum in the band $[0, E_0]$ and $V(\omega t)$ a time-periodic potential with period $T = 2\pi/\omega$. We show that the corresponding monodromy operator has in general both absolutely continuous spectrum and point spectrum (cyclic states) when the period is small enough. In fact, these cyclic states originate from the eigenstates of the static time-averaged Hamiltonian $\bar{H} = \frac{1}{T}\int_0^T dtH(t)$: they loose their stability as T increases and may be transformed into resonances. These mechanisms are illustrated on the explicitly solvable model of a periodically kicked rank-one perturbation.

1 Cyclic states in the high-frequency limit

Scattering of a quantum mechanical particle by a short range time periodic potential plays an important role in a variety of problems ranging from tunneling through a modulated barrier to atoms in laser fields. We consider a simple model of such a system described by a Hamiltonian of the form

$$H(t) = H_0 + V(\omega t), \quad \omega = \frac{2\pi}{T} \tag{1}$$

acting on a Hilbert space $\mathcal{H}$ with the following specifications. The free Hamiltonian $H_0 = \int EdP(E)$ is a self-adjoint operator on $\mathcal{H}$ with absolutely continuous spectral measure $dP(E)$ supported in the interval $[0, E_0]$. The interaction $V(t)$ is time periodic with period 2π so that $V(\omega t)$ has period T. $V(t)$ is assumed to be uniformly bounded and norm continuous in t. If $U(t_0, t)$ and $U_0(t_0, t)$ denote the full and the free evolution operators, we set simply

$$U_T = U(0, T), \quad U_{0T} = e^{-iH_0T} \tag{2}$$

for the corresponding evolutions over one period. The relevant informations on the system are contained in the spectral properties of U_T (the monodromy operator) and we like to analyze them as function of the period T. Throughout the paper we make the assumption that $U_T - U_{0T}$ belongs to the trace class for $T > 0$ so that the pair U_T, U_{0T} of unitary operators forms a scattering system with complete wave operators [1]. For $T < T_0 = \frac{2\pi}{E_0}$, U_{0T} has absolutely continuous spectrum on the unit circle in the sector $[0, \theta_0 = E_0 T]$ and the same is true for U_T under the above assumption. The question we address here is the possible emergence of point spectrum of U_T in the sector $(\theta_0, 2\pi)$ and its influence on scattering. We call an eigenvector ψ_T of U_T with eigenphase $\alpha_T = \epsilon_T T$

$$U_T \psi_T = e^{-i\epsilon_T T}\psi_T, \quad E_0 < \epsilon_T < \frac{2\pi}{T} \tag{3}$$

a cyclic state with quasi-energy ϵ_T. It represents a state of the particle that remains stable and localized under the evolution since it is left invariant by this evolution over one period (up to a phase). There is a large literature on properties of point spectrum of the monodromy operator for confining potentials (rotators, oscillators under time periodic perturbations [2]), but we are not aware of a proof for the existence of such states when the Hamiltonian has also absolutely continuous spectrum allowing for scattering processes. This is shown in the two following elementary propositions.

We like to study the dynamics generated by H(t) in the limit of high frequency. For this, we introduce the time averaged Hamiltonian over one period

$$\bar{H} = H_0 + \bar{V}, \quad \bar{V} = \frac{1}{2\pi}\int_0^{2\pi} dt V(t) \tag{4}$$

and show that the corresponding evolution $\bar{U}(t) = \exp(-i\bar{H}t)$ is a good approximation of $U(0,t)$ for finite time intervals when T is small[1].

Proposition 1

(i) For $t_0 > 0$, there exists c_1 such that

$$||U(0,t) - \bar{U}(t)|| \le c_1 T \tag{5}$$

uniformly with respect to t, $|t| \le t_0$.

(ii) For $t = T$,

$$||U_T - \bar{U}(T)|| \le c_2 T^2 \tag{6}$$

Proof Define $W(t) = V(t) - \bar{V}$ so that $H(t) = \bar{H} + W(\omega t)$ and set $W_1(t) = \int_0^t ds W(s)$. W_1 is periodic, $W_1(0) = W(2\pi) = 0$, and uniformly bounded in time. By integration by parts,

$$U(0,t) - \bar{U}(t) \quad = \quad -i\int_0^t ds U(s,t) W(\omega s)\bar{U}(s)$$

[1]The high-frequency limit for the scattering operator has been studied in [3].

$$
\begin{aligned}
&= -\frac{i}{\omega}\int_0^t dsU(s,t)\frac{dW_1(\omega s)}{ds}\bar{U}(s) \\
= -\frac{i}{\omega}W_1(\omega t)\bar{U}(t) \quad &- \frac{1}{\omega}\int_0^t dsU(s,t)H(s)W_1(\omega s)\bar{U}(s) \\
&+ \frac{1}{\omega}\int_0^t dsU(s,t)W_1(\omega s)\bar{H}\bar{U}(s)
\end{aligned} \tag{7}
$$

Since all operators are uniformly bounded in time, the norm of the first term of (7) is less than const.T for all times whereas the norms of the two other terms are less than const.$|t|T$, giving (i). Since $W_1(\omega T) = 0$ the first term vanishes at $t = T$ and the two other terms are bounded by const. T^2, giving (ii).

Proposition 2

Suppose that $\bar{H}$ has an isolated eigenvalue $\bar{E}$ of finite multiplicity m. Then for T small enough, U_T has m cyclic states with quasi-energies $\epsilon_T^{(j)}, \epsilon_T^{(j)} = \bar{E} + \mathrm{O}(T)$, $j = 1, \ldots, m$

Proof Set $t_0 = \pi||\bar{H}||^{-1}$. For $|t| < t_0$, $\bar{U}(t)$ has an isolated eigenvalue $\bar{\xi}(t) = \exp(-i\bar{E}t)$ on the unit circle. In view of (5), Rellich's theorem ensures that $U(0,t)$ has also m eigenvectors for T small enough and the corresponding eigenvalues $\xi^{(j)}(t)$ satisfy $|\xi^{(j)}(t) - \bar{\xi}(t)| = \mathrm{O}(T)$ uniformly in $|t| \le t_0$. For $t = T$, $\xi^{(j)}(T) = \exp(-i\epsilon_T^{(j)}T)$ is an eigenvalue of U_T with quasi-energy $\epsilon_T^{(j)}$ and (6) implies

$$|\exp(-i\epsilon_T^{(j)}T) - \exp(-i\bar{E}T)| = \mathrm{O}(T^2),$$

hence $\epsilon_T^{(j)} = \bar{E} + \mathrm{O}(T)$.

What is the fate of such cyclic states as T increases from 0 to T_0? Consider for simplicity the situation (illustrated in the next section) where the spectrum of U_T in the sector $(\theta_0, 2\pi)$ consists of a single eigenphase $\alpha_T = \epsilon_T T$ inherited from the time averaged Hamiltonian $\bar{H}$, as described in Prop. 2. As T increases, α_T moves on the circle, and for a certain period T_1' it may reach the threshold of the continuous spectrum, say, at $\theta = 2\pi$ where it looses stability and may be transformed into a resonance seen in the scattering amplitude. For a subsequent period $T_2 > T_1'$, α_T can emerge out of the continuum at $\theta = E_0T_2$ and migrate again until it is reabsorbed when $T = T_2' > T_2$ and so on. Thus there will be a succession of period intervals (T_i, T_i'), called stability intervals, where the cyclic state exists. When $T = T_0$, the gap between the two extremities of the absolutely continuous spectrum of U_T closes, and no state is expected to remain stable beyond this point.

This situation shows the phenomenon of stabilization of the dynamics under high-frequency motion: it is well known in classical mechanics (see for instance [4]) and it occurs also in atomic physics (see the concluding remarks).

2 A solvable model

2.1 Cyclic states and stability intervals

An explicit illustration of these mechanisms is given by a particle periodically kicked by a rank one perturbation (details can be found in [5]). The total Hamiltonian is

$$H(t) = H_0 + \lambda f(\omega t)|\,\phi\rangle\langle\phi\,| \tag{8}$$

The free Hamiltonian has simple absolutely continuous spectrum in the band $[0, E_0]$ and the kick function formally reads

$$f(t) = 2\pi \sum_{n=0}^{\infty} \delta(t-2\pi n), \qquad \frac{1}{2\pi}\int_0^{2\pi} dt f(t) = 1 \tag{9}$$

The normalized vector $|\,\phi\rangle$ is represented by an absolutely continuous function $\phi(E)$ that does not vanish on the spectrum of H_0, except possibly at its extremities 0 and E_0 (λ is a coupling constant). We assume there a power law behaviour

$$\begin{aligned} \varphi(E) &\sim \text{const.}\, E^{\nu}, \;\; E \to 0 \\ \varphi(E) &\sim \text{const.}\, (E-E_0)^{\nu}, \;\; E \to E_0 \end{aligned} \tag{10}$$

The evolution over one period corresponding to the interaction (8) is given by

$$U_T = U_{0T}\exp(-i\lambda T|\,\phi\rangle\langle\phi\,|) = U_{0T}(I + (\exp(-i\lambda T)-1)|\,\phi\rangle\langle\phi\,|) \tag{11}$$

Since U_T and U_{0T} differ by a rank one operator, there is a well defined scattering theory. The point spectrum of U_T is determined by the poles of its resolvent $R(z) = (U_T - z)^{-1}$, see [6]

$$R(z) = \left(I - \frac{1}{g(z)}|\,\varphi\rangle\langle\varphi\,|\right) R_0(z) - \frac{z}{g(z)} R_0(z)|\,\varphi\rangle\langle\varphi\,|R_0(z) \tag{12}$$

where

$$\begin{aligned} g(z) &= \frac{1}{1-\exp(i\lambda T)} + z\langle\varphi\,|R_0(z)|\,\varphi\rangle \\ &= \frac{i}{2}\left(\cot\left(\frac{\lambda T}{2}\right) + h(\alpha \pm i\eta, T)\right) \end{aligned} \tag{13}$$

and

$$h(\alpha \pm i\eta, T) = \frac{1}{T}\int_0^{\theta_0} d\theta \left|\varphi\left(\frac{\theta}{T}\right)\right|^2 \cot\left(\frac{\theta-\alpha \mp i\eta}{2}\right), \quad \theta_0 = E_0 T \tag{14}$$

One has set $z = e^{-i(\alpha \pm i\eta)}$ and used the formula $\frac{1}{e^{ix}-1} = -\frac{1}{2}(1 + i\cot(\frac{x}{2}))$; $\cot x$ is singular at the points $n\pi$, n integer, and behaves as $(x-n\pi)^{-1}$ near these points.

For $t < T_0 = \frac{2\pi}{E_0}$, $R(z)$ is holomorphic in the sector $z = e^{-i\alpha}$, $\theta_0 < \alpha < 2\pi$, except at the zeros of $g(z)$, i.e. if

$$h(\alpha, T) = -\cot\left(\frac{\lambda T}{2}\right) \tag{15}$$

Since $h(\alpha, T)$ is monotonously increasing with α, $\theta_0 < \alpha < 2\pi$, there is exactly

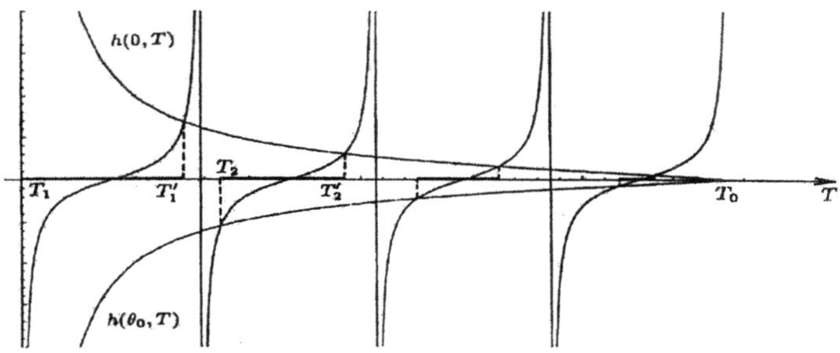

Figure 1: *Intervals of stability. Here* $E_0 = 1$, $T_0 = 2\pi$, $\varphi(E) = \sqrt{30}E(E-1)$, $\lambda = 4$. *One has the symmetry* $\varphi(E) = \varphi(1-E)$, $h(\theta_0, T) = -h(0, T)$.

one zero α_T if

$$h(\theta_0, T) < -\cot\left(\frac{\lambda T}{2}\right) < h(2\pi, T) \tag{16}$$

We distinguish two cases.

(i) The exponent ν in (10) is non positive implying $h(\theta_0) = -\infty$, $h(0) = \infty$. Then (15) has a solution for all $T < T_0$ and all coupling constants λ, provided that $\frac{\lambda T}{2} \neq n\pi$.

(ii) The exponent ν in (10) is positive so that $h(\theta_0)$ and $h(2\pi) = h(0)$ are both finite. Then for $T < T_0$, Eq. (15) has a solution if $\frac{\lambda T}{2}$ belongs to a certain range of values defined by the condition (16), consisting in general of an union of intervals (the stability intervals) because of the periodicity of $\cot(\frac{\lambda T}{2})$. Stability intervals are illustrated in the example of Fig.1.

In either of the two cases this produces a single pole in the resolvent (12), and thus an eigenstate ψ_T of U_T satisfying (3)

$$\psi_T(E) = a\frac{\varphi(E)}{e^{-i(\epsilon_T T - ET)} - 1} \tag{17}$$

where a is a normalization constant. It is easy to verify the findings of Prop. 2: in this case the time averaged Hamiltonian is simply $\bar{H} = H_0 + \lambda|\phi\rangle\langle\phi|$ and the eigenvalue equation (15) as well as ϵ_T and ψ_T reduce to the corresponding quantities for this static Hamiltonian $\bar{H}$ as $T \to 0$.

2.2 Scattering and resonances

Since $U_T - U_{0T}$ is trace class, we have complete wave operators[2]

$$\begin{aligned} \Omega_+ &= s\lim_{n\to\infty}(U^*)^n U_0^n \\ \Omega_- &= s\lim_{n\to\infty} U^n (U_0^*)^n \end{aligned} \tag{18}$$

with equal ranges $\mathcal{R} = \mathcal{R}(\Omega_-) = \mathcal{R}(\Omega_+)$. In (18) the infinite time limit is taken through a discrete set of multiples of the period T. In the model, the scattering operator $S = \Omega_+^*\Omega_-$ can be explicitly computed. For this it is useful to express the wave operators directly in terms of the "interaction" $W = U - U_0$ as represented by the difference of the full and free evolutions on one period. This can be done by deriving the equivalent of the usual Lippman-Schwinger equations (in operator forms)

$$\Omega_\pm = I - s\lim_{\eta\to 0_+}\int_0^{2\pi} R(e^{-i(\theta\mp i\eta)})W dF_0(\theta) \tag{19}$$

and on $\mathcal{R}$

$$\Omega_\pm^* = I + s\lim_{\eta\to 0_+}\int_0^{2\pi} R_0(e^{-i(\theta\mp i\eta)})W dF(\theta) \tag{20}$$

As before $R(z) = (U - z)^{-1}$ and $R_0(z) = (U_0 - z)^{-1}$ are the resolvents associated with the unitary operators U and U_0, and $dF(\theta)$ and $dF_0(\theta)$ are the corresponding spectral measures. Then the S operator is found from the usual manipulations

$$S - I = (\Omega_+^* - \Omega_-^*)\Omega_-$$

$$= s\lim_{\nu\to 0_+} s\lim_{\eta\to 0_+}\int_0^{2\pi}\{R_0(e^{-i(\theta-i\nu)}) - R_0(e^{-i(\theta+i\nu)})\}\mathcal{T}(e^{-i(\theta+i\eta)})dF_0(\theta) \tag{21}$$

where we have introduced the $\mathcal{T}$-operator by

$$\mathcal{T}(z) = W - WR(z)W \tag{22}$$

in analogy with its usual definition when the interaction is time-independent. Finally, writing any E in the spectrum of H_0 as $E = \epsilon + n\omega$ where ϵ, $0 \le \epsilon < \omega$, is the conserved quasi-energy, the S-operator can be reduced to the quasi-energy shell matrix $S(\epsilon)$

$$\langle \epsilon + n\omega\,|S|\,\epsilon' + n'\omega\rangle = \delta(\epsilon - \epsilon')\langle n\,|S(\epsilon)|\,n'\rangle \tag{23}$$

with $\langle n\,|S(\epsilon)|\,n'\rangle$ given by

$$\langle n\,|(S(\epsilon) - I)|\,n'\rangle = \frac{2\pi e^{i\epsilon T}}{T}\lim_{\eta\to 0}\langle \epsilon + n\omega\,|\mathcal{T}(e^{-i(\epsilon T + i\eta)})|\,\epsilon + n'\omega\rangle \tag{24}$$

[2]In this subsection we consider that the period is fixed and omit it from the notation.

with ϵ, n and n' such that both $\epsilon + n\omega$ and $\epsilon + n'\omega$ belong to the spectrum $[0, E_0]$ of H_0. The matrix elements $\langle n\,|S(\epsilon)|\,n'\rangle$ correspond to transitions where $(n-n')\omega$ energy quanta are emitted or absorbed in the course of the scattering process.

In our simple model, the $\mathcal{T}$-operator is easily found from (22) and (12)

$$\mathcal{T}(z) = \frac{1}{g(z)} U_0 |\,\varphi\rangle\langle\varphi\,| \tag{25}$$

Hence according to (24) and (25) the scattering matrix is

$$\langle n\,|S(\epsilon)|\,n'\rangle = \delta_{n,n'} + \frac{2\pi}{T}\left(\frac{\varphi(\epsilon+n\omega)\varphi^*(\epsilon+n'\omega)}{-A(\epsilon T)+iB(\epsilon T)}\right) \tag{26}$$

The denominator in (26) comes from the evaluation of $g(z)$ on the continuous part of the spectrum with the help of the Cauchy principle value formula i.e.

$$\lim_{\eta\to 0} g(e^{-i(\alpha\pm i\eta)}) = \mp A(\alpha) + iB(\alpha) \tag{27}$$

with

$$\begin{aligned} A(\alpha) &= \frac{\pi}{T}\left|\varphi\left(\frac{\alpha}{T}\right)\right|^2 \\ B(\alpha) &= \frac{1}{2}\left(\cot\left(\frac{\lambda T}{2}\right) + \frac{1}{T}\mathcal{P}\!\!\int_0^{\theta_0} d\theta \left|\varphi\left(\frac{\theta}{T}\right)\right|^2 \cot\left(\frac{\theta-\alpha}{2}\right)\right) \end{aligned} \tag{28}$$

If $T < T_0$, the energy is necessarily conserved and the only non vanishing S-matrix element is $S_{00}(\epsilon) \equiv \langle n=0\,|S(\epsilon)|\,n'=0\rangle$ which is a pure phase factor

$$S_{00}(\epsilon) = \frac{A(\epsilon T)+iB(\epsilon T)}{-A(\epsilon T)+iB(\epsilon T)} = e^{2i\delta(\epsilon)}, \quad 0 \le \epsilon \le E_0 \tag{29}$$

The phase shift $\delta(\epsilon)$ verifies Levinson's theorem

$$\delta(0) - \delta(E_0) = \begin{cases} \pi & \text{if } T \text{ belongs to an interval of stability} \\ 0 & \text{otherwise} \end{cases} \tag{30}$$

Moreover, if one has $\nu > \frac{1}{2}$ in (10), and T reaches the extremity of an interval of stability (T close, say, to $T_c \equiv T_i'$), the cyclic state with eigenphase α_T is transformed into a resonance at threshold. When $T = T_c$ the function $B(\alpha)$ vanishes at α_{T_c} and the implicit function theorem implies that $B(\alpha)$ has a close-by zero $\alpha_r = \epsilon_r T$ when T is slightly above T_c: $B(\epsilon T) \sim b(\epsilon - \epsilon_r)T$, $b > 0$. Then one finds from (29) the usual Breit-Wigner form

$$|S_{00}(\epsilon) - 1|^2 \sim \frac{2\Gamma_r^2}{(\epsilon-\epsilon_r)^2 + \Gamma_r^2} \tag{31}$$

when T is close to T_c ($T > T_c$), and $|\epsilon - \epsilon_r| \ll \epsilon_r$. The energy of the resonance is of the order const.$(T - T_c)$ and by (28) and (10) its width Γ_r (the inverse life time) behaves as

$$\Gamma_r = \frac{A(\epsilon_r T)}{bT} = \frac{\pi}{bT^2}|\varphi(\epsilon_r)|^2 \sim \text{const. } \epsilon_r^{2\nu} \sim \text{const. } (T - T_c)^{2\nu} \tag{32}$$

with $\nu > \frac{1}{2}$ and $T \to T_c$.

2.3 Concluding remarks

The model of the preceding section can describe an electron hopping on a d-dimensional lattice and submitted to a time periodic point interaction acting at a single site. The kinetic energy H_0 is then the finite difference Laplacian on the lattice, and $|\varphi(E)|^2$ is determined by the density of states which behaves at threshold as $(\sqrt{E})^{d-2}$. Therefore we have the case (i) of section 2.1 if $d = 1, 2$ and the case (ii) if $d = 3$.

If the spectrum of H_0 consists of a finite number of bands, cyclic states can appear in all the gaps of U_T at sufficiently high frequency, depending on the nature of the point spectrum of the time averaged Hamiltonian $\bar{H}$.

If the spectrum of H_0 is not bounded, the situation is more delicate. For instance if the spectrum of H_0 is the half axis $[0, \infty)$, U_T has also an absolutely part in its spectrum that wraps the unit circle infinitely many times. Then one expects that the point spectrum of $\bar{H}$ is transformed into resonances of U_T whose life times tend to ∞ as $T \to 0$. At the moment there is no example of a cyclic state with eigenvalue embedded into the continuous spectrum of U_T, but there exist no proof that this case cannot occur.

A case of physical interest (atom in strong high frequency laser field) is obtained when $H_0 = -$Laplacian on $\Re^3$ and $V(\omega t)$ is an atomic potential of the form $V(\mathbf{x}+\mathbf{a}\cos\omega t)$. After a suitable unitary transformation (the Kramers-Henneberger transformation), it represents the action of a linearly polarized monochromatic field on the atom, and the quantity $\mathbf{a}$ is determined by the ratio of the field amplitude to the frequency. The corresponding averaged Hamiltonian $\bar{H}$ is called the dressed atomic Hamiltonian and it may support more bound states than H [7]. A perturbative scheme in inverse powers of ω has been proposed [8,9] to compute their energies and lifetimes at high-frequency. Although there is a fairly large physical literature on this subject (see for instance [10] for recent references), we feel that it still deserves mathematical clarification.

References

[1] D.R. Yafaev, *Mathematical Scattering Theory*, Chap. 6, Mathematical Monographs **105**, Amer. Math. Soc. (1991)

[2] G. Casati and L. Molinari, *Progress in Theoretical Physics Supplement*, **98**, 287 (1989)

[3] Ph.A. Martin and M. Sassoli de Bianchi, *J. Phys. A: Math. Gen.*, **28**, 2403 (1995)

[4] V.I. Arnold, *Mathematical Methods of Classical Mechanics*, section 25 E, Springer (1989)

[5] T. Kovar and Ph.A. Martin,*J. Phys. A: Math. Gen.*, **31**, 385 (1998)

[6] M. Combescure, *J. Stat. Phys.*, **59**, 679 (1990)

[7] R. Bhatt, B. Piraux, and K. Burnett, *Phys. Rev. A*, **37**, 98 (1988)

[8] M. Gavrila and J.Z. Kaminski, *Phys. Rev. Lett.*, **52**, 613 (1984)

[9] *Atoms in Intense Laser Fields*, Ed. M. Gavrila, Academic Press, New-York (1992)

[10] H. Wiedemann, *Phys. Rev. A*, **50**, 2769 (1994)

Operator Theory:
Advances and Applications, Vol. 108

Some geometry related to decay properties of the resolvent of a class of symmetric operators

E. Mourre*

Abstract

We introduce in a simple way the notion of stability sets associated to the resolvent $\frac{1}{H-E+i\epsilon}$ of a class of symmetric operators; then we give theorems which estimate the decay properties of the resolvent, in terms of an appropriate distance defined by the operator H and stability sets. We also give the criteria of invariance of these sets under exponential changes of scales. Next, we give the example of the non-triviality of stability distances for ergodic families of one-dimensional operators. In section 4 we prove an algebraic coupling identity and inequalities associated with decoupling projectors, and relations with stability sets.

1 Introduction

1.1 Example of the Anderson model.

Let the Hilbert space $\mathcal{H}$ be described by $l^2(\mathbf{Z}^d)$, and consider the following operators:

$$H_\omega = \Delta_d + \sum_{n\in Z^d} \lambda\omega_n P_n \tag{1.1}$$

where $(P_n)_{n\in\mathbf{Z}^d}$ is the canonical family of the one dimensional projectors and

$$(\Delta_d\Psi)(n) = \sum_{|m-n|=1, m\in\mathbf{Z}^d} \Psi(m) \tag{1.2}$$

$$\omega = (\omega_n)_{n\in\mathbf{Z}^d} \in (-1,+1)^{\mathbf{Z}^d} = \Omega. \tag{1.3}$$

For simplicity we make Ω a probability space with the measure:

$$\mu = \prod_{n\in\mathbf{Z}^d} \frac{1}{2}dx, \tag{1.4}$$

*E-mail: mourre@cpt.univ-mrs.fr
Centre de Physique Théorique, CPT-CNRS, Luminy Case 907, F–13288 Marseille, Cedex 9, FRANCE.

dx being the Lebesgue measure on $(-1, +1)$. λ is a coupling constant.

1.2 The present results

Many interesting developments have been concerned with the study of random and quasi-periodic operators; for references, see the following books [1, 2] and articles [3, 4].

The results that are known are of the following kind:

a) $\sigma(H_\omega) = [-2d - \lambda, 2d + \lambda]$, almost surely in Ω.

b) For λ big enough these operators have almost surely in Ω dense pure point spectrum; this is known for all dimensions. In dimension 1, the pure point spectrum is true for any non zero λ.

Under the conditions for which the pure point spectrum is proved, we also have:

c) For any fixed $E \in [-2d - \lambda, +2d + \lambda]$ there exists a set $\Omega' \subset \Omega$ of measure 1 such that, for any $\omega \in \Omega'$, there exists a constant C_ω such that:

$$||P_o \frac{1}{H_\omega - E + i\varepsilon} P_i|| \leq C_\omega e^{-\gamma(E) d(o;i)} \tag{1.5}$$

for any $\varepsilon \in \mathbf{R}$ and $i \in \mathbf{Z^d}$.

1.3 Question

How to understand that the resolvent of an operator has an exponential decay despite the fact that the spectrum for that operator is dense around E.

The easy way to get exponential decay is usually when E is not in the spectrum of H; in that case $d(E; \sigma(H)) \geq \alpha > 0$ and we can use the analyticity in β of the resolvent:

$$\frac{1}{H - E + \beta C} \tag{1.6}$$

for any operator C with $||C|| \leq 1$ and $|\beta| < \alpha$.

We shall see that in the above situations we may expect the existence of projectors:

$$P_{S(\omega,E)} = \sum_{i \in S(\omega,E)} P_i \tag{1.7}$$

where $S(\omega, E) \subset \mathbf{Z}^d$ and such that we still have non trivial analyticity in β for the operators:

$$\frac{1}{H_\omega - E + i\varepsilon + \beta P_{S(\omega,E)} C P_{S(\omega,E)}} \tag{1.8}$$

for any $\varepsilon \in \mathbf{R}$ and any operator C with $||C|| \leq 1$.

Because of the density of the spectrum of H_ω around E, the set $S(\omega, E)$ must be a porous set in $\mathbf{Z}^d$ and certainly strictly smaller.

2 Stability sets and distances

2.1 Some definitions

Let $\mathcal{H}$ be a Hilbert space and $(P_i)_{i\in I}$ a complete family of orthogonal projectors:

$$P_i = P_i^*, \quad P_iP_j = \delta_{ij}P_i, \quad \sum_{i\in I} P_i = \mathbb{1} \tag{2.1}$$

Definition. For a symmetric operator H, the H-neighbors of a point $i \in I$ are:

$$V_H(i) = \{j \in I | j \neq i, \text{ and } \|P_iHP_j\| \neq 0\} \tag{2.2}$$

Hypothesis and notation

$$\sharp(V_H(i)) \leq n_0 \quad \text{for any } i \in I, \tag{2.3}$$

$$h = \sup_{i\in I} \Big\{ \sum_{j\neq i} \|P_iHP_j\| \Big\}. \tag{2.4}$$

Notice that finiteness of h implies that the off diagonal part of H is a bounded operator.

Definition of the set Γ_H of H-paths: A path $\gamma \in \Gamma_H$ is a finite sequence in I: $\gamma = \{\gamma(0), \gamma(1), \ldots, \gamma(k), \gamma(k+1), \ldots, \gamma(N(\gamma))\}$, such that for $k \in \{0, 1, 2, \ldots, N(\gamma) - 1\}$ $\gamma(k+1) \in V_H(\gamma(k))$.

Definition of the distance induced by H and a set $S \subset I$.
For a set $S \subset I$, and any path γ we define:

$$l_H^S(\gamma) = \sharp\{k \in [0, N(\gamma) - 1] | \{\gamma(k), \gamma(k+1)\} \subset S\} \tag{2.5}$$

Then the induced distance is:

$$d_H^S(i,j) = \inf_{\gamma\in\Gamma_H(i,j)} l_H^S(\gamma) \tag{2.6}$$

2.2 Example of porous sets

On $l^2(\mathbf{Z}^2)$ for operators H of the form :

$$\Delta_2 + \sum_{n\in\mathbf{Z}^2} v(n)P_n \tag{2.7}$$

the neighbors of any point are the usual ones, and then the paths are the usual paths on $\mathbf{Z}^2$; $n_0 = 4$ and $h = 4$. For an increasing sequence $(r_n)_{n\in\mathbf{Z}}$ of radii we can construct a porous set in the following way:

$$S = \mathbf{Z}^2 \setminus \bigcup_n \{ \bigcup_{k\in\{C_n\}} B(k, r_n)\} \tag{2.8}$$

where C_n is the subset of centers of balls of radius r_n. Even in the case where the sequence of radii tends to infinity, if the sets C_n get sparser and sparser then for each fixed point o, the distance $d_H^S(o;m)$ may remain proportional to the Euclidean one, uniformly in m for sufficiently distant m .

2.3 Stability distances and exponential decrease of the resolvent

Definition: stability sets

For $C_o \in \mathbf{R}^+$ and $p \in [1,2]$ let us define the following stability set:

$$F(H-E,C_o,p) = \{i \in I | \sup_{\varepsilon \neq 0} \sum_{k \in I} ||P_i \frac{1}{H-E+i\varepsilon} P_k||^p \leq C_o^p\} \ . \tag{2.9}$$

Theorem 2.1 *Let H a symmetric operator, and $(P_i)_{i\in I}$ a complete family of orthogonal projectors;*

then for any $J \subset F(H-E,C_o,1)$, and for any $i,j \in I$, the following holds:

$$||P_J \frac{1}{H-E+i\varepsilon} P_J|| \leq C_o \tag{2.10}$$

$$||P_i \frac{1}{H-E+i\varepsilon} P_j|| \leq (||P_i \frac{1}{H-E+i\varepsilon} P_j|| + C_o \sharp\{J\}) e^{-\log(1+\frac{1}{2hC_o}) d_H^J(i,j)} \tag{2.11}$$

For the proof see [5].

Remark:

If the stability distance between i and j is equal to L, strictly positive, we get an explicit estimate on the resolvent by taking $J = \{k | d_H^{F(H-E,C_o,1)}(i,k) \leq 2L\}$ for the set J.

2.4 Criteria of invariance of stability sets under scaling

Definition: H-boundary of a set

Let $(P_i)_{i\in I}$ a complete family of orthogonal projectors, and H a symmetric operator; the H-boundary of a set $S \subset I$ is:

$$\partial_H(S) = S \cap V_H(I \setminus S)$$

where, for any set S', $V_H(S') = \cup_{i \in S'} V_H(i)$.

Theorem 2.2 *Let $(P_i)_{i\in I}$ a complete family of orthogonal projectors, and H a symmetric operator; let $S_o \subset S_1 \subset I$ and H_{S_o} and H_{S_1} the symmetric restrictions of the operator H on $P_{S_o}\mathcal{H}$ and $P_{S_1}\mathcal{H}$. Let*

$$i \in S_o \cap F(H_{S_o} - E, C_o, 1) \quad , \quad L = d_{H_{S_o}}^{F(H_{S_o}-E,C_o,1)}(i, \partial_{H_{S_1}}(S_o)) \tag{2.12}$$

and suppose that the following conditions hold:

$$L \log(1 + \frac{1}{2hC_o}) \geq 2 \tag{2.13}$$

$$d(E, \sigma(H_{S_1}) \geq e^{-\frac{L}{3}\log(1+\frac{1}{2hC_o})} \tag{2.14}$$

$$\sharp(S_o)\sharp(S_1) \leq e^{+\frac{L}{3}\log(1+\frac{1}{2hC_o})} \ , \tag{2.15}$$

then :

$$i \in F(H_{S_1} - E, C_1, 1) \quad \textit{with} \quad C_1 \leq C_o(1 + 3e^{-\{\frac{L}{3}\log(1+\frac{1}{2hC_o})\}}) \ . \tag{2.16}$$

For the proof see [5].

Remark:

This theorem concerning the invariance of stability subsets under exponential change of scale should relate for example the construction of stability sets in the Anderson model to classical probability estimates as large deviations, under some initial hypothesis of the following form:

For some C_o, and β_o there exist a box Λ_o, large enough, such that the probability of a regular event $\omega_o \in \Omega_{\Lambda_o}^{regular}(E, C_o, \beta_o)$ is larger than some minimal probability $p_m(d)$, where

$$\Omega_{\Lambda_o}^{regular} = \{\omega \in \Omega_{\Lambda_o} | d^{F(H_{\Lambda_o}(\omega)-E,C_o,1)}(i,j) \geq \beta_o|i-j| \, , \text{for } |i-j| \text{ big enough}\} \ . \tag{2.17}$$

For some reference for that kind of multiscale analysis (in dimension 1) see [5].

Remark:

Moreover the above argument with the theorem 2.1 suggest, and section 3 proves (in dimension 1) the existence of decoupling sets and decoupling projectors:

Definition. Decoupling sets and projectors :

A set S is called a decoupling set which defines a decoupling projector when :

$P_{I\setminus S} H P_{I\setminus S}$ becomes a direct sum of operators :

$$P_{I\setminus S} H P_{I\setminus S} = \sum_{\alpha\in A} P_{S_\alpha} H P_{S_\alpha} \quad \text{with} \quad \mathrm{S}_\alpha \cap \mathrm{S}_{\alpha'} = \emptyset \quad \text{for} \quad \alpha \neq \alpha'$$

and such that $P_S \frac{1}{H-E+i\varepsilon} P_S$ are uniformly bounded operators for $\varepsilon \in \mathbf{R}$.

These decoupling sets are interesting by themselves and may be useful to construct stability sets in the first step of multiscale analysis (see section 4).

3 Example of non triviality of stability distance in dimension 1

Let $\mathcal{H} = l^2(\mathbf{Z})$ and H of the form:

$$H_\omega = \Delta_1 + \sum_{i\in\mathbf{Z}} v_\omega(i) P_i \tag{3.1}$$

Then we have the following theorem which is a version of the general theorem 2.1, but concerns stability sets $F(H-E, C_o, 2)$ and dimension 1:

Theorem 3.1 *There exist two functions γ and D such that, for any $C_o \geq 0$, the following holds:*

a) For any subset $J \subset F(H_\omega - E, C_o, 2)$ with the property that if $i \in J$ then $i+1 \in F(H_\omega - E, C_o, 2)$ we have:

$$||P_{\bar{J}} \frac{1}{H_\omega - E + i\varepsilon} P_{\bar{J}}|| \leq D(C_o) \tag{3.2}$$

where $\bar{J} = \cup_{i\in J}\{i, i+1\}$,

b) and for any $i, j \in \mathbf{Z}$ we have:

$$\begin{aligned} ||P_i \tfrac{1}{H_\omega - E + i\varepsilon} P_j|| \quad &\leq \quad \left(||P_i \tfrac{1}{H_\omega - E + i\varepsilon} P_j|| + 2C_o d^{F(H_\omega - E, C_o, 2)}(i,j)\right) \times \\ &\times e^{-\left(\gamma(C_o) d^{F(H_\omega - E, C_o, 2)}(i,j)\right)} . \end{aligned} \tag{3.3}$$

For the proof see [5].

Now we are going to make the following hypotheses on the family of the one-dimensional operators

$$H_\omega = \Delta + \sum_{i\in\mathbf{Z}} v_\omega(i) P_i$$

Hypotheses
Let $\omega \in \Omega$, where Ω is a probability space on which an invertible ergodic transformation T exists which preserves the measure and such that :

$$v_\omega(i+1) = v_{T(\omega)}(i) \quad \forall i \in \mathbf{Z}$$

Then we have the following theorem:

Theorem 3.2 *Assume that for some fixed $E \in \mathbf{R}$:*

a) E is not an eigenvalue of H_ω for a.e ω

b) for a.e. ω there exist solutions u_ω^+ and u_ω^- , of the equation :

$$(H_\omega - E)u = 0 \tag{3.4}$$

which are respectively in $l^2(\mathbf{Z}^+)$ and in $l^2(\mathbf{Z}^-)$

Then there exist C_o , $\alpha > 0$ and a set $\Omega' \subset \Omega$ of measure 1 such that $\forall \omega \in \Omega'$

$$\liminf_{N\to\infty} \quad \frac{d^{F(H_\omega - E, C_o, 2)}(0, N)}{N} \geq \alpha \tag{3.5}$$

Proof of theorem 3.2

If for some E and ω hypotheses a) and b) are true, then by a nice theorem (see [4]: Spectral analysis of rank one perturbations and applications) one has:

$$\sup_\varepsilon \|P_o \frac{1}{(H_\omega - E)^2 + \varepsilon^2} P_o\| < \infty \tag{3.6}$$

Then for a fixed E this holds by assumption on a set of ω of measure 1; by the invariance of the measure under T we deduce that the function

$$dl(\omega) = \frac{1}{1 + \|P_o \frac{1}{(H_\omega - E)^2} P_o\| + \|P_1 \frac{1}{(H_\omega - E)^2} P_1\|} \tag{3.7}$$

is non-zero on a set of ω of measure 1. It is a bounded, measurable function and we have:

$$E_\Omega(\frac{1}{1 + \|P_o \frac{1}{(H_\omega - E)^2} P_o\| + \|P_1 \frac{1}{(H_\omega - E)^2} P_1\|}) = 2\alpha > 0. \tag{3.8}$$

We can apply the Birkhoff ergodic theorem to the function $dl(\omega)$ and thus deduce the existence of a set Ω' of measure 1, such that for any $\omega \in \Omega'$ we have:

$$\lim_{N\to\infty} \quad \frac{1}{N} \sum_{k \in [1,2,..,N]} \frac{1}{1 + \|P_k \frac{1}{(H_\omega - E)^2} P_k\| + \|P_{k+1} \frac{1}{(H_\omega - E)^2} P_{k+1}\|} = 2\alpha \tag{3.9}$$

Then for any $\alpha' < \alpha$, and N big enough we get:

$$\sharp\left\{ k \in [1,..,N] \quad | \quad \frac{1}{1 + \|P_k \frac{1}{(H_\omega - E)^2} P_k\| + \|P_{k+1} \frac{1}{(H_\omega - E)^2} P_{k+1}\|} > \alpha \right\} \geq N\alpha' \tag{3.10}$$

Theorem 3.2 is then proved with $C_o = \sqrt{1/\alpha}$.

The theorem 3.1 and 3.2 prove the equivalence of the positivity for the Lyapounov exponents for ergodic family of one-dimensional type operators, with the asymptotic proportionality of stability distances to the Euclidean one.

Remark:

A similar theorem can not be expected in more than one dimension, under such general hypotheses, since usually the ergodicity may not distinguish between situations with and without a percolation.

4 An algebraic expression of the resolvent of an operator in term of reduced resolvents, and operatorial decoupling

Proposition 4.1 *Let T be a bounded operator on an Hilbert space, and P and Q two projectors with:*

$$P^2 = P,\ Q^2 = Q,\ PQ = 0,\ P + Q = 1 \tag{4.1}$$

and we assume that T and $PTP + QTQ$ have bounded inverses, then we have the following identities:

$$P\frac{1}{T}P = \frac{1}{PTP - PTQ\frac{1}{QTQ}QTP} \tag{4.2}$$

$$P\frac{1}{T} = P\frac{1}{T}P\left(P - PTQ\frac{1}{QTQ}Q\right). \tag{4.3}$$

The identity (4.2) is the effective operator formula which says that $P\frac{1}{T}P$ is invertible if and only if

$$PTQ\frac{1}{QTQ}QTP$$

is a bounded operator, and gives an explicit form of its inverse.

Proof of proposition 4.1

By the resolvent equation associated with T and $PTP + QTQ$ we get:

$$P\frac{1}{T} = P\frac{1}{PTP} - P\frac{1}{PTP}PTQ\frac{1}{T}. \tag{4.4}$$

Using once more the resolvent equation, we get:

$$\left(P - P\frac{1}{PTP}PTQ\frac{1}{QTQ}QTP\right)P\frac{1}{T} = P\frac{1}{PTP}\left(P - PTQ\frac{1}{QTQ}\right) \tag{4.5}$$

multiplying on the left both sides of equation (4.5) by PTP we obtain:

$$\left(PTP - PTQ\frac{1}{QTQ}QTP\right)P\frac{1}{T} = \left(P - PTQ\frac{1}{QTQ}\right). \tag{4.6}$$

Then multiplying on the right the equation (4.6) by P one first gets equation (4.2), and then using that equation hols in (4.6) we get (4.3).

Proposition 4.2 *Assume $T_\varepsilon, \varepsilon \in (0,1]$, satisfy hypothesis of proposition 4.1 and converge (in the operator norm) to T;*
Then $P\frac{1}{T}P$ is non invertible if and only if :

$$\lim_{\varepsilon\to 0} \|PT_\varepsilon Q\frac{1}{QT_\varepsilon Q}QT_\varepsilon P\| = \infty. \tag{4.7}$$

Proposition 4.3 *Let T be a bounded and invertible operator and assume that $P\frac{1}{T}P$ is invertible (on the range of P),*

then we have:

$$\frac{1}{T} = Q\frac{1}{QTQ}Q + \frac{1}{T}P\frac{1}{P\frac{1}{T}P}P\frac{1}{T}. \tag{4.8}$$

Proof of proposition 4.3
Under the hypothesis that $P\frac{1}{T}P$ is invertible, for large enough λ, the following operator is well defined:

$$\frac{1}{T} - \frac{1}{T}P\frac{1}{\frac{1}{\lambda} + P\frac{1}{T}P}P\frac{1}{T}. \tag{4.9}$$

By multiplying it on the left or right with $T + \lambda P$ we verify that they are inverse of each other and

$$\frac{1}{T+\lambda P} = \frac{1}{T} - \frac{1}{T}P\frac{1}{\frac{1}{\lambda} + P\frac{1}{T}P}P\frac{1}{T}. \tag{4.10}$$

Now we let λ tend to infinity and remark that under the hypothesis, $P\frac{1}{T}P$ is invertible in the range of P. Hence both sides of equation (4.10) converge in operator norm, and in particular it is easy to see that:

$$\lim_{\lambda\to\infty} \frac{1}{T+\lambda P} = Q\frac{1}{QTQ}Q. \tag{4.11}$$

This proves the proposition 4.3 .

Coupling identity (4.12), and decoupling projectors

Theorem 4.4 *Let P and Q two orthogonal projectors on $\mathcal{H}$ with $P+Q=1$. Assume that T and $PTP+QTQ$ are bounded invertible operators,*

a) then we have the following identities:

$$\frac{1}{T}=Q\frac{1}{QTQ}Q+\left(P-Q\frac{1}{QTQ}QTP\right)P\frac{1}{T}P\left(P-PTQ\frac{1}{QTQ}Q\right), \tag{4.12}$$

$$P\frac{1}{T}P=\frac{1}{PTP-PTQ\frac{1}{QTQ}QTP}. \tag{4.13}$$

b) Assume furthermore that P is decoupling, namely : $\|P\frac{1}{T}P\|\leq C$ and that QTQ is a direct sum :

$$QTQ=\sum_{\alpha\in A}Q_\alpha TQ_\alpha,$$

where Q_α are orthogonal projectors; let $Q^r(W)$ be defined by :

$$Q^r(W)=\sum_{\alpha\in B\subset A}Q_\alpha \text{ with for any } \alpha\in \mathrm{B}\,, \|\mathrm{Q}_\alpha\frac{1}{\mathrm{Q}_\alpha\mathrm{T}\mathrm{Q}_\alpha}\mathrm{Q}_\alpha\|\leq \mathrm{W}$$

then we have:

$$\|\left(Q^r(W)+P\right)\frac{1}{T}\left(Q^r(W)+P\right)\|\leq W+C(1+W\|QTP\|)^2. \tag{4.14}$$

To prove the theorem 4.4, we first use proposition 4.3, and we express $P\frac{1}{T}$ by equation (4.3) and simultaneously the transposed expression for $\frac{1}{T}P$; thus we have proved the part a) from which part b) easily follows.

Decoupling projectors and construction of stability sets

Theorem 4.5 *Let $(P_i)_{i\in I}$ a complete family of orthogonal projectors, and H a bounded symmetric operator, E and $\varepsilon\in\mathbf{R}$. Let $S\subset I$ be a decoupling set for the resolvent of $(H-E)^2+\varepsilon^2$:*

a) $\|P_S\frac{1}{(H-E)^2+\varepsilon^2}P_S\|\leq C$ for any non zero ε

b) $P_{I\setminus S}(H-E)^2P_{I\setminus S}$ is a direct sum:

$$P_{I\setminus S}(H-E)^2P_{I\setminus S}=\sum_{\alpha\in A}P_{S_\alpha}(H-E)^2P_{S_\alpha}. \tag{4.15}$$

Let :

$$S^r(W)=\bigcup_{\alpha\in B\subset A}S_\alpha\ , \tag{4.16}$$

where for any

$$\alpha \in B,\ \|P_{S_\alpha} \frac{1}{P_{S_\alpha}((H-E)^2+\varepsilon^2)P_{S_\alpha}} P_{S_\alpha}\| \leq W. \tag{4.17}$$

Let $J = S^r(W) \cup S$; *then the following holds:*

$$\|P_J \frac{1}{(H-E)^2+\varepsilon^2} P_J\| \leq C_0^2 = W + C(1 + W\|P_{I\setminus S}(H-E)^2 P_S\|)^2 \tag{4.18}$$

and

$$\|P_J \frac{1}{H-E+i\varepsilon} P_J\| \leq C_0 = (W + C(1 + W\|P_{I\setminus S}(H-E)^2 P_S\|)^2)^{\frac{1}{2}}. \tag{4.19}$$

As a consequence of the fact that $J \subset F(H-E, C_0, 2)$, *we obtain by the tools used in the proof of theorem 2.1 by and Schwartz inequality for any* $j \in J$ *and for any* $k \in I$:

$$\|P_j \frac{1}{H-E+i\varepsilon} P_k\| \leq 2C_0 e^{-log(1+\frac{1}{2hC_0})d_H^J(j,k)}. \tag{4.20}$$

But also:

$$\sum_{k\in I} \|P_j \frac{1}{H-E+i\varepsilon} P_k\| \leq 2C_0 \left(\sum_{k\in I} e^{-2log(1+\frac{1}{2hC_0})d_H^J(j,k)} \right)^{1/2}, \tag{4.21}$$

where h *and* d_H^J *are defined by formulas (2.4) and (2.6).*

We have to remark that decoupling projectors for the resolvent of $(H-E)^2$ which may be found and estimated using formula (4.13) with $T = (H-E)^2 + \varepsilon^2$ may allow us to construct strong estimates (4.17). Weaker estimates are that $J \subset F(H-E, C_0, 2)$ and $\|P_J \frac{1}{H-E+i\varepsilon} P_J\| \leq C_0$, from which we can construct by (4.20) stability sets of the form $F(H-E, C', 1)$, which are convenient for multiscale-analysis.

Acknowledgments: Many thanks are due to P. Briet for numerous and interesting discussions.

References

[1] L. Pastur, A. Figotin, *Spectra of random and almost periodic operators*, Springer Verlag Berlin (1992).

[2] R. Carmona, J. Lacroix, *Spectral theory of random Schrödinger operators*, Birkhauser,Boston (1990).

[3] M. Aizenman, S. Molchanov, *Localization at large disorder and at extreme energies: an elementary derivation*, Commun. Math. Phys. 157 (1993), 245–278.

[4] B. Simon, *Spectral analysis of rank one perturbations and applications*, CDM Proc. Lecture Notes 8 (1995), 109–149.

[5] E. Mourre, *Géométrie des ensembles de stabilité et propriétés de décroissance de la résolvante d'un opérateur self adjoint.*, preprint: CPT-97 P3468.

Operator Theory:
Advances and Applications, Vol. 108

On Some Asymptotic Formulas in the Strong Localization Regime of the Theory of Disordered Systems

L.Pastur *†

Abstract

We present certain results, both heuristic and rigorous, on the asymptotics study of the simplest interesting observables of the theory of disordered systems: the density of states (the DOS) and the conductivity. First we outline heuristic arguments dating back to I.Lifshitz and allowing us to write the leading term of the low-energy asymptotics of the logarithm of the DOS for the non-positive Poisson potential, less studied so far than, say, the alloy-type potential and modeling chaotically distributed attractive impurities. Second we formulate recent rigorous results ([16]) that justify and develop this heuristics in the case of singular one-impurity potentials, the screened Coulomb in particular. At last we presents heuristic arguments ([9]) that, we believe, make more precise and detailed the arguments of I. Lifshitz and N. Mott and allowing us to obtain the known Mott formula for the low-frequency conductivity of disordered systems. The arguments are based on an Ansatz describing the form of relevant realizations of the random potential. We believe that the Ansatz is a natural consequence of the studies of the strong localization regime carried out so far and can be used to write a number of other observables in the strong localization regime.

1 Introduction

In recent decades there has been considerable progress in spectral theory of random selfadjoint operators, differential and finite-difference operators in particular. This has been achieved via synthesis of the spectral analytic and probabilistic ideas and was strongly motivated by theoretical physics. The result of this progress is a rather complete spectral analysis of finite-difference and differential operators with random coefficients in the one dimensional case and in the strong localization regime of the multidimensional case. In all these cases it has been proved that if a random potential in the Schrödinger equation is shortly enough correlated and

*UFR de Mathematiques, Université Paris 7, 75252, Paris, France
†Mathematical Division, Institute for Low Temperature Physics, 310164, Kharkov, Ukraine

smoothly distributed, then all the states in these cases are exponentially localized and corresponding levels are non-degenerated and dense.

There are, however, rather serious questions to be understood even in the one-dimensional case and in the strong localization regime, having the pure point spectrum, not to mention other regimes and spectral types. First of all it is the development of powerful and general enough techniques of the study of the strong localization regime for broad enough classes of partial differential operators with random coefficients, periodically random operators, and operators of the divergent form in particular. It seems that new rather important ingredients have to be added to the toolkit of the localization theory. There is one more problem, serious and important in our opinion. Spectral theory of random and almost periodic operators has strong theoretical-physics motivation, solid-state-theory motivation first of all. However the goal of the latter is not just to establish the character of the spectrum but to compute certain moments of the Green function, the evolution operator and the spectral kernel, which represent the observables of the theory. It turns out that the road from the knowledge of the spectrum (eigenvalues and eigenfunctions) of random operators to the formulas and properties of observables is rather hard and the respective difficulties are far from being just technical. On the other hand, it seems that these topics are certainly among the most important in the present-time spectral theory and mathematical physics of disordered systems.

In this paper we communicate certain results, both heuristic and rigorous, obtained recently in [16, 9] in this direction.

2 Density of States

2.1 History and Heuristics

The Density of States (DOS) and its once integrated form known as the integrated density of states (IDS) are the simplest spectral characteristics of a random Schrödinger operator. Among numerous problems related to the IDS, the problem of its asymptotic behavior near the edges (i.e. edges of the spectrum) is well known and fairly studied. Results of these studies can be summarized as follows (see e.g. [8, 20, 23]). One has to distinguish two types of spectral edges: stable and fluctuational. The latter are special for shortly correlated random potentials. The simplest case is the lower edge of the spectrum. It is determined by the absolute minimum of the potential since the spectrum in a neighborhood of this edge exists only because of those fluctuations of the potential where it is arbitrarily close to the minimum. This is why the edge is called the fluctuational edge ([20]). By using the quantum mechanical terminology one can call these portions of the realizations the localization wells.

A heuristic derivation of the fluctuational asymptotics was proposed by I. Lifshitz in the early 60's ([18, 19]). The asymptotics is given by the probability to have a potential well whose ground state energy is close enough to the spectral edge.

Since the probability of these realizations having the form of very broad and/or deep potential wells (known as optimal fluctuations) on distances much bigger than their size is usually exponentially small, one has to deal here with a version of the large deviation technique in the spectral context. In particular, to determine the asymptotic formula for the DOS one has to be able to give rather detailed description of the statistics of these special realizations. This is why precise and explicit asymptotic formulae are known only for comparatively restricted classes of random potentials.

One of physically important random potentials is the Poisson potential having the form

$$V_\omega(x) = \sum_j V(x - x_j) \tag{1}$$

where $\{x_j\}$ is the Poisson point field of the density μ in $\mathbf{R}^d$ and $V(x)$, the one-impurity potential, is a function decaying sufficiently fast at infinity. The Poisson potential is of considerable interest both in the spectral theory and in theoretical physics of disordered systems. It possesses a number of nontrivial regimes and properties, a part of which has been studied only so far. In particular, it has not been proved yet rigorously that in the multidimensional case the spectrum near the fluctuational edges is pure point.

Speaking on this potential one has to mention first the case of the nonnegative one-impurity potential $V(x)$ of a compact support. In this case $E = 0$ is a fluctuational edge and according to I. Lifshitz [18, 19] the 'IDS $N(E)$ has the following asymptotic form

$$N(E) \simeq \exp(-const \cdot E^{-d/2}), \quad E \to +0. \tag{2}$$

Here and below we use the symbol "$\simeq$" to denote the asymptotic equivalence without indicating explicitly the order of the remainder and respective constants.

The r.h.s. of this formula is just the probability to have a well (a region of $\mathbf{R}^d$ free of x_j''s) of width $L \simeq E^{-1/2}$; the latter relation is due to the uncertainty principle. In other words, the asymptotics of the IDS in this case is determined by an optimization procedure, balancing the competition of the quantum and the probabilistic components of the problem. This procedure singles out the exponentially small portion of realizations that nevertheless make the leading contribution to the IDS. Following I. Lifshitz we will call these realizations optimal. Formula (2) is often called the quantum Lifshitz tail. Rigorous derivations of various versions of (2) (e.g. its logarithmic or even its the double logarithmic forms) have required a number of rather sophisticated probabilistic and spectral techniques (see e.g. [3, 22, 23, 25, 26, 14] for results and references).

Other asymptotic regimes of the IDS for the potential (1) correspond to the case when the one-impurity potential has a non-positive part, i.e. $\inf_{x\in\mathbf{R}^d} V(x) < 0$, so that the lower edge of the spectrum is $E = -\infty$. In this case one has to distinguish two asymptotic regimes, called usually the quantum and the classical one. We will present the respective asymptotic formulae by using a version of Lifshitz's arguments adapted to this case.

Recall the definition of the IDS (implicit also in [18, 19]): it is the limit as $\Lambda \to +\infty$ of the expectation of the normalized counting function of eigenvalues of the Schrödinger operator H_Λ, where H_Λ is the restriction of $H_\omega = -\Delta + V_\omega$ to $L^2(\Lambda)$ (see e.g. [23, 8]). Here Λ is the cube of the $\mathbf{R}^d$ of center zero and of side length L. The definition shows that the IDS can be regarded as the probability to find an eigenvalue of H_Λ lying below a given energy E. For $E \to -\infty$ these eigenvalues are produced by very deep potential wells, created by large clusters of k Poisson points x_j's, confined to a sufficiently small regions of the space, say, a cube of the side length $l << L$. The respective probability is

$$e^{-\mu l^d}(\mu l^d)^k/k!\,. \tag{3}$$

Assume first that l is independent of k. This corresponds to relatively (with respect to the exponential decay radius of the one-impurity potential) slowly varying one-impurity potentials, i.e. to relatively broad optimal fluctuations. Then the IDS has to be asymptotically equivalent to the probability (3) in which k is determined by the equation $E(k) = E$ where $E(g)$ is the ground state energy of the well $gV(x)$. In other words, for a non-positive one-impurity potential $V(x)$ optimal fluctuations are potential wells whose form is roughly $kV(x)$. Thus, we have to find the ground state energy $E(g)$ of the well $gV(x)$ as the function of its coupling constant g and then the asymptotic form of the IDS will be given by (3) in which $k = g(E)$ where $g(E)$ is the asymptotic inverse of $E(g)$ for $E \to -\infty$ as $g \to +\infty$ and the leading contribution is given by the factor $1/k!$. Thus the asymptotic formula should have the form

$$N(E) \simeq \exp(-g(E)\log g(E)), \quad E \to -\infty \tag{4}$$

or the less precise form

$$\log N(E) \simeq -g(E)\log g(E), \quad E \to -\infty. \tag{5}$$

The simplest case when this formula can be made rigorous is when $V(x)$ is continuous in its (negative) absolute minimum. In this case is easy to show that

$$E(g) = gV(0)(1 + o(1)), \quad E \to -\infty. \tag{6}$$

The respective rigorous formula

$$\log N(E) = -\frac{E}{V(0)}\log|E|(1 + o(1)), \quad E \to -\infty \tag{7}$$

was proved in [22] using a combination of variational and the Wiener integral techniques. The formula is obviously classical because it includes only the characteristics of the potential but not the Laplacian. In addition it can be easily shown that the r.h.s. of (4) is the asymptotic form of the logarithm of the probability $P(V_\omega(x) < E)$, i.e.

$$\log N(E) = \log P(V(0) \leq E)(1 + o(1)), \quad E \to -\infty \tag{8}$$

This provides one more evidence of the classical nature of the formula.

Situation is less satisfactory for the case of singular one-impurity potentials, where the role of the Laplacian is the same as that of the potential. In other words, in this case because of quantum effects the optimal fluctuations are more narrow than in the classical case and the second factor in (3) is not negligible. Despite a considerable amount of physical results in this case (see e.g. [20, 4]) rigorous results are almost absent, except the one-dimensional case of the point one-impurity potentials

$$V(x) = -g_0\delta(x), \quad g_0 > 0 \tag{9}$$

where $\delta(x)$ is the Dirac delta-function. In this case by using the special Markov processes technique one can show (see [20]) an asymptotic of the form

$$\log N(E) = -2\sqrt{\frac{E}{E_0}} \log \sqrt{\frac{E}{E_0}}(1 + o(1)), \quad E \to -\infty \tag{10}$$

where $E_0 = -g_0^2/4$ is the ground state energy of the well (9).

We see that formula (5) is now valid up to the factor 2. This difference is the result of a tunneling phenomenon related to the question how close to each other should be k potentials (9) in order to be regarded as the potential $k^*(k)\delta(x)$, i.e. the potential of same shape as (9) and of amplitude $k^*(k)$. In other words, in this case optimal fluctuations are more narrow then in the classical case. For many interesting one-impurity potentials l behaves as $k^{-\alpha^*}$ for some $\alpha^* > 0$ and the second factor in (3) also contributes to the asymptotics of the IDS. The study of this phenomenon is one of the topics dealt with in the paper [16].

Following this paper we present rigorous asymptotic formulas for $\log N(E)$ corresponding to singular one-impurity potentials (mainly with power-law singularities) obtained in the framework of the scheme outlined above. We find the precise form of $k^*(k)$ (see Theorem 4 and 6 below). This became possible owing to an improvemenet of one of the widely used techniques in the field, based on approximations of the Schrödinger operator in the whole space by the operator with the same potential but defined in a finite box whose size is properly chosen as a function of energy. Previous versions of this technique were based on the so-called Neumann-Dirichlet bracketing where the boxes with the Neumann and Dirichlet boundary conditions were used to construct the upper and the lower bounds for the IDS. The error in these bounds is of the order $O(L^{-1})$ where L is the size of the box and this is not sufficient to treat the quantum case. In [16] we approximate the IDS of the random model by the IDS of some well chosen periodic Schrödinger operators and obtain much more precise bounds. This method was proposed in

[14] and has been used to solve several problems in the field ([13, 12]). We obtain the once logarithmic versions of (4), i.e. (5) with explicit $g(E)$ and constants in front of $g(E)\log g(E)$ (see Theorems 5 and 6).

Let us give an example of the obtained results (for other examples see the next subsection and [16]). Assume that the one-impurity potential is the 3-dimensional attractive screened Coulomb potential

$$V(x) = -\frac{g_0 e^{-|x|}}{|x|} \tag{11}$$

important in the semiconductor physics. In this case we prove that (cf. (10))

$$\log N(E) = -2\sqrt{\frac{E}{E_0}}\log|E|(1+o(1)), \quad E \to -\infty, \tag{12}$$

where $E_0 = -g_0^2$ is the ground state energy of the well (11) (see Theorem 6 and the discussion following it).

The role of the IDS in the spectral theory and theoretical physics of disordered systems is well known and appreciated. However, there seems to be one more reason to study this quantity. Since the pioneering papers of I. Lifshitz, the study of the IDS was providing the first important step in understanding and study of more complicated properties and quantities in a respective version of the strong localization regime. In particular, in a subsequent paper ([15]) we use the technique developed here in order to find the large-deficit asymptotic behavior of the interband light absorption coefficient.

2.2 Rigorous Results.

Assume that the one-impurity potential $V : \mathbf{R}^d \to \mathbf{R}$ in (1) is such

$$V = V_1 + V_2$$

where

H1: for some $C > 0$ and any $x \in \mathbf{R}^d$, $|V_1(x)| \le Ce^{-|x|/C}$;

H2: V_2 is a compactly supported function satisfying $V_2 \in L^p(\mathbf{R}^d)$ where $p > p(d)$ and $p(d) = 2$ if $d \le 2$ and $p(d) = d/2$ if $d \ge 3$;

H3: for some set $Q \subset \mathbf{R}^d$ of a positive measure, $V\big|_Q < 0$.

Consider the random Schrödinger operator

$$H_\omega = -\Delta + V_\omega. \tag{13}$$

One has

Theorem 1 ([8]) *Under the assumptions made above, H_ω is essentially self-adjoint on $C_0^\infty(\mathbf{R}^d)$ ω-almost surely.*

Under our assumptions on V, we know that the almost sure spectrum of H_ω is $\Sigma = \mathbf{R}$([23, 8]).

Let Λ be a cube centered at 0 in $\mathbf{R}^d$. We define $H^D_{\omega,\Lambda}$ to be the Dirichlet restriction of H_ω to Λ. Pick $E \in \mathbf{R}$ and consider the quantity

$$N_{\omega,\Lambda}(E) = \frac{1}{|\Lambda|}\sharp\{\text{eigenvalues of } H^D_{\omega,\Lambda} \text{ smaller than or equal to } E\}. \tag{14}$$

Then one has

Theorem 2 ([8]) *Under the assumptions made above, there exists a non random, non decreasing, non negative, right continuous function $N(E)$ such that, ω-almost surely, for all $E \in \mathbf{R}$, E a continuity point of N, $N_{\omega,\Lambda}(E)$ converges to $N(E)$ as Λ exhausts $\mathbf{R}^d$.*

$N(E)$ is the *integrated density of states* (IDS) of H_ω.

As N is non decreasing, one can define its distributional derivative dN. It is a positive measure and is supported on the almost sure spectrum of H_ω (see [8, 23]). One has the following result

Theorem 3 *For $\varphi \in C_0^\infty(\mathbf{R})$, we have*

$$(\varphi, dN) = \mathbf{E}(\mathit{Tr}(\mathbf{1}_{C(0,1)}\varphi(H_\omega)\mathbf{1}_{C(0,1)})). \tag{15}$$

where $C(0,1)$ is the cube of center 0 and side length 1 and the symbol $\mathbf{E}(...)$ denotes the expectation with respect to realizations of the random potential.

Formula (15) is well known under more restrictive assumptions on the potential V_ω, i.e. for a less singular one-impurity potential V (see [23]).

It is clear from the above that to describe the asymptotic behavior of $N(E)$ near $-\infty$, we will need to define an auxiliary operator. For $g \in \mathbf{R}$, define

$$H(g) = -\Delta + gV. \tag{16}$$

Under our assumptions V is relatively form-bounded with respect to $-\Delta$ with the relative bound 0. Hence $H(g)$ admits a unique self-adjoint extension. Let $\sigma(H(g))$ denote its spectrum. It is lower semi-bounded. The infimum of $\sigma(H(g))$, i.e. the ground state energy of $H(g)$, will be denoted by $E(g)$. Let φ_g be the respective ground state, i.e. the unique positive normalized eigenfunction of $H(g)$ associated to energy $E(g)$ ([24]). In the sequel it will be often more convenient to work with $E_-(g) = -E(g)$ instead of $E(g)$ itself. Our assumption **H1** implies that $E_-(g) \to +\infty$ when $g \to +\infty$. Moreover E_- is strictly increasing in a neighborhood of $+\infty$. Let g be the inverse of E_- in a neighborhood of $+\infty$.

In the regular (classical) case, it was found that g is governing the first-term asymptotic of $\log N$ (cf [23, 22]). In the singular (quantum) case, the singular set of V will play a special part in the asymptotics. To measure this role, we introduce the notion of *asymptotic ground state*, i.e.

Definition 1 *Let $g \in (1, +\infty) \to \psi_g \in H^1(\mathbf{R}^d)$. We will say that ψ_g is an asymptotic ground state if and only if*

- *the vector ψ_g is normalized.*
- *$\exists g_0 > 1$, $l_0 > 0$ such that $\forall g \geq g_0$, $\operatorname{supp} \psi_g \subset C(0, l_0)$ (where $C(x, l)$ denotes the cube of center x and side length l).*
- $$\frac{|\langle (H(g) - E(g))\psi_g, \psi_g\rangle|}{|E(g)|} \to 0, \quad g \to +\infty. \tag{17}$$

We prove the existence of an asymptotic ground state. For $a \in \mathbf{R}^d$, we define the translation τ_a by $\tau_a V(x) = V(x - a)$ and

$$\mathcal{A}_{\psi_g} = \left\{ \alpha > 0;\ \lim_{g \to +\infty} \sup_{|a| \leq g^{-\alpha}} \left[\frac{g|\langle (\tau_a V - V)\psi_g, \psi_g\rangle|}{E_-(g)} \right] = 0 \right\}.$$

If $\mathcal{A}_{\psi_g} \neq \emptyset$, then we define $\alpha^*(\psi_g) := \inf \mathcal{A}_{\psi_g}$. Moreover, we define $\mathcal{A}$ to be the union of all $\mathcal{A}_{\psi_g}$. It can be proved that $\mathcal{A} \neq \emptyset$. We define

$$\alpha^* := \inf \mathcal{A}. \tag{18}$$

Roughly speaking $g^{-\alpha^*}$ is the form of dependence of the radius of the exponential decay of the one-impurity ground state on the coupling constant. It determines the size l of optimal fluctuations whose order now is $k^{-\alpha^* d}$.

Then we prove

Theorem 4 *Under the assumptions made above and for sufficiently large E, one has*

$$-(1 + \alpha^* d) g(E) \log g(E)(1 + o(1)) \leq \log N(-E) \leq -g(E) \log g(E)(1 + o(1)). \tag{19}$$

One may complain that Theorem 4 is not precise enough somewhat in that it only gives a two-sided estimate. But, as we will see below, this is in some way unavoidable as the true asymptotics depends not only on g but also on the singular set of the negative part of V. More precisely, as can be seen from Theorem 6 below (and from the proof of Theorem 4), the asymptotics of the IDS depends on the way the eigenfunction associated to the lowest eigenvalue for the operator $-\Delta + gV$ concentrates near the singular set of the negative part of V as g becomes large. In general the correction also depends on the geometry of the singular set. For example, if the singular set is a segment (e.g. a dislocation), then one can show that neither the lower nor the upper bound given by Theorem 4 are sharp. The two-sided estimate (19) can be made more precise if we know more about V.

The first and simplest example we give is the case when V is bounded from below, reaches its minimum at a single point, say 0, and is continuous near 0. Then one easily proves that $\alpha^* = 0$ and the upper and lower bounds in (19) coalesce to give (7).

We will now give other results that, we think, enclose most of the physically relevant examples. Let v_- be the essential infimum of V and assume that V is bounded from below, say satisfies

H1': $-\infty < v_- < 0$.

It is easy to show that $g(E) \sim E/v_-$ when $E \to -\infty$. We obtain

Theorem 5 *Under assumptions* **H1**, **H2** *and* **H1'** *one has*

$$\log N(E) \sim -g(E)\log g(E) \sim \frac{E}{v_-}\log|E|, \quad E \to -\infty. \tag{20}$$

Here and in the rest of the paper, $a \sim b$ *will always mean* $a = b(1 + o(1))$.

This extends (7) by removing the continuity assumption near the minimum.

Consider now a bit more general example that could be regarded as a model of a line defect (charged dislocation). In this case, $d = 2$ and

$$V_2(x) = \log_- |x|, \quad x \in \mathbf{R}^2$$

where, for $a \geq 0$, $\log_- a = \min\{\log a, 0\}$. Using the inequality $\log_- |x| + \log R \leq \log_- R|x| \leq \log|x|$ for $0 < R < 1$ and the variational principle for the ground state energy, one shows that in this case $E_-(g) \sim \frac{1}{2} g \log g, \quad g \to +\infty$, hence $g(E) \sim 2E/\log E, \quad E \to -\infty$. One also shows that $\alpha^* = 0$ for this one-impurity potential. Therefore, Theorem 4 tells us that

$$\log N(E) \sim -g(E)\log g(E) \sim -2|E|, \quad E \to -\infty.$$

Hence asymptotics (5) is also valid for certain mildly singular potentials.

Another case where one can find an asymptotics for $\log N$ is when V has only power-law singularities. Let q be a positive integer and pick q positive exponents $(\nu_i)_{i=1,\dots,q}$ and q functions $\{h_i(\theta)\}_{i=1,\dots,q}$ continuous on the sphere $\mathbf{S}^{d-1}$. For $1 \leq i \leq q$, consider the potentials

$$V_i(x) = \frac{h_i(\theta(x))}{|x|^{\nu_i}} \text{ where } \theta(x) = \frac{x}{|x|} \tag{21}$$

Assume that

$$0 < \nu_i < \begin{cases} 2, & \text{if } d = 1, 2; \\ d/2, & d \geq 3. \end{cases} \tag{22}$$

Then V_i is relatively form-bounded with respect to $-\Delta$ with the relative bound 0 and we can consider the operators $H_i = -\Delta + V_i$ with the form domain $H^1(\mathbf{R}^d)$. For $1 \leq i \leq q$, E_i denotes the ground state energy of H_i. Now we assume that

H1":
- there exist q distinct points $(x_i)_{i=1,\dots,q}$ in $\mathbf{R}^d$ and q continuous compactly supported functions $(W_i)_{i=1,\dots,q}$ such that $W_i(0)=1$ and

$$V_2(x)=\sum_{i=1}^{q} W_i(x-x_i)V_i(x-x_i)=\sum_{i=1}^{q}\tau_{x_i}(W_iV_i)(x) \tag{23}$$

where the exponents $(\nu_i)_{1\le i\le q}$ satisfy (22);
- for some $1\le i_0\le q$, we have $E_{i_0}<0$.

Notice that assumption **H1**$''$ implies assumptions **H2** and **H3**. Define

$$\nu^\dagger=\sup\{\nu_i;\ 1\le i\le q \text{ such that } E_i<0\},$$

$$E_-=\sup\left\{|E_i|;\ 1\le i\le q \text{ such that } E_i<0 \text{ and } \nu_i=\nu^\dagger\right\}$$

$$\alpha^\dagger=\frac{1}{2-\nu^\dagger}$$

Then by using scaling arguments one can show that $E_-(g)\sim -E_-g^{2\alpha^\dagger}$ when $g\to+\infty$. In addition, using the periodic approximation scheme in this case, we can find an upper bound having the same form as the lower bound in (19) where α^* is replaced by $\alpha^\dagger$. On the other hand, we can prove that $\alpha^*\le\alpha^\dagger$. From these two facts and from Theorem 4, we deduce $\alpha^*=\alpha^\dagger$ and

Theorem 6 *Under assumptions* **H1** *and* **H1**$''$ *one has*

$$\begin{aligned}\log N(E) &\sim -(1+\alpha^\dagger d)g(E)\log g(E) \\ &\sim -\left(1+\frac{d-\nu^\dagger}{2}\right)\left(\frac{E}{E_-}\right)^{1-\nu^\dagger/2}\log\left(\frac{E}{E_-}\right),\quad E\to-\infty\end{aligned} \tag{24}$$

We see that the lower bound in (19) is rather universal.

There exist also other methods to obtain this bound, for instance a version of the variational method of [20, 23] (see also [15]). This method also gives a sharp lower bound for practically all asymptotic formulae for $\log N(E)$ known so far. This is true in particular for the Poisson potential with a non-negative one-impurity potential. Thus it is the upper bound that has been requiring different techniques depending on the specific potential (see e.g. [3, 22, 25, 26, 14])

Assumptions **H1** and **H1**$''$ include most physically interesting cases as, for example, the 3-dimensional attractive screened Coulomb potential (11). In this case we have $\alpha^*=\alpha^\dagger=1$ and $E_-=1$ (see [17]) and we obtain (12).

There is another physically interesting case that has not been discussed here: it is the case of point potentials. Such potentials will be studied in the sequel [15] to this paper.

3 Conductivity

3.1 History

It is widely accepted and proved in many cases that elementary excitations in disordered systems are localized if the disorder is strong enough or if the energy of the excitations is close enough to the spectrum edges, i.e. that the spectrum of the respective one-particle Hamiltonian under these conditions is pure point and the eigenfunctions are exponentially decaying.

However, mathematical-physics aspects of disordered systems are less advanced. By mathematical-physics aspects we mean the study (computation, asymptotical analysis, etc.) of expressions for relevant physical quantities, usually having the form of moments of the spectral kernel of the Schrödinger operator H, i.e. of the kernel (the matrix) $\rho_E(x, y)$ of the operator $\delta(H - E)$, called often the local density of states. The most known and simple of these quantities is the density of states (DOS) $\rho(E) = \mathbf{E}(\rho_E(x, x))$ (see (15) as one of rigorous versions of this relation). But even for this simplest quantity there are several rather nontrivial asymptotic regimes as we discussed above.

The role of the DOS in spectral theory is well known. From the mathematical-physics point of view the DOS determines the equilibrium properties of disordered systems, i.e. of the ideal gas of elementary excitation (electrons, phonons, spin waves etc.) in the random environment. For the sake of definiteness we will discuss below the electron gas. The study of the kinetic properties of this gas and of interaction effects requires the knowledge of higher moments of the local density, the second moments in the first place. Two important quantities that can be expressed via the second moments are the density-density correlator and the current-current correlator [20]. The knowledge of these correlators allows one to answer a number of relevant questions concerning the existence and the nature of the localization and the behavior of observables in various situations.

Among them the low-frequency conductivity is of considerable interest. Denote by T the temperature, by ν the frequency of an external electric field, homogeneous in coordinates and alternating with time, and by E_F the Fermi energy by the electron gas. Then, according to N. Mott [21] the asymptotic form of the alternating current (ac) conductivity $\sigma(\nu, E_F)$ for $T << \nu << E_F$ is

$$\sigma(\nu, E_F) = A\rho^2(E_F)\nu^2[\log \nu_0/\nu]^{(d+1)}, \tag{25}$$

where A is a quantity determined by fundamental physical constants and by characteristics of the system, and ν_0 is determined by the random potential.

Defining the conductivity as the rate of the absorption of energy for the unit field Mott argued that the main contribution to this quantity in the sufficiently well developed localization regime and at low frequencies and temperatures has to be due to the resonance tunneling between pairs of the localization well having an optimum separation. Thus again as in the case of the DOS we have an optimization procedure that singles out the most relevant realizations of the random potential.

Formula (25) was discussed in many works devoted to disordered systems (see e.g. review works [21, 4, 5, 20]). Since rigorous results on this subject are absent we mention here several theoretical physics works that deal with the "first principles" derivation of (25) , i.e. derivations that start from the Kubo formula (see formula (40) below) or its analogs and in which the average of the product of two local densities in the corresponding asymptotic regime is computed by using certain approximations. In [2] the Mott formula was derived in one dimension for the white-noise Gaussian potential $\mathbf{E}(V(x)) = 0, \quad \mathbf{E}(V(x)V(y)) = l^{-3}\delta(x-y)$ in its weak disorder regime $E_F > 0, l^2 E_F >> 1$ (see also [1, 20]). In [6] the strong disorder regime ($E_F < 0, l^2|E_F| >> 1$) of the same potential was considered by using the Grassman integration technique. The results of this paper are the closest to those of this paper. In [7] the multidimensional case was considered in the framework of the instanton approach. Authors discuss the two regimes, that they call the weak localization regime and the hydrodynamic regime and obtain (25) in the latter regime but not in the first.

3.2 Effective Potential

As we saw in the previous sections the spectrum near band edges is due to rather special realizations of the random potential having the form of broad or deep (for a potential bounded or unbounded below, respectively) potential wells that are so rare and whose form is so random that the quantum mechanical probability for the electron to tunnel away vanishes for microscopic bodies. This is why it is enough to deal with only one localizing well in order to compute the leading terms of asymptotic behavior of the DOS.

For a below unbounded potential (like Poisson random potential of the previous section) the large parameter of the theory is the energy. Analogous realizations exist in the case of large disorder, i.e. when the large parameter of the theory is the amplitude of the random potential. These two cases of the strong localization regime of the localization theory are two manifestations of the simplest localization mechanism: capturing an electron in strong and rare fluctuations of the random potential[1].

It is clear that the properties of the random potential that facilitate localization are its short ranged correlations and sufficiently smooth probability distribution. These properties result in suppressing tunneling phenomena and in bounding excitation in spatial regions where the strong fluctuations of the random environment are localized.

Results of extensive study of the strong localization regime make reasonable the Ansatz, according to which relevant properties and phenomena can be un-

[1] We mention in passing another mechanism of localization: enhanced backscattering. It is widely believed that this mechanism is responsible for localization in the middle of the band for phonons and for high energies for electrons in one dimension for arbitrary disorder, and in the middle of the acoustic band for phonons in three dimensions for large disorder. This mechanism is much less studied.

derstood and computed assuming that in this regime any short correlated and smoothly distributed random potential can be replaced by an (effective) random potential of the form

$$V_{eff} = \sum_j V_j(x - x_j) \tag{26}$$

where $\{x_j\}$ are the Poisson random points of the density μ in $\mathbf{R}^d$ modeling the centers of localizing wells and $\{V_j(x)\}$ are random functions independent mutually and of $\{x_j\}$, modeling the random form of the localizing wells. Thus they should be non-positive. We assume for simplicity that they are even and have a compact support. The density μ of localization centers is not known a priori and should be found self-consistently. It may depend on an energy interval under consideration. The same is true for $\{V_j(x)\}$. It is reasonable to expect that final results do not depend strongly on a concrete form of these functions. Thus one can try the simplest possible form like

$$V_j(x) = g_j V(\sqrt{g_j}x) \tag{27}$$

where $\{g_j\}$ are independent mutually and of $\{x_j\}$, and identically distributed random variables whose probability distribution has smooth enough density $p(g)$ with a non-negative unbounded support to model deep and narrow wells of the large disorder regime (like we had in the previous section). The function V is non-positive, even, and has a compact support. Thus mathematically our effective random potential is a randomized (by g_j's) Poisson random field.

By using the terminology dating back to I. Lifshitz we can call our effective potential optimal realizations, i.e. realizations that produce leading contributions in relevant quantities in the strong localization regime.

3.3 Concentration Expansion

We describe now our main technical tool. Let $\{F_l(x_1, ..., x_l)\}_{l=1}^{\infty}$ be a system of functions symmetric with respect permutations of their arguments that are points of $\mathbf{R}^d$. We denote for brevity the set $(x_1, ..., x_l)$ of arguments of F_l by X. We suppose that these functions satisfy two following properties:

(i) translational invariance: $F_l(X) = F_l(X + a), \quad \forall a \in \mathbf{R}^d, \quad \forall l$ where $X + a = (x_1 + a, ..., x_l + a)$;

(ii) additive clustering: $F_l(X \cup Y + a) \to F_l(X) + F_l(Y), \quad a \to \infty$.

In the latter case we assume that the convergence is fast enough to guarantee convergence of all the integrals below.

For any system of such functions we can write the combinatorial identity

$$F_N(X) = \sum_{Y \subset X} \sum_{Z \subset X} (-1)^{\sharp(Y \setminus Z)} F(Z) \tag{28}$$

that can be proved by using for example the exclusion-inclusion principle. Assume now that all x_j, $j = 1, 2, ..., N$ are in the cube Λ of the side length L centered at the origin. Then, integrating with respect to all $x_j \in \Lambda$ and dividing by $|\Lambda|^N$ we obtain

$$|\Lambda|^{-N} \int_{\Lambda^N} F_N(x_1, ..., x_N) dx_1 ... dx_N = \frac{N}{|\Lambda|} \int_\Lambda F_1(x_1) dx_1 + \frac{N(N-1)}{2|\Lambda|^2} \int_{\Lambda^2} (F_2(x_1, x_2) - F_1(x_1) - F_1(x_2)) dx_1 dx_2 + ... \tag{29}$$

We can consider a bit more general situation when the role of x_j's play the pairs (x_j, g_j)'s where $\{g_j\}$ are the i.i.d random variable independent also of $\{x_j\}$ and having a probability density $p(g)$. Then we obtain an analogous identity containing now the appropriate number of integrations with respect to g's, each with the weight $p(g)$.

Recall now that the Poisson random point field x_j of the density μ can be obtained as the weak limit of the point field of N random point uniformly and independently distributed inside the cube Λ as $N \to \infty, L \to \infty, \frac{N}{|\Lambda|} \to \mu$. By using this fact and the translational invariance of F_l's in x's we obtain from the last identity

$$\lim_{\frac{N}{|\Lambda|} \to \mu} |\Lambda|^{-1} \mathbf{E}(F_N) = \int_{\mathbf{R}} F_1(g_1) \mu(g_1) dg_1 + \frac{1}{2} \int_{\mathbf{R}^d} dx \int_{\mathbf{R}^2} (F_2(x, g_1, g_2) - F_1(g_1) - F_1(g_2)) \mu(g_1) \mu(g_2) dg_1 dg_2 + ... \tag{30}$$

where we assume that the (thermodynamic) limit in the l.h.s. of (30) exists and we use the notations

$$\mu(g) = \mu p(g). \tag{31}$$

Formula (30) provides the formal concentration expansion of random functions defined on realizations of the randomized Poisson field (26). For its various theoretical physical applications see ([20]). Its mathematical aspects are considered in [10] and in [27].

3.4 Concentration Expansions of the DOS

Now we are going to apply the concentration expansion to the density of states (DOS) $\rho(E)$. We use the well known fact employed already above that the DOS can be regarded as a formal density of the measure (the Integrated Density of States) that is the weak limit of the normalizing counting measures of the Schrödinger operator H_Λ defined in the cube Λ by the Schrödinger equation with the same potential and by some selfadjoint conditions on the cube boundary, i.e.

$$\rho(E) = \lim_{\Lambda \to \infty} \mathbf{E}(\frac{1}{|\Lambda|} \sum_i \delta(E - E_i)) \tag{32}$$

where E_i are the eigenvalues of H_Λ. Comparing (30) and (32) we see that if we want to apply the formula (30) to the DOS we have to use as functions F_l the expressions

$$\sum_i \delta(E - E_i^{(l)}(x_1, ..., x_l; g_1, ..., g_l)) \tag{33}$$

where $E_i^{(l)}(x_1, ..., x_l; g_1, ..., g_l)$ are the eigenvalues of the l-well Hamiltonian

$$H^{(l)}(x_1, ..., x_l; g_1, ..., g_l) = -\Delta + \sum_{j=1}^{l} g_j V(\sqrt{g_j}(x - x_j)). \tag{34}$$

Then applying (30) to $\rho(E)$ and taking into account that since we are interested in negative energies large in modulus, and since $H^{(0)} = -\Delta$ has no negative spectrum, we find that the term with $l = 0$ will be absent in the expansions of the DOS and that the one-well contribution $\rho^{(1)}(E)$ (the term corresponding to $l = 1$) is

$$\rho^{(1)}(E) = \sum_i \int_{\mathbf{R}} \delta(E - E_i^{(1)}(g))\mu(g)dg. \tag{35}$$

For a well of form (27)

$$E_i^{(1)}(g) = g\varepsilon_i \tag{36}$$

where ε_i are the negative eigenvalues of the one-well Hamiltonian $H^{(1)} = -\Delta + V(.)$. Thus

$$\rho^{(1)}(E) = \sum_i \mu(-\frac{E}{\varepsilon_i})\frac{1}{|\varepsilon_i|}. \tag{37}$$

According to the spirit of our approach the density $p(g)$ should decay very fast (at least exponentially in the case of the non-positive Poisson potential of the previous Section, see (3)). Thus the leading contribution to $\rho^{(1)}(E)$ is due to the first term of the sum in r.h.s of (37), i.e.

$$\rho^{(1)}(E) = \mu(-\frac{E}{\varepsilon_i})\frac{1}{|\varepsilon_i|}. \tag{38}$$

Normalizing V by the condition $\varepsilon_1 = -1$ we can write

$$\rho^{(1)}(E) = \mu(E). \tag{39}$$

The last formula can be regarded as a version of the classical asymptotic formula (8). By choosing as a randomizing parameter of the localizing wells $V_j(x)$ their ground state energies and assuming that they are i.i.d. random variables with the common density $p_1(E)$, we can show that an analogue of (39) allow us to obtain quantum versions of asymptotic formulas for the DOS.

It can be also shown that the two-well contribution to the DOS has the order $O((\mu(E))^2)$. We postpone the respective arguments to the next subsection where it will be more natural to present them. Thus the two-well contribution is also negligible with respect to the r.h.s of (39).

We conclude that the unknown (and small) function $\mu(g)$ determining our effective potential and having the sense of the probability density to find a well of the amplitude between g and $g+dg$ with the center in an infinitesimal neighborhood of a given x, coincides in our approximation with the DOS of the Schrödinger operator. This important conclusion first makes our scheme self-consistent and second correspond to the basic ingredient of the Lifshitz's approach, according to which the DOS is the probability density of the localizing wells.

3.5 A. C. Conductivity.

According to the Kubo formula the conductivity $\sigma(\nu, E_F)$ at zero temperature is in proper units

$$\sigma(\nu, E_F) = \nu^2 \lim_{\Lambda\to\infty} \mathbf{E}(\frac{1}{|\Lambda|}\sum_{i\neq j}|x_{ij}|^2\delta(E_F+\nu-E_i)\delta(E_F-E_j)) \tag{40}$$

where

$$x_{ij} = \int_\Lambda x\psi_i\psi_j dx \tag{41}$$

and ψ_i and E_j are eigenvalues and eigenfunction of the Hamiltonian H_Λ. Comparing (30) and (40) we see that the role of functions F_l is in this case played by the expression

$$\nu^2 \sum_{i\neq j}|x_{ij}^{(l)}|\delta(E_F+\nu-E_i^{(l)})\delta(E_F-E_j^{(l)}) \tag{42}$$

where $\psi_i^{(l)}$ and $E_j^{(l)}$ are eigenvalues and eigenfunction of the l-well Hamiltonian $H^{(l)}$.

For the same reasons as for the DOS the zero well contribution in the conductivity is absent. Let us show that the one-well contribution $\sigma^{(1)}$ is also absent. We use the same form (27) of the single-well potential and the form (36) of its spectrum. We obtain

$$\sigma^{(1)} = \nu^2 \int_{\mathbf{R}} \sum_{i\neq j}|x_{ij}^{(1)}|^2\delta(E_F+\nu-g_i\varepsilon_i)\delta(E_F-g_j\varepsilon_j)\mu(g)dg. \tag{43}$$

Nonzero contribution to this expression comes from the pairs (i,j) satisfying the condition $g(\varepsilon_i-\varepsilon_j)=\nu$. Since ν is small and ε_i's do not depend of g, this expression is zero as $g\to\infty$.

The two-well contribution in this case is

$$\sigma^{(2)} = \nu^2 \int_{\mathbf{R}^d} dx \int_{\mathbf{R}^2} \sum_{i\neq j}|x_{ij}^{(2)}(x,g_1,g_2)|^2 \times \tag{44}$$
$$\delta(E_F+\nu-E_i(x;g_1,g_2))\delta(E_F-E_j(x;g_1,g_2))\mu(g_1)\mu(g_2)dg_1dg_2.$$

We assume again that the single-well potential has the form (26)-(27) where the well $V(.)$ is spherically symmetric and produces only one negative eigenvalue which we again set to -1. Then according to the basic qualitative picture of the quantum mechanical tunneling (see e.g. [17]) the spectrum of the two-well Hamiltonian will consist of two eigenvalues $E^{(2)}_{1,2}(x;g_1,g_2)$ that up to the exponentially small corrections with respect to the distance between the well are

$$E^{(2)}_{2,1} = g \pm \sqrt{\delta^2 + I^2(x)} \tag{45}$$

where

$$g = \frac{g_1+g_2}{2}, \quad \delta = \frac{g_1-g_2}{2} \tag{46}$$

and

$$I(x) = \int_{\mathbf{R}^d} \psi^{(1)}(y)\psi^{(1)}(y+x)dy = \tag{47}$$

$$(g_1g_2)^{d/4}\int_{\mathbf{R}^d} \phi(g_1^{1/2}y)\phi(g_2^{1/2}(y+x))dy$$

is known as the overlap (resonant) integral and $\phi(x)$ is the eigenfunction corresponding to the eigenvalue $\varepsilon_1 = -1$. Under our assumptions $\phi(x)$ and $I(x)$ depend only on $|x| = r$, and since $\phi(x)$ is exponentially decaying with the rate 1, the overlap integral behaves as

$$I(r) \simeq I_0 e^{-2gr}, \quad r \to \infty \tag{48}$$

In addition the matrix element of the coordinate $x^{(2)}_{12}(x;g_1,g_2)$ in this case is

$$x^{(2)}_{12}(x;g_1,g_2) = \frac{|x|I(x)}{\sqrt{\delta^2+I^2(x)}} \tag{49}$$

We see that for small ν the nonzero contribution to $\sigma^{(2)}$ is due to large x and g_1 and g_2 close to each other. Thus, passing to the variables g and δ defined in (46) and taking into account that $\mu(g)$ is smooth and coincides in our approximation with the DOS ρ we obtain that

$$\sigma^{(2)}(\nu, E_F) = \frac{\nu\rho^2(E_F)}{2}\int_{r(\nu)}^{\infty} \frac{I(r)r^{d+1}dr}{\sqrt{\nu^2 - 4I^2(r)}}. \tag{50}$$

where $r(\nu)$ is the solution of the equation $I(r) = \nu/2$ which is called the tunneling distance [21]. In view of (48) we have that asymptotically for small ν (and large r)

$$r(\nu) \sim \frac{1}{2g}\log\frac{I_0}{\nu}$$

Thus it is easy to see that for small ν the last expression has asymptotic form (25).

3.6 Corrections

We comment now on the corrections to (39) and (50). We will not give a proof of the convergence of the respective expansions. We simply argue that they should be asymptotic, i.e. that their terms should be small in successive powers of $\rho(E)$.

It is easy to see that the next term in the expansion of the DOS have the form

$$\rho^{(2)} = \int_{\mathbf{R}^d} dx \int_{\mathbf{R}^2} \{[\delta(E - \varepsilon_2^{(2)} - \Delta^{(2)}(\delta, I(x))) - \delta(E - \varepsilon_2^{(2)})]+ \\ [\delta(E - \varepsilon_1^{(2)} + \Delta^{(2)}(\delta, I(x))) - \delta(E - \varepsilon_2^{(2)})]\}\mu(g_1)\mu(g_2)dg_1dg_2, \tag{51}$$

where $\varepsilon_1^{(2)} = -\max(g_1, g_2)$, $\varepsilon_2^{(2)} = -\min(g_1, g_2)$ and $\Delta^{(2)} = \sqrt{\delta^2 + I^2} - \delta$. Recall that we assume $\mu(g)$ sufficiently fast decaying for large g. Thus $\rho^{(2)}$ will be of the order $O(\mu)^2$ if the integral in the relative distance x between the wells will be convergent. This follows from the inequality $0 \leq \Delta^{(2)} \leq I$, the exponential decay of $I(x)$ (see (48)), and the smoothness of $\mu(g)$, allowing us to transfer derivatives of delta functions on μ's.

In the general case of the correction of order l the respective integrals in relative distances between wells will be convergent because of the subtractions of expressions (33) of lower orders $k < l$ from that of the order l, the sufficiently fast splitting (additive clustering) of negative eigenvalues $E^{(l)}$ of the l-well problem into the sums of negative eigenvalues $E^{(k)}$ of the k-well problems $k < l$ and of the smoothness of $\mu(g)$.

The situation is less simple in the case of conductivity as we have seen already in the case $l = 2$. This is because of the presence of families of tunneling configurations for any number of wells (for example, for $l = 3$ there are two families: the isosceles triangles of an arbitrary size and three equidistant points on a straight line also with of an arbitrary distance between them). These configurations are responsible for the absence of the decay (and even for the polynomial growth) in distances between wells of the coordinate matrix elements x_{ij}^l on respective resonant submanifolds and for the resulting powers of $\log 1/\nu$. However, since the dimension of these resonant manifolds grows slower in l than l, these powers of $\log 1/\nu$ will be always multiplied by growing powers of ν, giving the dimension of the manifold transversal to the resonant ones. This is why the higher terms in the expansion of the low frequency conductivity should be small relative to the Mott's formula (25).

Acknowledgement. I am grateful to W.Kirsch, F.Klopp, and F.Perez with whom the results presented above were obtained.

References

[1] A. Abrikosov and I. Ryzhkin. Conductivity of quasi-one-dimensional metal systems. *Advances in Physics*, 27:147-251, 1975.

[2] V. Berezinskii. Kinetics of a quantum particle in a one-dimensional random potential. *Soviet Physics JETP*, 38,620-627,1974.

[3] M. Donsker and S.R.S. Varadhan. Asymptotics for the Wiener sausage. *Communications on Pure and Applied Mathematics*, 28:525–565, 1975.

[4] A. Efros and B. Shklovski. *Electronic properties of doped semi-conductors.* Springer Verlag, Heidelberg, 1984.

[5] L. Gorkov. Disorder and interaction in the system of quasi one-dimensional electrons. In A.L. Efros and M. Polak, editors, *Electron-Electron Interaction in Disordered Systems*, New-York, 1985. Elsevier.

[6] R. Hayne and W. John. Instanton approach to the conductivity of a disordered solid. *Nuclear Physics*, B348:766-786, 1991.

[7] A. Haughton, L. Schaffer and F. Wegner. Two-particle spectral function and ac conductivity of an amorphous system far below the mobility edge: a problem of interacting instantons. *Phys.Rev.*, B22:3599-3608, 1980.

[8] W. Kirsch. Random Schrödinger operators. In A. Jensen, H. Holden, editors, *Schrödinger Operators*, number 345 in Lecture Notes in Physics, Berlin, 1989. Springer Verlag. Proceedings, Sonderborg, Denmark 1988.

[9] W. Kirsch, L. Pastur and F. Perez. To be submitted.

[10] F. Klopp. An asymptotic expansion for the density of states of a random Schrödinger operator with Bernoulli disorder. *Random Operators and Stochastic Equations*, 3(4):315–332, 1995.

[11] F. Klopp. A low concentration asymptotic expansion for the density of states of a random Schrödinger operator with Poisson disorder. *Journal of Functional Analysis*, 145:267–295, 1995.

[12] F. Klopp. Precise high energy asymptotics for the integrated density of states of an unbounded random Jacobi matrix. Preprint, Universite Paris-Nord, Villetaneuse, 1997.

[13] F. Klopp. Band edge behaviour for the integrated density of states of random Jacobi matrices in dimension 1. *Journal of Statistical Physics*, 90(3-4):927–947, 1998.

[14] F. Klopp. Internal Lifshitz tails for random perturbations of periodic Schrödinger operators. *Duke Mathematical Journal*, 1998. to appear.

[15] F. Klopp and L. Pastur. In progress.

[16] F. Klopp and L. Pastur. Lifshitz tails for random Schrödinger operator with negative singular Poisson potential. To be submitted.

[17] L. D. Landau and E. M. Lifshitz. *Mécanique quantique, théorie non-relativiste*. Editions MIR, Moscou, 1966.

[18] I. M. Lifshitz. Structure of the energy spectrum of impurity bands in disordered solid solutions. *Soviet Physics JETP*, 17:1159–1170, 1963.

[19] I. M. Lifshitz. Energy spectrum structure and quantum states of disordered condensed systems. *Soviet Physics Uspekhi*, 7:549–573, 1965.

[20] I.M. Lifshitz, S.A. Gredeskul, and L.A. Pastur. *Introduction to the Theory of Disordered Systems*. Wiley, New-York, 1988.

[21] N.F. Mott and E.A. Davis. *Electronic properties of in Non-Crystalline Materials*. Clarendon, Oxford, 1971.

[22] L. Pastur. Behaviour of some Wiener integrals as $t \to +\infty$ and the density of states of the Schrodinger equation with a random potential. *Teor.-Mat.-Fiz*, 32:88–95, 1977. (in russian).

[23] L. Pastur and A. Figotin. *Spectra of Random and Almost-Periodic Operators*. Springer Verlag, Berlin, 1992.

[24] M. Reed and B. Simon. *Methods of Modern Mathematical Physics, Vol IV: Analysis of Operators*. Academic Press, New-York, 1978.

[25] B. Simon. Lifshitz tails for the Anderson model. *Journal of Statistical Physics*, 38:65–76, 1985.

[26] A. Sznitman. Lifshitz tails and Wiener sausages. I. *Journal of Functional Analysis*, 94:223–246, 1990.

[27] W. von Waldensfels. Taylor expansion of a Poisson mesure. *Lecture Notes in Mathematics*, 381:303–375, 1974.

Operator Theory:
Advances and Applications, Vol. 108

SPECTRAL MEASURES AND CATEGORY

Rafael del Rio
IIMAS-UNAM
Circuito Escolar, Ciudad Universitaria
C.P. 04510, México, D.F.
México
delrio@servidor.unam.mx

Alexei Poltoratski
Department of Mathematics
College Station, Texas 77843–3368
U.S.A
alexei.poltoratski@math.tamu.edu

Abstract

We analyze spectral properties of self-adjoint operators under rank one perturbations. We show that for any self-adjoint operator with dense spectrum and any standard family of its rank one perturbations there exists a first category set supporting the spectral measures of all the operators in the family. An analogous result is true for the spectra of a family of Sturm–Liouville operators corresponding to various boundary conditions.

Introduction

This note analyzes spectral behavior of a selfadjoint operator under a rank one perturbation and shows how the same methods can be applied to study the perturbations of spectra of ordinary differential expressions when the boundary condition at one end changes.

Consider a selfadjoint operator A in a separable Hilbert space, a vector φ and the corresponding family of selfadjoint rank one perturbations of A:

$$A + \lambda(\varphi, \cdot)\varphi, \ \lambda \in \mathbf{R}.$$

In many applications of perturbation theory, operator A satisfies an additional condition that its spectrum contains an interval $I \subset \mathbf{R}$. It is well known that then there exists a first category set (see [10] for precise definitions), which contains all the embedded eigenvalues on I of all the operators A_λ, $\lambda \in \mathbf{R}$, see [2], [7]. This fact was used in particular, to prove existence of singular continuous spectrum for large sets of boundary conditions for ordinary differential expressions and to answer some questions about the perturbational behavior of the essential spectrum.

In regard of this result, it is natural to ask whether the other parts of the spectra have similar properties. In this paper we give a complete answer to this question. Namely, we prove that there exists a first category set which supports all spectral measures of all the operators A_λ, $\lambda \in \mathbf{R}$ on I. Our result implies the before mentioned property of embedded eigenvalues as well as some other known results on the perturbations of spectra. We shall give a general argument which includes the cases of rank one perturbations and Sturm–Liouville operators. The unified approach developed in [9], [5] which needs the negative part of the potential to be infinitesimally form bounded with respect to $-\frac{d^2}{dx^2}$, will not be used. We will need only elementary calculations.

Our results could be considered as a contribution to the understanding of the rich interplay between the concepts of measure and category in spectral theory. We proceed as follows. In Section 1 we introduce the families of measures that we are going to study. These diagonalize the corresponding operators and appear in the Poisson integral representation of harmonic functions related to the resolvent operator. In Section 2 we prove our main results about the existence of a common support of first category. In Section 3 we give explicit example where such support can be constructed. This support will be the set of points where there are subordinate solutions, a concept generalizing the idea of eigenfunctions.

Section 1

Let ϕ be a nonconstant analytic function in the upper half plane $\mathbf{C}_+$ such that $|\phi| \leq 1$. Then for any α with $|\alpha| = 1$ the function

$$i\frac{\alpha+\phi}{\alpha-\phi}$$

has positive imaginary part in $\mathbf{C}_+$.

Thus, according to a well known representation theorem, see for example [13] page 83, there exists a measure μ_α such that

$$Im\left(i\frac{\alpha+\phi(z)}{\alpha-\phi(z)}\right) = c\,y + \frac{1}{\pi}\int_{\mathbf{R}} \frac{y}{(t-x)^2+y^2}d\mu_\alpha(t) \tag{1}$$

and

$$\int_{\mathbf{R}} \frac{d\mu_\alpha(t)}{1+t^2} < \infty \quad , \quad c \geq 0 \quad z = x+iy.$$

In what follows we shall study two cases where families of measures of the above type appear as spectral measures of self–adjoint operators.

Case a) Let $\mathcal{H}$ be a Hilbert space, $\mathcal{H}_0$ a self–adjoint operator on $\mathcal{H}$ which may or may not be bounded and P a self–adjoint projection on a fixed normalized element k. For all $x \in \mathcal{H}$ we have $Px = (x,k)k$. Consider the family of operators $H_\lambda = H_0 + \lambda P$ where λ is real.

Let us introduce the resolvent operator

$$R_z^\lambda = (H_\lambda - zI)^{-1}$$

and the functions

$$F_\lambda(z) = (R_z^\lambda k, k) = \int_{\mathbf{R}} \frac{d\nu_\lambda(t)}{t-z}$$

which are analytic in the upper and lower half–planes and have a positive imaginary part in the upper half–plane. The measure ν_λ is the spectral measure corresponding to k given by the spectral theorem.

The following relation holds. See [4], [15] (1.13)

$$F_\lambda(z) = \frac{F_0(z)}{1+\lambda F_0(z)}.$$

We can write

$$F_0(z) = i\frac{1+\phi(z)}{1-\phi(z)}$$

where

$$\phi(z) = \frac{F_0(z)-i}{F_0(z)+i}.$$

The function ϕ is analytic and satisfies $|\phi(z)| \leq 1$.

A simple calculation gives the following equality

$$(1+\lambda^2)\frac{i\frac{1+\phi}{1-\phi}}{1+i\lambda\frac{1+\phi}{1-\phi}} - \lambda \quad = \quad i\frac{\frac{1+i\lambda}{1-i\lambda}+\phi}{\frac{1+i\lambda}{1-i\lambda}-\phi}. \tag{2}$$

Notice that $\left|\frac{1+i\lambda}{1-i\lambda}\right| = 1$.

The relation between the spectral measures ν_λ and the measures μ_α which appeared in the integral representation (1) is then given by

$$(1+\lambda^2)d\nu_\lambda = d\mu_\alpha$$

where

$$\alpha = \frac{1+i\lambda}{1-i\lambda}.$$

Case b) Let us consider now operators generated by the differential expression

$$lu = -u'' + q(x)u \qquad 0 \leq x < \infty$$

(where q is real valued, locally integrable function defined in $[0,\infty)$) and the boundary conditions at zero

$$u(0)\cos\theta + u'(0)\sin\theta = 0 \qquad 0 \le \theta < \pi.$$

We assume that the limit point case occurs at infinity.

For each θ there is a function $m_\theta(z)$ which is analytic and with positive imaginary part in the upper half plane, called the Weyl–m function.

The imaginary part of $m_\theta(z)$ has the following integral representation

$$Im\ m_\theta(z) = \int_{\mathbf{R}} \frac{\mu}{(t-\xi)^2+\mu^2} d\rho_\theta(t) \qquad z = \xi + i\mu.$$

The measures ρ_θ are the Weyl spectral measures of the boundary value problem. These measures diagonalize the corresponding operators and contain all the information about the spectrum. The following relation is satisfied:

$$m_\theta(z) = \frac{-\sin(\theta-\beta) + m_\beta(z)\cos(\theta-\beta)}{\cos(\theta-\beta) + m_\beta(z)\sin(\theta-\beta)}. \tag{3}$$

See [1], [15].

After some elementary calculations this equality can be rewritten, for $\beta = 0$, as

$$i\frac{e^{2i\theta}+\psi(z)}{e^{2i\theta}-\psi(z)} = m_\theta(z)$$

where $\psi(z) = \frac{m_0(z)-i}{m_0(z)+i}$. Observe that $|\psi(z)| \le 1$.

Therefore $\rho_\theta = \mu_\alpha$ where $\alpha = e^{2i\theta}$ and μ_α are the measures in the expression (1).

Section 2

Using equality (2) we obtain

$$Im\left(i\frac{\alpha+\phi}{\alpha-\phi}\right) \le \frac{1+\lambda^2}{\lambda^2}\cdot\frac{1}{\varepsilon}$$

if

$$Im\left(i\frac{1+\phi}{1-\phi}\right) > \varepsilon \quad , \quad \lambda \ne 0$$

where

$$\alpha = \frac{1+i\lambda}{1-i\lambda}$$

and from (1) we get then,

$$cy + \frac{1}{\pi}\int_{-\infty}^{\infty} \frac{y}{(t-x)^2+y^2} d\mu_\alpha(t) \le \frac{1+\lambda^2}{\lambda^2}\cdot\frac{1}{\varepsilon}. \tag{4}$$

If we define for $z \in \mathbf{C}, z = x + iy$ the set

$$V_z = \left\{ p \in \mathbf{R} : |x - p| < \frac{1}{2} y \right\}$$

we can obtain the inequalities

$$\begin{aligned} cy \quad + \quad & \frac{1}{\pi} \int_{-\infty}^{\infty} \frac{y}{(t-x)^2 + y^2} d\mu_\alpha(t) \geq \\ & \frac{1}{\pi} \int_{-\infty}^{\infty} \frac{y}{(t-x)^2 + y^2} d\mu_\alpha(t) \geq \\ & \frac{1}{\pi} \int_{V_z} \frac{|V_z|}{(t-x)^2 + |V_z|^2} d\mu_\alpha \geq \\ & \frac{4}{5\pi} \frac{\mu_\alpha(V_z)}{|V_z|} \qquad z = x + iy. \end{aligned}$$

Therefore together with inequality (4) we get

$$\frac{4}{5\pi} \frac{\mu_\alpha(V_z)}{|V_z|} \leq \frac{1 + \lambda^2}{\lambda^2} \cdot \frac{1}{\varepsilon}$$

that is

$$\mu_\alpha(V_z) \leq |V_z| \left(\frac{5\pi}{4} \frac{1 + \lambda^2}{\lambda^2} \right) \cdot \frac{1}{\varepsilon} \tag{5}$$

if $Im\, i\frac{1+\phi}{1-\phi} > \varepsilon$ for $\alpha \neq 1$.

We say that S is a support of a measure μ or that μ is supported on S if

$$\mu(\mathbf{R} \backslash S) = 0.$$

A set S is a closed support if it is closed and it is a support.

Lemma 1 Assume that μ is a $\sigma -$ finite measure on $\mathbf{R}$ and that S is a closed support of μ without isolated points. Then there exists $\tilde{S} \subset S$ such that $\tilde{S}$ is of first category in S and $\mu(\mathbf{R} \backslash \tilde{S}) = 0$.

Proof Let $P = \{p \in S / \mu(p) > 0\}$ Since μ is $\sigma -$ finite , P is at most countable. We are assuming that S is perfect (closed and without isolated points), therefore the set $S \backslash P$ is dense in S. Since any subset of $\mathbf{R}$ is separable, we can choose a countable dense subset $\{a_1, a_2, \ldots\}$ of $S \backslash P$. This set is dense in S.

For each $n, m \in \mathbf{N}$ we choose an open interval I_{nm} with center a_n, such that

$$\mu(I_{nm}) < \frac{1}{2^{n+m}}.$$

This can be done because $\mu(a_n) = 0$ and, for any nested sequence of intervals I_i,

$$\mu(I_i) \to \mu(\bigcap_{j=1}^{\infty} I_j) \qquad \text{as } i \to \infty$$

holds.

Let

$$G_m = \bigcup_{n=1}^{\infty} (I_{nm} \cap S).$$

Then

$$\mu(G_m) \le \sum_{i=1}^{\infty} \mu(I_{im}) \le \frac{1}{2^m} \overset{m\to\infty}{\longrightarrow} 0\,.$$

Therefore,

$$\mu(\bigcap_{m=1}^{\infty} G_m) = 0$$

and the lemma follows with $\tilde{S} = \mathbf{R} \setminus \bigcap_{m=1}^{\infty} G_m$. □

Inequality (5) now allows to prove the following result which shows the existence of a "small" support for all measures μ_α.

Theorem 1 *Assume that $I \subset \mathbf{R}$ is an interval such that for every subinterval $J \subset I$ we have $\mu_{\alpha_0}(J) > 0$, for some $\alpha_0 \in \partial D = \{z \in \mathbf{C} : |z| = 1\}$.*

Then there exists a set $S \subset I$, S of first category such that

$$\mu_\alpha(I \setminus S) = 0$$

holds for every $\alpha \in \partial D$.

Proof Without loose of generality we can choose $\alpha_0 = 1$.

First we will show the existence of a countable set $\{\lambda_n\}_{n=1}^{\infty}$, dense in I such that

$$Im\, i \frac{1 + \phi(\lambda_n + i0)}{1 - \phi(\lambda_n + i0)} > \varepsilon_n > 0$$

where

$$\phi(\lambda + i0) := \lim_{y \to 0_+} \phi(\lambda + iy).$$

Recall that (see [11] p. 44)

$$Im\, i \frac{1 + \phi(\lambda + i0)}{1 - \phi(\lambda + i0)} = \pi \lim_{\varepsilon \to 0_+} \frac{\mu(\lambda - \varepsilon, \lambda + \varepsilon]}{2\varepsilon}$$

(where $\mu := \mu_{\alpha=1}$).

From a theorem of De la Vallée Poussin we know that

$$\mu\left(\left\{\lambda / \lim_{\varepsilon\to 0_+} \frac{\mu(\lambda-\varepsilon,\lambda+\varepsilon]}{2\varepsilon} = 0\right\}\right) = 0, \tag{6}$$

See [14] p. 127.

If $\mu(J) > 0$ then according to (6), there exists $\lambda_0 \in J$ such that

$$Im\; i\frac{1+\phi(\lambda_0+i0)}{1-\phi(\lambda_0+i0)} > \varepsilon_0 > 0.$$

Since this happens for every subinterval $J \subset I$ we can construct the set $\{\lambda_n\}_{n=1}^{\infty}$ mentioned above.

Now for $n, m \in \mathbf{N}$ define

$$z_n^m = \lambda_n + i\; y_n^m \in \mathbf{C}$$

where we choose y_n^m such that

$$y_n^m < \varepsilon_n \frac{1}{2^{n+m}}$$

and

$$Im\; i\frac{1+\phi(z_n^m)}{1-\phi(z_n^m)} > \varepsilon_n > 0.$$

Let us define

$$G_m = \bigcup_{n=1}^{\infty} (V_{z_n^m} \cap I)$$

the set G_m is open and dense in I.

We have then

$$\mu_\alpha(G_m) \le \mu_\alpha\left(\bigcup_{n=1}^{\infty} V_{z_n^m}\right) \le \sum_{n=1}^{\infty} \mu_\alpha(V_{z_n^m}).$$

Using inequality (5) we get

$$\begin{aligned}\sum_{n=1}^{\infty} \mu_\alpha(V_{z_n^m}) &\le \frac{5\pi}{4}\frac{1+\lambda^2}{\lambda^2}\sum_{n=1}^{\infty}\frac{|V_{z_n^m}|}{\varepsilon_n} \\ &< \frac{5\pi}{4}\frac{1+\lambda^2}{\lambda^2}\sum_{n=1}^{\infty}\frac{1}{2^{n+m}} = \frac{5\pi}{4}\frac{1+\lambda^2}{\lambda^2}\frac{1}{2^m} \overset{m\to\infty}{\to} 0\end{aligned}$$

therefore $\mu_\alpha(G_m) \overset{m\to\infty}{\to} 0$ and $\mu_\alpha\left(\bigcap_{m=1}^{\infty} G_m\right) = 0$ for every $\alpha \neq 1$.

Hence μ_α is supported in the set $I \setminus \bigcap_{m=1}^{\infty} G_m$ which is of first category *i.e.* countable union of nowhere dense sets (see [10]).

Since any single σ−finite measure may be supported on a first category set, according to Lemma 1, we have the desired support for $\mu_\alpha, \alpha = 1$. Taking the union of this set and the common support for all other $\alpha's$, the result follows. □

From here it follows in particular that all the eigenvalues of rank one perturbations live in a first category set.

Corollary Let A be a selfadjoint operator whose closed spectrum contains an interval $I \subset \mathbf{R}$. Let φ be cyclic vector for A and $A_\lambda = A + \lambda(\varphi, \cdot)\varphi, \lambda \in \mathbf{R}$, be the corresponding rank one perturbations of A. Then there exits a first category subset F of I such that for any operator A_λ the part of its spectral measure ν_λ in I, is supported on F.

Proof Straightforward from Theorem 1 and the relation

$$(1 + \lambda^2)d\nu_\lambda = d\mu_\alpha$$

mentioned at the end of case a) in Section 1.

□

Remark A simple example of a family of measures *not* supported on a first category set is given by $\{\mu_p\}_{p\in[0,1]}$ where

$$\mu_p(A) = \begin{cases} 1 & \text{if } p \in A \\ 0 & \text{if } p \notin A \end{cases} .$$

In some cases, a description of the first category set which supports the measures μ_θ can be given, as we will see in the next section.

Section 3

The next theorem shows that, when we have only singular spectrum in an interval I, the set of points corresponding to subordinate solutions is a first category support for all the measures μ_θ. This is not true if the spectrum has empty interior. See remark at the end of this section. We follow the notation introduced in Section 1 case b).

We recall the following (see [6]):

Definition A solution $u_s(r, z)$ of $lu = zu$ is said to be subordinate at infinity if, for every linearly independent solution $u(r, z)$,

$$\lim_{N\to\infty} \frac{||u_s(r, z)||_N}{||u(r, z)||_N} = 0$$

where $\| f(r) \|_N$ denotes $(\int_0^N |f(r)|^2 dr)^{1/2}$.

We shall need the following:

Lemma 2 Assume $S \subset \mathbf{R}$. If there exists $\varepsilon > 0$ such that

$$S_\varepsilon = \{x \in S \mid Im\ m_\beta(x+i0) > \varepsilon\}$$

is dense in S, then

$$B = \{x \in S \mid Im\ m_\beta(x+i0) = 0\}$$

is of first category in S.

Proof Let

$$A_{nk} = \left\{x \in S \mid Im\ m_\beta\left(x + i\frac{1}{k}\right) \leq \frac{1}{n}\right\}.$$

Then we have, for every $N \in \mathbf{N}$,

$$B \subset \bigcap_{n=N}^{\infty} \left[\bigcup_{j=1}^{\infty} \left(\bigcap_{k \geq j} A_{nk}\right)\right]. \tag{7}$$

We shall prove that the set on the right is of first category in S.
Let $N \in \mathbf{N}$ be such that $N > \frac{1}{\varepsilon}$.
Then, for any $J \in \mathbf{N}$,

$$S_\varepsilon \subset \bigcup_{k \geq J} A_{nk}^c \quad \text{if } n > N$$

where we take the complement of A_{nk} with respect to S, i.e.,

$$A_{nk}^c = \left\{x \in S \mid Im\ m_\beta\left(x + i\frac{1}{k}\right) > \frac{1}{n}\right\}.$$

Since S_ε is dense in S, we know that $\bigcap_{n \geq J} A_{nk}$ has empty interior with respect to S. Since, moreover, $\bigcap_{k \geq J} A_{nk}$ is closed with respect to S, then it is nowhere dense with respect to S and therefore the set on the right of (7) is of first category with respect to S and so is B. □

Let $\sigma(L_\beta)$ and $\sigma_{ac}(L_\beta)$ denote the spectrum and the absolutely continuous spectrum of L_β respectively. See [8] for precise definitions. Recall that μ_β is purely singular if and only if σ_{ac} is empty.

Theorem 2 Let $I \subset \mathbf{R}$ be an interval. Assume $I \subset \sigma(L_\beta)$ and $\sigma_{ac}(L_\beta) \cap I = \emptyset$.
Let

$$S = \{x \in I \mid \text{ a subordinate solution of } lu = xu \text{ exists }\}.$$

Then S is of first category with respect to I and

$$\mu_\theta(I \backslash S) = 0 \qquad \forall \theta.$$

Proof Let us consider the sets

$$B = \{x \in I \mid Im\ m_\beta(x + i0) = 0\}$$

$$C = \{x \in I \mid |m_\beta(x + i0)| = \infty\}.$$

According to Theorem 1 Section 6 of [6] we have that $S = B \cup C$. From Theorem 3(iii) of [6] it follows that

$$\mu_\theta(I \backslash S) = 0 \qquad \forall \theta.$$

Therefore, it is enough to prove that $B \cup C$ is of first category with respect to I.

To prove that B is of first category, recall that the set

$$\{x \in I \mid Im\ m_\beta(x + i0) = \infty\}$$

is dense in I, since it is a support of the singular part of the spectral measure (see Proposition 1(iii) of [6]).

Applying now Lemma 2 it follows that B is of first category in I. From formula (3) we know that

$$C \subset \{x \in I \mid Im\ m_\theta(x + i0) = 0\}$$

for $\theta \neq \beta$.

Since $\sigma_{ac}(L_\beta) \cap I = \emptyset$, we know that L_θ has only singular spectrum in I. It then follows that

$$\{x \in I \mid Im\ m_\theta(x + i0) = \infty\}$$

is dense in I.

Using Lemma 2 again, we obtain that

$$\{x \in I \mid Im\ m_\theta(x + i0) = 0\}$$

is of first category in I and, therefore, the set C is of first category with respect to I. □

Open Problem: Do we really need assumption $\sigma_{ac}(L_\beta) \cap I = \emptyset$ in Theorem 2?

It is hard to believe that existence of absolutely continuous spectrum can turn the set S into second category.

Remark If $\sigma(L_\beta)$ has empty interior, then it is possible to have for every point of an interval I a subordinate solution, even if $\sigma_{ess}(L_\beta) \cap I$ is "big", (where $\sigma_{ess}(L_\beta)$ denotes the set of accumulation points of $\sigma(L_\beta)$). See [3], [12].

Acknowledgment: Very useful conversations with S. Jitomirskaya, C. Remling and B. Simon are gratefully acknowledged.

References

[1] Aronszajn, N., *On a problem of Weyl in the theory of Sturm–Liouville equations.* Amer. J. Math. **79** (1957), 597–610.

[2] del Rio, R., Makarov, N., Simon, B., *Operators with Singular Spectrum: II. Rank One Operators.* Commun. Math. Phys. **165** (1994), 59–67.

[3] del Rio, R., *On a Problem of P. Hartman and A. Wintner.* Aportaciones Matemáticas SMM. Serie Comunicaciones **16** (1995), 119–123.

[4] Donoghue, W., *One the perturbation of spectra.* Commun. on Pure and Applied Mathematics., **18**, 559–579 (1965).

[5] Gesztesy, F., Simon, B., *Rank one perturbations at infinite coupling.* J. Funct. Anal. **128**, 245–252 (1995).

[6] Gilbert, D. J., Pearson, D., *On Subordinacy and Analysis of the Spectrum of One–Dimensional Schrödinger Operators.* J. Math. Anal. Appl. **128** (1987), 30–56.

[7] Gordon, A., *Pure point spectrum under 1–parameter perturbations and instability of Anderson localization.* Comm. Math. Phys. **164** (1994), 489–505.

[8] Kato, T., *Perturbation Theory for Linear Operators* Springer. Berlin, New York 1976.

[9] Kiselev, A., Simon, B., *Rank one perturbations with infinitesimal coupling.* J. Funct. Anal. **130**, (1995) 345–356.

[10] Oxtoby, J.C., *Measure and Category.* Springer. New York, Heidelberg 1980.

[11] Pearson, D.B., *Quantum scattering and spectral theory.* Academic Press. London, San Diego 1988.

[12] Remling, C., *Essential Spectrum and L_2–solutions of one–dimensional Schrödinger Operators.* Proceedings of the AMS **124** (1996), 2097–2100.

[13] Rosenblum, M., Rovnyak, J., *Topics in hardy classes and univalent functions.* Birkhäuser. Basel, Boston 1994.

[14] Saks, S., *Theory of the integral* Dover Publications, Inc. New York 1964.

[15] Simon, B., *Spectral analysis of rank one perturbations and applications.* Centre de Recherches Mathématiques CRM Proceedings and Lecture Notes. **8**, (1995), 109–149.

Operator Theory:
Advances and Applications, Vol. 108

Quantum dots

A survey of rigorous results

Jakob Yngvason
Institut für Theoretische Physik, Universität Wien
Boltzmanngasse 5, A 1090 Vienna, Austria

1 Introduction

Modern semiconductor technology has in recent years made it possible to fabricate ultrasmall structures that confine electrons on scales comparable to their de Broglie wavelength. If the confinement is only in one spatial direction such systems are called *quantum wells.* In *quantum wires* the electrons can move freely in one dimension but are restricted in the other two. Structures that restrict the motion of the electrons in all directions are called *quantum dots.* The number of electrons, N, in a quantum dot can range from zero to several thousand. The confinement length scales R_1, R_2, R_3 can be different in the three spatial dimensions, but typically $R_3 \ll R_1 \approx R_2 \approx 100$ nm. In models of such dots R_3 is often taken to be strictly zero and the confinement in the other two dimensions is described by a potential V with $V(x) \to \infty$ for $|x| \to \infty$, $x = (x^1, x^2) \in \mathbf{R}^2$. A parabolic potential, $V = \frac{1}{2}\omega|x|^2$, is often used as a realistic and at the same time computationally convenient approximation.

Quantum dots have potential applications in microelectronics and have been extensively studied both experimentally and theoretically. Apart from possible practical uses they are of great interest for basic quantum physics. Their parameters (strength and shape of the confining potential, magnetic field strength, number of electrons) can be varied in a controlled way and their properties can be studied by clever experimental techniques. This offers many possibilities to confront theoretical predictions with experimental findings. There exist by now many excellent reviews on the physics of quantum dots, e.g. [1]–[8]. In the present contribution the focus will be on some theoretical aspects that are only partly covered by these reviews, in particular on rigorous limit theorems [9], [10] which apply to dots in high magnetic fields and/or with high electron density.

A quantum dot with N electrons is usually modeled by a Hamiltonian of the following form, acting on the Hilbert space

$$\mathcal{H}_N = \bigwedge_1^N L^2(\mathbf{R}^2; \mathbf{C}^2) \qquad (1)$$

appropriate for two dimensional Fermions of spin 1/2:

$$H_N = \sum_{j=1}^{N} H_1^{(j)} + \sum_{1\leq i<j\leq N} W(x_i - x_j), \tag{2}$$

where $x_i \in \mathbf{R}^2$, $i = 1,\dots,N$ and $H_1^{(j)} = 1 \otimes \cdots \otimes H_1 \otimes \cdots \otimes 1$ (H_1 in the j-th place) with the one-body hamiltonian

$$H_1 = \frac{\hbar^2}{2m_*}\left(\mathrm{i}\nabla - \frac{e}{\hbar c}A(x)\right)^2 + V(x) + g_*\left(\frac{\hbar e}{2m_e c}\right)S_3 B - CB. \tag{3}$$

Here $A(x) = \frac{1}{2}(-Bx^2, Bx^1)$ is the vector potential of a homogeneous magnetic field of strength B in the x^3-direction, V is the confining potential, assumed to be continuous with $V(x) \to \infty$ for $|x| \to \infty$, and S_3 is the spin operator in x^3-direction. The parameters m_* and g_* are respectively the effective mass and the effective g-factor of the electrons, while m_e and e are the bare values of the electron mass and electric charge, and $\hbar$ and c have their usual meanings. The constant

$$C = \left(\frac{\hbar e}{2m_e c}\right)\left(\frac{m_e}{m_*} - \frac{|g_*|}{2}\right) \tag{4}$$

has been introduced in (3) for convenience: Subtraction of CB has the effect that the spectrum of the kinetic energy operator (including spin) $H_1^{\mathrm{kin}} = H_1 - V$ starts at zero for all B, even if $m_* \neq m_e$ and $g_* \neq 2$.

The interaction potential W represents the Coulomb repulsion between the electrons, modified by the properties of the surrounding medium. Usually it is simply taken to be

$$W(x_i - x_j) = e_*^2 |x_i - x_j|^{-1} \tag{5}$$

where $e_* = e/\sqrt{\epsilon}$ with ϵ the dielectric constant, but some regularization of the bare Coulomb potential, e.g.,

$$W(x_i - x_j) = e_*^2\left[(|x_i - x_j|^2 + \delta_+^2)^{-1/2} - (|x_i - x_j|^2 + \delta_-^2)^{-1/2}\right] \tag{6}$$

with $\delta_- > \delta_+ > 0$ [11], or even a potential that depends not just on the differences $x_i - x_j$, may sometimes fit the effective interaction better. For the proof of some of the theorems below the important property of (5) is that W is repulsive, of positive type and tends to zero at infinity; these features are shared by (6).

Writing the Hamiltonian in the above form is, of course, an approximation, because the effect of the medium on the electrons is only taken into account through the modification of the parameters from their bare values. The size of a quantum dot ($\approx$ 100 nm) is however usually much larger than the lattice constant of the medium where it resides ($<$ 0.5 nm), so this approximation is usually a good one.

Quantum dots, especially such with few electrons, are sometimes referred to as ***artificial atoms*** with V playing the role of the attractive nuclear potential in real atoms. The analogy is not perfect, however, because V is regular around the origin in contrast to the potential from an atomic nucleus, and also because the electron interaction is the three dimensional Coulomb potential (5) (or modified Coulomb potential (6)), while the motion is (essentially) restricted to two dimensions. But in many respects quantum dots can indeed be regarded as artificial atoms, with an important additional aspect: The effective parameters are to a certain extent tunable and may differ appreciably form their counterparts in real atoms.

In a quantum dot the natural atomic unit of length is $a_* = \epsilon\hbar^2/(m_* e^2)$. Compared with the usual Bohr radius, $a_0 = \hbar^2/(m_e e^2) = 0.53 \times 10^{-1}$ nm, the length a_* is typically large, e.g., $a_* \approx 185\ a_0 \approx 10$ nm in GaAs. The natural energy unit is $E_* = e_*^2/a_* = e_*^4 m_*/\hbar^2$, and in GaAs $E_* \approx 12$ meV, which should be compared with $E_0 = e^2/a_0 = e^4 m_e/\hbar^2 = 27.2$ eV, i.e., $E_* \approx 4 \times 10^{-4} E_0$.

The natural unit, B_*, for magnetic field strength is the field at which the magnetic length $\ell_B = \hbar e/(B^{1/2}c)$ equals a_*, or equivalently, at which the cyclotron energy $\hbar e B_*/m_* c$ equals E_*. Hence $B_* = (a_0/a_*)^2 B_0$, where $B_0 = e^3 m_e^2 c/\hbar^3 = 2.35 \times 10^5$ T is the value corresponding to free electrons. If a_0/a_* is small, B_* can be much smaller than B_0. Thus $B_* \approx 7$ T in GaAs. This opens the very interesting possibility to study in the laboratory magnetic effects, whose analog for real atoms require field strengths prevailing only on neutron stars.

On the experimental side the main techniques for studying quantum dots are *charge transport and capacitance spectroscopy* ([12], [6], [13]) and *optical far infrared spectroscopy* ([14], [15]). The former is in particular suited for measuring the N dependence of ground state energies, but excited states can be investigated as well. The applicability of optical spectroscopy is to a certain extent limited by Kohn's theorem, to be discussed below, but refined techniques allow also to infer many properties by this method. Altogether it is fair to say that the energy spectrum of quantum dots and even some aspects of the corresponding wave functions are experimentally accessible with considerable precision. See also [16] for recently discovered effects in charge transport spectroscopy that wait for an adequate explanation.

On the theoretical side the spectral properties of the Hamiltonian (2) have been studied by a variety of methods, which can be roughly divided into the following categories:

- Exact analytic solutions
- Rigorous limit theorems
- Numerical diagonalizations
- Hartree and Hartree-Fock approximations
- Variational calculations

- Density functional methods

In the condensed matter literature the last four categories are by far the most prominent, but the present survey is mostly concerned with the first two, which lie within the realm of mathematical physics. It is impossible here to do any justice to the extensive physics literature on the theory of quantum dots, but an annotated bibliography of some representative references will be given in the last section.

2 Exact solutions

2.1 The Fock-Darwin spectrum

From now on units are chosen such that $\hbar = e_* = m_* = B_* = 1$. The Hamiltonian (3) can then be written

$$H_1 = \frac{1}{2}\left(\mathrm{i}\nabla - A(x)\right)^2 + V(x) + \gamma S_3 B - \tfrac{1}{2}(1-|\gamma|)B \tag{7}$$

with $\gamma = g_* m_*/(2m_e)$. For a confining potential of the harmonic oscillator form

$$V(x) = \tfrac{1}{2}\omega|x|^2, \tag{8}$$

the eigenvalues and eigenfunctions of

$$H_1^{\mathrm{orb}} = \tfrac{1}{2}\left(\mathrm{i}\nabla - A(x)\right)^2 + \tfrac{1}{2}\omega^2|x|^2 \tag{9}$$

were determined by Fock already in 1928 [17], and also by Darwin in 1930 [18]. The Fock-Darwin spectrum of (9) consists of the eigenvalues

$$\varepsilon_{k,l}^{\mathrm{FD}} = (2k + |l| + 1)\Omega - \tfrac{1}{2}lB \tag{10}$$

with

$$\Omega = \left(\tfrac{1}{4}B^2 + \omega^2\right)^{1/2}, \tag{11}$$

$k = 0, 1, 2\ldots$, $l = 0, \pm1, \pm2, \ldots$. The corresponding eigenfunctions are

$$\psi_{k,l}^{\mathrm{FD}}(x) = (\mathrm{const.})\exp(\mathrm{i}l\varphi)r^{|l|}\exp(-\Omega r^2/2)L_k^{|l|}(\Omega r^2) \tag{12}$$

where $x = (r\cos\varphi, r\sin\varphi)$ and $L_k^{|l|}$ is an associated Laguerre polynomial. These functions are also eigenfunctions of the angular momentum L_3 in the x^3-direction with eigenvalue l. Eigenfunctions with the same value of

$$n = k + \tfrac{1}{2}(|l| - l) \tag{13}$$

are grouped together in a *Fock-Darwin level* (FDL). In the limit $\omega/B \to 0$ the eigenvalues in a FDL coalesce and a FDL becomes identical to a *Landau level* (LL) with the eigenvalues

$$\varepsilon_n^{\mathrm{L}} = (n + \tfrac{1}{2})B \tag{14}$$

and eigenfunctions

$$\psi^{\mathrm{L}}_{k,l}(x) = (\text{const.}) \exp(\mathrm{i}l\varphi) r^{|l|} \exp(-Br^2) L_k^{|l|}(2Br^2) \tag{15}$$

The degeneracy of a LL per unit area is $B/(2\pi)$.

If the interaction W between the electrons is ignored the FD spectrum together with the Pauli principle, taking spin into account, completely solves the eigenvalue problem for H_N in the case of the quadratic potential (8). This approximation even fits some experimental data quite well [1], [2]. When the interaction W is taken into account this picture has to be modified, of course. For a quadratic confining potential, however, the FD spectrum continues to apply to the motion of the center of mass, independently of the interaction. This simple, but important fact [19], [20] goes under the heading *Kohn's theorem*. The proof is essentially contained in the identity

$$N \sum_{j=1}^{N} x_j^2 = \left(\sum_{j=1}^{N} x_j \right)^2 + \sum_{i<j} (x_i - x_j)^2, \tag{16}$$

for it implies that H_N can be written

$$H_N = H_N^{\mathrm{CM}} + H_N^{\mathrm{spin}} + H_N^{\mathrm{rel}} \tag{17}$$

where H_N^{rel} operates only on the relative coordinates $x_i - x_j$, while all dependence on the center of mass coordinate $X = (x_1 + \cdots + x_N)/N$ is contained in

$$H_N^{\mathrm{CM}} = \tfrac{1}{2N} \left(\mathrm{i}\nabla_X - NA(X) \right)^2 + \tfrac{N}{2} \omega^2 |X|^2. \tag{18}$$

The spectrum of H_N^{CM} is exactly the same as that of H_1^{orb}, independent of N and W. In the dipole approximation the radiation field couples only to X and in this approximation optical FIR spectroscopy thus probes only the FD spectrum.

There is a further instance where the FD eigenfunctions play a role even for $W \neq 0$. Let Π_0^{FD} denote the projector on the subspace of $\mathcal{H}_N$ generated by eigenfunctions with FD index $n = 0$ and complete polarization, i.e., spin magnetic moment in the direction of the field. Consider the operator

$$H_N' = \Pi_0^{\mathrm{FD}} H_N \Pi_0^{\mathrm{FD}}. \tag{19}$$

For large B the spectral properties of this operator can be expected to approximate those of H_N. If $W = 0$ the ground state of this operator has the orbital wave function

$$\Psi^{\mathrm{mdd}} = \psi_{0,0}^{\mathrm{FD}} \wedge \psi_{0,1}^{\mathrm{FD}} \wedge \ldots \wedge \psi_{0,N-1}^{\mathrm{FD}}. \tag{20}$$

This eigenfunction is called the *maximum density droplet* [21] because the electrons are as "tightly packed" as possible around the origin. It is an amusing observation that Ψ^{mdd} remains an *exact eigenfunction* of H_N' also for $W \neq 0$. The proof is very simple: H'_N and the angular momentum operator L_3 on $\mathcal{H}_N$ commute and have discrete spectrum. Hence for every eigenvalue of L_3 there is a corresponding eigenfunction that is simultaneously an eigenfunction of H'_N. But the lowest eigenvalue, $N(N-1)/2$, of L_3 in the subspace of $\mathcal{H}_N$ generated by the FD functions with $n=0$ is nondegenerate, and the corresponding eigenfunction $\Psi^{\rm mdd}$ must hence also be an eigenfunction of H'_N. There is numerical ([21], [22], [23]) and even experimental [24] evidence that $\Psi^{\rm mdd}$ is a ground state of H'_N for some range of values of B.

2.2 Analytic solutions for $N > 1$

For $N=2$ the Hamiltonian for the relative coordinate $x = x_1 - x_2$ is

$$H_2^{\rm rel} = (\mathrm{i}\nabla - \tfrac{1}{2}A(x))^2 + \tfrac{1}{4}\omega^2|x|^2 + W(x) \tag{21}$$

In the case of a pure Coulomb interaction, $W(x) = 1/|x|$, explicit formulas for eigenfunctions and eigenvalues of (21) have been found by Taut [25]. His approach is based on an ansatz for the wave functions of the form

$$\psi(x) = \exp(\mathrm{i}m\varphi)\exp(-\rho^2/2)\rho^{|m|}P(\rho) \tag{22}$$

where P is a polynomial in $\rho = (\Omega/2)^{1/2}r$, with $\Omega = \left(\tfrac{1}{4}B^2 + \omega^2\right)^{1/2}$, φ is the angular variable and $m \in \mathbf{Z}$. Solutions of the form (22) with P a polynomial do not exist for arbitrary values of B, but Taut's method produces at least eigenfunctions and eigenvalues for a countable infinity of values of Ω which accumulate at 0. In fact, an ansatz like (22) in the eigenvalue equation $H_2^{\rm rel}\psi = E\psi$ with P a power series,

$$P(\rho) = \sum_{\nu=0} a_\nu \rho^\nu, \tag{23}$$

leads to

$$a_\nu = F(|m|, \nu, E')a_0 \tag{24}$$

with a certain recursively computable function F and where E' is related to E by

$$E = \tfrac{1}{4}\Omega E' - \tfrac{1}{2}mB. \tag{25}$$

The condition that a_ν vanishes for all ν larger than some positive integer n is equivalent to the two equations

$$F(|m|, n, E', \Omega) = 0 \tag{26}$$

and

$$E' = 2(|m| + n). \tag{27}$$

For given n and m this gives one or more acceptable values for Ω and corresponding energy values

$$E = \tfrac{1}{2}(n + |m|)\Omega - \tfrac{1}{2}mB. \tag{28}$$

The solutions found in this way are not necessarily ground states of (21) but the position of E in the spectrum can be inferred from the number of nodes of the corresponding wave function.

The solutions of Taut seem so far to be the only known exact solutions for $N = 2$ and the Coulomb interaction (5). They are limited to the special values of Ω defined by (26) and (27). For W of the inverse square form

$$W(x) = \alpha|x|^{-2} \tag{29}$$

on the other hand, the Hamiltonian (21) can be exactly diagonalized for all B [26]. In fact, addition of (29) merely modifies the centrifugal term in the radial part of the FD Hamiltonian (9) and we obtain as eigenvalues of (21)

$$E = [2n + \mu + 1]\Omega - \tfrac{1}{2}mB \tag{30}$$

with $\mu = (m^2 + \alpha)^{1/2}$ not necessarily an integer, and the eigenfunctions

$$\psi(x) = \exp(\mathrm{i}m\varphi)\exp(-\rho^2/2)\rho^{\mu}L_n^{\mu}(\rho^2). \tag{31}$$

It should be noted that the inverse square form (29) for the effective interaction between the electrons is not necessarily less realistic than the pure Coulomb repulsion (5). In fact, the form (6) of the interaction, that is motivated by the situation in real dots [11], has an inverse square decrease for large separation. At small separation, on the other hand, the effective interaction may be less singular than (5). It is therefore not entirely academic to consider also a harmonic interaction [27] of the form

$$W(x_i - x_j) = 2W_0 - \tfrac{1}{2}\beta|x_i - x_j|^2 \tag{32}$$

with positive parameters W_0 and β. For this case one can even solve the problem for *all* N exactly, using (16) to decouple the oscillators. The result for the ground state energy $E^{\mathrm{Q}}(N,B)$ of H_N with $N \geq 2$ and $\gamma = 0$ (for simplicity) is [27]

$$E^{\mathrm{Q}}(N,B) = \Omega + \tfrac{1}{2}(N-1)(N-2)\Omega_0(N) - \tfrac{1}{4}N(N+1)B + N(N-1)W_0 \tag{33}$$

with $\Omega_0(N) = (\Omega^2 - N\beta^2)^{1/2}$. It is assumed that $\omega \geq N^{1/2}\beta$, so $\Omega_0(N) \in \mathbf{R}_+$ for all B. The corresponding wave function for the relative motion is

$$\psi(x) = \prod_{i<j}[z_{ij}\exp(-\Omega_0|z_{ij}|^2/2N)] \tag{34}$$

where $z_{ij} = (x_i^1 - x_j^1) - \mathrm{i}(x_i^2 - x_j^2)$ are the relative coordinates regarded as points in $\mathbf{C}$. The form (32) of the repulsion, of course, quite wrong for large separation, but this error is to some extent counterbalanced by the confining potential. For a comparison of the solutions for different W's and their confrontation with numerical calculations and experiments we refer to [4].

3 Rigorous limit theorems

While exact solutions of the eigenvalue problem for H_N with W of the form (5) or (6) are not available for $N > 2$, it is possible to analyze at least the ground state properties for large B or large N by minimizing simple functionals of the electron density. In this analysis, which implies a drastic reduction of the quantum mechanical N-body problem, it is not necessary that the confining potential V has the quadratic form (8). From now on V will stand for an arbitrary continuous function of $x \in \mathbf{R}^2$ tending to ∞ for $|x| \to \infty$; when additional properties are required these will be explicitly stated. It is no restriction to assume that $V \geq 0$. To be able to consider variations of the strength of the potential at fixed shape, we write

$$V(x) = Kv(x) \tag{35}$$

with a coupling constant K. The interaction W will for definiteness be assumed to be pure Coulomb with $e_* = 1$, i.e.,

$$W(x-y) = |x-y|^{-1}, \tag{36}$$

but other repulsive potentials of positive type could be treated similarly. The quantum mechanical ground state energy is

$$E^{\mathrm{Q}}(N,B,K) = \langle \Psi_0, H_N \Psi \rangle, \tag{37}$$

where Ψ_0 is a normalized ground state of H_N. The corresponding ground state electron density is

$$\rho^{\mathrm{Q}}_{N,B,K}(x) = \sum_{\text{spins } \sigma_k = \pm 1/2} \int |\Psi_0(x,\sigma_1;x_2,\sigma_2;\ldots,x_N,\sigma_N)|^2 dx_2 \cdots dx_N. \tag{38}$$

We are concerned with the asymptotics of these quantities when one or more of the parameters N, K and B tends to ∞ with v fixed.

The large B limit at fixed N and K is easiest and will be considered first.

3.1 High field limit

In the lowest Landau level (i.e., for $n = 0$) the wave functions (15) are essentially localized on scale $B^{-1/2}$, and the quantum mechanical kinetic energy vanishes after the spin contribution and the subtraction of $\frac{1}{2}(1-|\gamma|)$ in (7) have been taken into account. In the limit $B \to \infty$ it can therefore be expected that a classical model of N point particles with the energy function

$$\mathcal{E}^{\mathrm{P}}[x_1,\ldots,x_N] = \sum_{i=1}^{N} V(x_i) + \sum_{i<j} |x_i - x_j|^{-1} \tag{39}$$

describes the ground state energy correctly. This is indeed the case: Defining

$$E^{\mathrm{P}}(N,K) = \inf \mathcal{E}^{\mathrm{P}}[x_1,\dots,x_N] \tag{40}$$

we have

Theorem 3.1 (High field limit.) *For N and K fixed,*

$$\lim_{B\to\infty} E^{\mathrm{Q}}(N,B,K) = E^{\mathrm{P}}(N,K). \tag{41}$$

Proof: The lower bound for E^{Q} in the proof of this theorem is trivial: Since the kinetic energy part of (2) is ≥ 0, $E^{\mathrm{P}} \leq E^{\mathrm{Q}}$. The upper bound is a simple variational calculation: Since $V(x)$ tends to ∞ for $|x| \to \infty$ and $|x_i - x_j|^{-1} \to \infty$ for $|x_i - x_j| \to 0$, the infimum of $\mathcal{E}^{\mathrm{P}}$ is obtained at some point $(\bar{x}_1,\dots,\bar{x}_N) \in \mathbf{R}^{2N}$ with $\bar{x}_i \neq \bar{x}_j$ for $i \neq j$. We test the Hamiltonian with the wave function

$$\Psi(x_1\dots,x_N) = \exp(\mathrm{i}x_1\times\bar{x}_1/B)\psi^{\mathrm{L}}_{0,0}(x_1-\bar{x}_1)\wedge\dots\wedge\exp(\mathrm{i}x_N\times\bar{x}_N/B)\psi^{\mathrm{L}}_{0,0}(x_N-\bar{x}_N) \tag{42}$$

$\wedge$ indicates antisymmetrization in the coordinates $(x_1,\dots,x_N)$, and the vector product $\times$ of two points in $\mathbf{R}^2$ is regarded as $\mathbf{R}$ valued. The spins are totally aligned and are therefore omitted in the notation. The kinetic energy of Ψ is zero, because the wave functions $\exp(\mathrm{i}x \times \bar{x}_j/B)\psi^{\mathrm{L}}_{0,0}(x - \bar{x}_j)$ belong to the lowest Landau level. The potential energy part approximates $\mathcal{E}^{\mathrm{P}}[\bar{x}_1,\dots,\bar{x}_N]$ arbitrarily well as $B \to \infty$, because $|\psi^{\mathrm{L}}_{0,0}(x - \bar{x}_j)|^2$ is essentially localized within a radius $\sim B^{-1/2}$ around $\bar{x}_j$ and tends to a delta function as $B \to \infty$.

3.2 Large N limits

Before stating the limit theorems formally let us briefly discuss their heuristic basis.

For $B = 0$ each of the N electrons in a dot of radius R occupies a "private room" of spatial extension $R/N^{1/2}$, because of the Pauli principle. Hence the kinetic energy of an electron is $\varepsilon_{\mathrm{kin}} \sim N/R^2$. The potential energy due to the confining potential is $\varepsilon_{\mathrm{conf}} \sim Kv(R)$ and the repulsive energy due to the other electrons $\varepsilon_{\mathrm{rep}} \sim N/R$. The radius R can be estimated by minimizing the sum $\varepsilon_{\mathrm{kin}} + \varepsilon_{\mathrm{conf}} + \varepsilon_{\mathrm{rep}}$. If K/N is kept fixed while $N \to \infty$, all these terms are proportional to N. The radius is therefore *independent of* N an the mean distance between the electrons is $\sim RN^{-1/2}$. In this limit the density tends to infinity. The ground state energy of the dot, i.e., the total energy E^{Q} of the N electrons, is $\sim N^2$, and this is large compared to the exchange/correlation energy $\sim N \cdot (N^{-1/2})^{-1} = N^{3/2}$. One may therefore expect that a Thomas-Fermi theory [28] captures the leading asymptotics for $N, K \to \infty$ with K/N fixed.

A magnetic field B will not influence the asymptotics as long as $B \ll N$, because the energy differences between Landau levels are much smaller than the other energy contributions. If $B \sim N$, on the other hand, all Landau levels have to be taken into account, and the Thomas-Fermi theory has to be modified ([11], [29]). For $B \gg N$ the electrons will essentially all sit in the lowest Landau level and the kinetic energy contribution vanishes. The radius is essentially independent of B and N (as long as $B \gg N$) and slightly smaller than in the $B = 0$ case, because the positive kinetic energy term $\varepsilon_{\text{kin}} \sim N/R^2$ is now missing. The ground state energy is $\sim N^2$ as before. Theorem 3.1 applies, but the large N limit leads to an additional simplification: The point charges can be replaced by a continuous distribution [28], because the interelectronic distance $\sim RN^{-1/2}$ tends to zero. The asymptotics is described by the energy functional of a charged fluid in the confining potential V.

For a homogeneous potential V of degree $s \geq 1$, i.e., $V(\lambda x) = \lambda^s V(x)$, this last mentioned "classical" energy functional describes also the large N limit for arbitrary B if $K/N \to 0$, in particular if K is fixed while $N \to \infty$. The radius $R \sim (N/K)^{1/(s+1)}$ tends in this case to ∞. The density may go to zero or to infinity. The kinetic energy per particle $\sim N/R^2 \sim N(K/N)^{2/(s+1)}$ is in both cases small compared to the other contributions $\sim KR^s \sim K(N/K)^{s/(s+1)}$. Smearing out the point charges brings in an error $\sim N^{1/2}/R$ per particle, but it is small compared to the total Coulomb repulsion $\sim N/R$.

These heuristic considerations will now be turned into precise statements. We define three functionals of the electron density $\rho \in L_1(\mathbf{R}^2, dx)$, $\rho \geq 0$ as follows.

The **2D Thomas-Fermi functional** is

$$\mathcal{E}^{\text{TF}}[\rho; K] = \pi \int \rho(x)^2 dx + \int V(x)\rho(x)dx + D(\rho, \rho) \tag{43}$$

with $D(\rho,\rho) = \frac{1}{2}\int\int \rho(x)|x-y|^{-1}\rho(y)dxdy$. All integrals are over $\mathbf{R}^2$. The $\pi\rho^2$ in the first term is just the kinetic energy density of a two dimensional noninteracting electron gas of density ρ at $B = 0$.

The **2D magnetic Thomas-Fermi functional** is defined as

$$\mathcal{E}^{\text{MTF}}[\rho; B, K] = \int j_B(\rho(x))dx + \int V(x)\rho(x)dx + D(\rho,\rho) \tag{44}$$

where j_B is a piece wise linear function representing the kinetic energy density of the electrons in a magnetic field B, taking all Landau levels (including spin) into account. If γ in (7) were 0, the derivative $j_B' = dj_B/d\rho$ would just be the step function

$$j_B'(\rho) = B[2\pi\rho/B] \tag{45}$$

where $[t]$ denotes the integer part of a real number t. This, together with $j_B(0) = 0$ fixes $j_B(\rho)$. As $B \to 0$, $j_B(\rho) \to \pi\rho^2$. If $\gamma \neq 0$, then j_B is a slightly more complicated step function, cf. Eq. (2.3) in [9], but since the explicit form is not significant we refrain form stating it here. When a concrete j_B is needed for discussion purposes we shall stick to the simplest case (45).

Finally, the **classical energy functional** is defined as

$$\mathcal{E}^{\mathrm{C}}[\rho; K] = \int V(x)\rho(x)dx + D(\rho, \rho). \tag{46}$$

All three functionals are convex, and for each there is a unique nonnegative density that minimizes the functional under the constraint $\int \rho(x)dx = N$; this is discussed in [9].

We denotes the minimizing densities and the corresponding energies by $\rho^{\mathrm{TF}}_{N,K}$ and $E^{\mathrm{TF}}(N,K)$ for the TF functional (43), by $\rho^{\mathrm{MTF}}_{N,B,K}$ and $E^{\mathrm{MTF}}(N,B,K)$ for the magnetic TF functional (44), and by $\rho^{\mathrm{C}}_{N,K}$ and $E^{\mathrm{C}}(N,K)$ for the classical functional (46). In the last case some regularity of V is needed in order to ensure that $\rho^{\mathrm{C}}_{N,K}$ is a function and not just a positive measure; a sufficient conditions is that V satisfies locally an estimate of the form

$$|\nabla V(x) - \nabla V(y)| \leq (\text{const.})\, |x-y|^{\alpha} \tag{47}$$

with some $\alpha > 0$.

The TF functional (43) and the classical functional (46) are both limiting cases of the MTF functional (44), for $B \to 0$ and $\to \infty$ respectively. More precisely,

$$\lim_{B\to 0} E^{\mathrm{MTF}}(N,B,K) = E^{\mathrm{TF}}(N,K) \tag{48}$$

$$\lim_{B\to 0} \rho^{\mathrm{MTF}}_{N,B,K} = \rho^{\mathrm{TF}}_{N,K} \tag{49}$$

and

$$\lim_{B\to \infty} E^{\mathrm{MTF}}(N,B,K) = E^{\mathrm{C}}(N,K) \tag{50}$$

$$\lim_{B\to \infty} \rho^{\mathrm{MTF}}_{N,B,K} = \rho^{\mathrm{C}}_{N,K}. \tag{51}$$

The limit for the densities should be understood in the weak L_1 sense, but for special V much stronger convergence may hold. For instance, if V is monotonically increasing with $|x|$, $\rho^{\mathrm{C}}_{N,K}$ is a bounded function, and $\rho^{\mathrm{C}}_{N,K} = \rho^{\mathrm{MTF}}_{N,B,K}$ for sufficiently large B, because $j_B(\rho^{\mathrm{C}}_{N,K}) = 0$ for $B > 2\pi\|\rho^{\mathrm{C}}\|_\infty$.

The MTF theory has two nontrivial parameters because of the **scaling relations**

$$E^{\mathrm{MTF}}(N,B,K) = N^2 E^{\mathrm{MTF}}(1, B/N, K/N) \tag{52}$$

$$\rho^{\mathrm{MTF}}_{N,B,K}(x) = N\rho^{\mathrm{MTF}}_{1,B/N,K/N}(x). \tag{53}$$

Corresponding relations (without B) hold for the TF theory and the classical theory, and also for E^{P}.

A further important property of the densities is their compact support: For fixed K/N the minimizers of $\mathcal{E}^{\mathrm{TF}}$, $\mathcal{E}^{\mathrm{MTF}}$, $\mathcal{E}^{\mathrm{C}}$ and also of $\mathcal{E}^{\mathrm{P}}$ have support in a disc whose radius is uniformly bounded in N and B (Lemma A.1 in [9]).

Each minimizer satisfies a variational equation, which in the case of the MTF theory has an unusual form, since it consists really of inequalities. To state it compactly it is convenient to modify the definition (45) slightly and regard j'_B as an *interval valued* function if $2\pi\rho/B$ is an integer, namely, if $2\pi\rho/B = n$, then j'_B is the closed interval $[(n-1)B, nB]$. The **MTF equation** that is satisfied by ρ^{MTF} can then be written

$$\mu - V(x) - \rho * |x|^{-1} \begin{cases} \in j'_B(\rho(x)) & \text{if} \quad \rho(x) > 0 \\ \leq 0 & \text{if} \quad \rho(x) = 0 \end{cases} \tag{54}$$

with a unique $\mu = \mu(N, B, K)$. Such generalized variational equations have been studied by Lieb and Loss [30].

If the potential is quadratic, $V(x) = K|x|^2$, there is an explicit formula ([28], [9]) for the minimizer for $\mathcal{E}^{\mathrm{C}}$, which is equal to ρ^{MTF} for B sufficiently large:

$$\rho^{\mathrm{C}}_{N,K}(x) = \begin{cases} \frac{3}{2\pi} N\lambda\sqrt{1-\lambda|x|^2} & \text{if} \quad |x| \leq \lambda^{-1} \\ 0 & \text{if} \quad |x| > \lambda^{-1} \end{cases} \tag{55}$$

with $\lambda = (8K/3\pi N)^{2/3}$. The density profile has the shape of a half ellipsoid with a maximum at $x = 0$. Note the difference between the two dimensional case considered here, and three dimensional electrostatics: In three dimensions the density would be homogeneously distributed in a ball.

The criterion for $\rho^{\mathrm{MTF}} = \rho^{\mathrm{C}}$ is that $j_B(\rho^{\mathrm{C}}(0)) = 0$, which holds if

$$B \geq (6/3^{2/3}\pi^{5/3})K^{2/3}N^{1/3}. \tag{56}$$

Numerically computed profiles of the minimizers ρ^{MTF} and the corresponding effective potentials

$$V_{\mathrm{eff}}(x) = V(x) + \rho^{\mathrm{MTF}} * |x|^{-1} \tag{57}$$

with $V(x) = K|x|^2$ are shown in Fig. 1. The computations were carried out by Kristinn Johnsen.

At the highest value of the field (Fig. 1(a)) condition (56) is fulfilled and ρ^{MTF} has the form (55). On the support of $\rho^{\mathrm{MTF}} = \rho^{\mathrm{C}}$ we have $V_{\mathrm{eff}}(x) =$constant$= \mu$.

When the field is gradually turned down the maximal density $B/(2\pi)$ of electrons that can be accommodated in the lowest Landau level goes down also. Condition (56) no longer holds, i.e., the density ρ^{C} near the center is higher than $B/(2\pi)$ and charges have to be moved into other states in phase space. If B is only slightly smaller than the value given by (56) (Fig. 1(b)) it would cost too much energy to bring the electrons near the origin into the next Landau level and it pays to move them spatially away from the center, because the potential energy increase is less than B. Hence in a certain range of B values, the density near the center is locked at the value $B/(2\pi)$ ("incompressible" domain). The effective potential is no longer constant in this domain. In the complementary "compressible" domain, on the other hand, the density is below the critical value $B/(2\pi)$, and tends to zero in such a way that the effective potential stays constant. Reducing the field strength

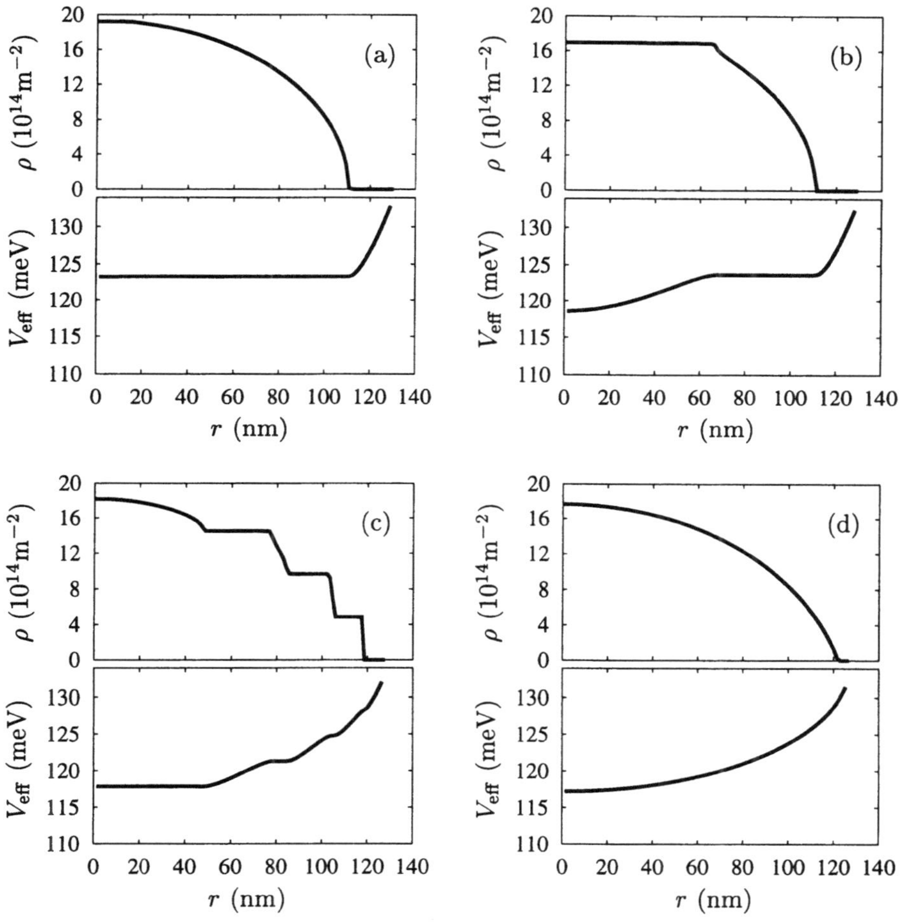

Figure 1: Density profiles and effective potentials for the MTF theory at different magnetic field strengths, calculated for $N = 50$ and $V(x) = K|x|^2$ with $K = 1,7$ meV and the material parameters of GaAs. (a) $B = 8$ T, (b) $B = 7$ T, (c) $B = 2$ T, (d) $B = 0$ T.

further brings more Landau levels into play (Fig. 1(c)). Incompressible domains, where the density is an integer multiple of $B/(2\pi)$, alternate with compressible domains, where the effective potential has a constant value. When $B \to 0$ the profile becomes indistinguishable from the smooth profile of $\rho^{\rm TF}$ (Fig. 1(d)). It is interesting to note that the alternation of compressible and incompressible domains in moderate magnetic fields may account for some fine structure in the charge transport spectroscopy of quantum dots with a large number of electrons [31].

The basic **limit theorem** [9] that relates the energy functionals (43)-(46) to the quantum mechanical ground state of H_N is as follows:

Theorem 3.2 (High density limit.) *Let $N \to \infty$ with K/N fixed. Then, uniformly in B/N,*

$$E^{\rm Q}(N,B,K)/E^{\rm MTF}(N,B,K) \to 1 \tag{58}$$

and

$$N^{-1}\rho^{\rm Q}_{N,B,K}(x) \to \rho^{\rm MTF}_{1,B/N,K/N}(x) \tag{59}$$

in weak L^1 sense. Moreover, if $B/N \to 0$, then

$$E^{\rm Q}(N,B,K)/E^{\rm TF}(N,B,K) \to 1 \tag{60}$$

$$N^{-1}\rho^{\rm Q}_{N,B,K}(x) \to \rho^{\rm TF}_{1,B/N,K/N}(x), \tag{61}$$

and if $B/N \to \infty$, then

$$E^{\rm Q}(N,B,K)/E^{\rm C}(N,B,K) \to 1 \tag{62}$$

$$N^{-1}\rho^{\rm Q}_{N,B,K}(x) \to \rho^{\rm C}_{1,B/N,K/N}(x). \tag{63}$$

According to this theorem there are thus three asymptotic regimes for quantum dots as N and K tend to ∞ with K/N fixed: $B \ll N$, $B \sim N$ and $N \ll B$. This should be compared with the more complex situation for three dimensional natural atoms in strong magnetic field, where there are five regimes [32], [33] for $N \to \infty$ with Z/N fixed (Z = nuclear charge): $B \ll N^{4/3}$, $B \sim N^{4/3}$, $N^{4/3} \ll B \ll N^3$, $B \sim N^3$, $N^3 \ll B$.

For homogeneous potentials a stronger asymptotic result holds, for K/N may tend to zero as $N \to \infty$.

Theorem 3.3 (Homogeneous potentials.) *Assume that V is homogeneous of degree $s \geq 1$, i.e.,*

$$V(\lambda x) = \lambda^s V(x). \tag{64}$$

Then

$$\lim_{N\to\infty} E^{\rm Q}(N,B,K)/E^{\rm MTF}(N,B,K) = 1 \tag{65}$$

uniformly in B and in K as long as K/N is bounded above. Moreover, if $K/N \to 0$ as $N \to \infty$, then

$$\lim_{N\to\infty} E^{\rm Q}(N,B,K)/E^{\rm C}(N,K) = 1 \tag{66}$$

uniformly in B.

We shall now discuss briefly the main techniques used for the proof of these theorems. As usual it is sufficient to prove the limit theorems for the energy, because the corresponding results for the density can be obtained by variation with respect to the potential V. The basic result is thus Eq. (58); the other limit theorems follow by (48)–(51). One has to prove upper and lower bounds for the quantum mechanical energy E^{Q} in terms of the energy E^{MTF}, with controllable errors.

The upper bound is obtained, using the variational principle of [34], by testing H_N with a suitable one particle density operator. Its kernel in the space and spin variables x, σ has the form

$$\mathcal{K}(x,\sigma;x',\sigma') = \sum_{\nu} f_\nu(u)\Pi_{\nu u}(x,\sigma;x',\sigma')d^2u \tag{67}$$

where the sum is over all Landau levels and $f_\nu(u)$ is the filling factor of the ν-th Landau level at point u when the density is $\rho^{\mathrm{MTF}}(u)$. The kernel $\Pi_{\nu u}(x,\sigma;x',\sigma')$ is obtained from the kernel $\Pi_\nu(x,\sigma;x',\sigma')$ of the projector on the ν-th Landau level by localizing around u with a smooth function g of compact support, i.e.,

$$\Pi_{\nu u}(x,\sigma;x',\sigma') = g(x-u)\Pi_\nu(x,\sigma;x',\sigma')g(x'-u). \tag{68}$$

This operator is positive and approximately a projector, localizing simultaneously in space, i.e., around u, and in the Landau level index ν. By letting the support of g shrink with N more slowly that the average electron spacing $N^{-1/2}$, the error terms in the estimate above for $E^{\mathrm{Q}} - E^{\mathrm{MTF}}$ are of lower order than N^2, which is the order of E^{MTF}.

The lower bound for E^{Q} is proved separately for large B and for small B. For large B, i.e., $B \gg N$, one starts with the obvious estimate $E^{\mathrm{Q}} \geq E^{\mathrm{P}}$. One then has to compare E^{P} with E^{C}, i.e., the energy of point charges with those of smeared charges. Since the electron distance is $\sim N^{-1/2}$ the self energy of a smeared unit charge is $\sim N^{1/2}$. Hence an estimate

$$E^{\mathrm{P}}(N,K) \geq E^{\mathrm{C}}(N,K) - bN^{3/2} \tag{69}$$

with b depending only on K/N is to be expected, and this can indeed be proved, using an electrostatic lemma of Lieb and Yau [35].

The lower bound for small B, i.e., $B \ll N$ or $B \sim N$, requires an estimate on the indirect Coulomb energy, that is derived in essentially the same way as a corresponding inequality in [38], using the positive definiteness of the Coulomb interaction 5, cf. also [39].

Lemma 3.1 (Exchange inequality in 2 dimensions.)

$$\sum_{\text{spins}} \int_{\mathbf{R}^{2N}} |\Psi|^2 \sum_{i<j} |x_i - x_j|^{-1} \geq D(\rho_\Psi,\rho_\Psi) - 192(2\pi)^{1/2} \int_{\mathbf{R}^2} \rho_\Psi^{3/2}. \tag{70}$$

In order to control negative term $\sim \int \rho_\Psi^{3/2}$ on the right side of (70) a lower bound on the kinetic energy is needed. This in turn is derived from a two dimensional magnetic Lieb-Thirring inequality, which has to be proved in a slightly different way from the corresponding inequality in [33], because there is no kinetic energy associated with a motion in the x^3 direction. The following inequality is adequate for the present purpose, but sharper Lieb-Thirring type inequalities that hold even for inhomogeneous fields have been derived by Erdős and Solovej [36], [37].

Theorem 3.4 (Lieb-Thirring inequality in 2 dimensions.) *Let U be locally integrable and let $e_1(U), e_2(U), \ldots$ denote the negative eigenvalues (if any) of the Hamiltonian $\frac{1}{2}(\mathrm{i}\nabla - \mathbf{A})^2 + S_3B - U$. Define $|U|_+(x) = \frac{1}{2}[|U(x)| + U(x)]$. For all $0 < \lambda < 1$ we have the estimate*

$$\sum_j |e_j(U)| \leq \lambda^{-1}\frac{B}{2\pi}\int_{\mathbf{R}^2} |U|_+(x)dx + \tfrac{3}{4}(1-\lambda)^{-2}\int_{\mathbf{R}^2} |U|_+^2(x)dx. \tag{71}$$

By a Legendre transformation with respect to U it follows from (71) that for all $0 < \lambda < 1$ the kinetic energy T_Ψ of a state Ψ is bounded below by $\frac{1}{3}(1-\lambda)^2\int[\rho_\Psi - \lambda^{-1}B/2\pi]_+^2$. It is then possible, for B/N smaller than a certain critical value depending on v, to chose an N dependent $\varepsilon > 0$ in such a way that $\varepsilon \to 0$ as $N \to \infty$, but $\varepsilon T_\psi - (\text{const}) \int \rho_\Psi^{3/2} \geq 0$ for all N-particle states Ψ.

4 Other approaches

The rigorous results presented above concern mainly (but not exclusively) the extreme cases of very few ($N = 1$ or $N = 2$) or very many ($N \to \infty$) electrons. These cases play a similar role as the hydrogen atom and the Thomas-Fermi atom do in ordinary atomic physics, i.e., they set a standard that can be used as a starting point of various approximation schemes, or as a test for such schemes whose connection with the original Hamiltonian (2) may not be entirely clear. The physics literature on quantum dots is by now quite extensive, and many approaches have been used for gaining insight where rigorous results are not yet available. Here it is only possible to mention the main methods and give a list of some references that are representative for the approaches of condensed matter physicists to these problems and from it further sources can be traced. See also [8] for a more comprehensive list.

For small N a direct numerical diagonalization of H_N is possible and has been carried out, e.g., in [40], [41], [42], [44], [45], for various values of $N \leq 8$. It is, of course, necessary to restrict H_N to a finite dimensional subspace of the full Hilbert space, and the error made in this step is seldom estimated rigorously. By the mini-max principle, however, the computed values give at least upper bounds to the true eigenvalues. One of the features studied by this method is the orbital angular momentum and spin of the ground state, and analogs of Hund's rules from atomic physics, as well as oscillations between triplet and singlet states for $N = 2$ as B is varied have been seen in the calculations [1], [41].

For $N > 10$ numerical diagonalization of the Hamiltonian is at present hardly feasible and resource is taken to other techniques of many body theory like Hartree and Hartree-Fock approximations (e.g., [42], [43], [22], [15]), perturbation theory [46], variational methods (e.g. [47]), quantum Monte Carlo methods ([48], [49]), and (current) density functional theory ([50], [23], [51], [52]). In quantum dots with a moderate electron number correlations play a much greater role than in natural atoms because the confining potential is usually quite shallow around the origin and the density may be low. (By contrast, one of the main steps in the proof of Theorem 3.2 is to show that exchange and correlation effects vanish in the high density limit for arbitrary magnetic fields.) Strong correlations together with strong dependence on magnetic fields make up much of the special flavor of quantum dot physics and it remains a challenge for mathematical physics to capture these effects in a rigorous way.

References

[1] T. Chakraborty, *Physics of the Artificial Atoms: Quantum Dots in a Magnetic Field*, Comments Cond. Mat. Phys. **16**, 35-68 (1992)

[2] M.A. Kastner, *Artificial atoms*, Phys. Today **46**, 24–31 (1993)

[3] D. Heitmann and J. Kotthaus, *The spectroscopy of quantum dot arrays*, Phys. Today **46**, 56–63 (1993)

[4] N. F. Johnson, *Quantum dots: few-body, low dimensional systems*, J. Phys.: Condens. Matter **7**, 965-989 (1995)

[5] M.A. Kastner, *Mesoscopic Physics with Artificial Atoms*, Comments Cond. Mat. Phys. **17**, 349–360 (1996)

[6] R.C. Ashori, *Electrons on artificial atoms*, Nature **379**, 413–419 (1996)

[7] J. H. Jefferson and W. Häusler, *Quantum dots and artificial atoms*, Molecular Physics Reports, **17** 81-103 (1997)

[8] L. Jacak, P. Hawrylak, A. Wójs, *Quantum Dots*, Springer, Berlin etc., 1998

[9] E.H. Lieb, J.P. Solovej and J. Yngvason, *The Ground States of Large Quantum Dots in Magnetic Fields*, Phys. Rev. B **51**, 10646–10665 (1995)

[10] E.H. Lieb, J.P. Solovej and J. Yngvason, *Quantum Dots*, in: Proceedings of the Conference on Partial Differential Equations and Mathematical Physics, University of Alabama, Birmingham, 1994, I. Knowles, ed., pp. 157–172, International Press 1995

[11] P.L. McEuen, E.B. Foxman, J. Kinaret, U. Meirav, M.A. Kastner, N.S. Wingreen and S.J. Wind, *Self consistent addition spectrum of a Coulomb island in the quantum Hall regime*, Phys. Rev. B **45**, 11419–11422 (1992)

[12] C. W. J. Beenakker, *Theory of Coulomb-blockade oscillations in the conductance of a quantum dot*, Phys. Rev. B **44** 1646–1656 (1991)

[13] L. P. Kouwenhouven, T.H. Osterkamp, M. W. S. Danoesastro, M. Eto, D. G. Austing, T. Honda and S. Tarucha, *Excitation Spectra of Circular Few-Electron Quantum Dots*, Science **278** 1788–1792 (1997)

[14] U. Merkt, *Far-infrared spectroscopy of quantum dots*, Physica B **189**, 165–175 (1993)

[15] D. Heitmann, K. Bollweg, V. Gudmundsson, T. Kurth, S. P. Riege, *Far-infrared spectroscopy of quantum vires and dots, breaking Kohn's theorem*, Physica E **1**, 204–210 (1997)

[16] N. B. Zhintev, R. C. Ashoori, L. N. Pfeiffer and K. W. West, *Periodic and Aperiodic Bunching in the Addition Spectra of Quantum Dots*, Phys. Rev. Lett. **79**, 2308–2311 (1997)

[17] V. Fock, *Bemerkung zur Quantelung des harmonischen Oszillators in Magnetfeld*, Z. Phys. **47**, 446–448 (1928)

[18] C.G. Darwin, *The Diamagnetism of the Free Electron*, Proc. Cambr. Philos. Soc. **27**, 86–90 (1930)

[19] W. Kohn, Cyclotron Resonance and the de Haas-van Alphen Oscillations of an Interacting Electron Gas, Phys. Rev. **123**, 1242–1244 (1961)

[20] A. O. Govorov and A. V. Chaplik, *Magnetoabsorption at quantum points*, JETP Lett. **52**, 31-33 (1990)

[21] A. H. MacDonald, S. R. Yang, M. D. Johnson, *Quantum dots in strong magnetic fields: Stability criteria for the maximum density droplet*, Australian J. Phys. **46**, 345 (1993)

[22] S. E. Koonin and H. M. Mueller, *Phase-Transitions in Quantum Dots*, Phys. Rev. B **54**, 14532–14539 (1996)

[23] M. Ferconi and G. Vignale, *Density functional theory of the phase diagram of maximum density droplets in two dimensional quantum dots in a magnetic field*, Phys. Rev. B **56**, 12108–12111 (1997)

[24] T. H. Osterkamp, J. W. Janssen, L. P. Kouwenhouven, D. G. Austing, T. Honda and S. Tarucha, *Stability of the maximum density drop in quantum dots at high magnetic fields*, Preprint, http:// vortex. tn.tudelft.nl/mensen/leok/papers/ (1998)

[25] M. Taut, *Two electrons in a homogeneous magnetic field: Particular analytical solutions*, J. Phys. A **27**, 1045–1055; Corrigendum **27**, 4723–4724 (1994)

[26] L. Quiroga, D. R. Ardila and N. F. Johnson, *Spatial Correlation of Quantum Dot Electrons in a Magnetic Field*, Solid State Comm. **86**, 775–780 (1993)

[27] N. F. Johnson and M. C. Payne, *Exactly Solvable Models of Interacting Particles in a Quantum Dot*, Phys. Rev. Lett. **67**, 1157–1160 (1991)

[28] V. Shikin, S. Nazin, D. Heitmann and T. Demel, *Dynamical response of quantum dots*, Phys. Rev. B **43**, 11903–11907 (1991)

[29] P.L. McEuen, N.S. Wingreen, E.B. Foxman, J. Kinaret, U. Meirav, M.A. Kastner, and Y. Meir, *Coulomb interactions and the energy-level spectrum of a small electron gas*, Physica B **189**, 70–79 (1993)

[30] E.H. Lieb and M. Loss, unpublished section of a book on stability of matter.

[31] N.C. van der Vaart, M.P. de Ruyter van Steveninck, L.P. Kouwenhoven, A.T. Johnson, Y.V. Nazarov, and C.J.P.M. Harmans, *Time-Resolved Tunneling of Single Electrons between Landau Levels in a Quantum Dot*, Phys. Rev. Lett. **73**, 320–323 (1994)

[32] E.H. Lieb, J.P. Solovej and J. Yngvason, *Asymptotics of Heavy Atoms in High Magnetic Fields: I. Lowest Landau Band Regions*, Commun. Pure Appl. Math. **47**, 513–591 (1994)

[33] E.H. Lieb, J.P. Solovej and J. Yngvason, *Asymptotics of Heavy Atoms in High Magnetic Fields: II. Semiclassical Regions*, Commun. Math. Phys **161**, 77-124 (1994)

[34] E.H. Lieb, *A Variational Principle for Many-Fermion Systems*, Phys. Rev. Lett. **46**, 457–459; Erratum **47**, 69 (1981)

[35] E.H. Lieb and H.-T. Yau, *The stability and instability of relativistic matter*, Commun. Math. Phys. **118**, 177–213 (1988)

[36] L. Erdős and J.P. Solovej, *Semiclassical eigenvalue estimates for the Pauli operator with strong non-homogeneous magnetic fields. I. Non-asymptotic Lieb Thirring estimates*, Duke Math. J., to appear

[37] L. Erdős and J.P. Solovej, *Semiclassical eigenvalue estimates for the Pauli operator with strong non-homogeneous magnetic fields. II. Leading order asymptotic estimates*, Comm. Math. Phys. **188**, 599–656 (1997)

[38] E. H. Lieb, *A lower bound for Coulomb energies*, Phys. Lett. **70** A, 444–446 (1979)

[39] V. Bach, *Error Bound for the Hartree-Fock Energy of Atoms and Molecules*, Commun. Math. Phys. **147**, 527–548 (1992)

[40] S. R. Eric Yang, A. H. MacDonald and M. D. Johnson, *Addition Spectra of Quantum Dots in Strong Magnetic Fields*, Phys. Rev. Lett. **71**, 3194–3197 (1993)

[41] M. Wagner, U. Merkt and A. V. Chaplik, *Spin-singlet-spin-triplet oscillations in quantum dots*, Phys. Rev. B **45**, 1951–1954 (1992)

[42] D. Pfannkuche, V. Gudmundsson and P. A. Maksym, *Comparison of a Hartree, a Hartree-Fock, and an exact treatment of quantum dot helium*, Phys. Rev. B **47**, 2244–2250 (1993)

[43] D. Pfannkuche and S. E. Ulloa, *Selection Rules for Transport Spectroscopy of Few-Electron Quantum Dots*, Phys. Rev. Lett. **74**, 1194–1197 (1995)

[44] L. Meza-Montes, S. E. Ulloa and D. Pfannkuche, *Electron interactions, classical instability, and level statistics in quantum dots*, Physica E **1**, 274–280 (1997)

[45] M. Eto, *Electronic Structures of Few Electrons in a Quantum Dot under Magnetic Fields*, Jpn. J. Appl. Phys. **36**, 3924–3927 (1997)

[46] E. Anisimovas and A. Matulis, *Energy Spectra of Few-Electron Quantum Dots*, Jour. Phys. Cond. Mat. **10**, 601–615 (1998)

[47] M. Dineykhan and R. G. Nazmitdinov, *Two-Electron Quantum Dot in Magnetic Field: Analytical Results*, Phys. Rev. B **55**, 13707-13714 (1997)

[48] A. Harju, V. A. Sverdlov and R. M. Nieminen, *Variational wave function for a quantum dot in a magnetic field: A quantum Monte-Carlo study*, Europhys. Lett. **41**, 407-412 (1998)

[49] A. Harju, V. A. Sverdlov and R. M. Nieminen, *Many-Body Wave Function for a Quantum Dot in a Weak magnetic Field*, Preprint, Univ. of Helsinki (1998)

[50] M. Ferconi and G. Vignale. *Current density functional theory of quantum dots in a magnetic field*, Physical Review B **50**, 14722–14725 (1994)

[51] M. Pi, M. Barranco, A. Emperador, E. Lipparini and Ll. Serra, *Current Density Functional approach to large quantum dots in intense magnetic fields*, Phys. rev. B **57**, 14783–14792 (1998)

[52] O. Heinonen, J. M. Kinaret and M. D. Johnson, *Ensemble Density Functional Approach to Charge-Spin Textures in Inhomogeneous Quantum Systems*, Phys. Rev. B (1998)

Part II

Session talks

Operator Theory:
Advances and Applications, Vol. 108

A Simple Model of Concentrated Nonlinearity

Riccardo Adami
Dipartimento di Matematica, Università di Roma "La Sapienza", Italy

Alessandro Teta
Dipartimento di Matematica, Università di Roma "La Sapienza", Italy

Abstract

We study the nonlinear Schrödinger equation in dimension one with a nonlinearity concentrated in a point. We give results on local and global existence of the solution and find conditions for the blow-up. In the critical case an explicit blow-up solution is constructed.

Recent studies in Solid State Physics concerning resonant tunneling, interactions with impurities etc. have motivated the analysis of the nonlinear Schrödinger equation (NLSE) with a nonlinearity concentrated in a (small) region of space. Such an equation exhibits interesting qualitative properties which have been discussed in the physical literature ([J-LPS], [MA], [BKB]). Some rigorous mathematical results are given in [N], where the existence of steady states in the case of an open system with smooth short-range nonlinearities is proved.

In this note we approach what we consider the simplest case of concentrated nonlinearity, i.e. the one-dimensional Schrödinger equation with nonlinearity concentrated in a fixed point.

In spite of its simplicity, this model seems to be not completely trivial (for instance, it produces blow-up solutions).

We consider this work as the first step in a line of research devoted to the mathematical analysis of the Schrödinger equation in dimension d with a nonlinearity concentrated on a set of lower dimension.

The paper is organized as follows.

First we formulate the nonlinear evolution problem and give a local existence theorem.

The second step is the analysis of the conservation laws and the proof of a global existence theorem.

Finally we give conditions for the blow-up of the solution and, in the critical case, we explicitly construct a class of blow-up solutions.

We start with the formulation of the problem. For any fixed real function $\alpha(t)$, we denote by $H_{\alpha(t)}$ the Schrödinger operator with point interaction of strength $\alpha(t)$ placed at the origin. It is well known ([AGH-KH]) that such operator can be defined for all t as a self-adjoint operator in $L^2(\mathbb{R})$. In particular, its form domain is $H^1(\mathbb{R})$ and the operator domain consists of functions $u \in H^2(\mathbb{R}\setminus\{0\})$ satisfying the boundary condition at the origin $u'(0^+) - u'(0^-) = \alpha(t)u(0)$. Moreover the linear Schrödinger equation with time-dependent hamiltonian $H_{\alpha(t)}$ can also be correctly defined and it admits a unique solution for $\alpha(t)$ smooth.
We define the nonlinear evolution problem by replacing $\alpha(t)$ with a function of the value at the origin of the solution itself. More precisely, for $\gamma \in \mathbb{R}$, $\sigma > 0$, we consider

$$i\frac{\partial \psi_t}{\partial t} = H_{\alpha(t)}\psi_t, \quad \alpha(t) = \gamma|\psi_t(0)|^{2\sigma}, \quad \psi_{t=0} = \psi_0 \tag{1}$$

For the sake of simplicity we have considered in (1) a power law nonlinearity; more general nonlinearities can also be introduced.
Problem (1) corresponds to an evolution generated by the laplacian in $\mathbb{R}$ with a nonlinear boundary condition at the origin, namely

$$\psi_t'(0^+) - \psi_t'(0^-) = \gamma|\psi_t(0)|^{2\sigma}\psi_t(0) \tag{2}$$

It is convenient to reformulate problem (1) in integral form. One obtains

$$\psi_t(x) + i\gamma \int_0^t dsU\left(t-s,|x|\right)|\psi_s(0)|^{2\sigma}\psi_s(0) = \left(U(t)\psi_0\right)(x) \tag{3}$$

where $U(t)$ denotes the free Schrödinger unitary group in dimension one. From (3) it is clear that the solution is entirely determined by its value at the origin. Evaluating equation (3) in $x = 0$ one has a closed equation for $\psi_t(0)$

$$\psi_t(0) + i\gamma \int_0^t ds \frac{1}{\sqrt{4\pi i(t-s)}}|\psi_s(0)|^{2\sigma}\psi_s(0) = \left(U(t)\psi_0\right)(0) \tag{4}$$

Equation (4) is a nonlinear Abel integral equation, involving only the time variable. The search for the solution of (3) is then reduced to the solution of (4).
As for the standard NLSE (see e.g. [GV]), the natural space where we study the equation (3) is $H^1(\mathbb{R})$, i.e. the space of finite energy. For technical reasons, in what follows we choose an initial datum in $H^2(\mathbb{R})$ but we are convinced that all the results remain valid under the more natural assumption $\psi_0 \in H^1(\mathbb{R})$.

Proposition 1. For any $\psi_0 \in H^2(\mathbb{R})$, there exists $\bar{t} > 0$ s.t. problem (3) has a unique solution $\psi_t \in H^1(\mathbb{R})$, $t \in [0,\bar{t})$.

Sketch of the proof. The r.h.s. of (4) is clearly continuous in t. Then, using the contraction principle, equation (4) can be uniquely solved for small times in the space of continuous functions ([M]). Substituting the solution $\psi_t(0)$ in (3), one easily sees that $\psi_t \in L^2(\mathbb{R})$. Now, exploiting the smoothing properties of the Abel operator ([GoVe]) and the regularity of ψ_0, one can prove that $\psi_t(0)$ is smooth enough to ensure $\psi_t \in H^1(\mathbb{R})$.

One of the main tools for the analysis of nonlinear evolution problems are conservation laws. In our case there are three conserved quantities: L^2-norm (denoted by $\|\cdot\|$), energy and parity.

Proposition 2. For any $\psi_0 \in H^2(\mathbb{R})$ we have

$$\|\psi_t\| = \|\psi_0\| \tag{5}$$

$$E(\psi_t) \equiv \int_{\mathbb{R}} dx |\psi_t'(x)|^2 + \frac{\gamma}{\sigma+1}|\psi_t(0)|^{2\sigma+2} = E(\psi_0) \tag{6}$$

$$\psi_0(x) = \pm\psi_0(-x) \text{ implies } \psi_t(x) = \pm\psi_t(-x) \tag{7}$$

Proof. We give some details for (6). The proof of (5) is similar while (7) is trivial. We define $q(t) \equiv |\psi_t(0)|^{2\sigma}\psi_t(0)$. The hypothesis $\psi_0 \in H^2(\mathbb{R})$ implies $q \in H^1((0,t))$ for all $t < \bar{t}$.

$$\begin{aligned}
\|\psi_t'\|^2 &= \int_{\mathbb{R}} dk\, k^2\, |\psi_t(k)|^2 \\
&= \int_{\mathbb{R}} dk\, k^2\, |\psi_0(k)|^2 - \sqrt{\frac{2}{\pi}}\, i\,\gamma\, Re \int_{\mathbb{R}} dk\, k^2 \overline{\tilde{\psi}_0(k)} \int_0^t ds\, e^{ik^2 s}\, q(s) \\
&\quad + \frac{\gamma^2}{2\pi}\int_{\mathbb{R}} dk\, k^2 \int_0^t ds \int_0^t ds'\, e^{ik^2(s'-s)}\overline{q(s)}q(s')
\end{aligned} \tag{8}$$

where $\tilde{\psi}$ denotes the Fourier transform of ψ.
After some calculations one obtains
$Re \int_{\mathbb{R}} dk\, k^2 \overline{\tilde{\psi}_0(k)} \int_0^t ds\, e^{ik^2 s}\, q(s)$

$$\begin{aligned}
&= \frac{\sqrt{\pi}}{\sqrt{2}\,(\sigma+1)}\left(|\psi_t(0)|^{2\sigma+2} - |\psi_0(0)|^{2\sigma+2}\right) \\
&\quad + \frac{\gamma}{2}\int_0^t \frac{ds}{\sqrt{t-s}}\left[Re(q(t)\overline{q(s)}) + Im(q(t)\overline{q(s)})\right] \\
&\quad - \frac{\gamma}{2}\int_{0\le s'\le s\le t} \frac{ds\, ds'}{\sqrt{s-s'}}\left[Re(\dot{q}(s)\,\overline{q(s')}) + Im(\dot{q}(s)\,\overline{q(s')})\right]
\end{aligned} \tag{9}$$

and
$\int_{\mathbb{R}} dk\, k^2 \int_0^t ds \int_0^t ds'\, e^{ik^2(s'-s)}\overline{q(s)}q(s')$

$$= \sqrt{2\pi}\int_0^t \frac{ds}{\sqrt{t-s}}\left[Re(q(t)\overline{q(s)}) + Im(q(t)\overline{q(s)})\right]$$

$$- \sqrt{2\pi} \int_{0 \le s' \le s \le t} \frac{ds\, ds'}{\sqrt{s - s'}} \left[Re(\dot{q}(s)\, \overline{q(s')}) + Im(\dot{q}(s)\, \overline{q(s')}) \right] \tag{10}$$

Then, from (8), (9) and (10), conservation of energy is proved.

Remark 1. Since $\psi_0^{odd}(0) = 0$, from (3) one has $\psi_t^{odd} = U(t)\psi_0^{odd}$ and ψ_t^{even} is given by equation (3), where ψ_0 is replaced by ψ_0^{even} (here we have denoted by f^{even} and f^{odd} respectively the even and the odd part of the function f). One can therefore restrict oneself to even initial data.

Using the conservation laws and the Sobolev inequality we easily obtain the following global existence theorem.

Proposition 3. Assume $\gamma > 0$ (case a)) or $\gamma < 0$, $\sigma < 1$ (case b)). Then for any $\psi_0 \in H^2(I\!R)$ and any $T > 0$ problem (3) has a unique solution $\psi_t \in H^1(I\!R)$, for $t \in [0, T]$.

Proof. In case a) the conservation of the energy directly provides an a priori estimate of $||\psi_t'||$, which implies that the solution is global in time.
In case b) we apply the Sobolev inequality ([A]) and the conservation of the L^2-norm to estimate the "potential" term in the energy

$$|\psi_t(0)|^{2\sigma+2} \le ||\psi_0||^{\sigma+1} ||\psi_t'||^{\sigma+1} \tag{11}$$

Combining (6) and (11) we have

$$E(\psi_0) \ge ||\psi_t'||^2 \left(1 - \frac{|\gamma|}{\sigma + 1} ||\psi_0||^{\sigma+1} ||\psi_t'||^{\sigma-1} \right) \tag{12}$$

Inequality (12) implies that $||\psi_t'||$ cannot diverge in finite time and this implies global existence.

Let us consider the problem of the existence of blow-up solutions of (3).
Following the usual definition, we say that ψ_t is a blow-up solution if there exists $t_0 < \infty$ s.t. $||\psi_t'|| \to \infty$ for $t \to t_0$.
Due to the conservation of the energy, it is evident that ψ_t is a blow-up solution if and only if $|\psi_t(0)|$ diverges for $t \to t_0$. This means that, in our model, a blow-up solution ψ_t can be equivalently defined by the condition $||\psi_t||_{L^\infty} \to \infty$ for $t \to t_0$.
Using standard arguments, i.e. the computation of the second time-derivative of the moment of inertia and the uncertainty principle ([RR]), we can prove existence of blow-up solutions.

Proposition 4. Assume $\gamma < 0$ and $\sigma \ge 1$. Then for any $\psi_0 \in H^2(I\!R)$ s.t. ψ_0 is an even function, $||x\psi_0|| < \infty$ and $E(\psi_0) < 0$ the solution ψ_t is a blow-up solution.

Proof. Let us define the moment of inertia (with respect to the origin) of a solution of (3)

$$I(t) \equiv \int_{\mathbb{R}} dx x^2 |\psi_t(x)|^2 \tag{13}$$

An explicit calculation yields

$$\dot{I}(t) = 4\, Im \int_{\mathbb{R}} dx \overline{\psi_t(x)} x \psi_t'(x) \tag{14}$$

$$\ddot{I}(t) = 8E(\psi_0) - 4|\gamma| \left(1 - \frac{2}{\sigma + 1}\right) |\psi_t(0)|^{2\sigma+2} \tag{15}$$

Since $\ddot{I}(t) < 0$, there exists $t_0 < \infty$ s.t. $I(t) \to 0$ for $t \to t_0$. Applying the uncertainty principle

$$||\psi_0||^2 \leq 2\, ||x\psi_t||\, ||\psi_t'|| \tag{16}$$

we conclude that ψ_t is a blow-up solution.

Remark 2. The initial datum in the above proposition is chosen to be even. For an arbitrary initial datum the conclusion of the theorem still holds if the condition $E(\psi_0) < 0$ is replaced by $E(\psi_0^{even}) < 0$.

From propositions 3, 4 it is clear that the critical exponent of the nonlinearity is $\sigma = 1$.
The critical case is of special interest because of the existence of an additional symmetry, the so-called pseudo-conformal invariance. This symmetry can be expressed as follows: if $u_t(x)$ is a solution then

$$v_\tau(y) \equiv \frac{e^{-i\frac{|y|^2}{4(t_0-\tau)}}}{[a(t_0-\tau)]^{\frac{1}{2}}}\, u_{\frac{1}{a^2(t_0-\tau)}} \left(\frac{y}{a(t_0-\tau)}\right) \qquad a, t_0 > 0 \tag{17}$$

is again a solution. Moreover, a stationary solution of the problem, i.e. a solution of the form $Q(x)e^{i\lambda t}$, is easily computed

$$u_t(x) \equiv e^{-\frac{|x|}{2} + i\frac{t}{4}} \tag{18}$$

Combining (17) and (18), we explicitly construct the blow-up solution

$$\psi_t(x) = \frac{1}{\sqrt{a(t_0-t)}} e^{-\frac{|x|}{2a(t_0-t)} - i\frac{|x|^2}{4(t_0-t)} + i\frac{1}{4a^2(t_0-t)}} \tag{19}$$

An advantage to have the explicit solution (19) is that one can directly check the properties of the solution near the blow-up (non existence of strong L^2-limit, L^2-concentration phenomenon, local behaviour, etc.).

On the other hand, the standard NLSE with critical power nonlinearity is also invariant under the pseudo-conformal transformation (17). This fact suggests that the behaviour of a solution that blows up at a single point can be well approximated by our solution (19) near the blow-up time. In this sense our model would exhibit a universal behaviour near the blow-up.
The heuristic motivation is the following. In the standard NLSE (critical case) the "nonlinear potential" is $|\psi_t(x)|^4$ and, near the blow-up, it behaves like the square of a δ-function (see e.g. [Me]).
In our case the "potential term" is $|\psi_t(0)|^2\delta(x)$ and it exhibits the same behaviour along the solution (19) for $t \to t_0$, since the square of the modulus of (19) is a δ-function approximation for $t \to t_0$.
In conclusion we point out that the natural generalization of the model discussed here is the case of NLSE with nonlinearities concentrated in n fixed points in dimensions one, two and three.
The two and three dimensional cases require some care in their formulation since the space of finite energy is strictly larger than H^1. This means that natural instruments like the Sobolev inequalities and the uncertainty principle cannot be directly used. Moreover, in the n points case the analysis of the blow-up solutions becomes more subtle because of the lack of conservation of parity (in dimension one) and angular momentum (in dimension two and three). We plan to approach these questions in a further work.

Acknowledgments. We thank prof. G.F. Dell'Antonio and prof. R. Figari for helpful discussions during the preparation of this work.

References

[A] Adams, R., *Sobolev Spaces*, Academic Press, New York, 1975.

[AGH-KH] Albeverio, S., Gesztesy, F., Högh-Krohn, R., Holden, H., *Solvable Models in Quantum Mechanics*, Springer-Verlag, New York, 1988.

[BKB] Bulashenko, O.M., Kochelap, V.A., Bonilla L.L., *Coherent Patterns and Self-Induced Diffraction of Electrons on a Thin Nonlinear Layer*, Phys. Rev. B, **54**, 3, 1996.

[GV] Ginibre, J., Velo, G., *On a Class of Nonlinear Schrödinger Equations. I. The Cauchy Problem, General Case*, J. Func. Anal., **32**, 1-32, 1979.

[GoVe] Gorenflo, R., Vessella, S., *Abel Integral Equations*, Springer-Verlag, Berlin Heidelberg, 1978.

[J-LPS] Jona-Lasinio, G., Presilla, C., Sjöstrand J., *On Schrödinger Equations with Concentrated Nonlinearities*, Ann. Phys., **240**, 1-21, 1995.

[M] Miller, R. K., *Nonlinear Volterra Integral Equations*, W. A. Benjamin Inc., 1971.

[Me] Merle, F., *Construction of Solutions with Exactly k Blow-up points for the Schrödinger Equation with Critical Nonlinearity*, Comm. Math. Phys., **129**, 223-240, 1990.

[MA] Malomed, B., Azbel, M., *Modulational Instability of a Wave Scattered by a Nonlinear Center*, Phys. Rev. B, **47**, 16, 1993.
[N] Nier, F., *The Dynamics of some Quantum Open Systems with Short-Range Nonlinearities*, preprint Ecole Polytechnique, 1998.
[RR] Rasmussen, J.J., Rypdal, K., *Blow-up in NLSE - A General Review*, Physica Scripta, **33**, 481-497, 1986.

Operator Theory:
Advances and Applications, Vol. 108

Anomalous electron trapping by magnetic flux tubes and electric current vortices

F. Bentosela,[a,b] P. Exner,[c,d] and V.A. Zagrebnov[a,b]

a) Centre de Physique Théorique, C.N.R.S., F–13288 Marseille Luminy
b) Université de la Mediterranée (Aix–Marseille II), F–13288 Luminy
c) Nuclear Physics Institute, Academy of Sciences, CZ–25068 Řež
d) Doppler Institute, Czech Technical Univ., Břehová 7, CZ-11519 Prague
bentosela@cpt.univ-mrs.fr, exner@ujf.cas.cz, zagrebnov@cpt.univ-mrs.fr

We consider an electron with an anomalous magnetic moment, $g > 2$, confined to a plane and interacting with a nonhomogeneous magnetic field B, and investigate the corresponding Pauli Hamiltonian. We prove a lower bound on the number of bound states for the case when B is of a compact support and the related flux is $N+\epsilon$, $\epsilon \in (0,1]$. In particular, there are at least $N+1$ bound states if B does not change sign. We also consider the situation where the magnetic field is due to a localized rotationally symmetric electric current vortex in the plane. In this case the flux is zero; there is a pair of bound states for a weak coupling, and higher orbital-momentum "spin-down" states appearing as the current strength increases.

Interaction of electrons with a localized magnetic field has been a subject of interest for a long time. It has been observed recently that a magnetic flux tube can bind particles with spin antiparallel to the field provided the latter have an anomalous magnetic moment, $g > 2$. Recall that this is the case for a free electron which has $g = 2.0023$. The effect was demonstrated first in simple examples [CFC, Mo], notably those of a circular tube with a homogeneous or a δ-shell field, and then extended to any rotationally invariant field $B(x)$ which is of a compact support and does not change sign [CC].

Our first aim here is to show that the last condition can be substantially weakened and the rotational–invariance requirement can be dropped altogether. We consider the standard two–dimensional Pauli electron Hamiltonian [Th],

$$H_P^{(\pm)}(A) = (-i\nabla - A(x))^2 \pm \frac{g}{2}B(x) = D^*D \pm \frac{1}{2}(g\pm 2)B(x), \qquad (1)$$

in natural units, $2m = \hbar = c = e = 1$; here $D := (p_1 - A_1) + i(p_2 - A_2)$ and the two signs correspond to two possible spin orientations. We are free to choose the magnetic flux direction; if it points conventionally up we will be concerned primarily with the operator $H_P^{(-)}(A)$ which describes electron with the spin antiparallel to the flux. The magnetic field $B = \partial_1 A_2 - \partial_2 A_1$ is supposed to be integrable and of a compact support Σ, with

$$F := \frac{1}{2\pi} \int_\Sigma B(x)\, d^2x = N + \epsilon\,, \tag{2}$$

where $\epsilon \in (0,1]$ and N is a non–negative integer. The quantity F, positive by assumption, is the total flux measured in the natural units $(2\pi)^{-1}$.

Recall further that by the theorem of Aharonov and Casher [AC, Th] the operator $H_P^{(-)}(A)$ without an anomalous moment, $g = 2$, has in this situation N zero energy eigenvalues. The corresponding eigenfunctions are given explicitly by

$$\chi_j(x) = e^{-\phi(x)}(x_1 + ix_2)^j, \quad j = 0, 1, \dots, N-1\,, \tag{3}$$

where

$$\phi(x) := \frac{1}{2\pi} \int_\Sigma B(y)\, \ln|x-y|\, d^2y\,. \tag{4}$$

Moreover, χ_N also solves the equation $H_P^{(-)}(A)\chi = 0$ representing a zero–energy resonance; this follows from the fact that $\chi_j(x) = o(|x|^{-F+j})$ as $|x| \to \infty$ — *cf.* [Th, Sec.7.2].

Theorem 1. Under the stated assumptions, the operator $H_P^{(-)}(A)$ has for $g > 2$ at least n_B negative eigenvalues, where n_B is the number of $j = 0, 1, \dots, N$ such that

$$\int_\Sigma B(x)\, e^{-2\phi(x)}\, r^{2j}\, d^2x \geq 0\,, \tag{5}$$

where $r := (x_1^2 + x_2^2)^{1/2}$. In particular, there are at least $n_B = N+1$ bound states if $B(x) \geq 0$.

Sketch of the proof: It is based on a variational argument. We employ the above mentioned zero–energy solutions to construct a family of trial functions ψ which make the quadratic form

$$(\psi, H_P^{(-)}(A)\psi) = \int_{I\!R^2} |D\psi|^2 d^2x - \frac{1}{2}(g-2) \int_{I\!R^2} B|\psi|^2 d^2x$$

negative. Specifically, we choose

$$\psi_j(x) := f_R(x)\chi_j(x) + \varepsilon h(x)\,, \tag{6}$$

where $h \in C_0^\infty(\Sigma)$ and $f_R(x) = f\left(\frac{r}{R}\right)$ for a suitable function $f : I\!R_+ \to I\!R$ such that $f(u) = 1$ for $u \leq 1$. It is then straightforward to compute the value of the energy form,

$$\begin{aligned}(\psi_j, H_P^{(-)}(A)\psi_j) &= \frac{1}{R^2}\int_{\mathbb{R}^2}\left|f'\left(\frac{r}{R}\right)\chi_j(x)\right|^2 d^2x + \varepsilon^2\int_\Sigma |(Dh)(x)|^2 d^2x \\ &- \frac{1}{2}(g-2)\left\{\int_\Sigma B(x)|\chi_j(x)|^2 d^2x\right. \\ &+ \left. 2\varepsilon\mathrm{Re}\int_\Sigma \bar{h}(x)B(x)\chi_j(x)\,d^2x + \varepsilon^2\int_\Sigma B(x)|h(x)|^2 d^2x\right\},\end{aligned}$$

where we have employed $D\chi_j = 0$ and the fact that h and $f'\left(\frac{\cdot}{R}\right)$ have disjoint supports. As we have said, $\chi_j \in L^2$ for $j = 0,\dots,N-1$. In this case we put $f = 1$ so the first term at the *rhs* is absent. If $\int_\Sigma B|\chi_j|^2 d^2x > 0$ we may set also $\varepsilon = 0$ to get a negative value. If B is non-negative, in particular, we obtain in this way $(\psi_j, H_P^{(-)}(A)\psi_j) < 0$ for $j = 0,\dots,N-1$.

For a sign-changing B the last integral might not be positive. If it is zero, a bound state still exists: it is always possible to choose h in such a way that $\mathrm{Re}\int_\Sigma \bar{h}B\chi_j\,d^2x \neq 0$. For small ε the linear term prevails over the quadratic ones and the form can be made negative by choosing properly the sign of ε. Finally, for $j = N$ the Aharonov–Casher solution has to be modified at large distances to produce a square integrable trial function. We choose, *e.g.*, $f \in C_0^\infty(\mathbb{R}_+)$ such that $f(u) = 0$ for $u \geq 2$. Using $|\chi_N(x)| = o(r^{-\epsilon})$ we estimate the first term at the *rhs* as

$$\frac{1}{R^2}\int_{\mathbb{R}^2}\left|f'\left(\frac{r}{R}\right)\chi_j(x)\right|^2 d^2x \leq C\|f'\|_\infty^2 R^{-2\epsilon}$$

for a positive C. If (5) is valid, one can achieve in the same way as above that the sum of the other terms is negative; it is then sufficient to set R large enough to make the whole *rhs* negative.

We have thus constructed n_B trial functions with the desired property. They are linearly independent, since the same is true for χ_j and the latter coincides with ψ_j in $\mathcal{B}_R \setminus \Sigma$. Consequently, the ψ_j's for which the requirement (5) is satisfied span an n_B-dimensional subspace in $L^2(\mathbb{R}^2)$. □

While the sufficient condition of Theorem 1 improves earlier results, it is still too restrictive. We postpone discussing how to optimize it to a subsequent paper.

The situation becomes more complicated when the total flux is zero. Here we will restrict ourselves to the particular case with a rotational symmetry; then (1) can be replaced by a family of partial wave Hamiltonians

$$H_\ell^{(\pm)} = -\frac{d^2}{dr^2} - \frac{1}{r}\frac{d}{dr} + V_\ell^{(\pm)}(r), \quad V_\ell^{(\pm)}(r) := \left(A(r) + \frac{\ell}{r}\right)^2 \pm \frac{1}{2}gB(r) \quad (7)$$

on $L^2(\mathbb{R}^+, r\,dr)$. The angular component $A(r)$ of the vector potential is now related to the magnetic field by $B(r) = A'(r) + r^{-1}A(r)$.

A typical situation with a vanishing flux arises when the field is generated by an electric current vortex in the plane. The physical appeal of such a problem stems in part from the fact that local current vortices are common in transport of charged particles [EŠSF]. In the following we shall discuss this example. We assume that the current is anticlockwise, $J(x) = \lambda J(r) e_\varphi$. Here r, φ are the polar coordinates, the total current is $\lambda \int_0^\infty J(r)\, dr$, and the positive parameter λ is introduced to control the vortex "strength".

It is necessary in this case to relax the compact–support requirement on the magnetic field. We suppose that J is C^2 smooth and non–negative, $J(r) \geq 0$, and has the following asymptotic behaviour,

$$J(r) = ar^2 + \mathcal{O}(r^3) \quad \text{and} \quad J(r) = \mathcal{O}(r^{-3-\epsilon}) \tag{8}$$

for some $\epsilon > 0$ at the origin and at large distances, respectively. The corresponding vector potential is easily evaluated [Ja],

$$A(r) \;=\; 4\lambda \int_0^\infty J(r')\, \frac{r'}{r_<} \left[K\left(\frac{r_<^2}{r_>^2}\right) - E\left(\frac{r_<^2}{r_>^2}\right) \right] dr' , \tag{9}$$

where $K,\, E$ are the full elliptic integrals of the first and the second kind, respectively, and the usual shorthands, $r_< := \min(r, r')$ and $r_> := \max(r, r')$ are employed. In view of the regularity of J the integral is finite for every r, because $E(\zeta)$ is regular at $\zeta = 1$ and $K(\zeta)$ has a logarithmic singularity there.

Let us label the Pauli Hamiltonian (1) with the vector potential (9) and its partial–wave components (7) by the current strength λ.

Theorem 2. Under the stated assumptions, $\sigma(H_\ell^{(\pm)}(\lambda)) = [0, \infty)$ for $\ell \neq 0$, while both $H_0^{(\pm)}(\lambda)$ exhibit a bound state if λ is small enough. On the other hand, each operator $H_\ell^{(-)}(\lambda)$ has a negative eigenvalue for a sufficiently large λ.

Sketch of the proof: By the regularity of J, the effective potentials (7) are C^1 smooth and

$$V_\ell^{(\pm)}(r) \;=\; \frac{\ell^2}{r^2} + \lambda m\, \frac{2\ell \mp g}{2r^3} + \mathcal{O}(r^{-3-\epsilon}) , \tag{10}$$

as $r \to \infty$, where $m := \pi \int_0^\infty J(r')\, r'^2 dr'$ is the dipole moment of the current for $\lambda = 1$. Consequently, the essential spectrum is not affected by the magnetic field. We rewrite the potentials into the form

$$V_\ell^{(\pm)}(r) \;=\; \left(\lambda A_1(r) + \frac{\ell}{r} \right)^2 \pm \frac{\lambda}{2}\, g B_1(r) , \tag{11}$$

where the indexed magnetic field refers to the value $\lambda = 1$. Since $H_\ell^{(\pm)}(\lambda)$ is nothing else than the s–wave part of the two–dimensional Schrödinger operator with the centrally symmetric potential (11), it is sufficient to find eigenvalues of the latter. If $\ell \neq 0$, the first term in (11) is below bounded by $\lambda h(r)$ for a suitably chosen positive function h of compact support. Since the second term does not contribute to $\int_0^\infty V_\ell^{(\pm)}(r)\, r\, dr$ which determines the weak-coupling behaviour, the result follows from the standard condition [Si] and the minimax principle.

While the above integral is positive in the case $\ell = 0$ as well for any $\lambda \neq 0$, this fact itself need not prevent binding. A more careful Birman–Schwinger analysis up to the second order in λ is required: it shows that a weakly coupled bound state exists if

$$\int_{\mathbb{R}^2} A(x)^2\, d^2x \,+\, \frac{g^2}{8\pi} \int_{\mathbb{R}^2\times\mathbb{R}^2} B(x) \ln|x-x'|\, B(x')\, d^2x\, d^2x' \,<0\,. \tag{12}$$

Evaluating the last integral, we find that the condition is satisfied for $g > 2$. This rectifies an incorrect claim made in [BEZ]; a more detailed discussion on that point will be presented in a forthcoming publication. The asymptotic behaviour of the bound state energy (in the sense of [Si]) is

$$\epsilon(\lambda) \,\approx\, -\,\exp\left\{ -\left(\frac{\lambda^2}{8}(g^2-4) \int_{\mathbb{R}^2} A(r)^2\, r\, dr \right)^{-1} \right\} \tag{13}$$

for both spin orientations (since $g \neq 2$, the second theorem of [AC] does not apply and the degeneracy may be lifted in the next order).

On the other hand, the existence of the "antiparallel" bound states in a strong vortex follows from the behaviour of the effective potential around the origin. We have

$$A(r) \,=\, \lambda\mu r + \alpha_0(r)\,, \qquad \mu \,:=\, \int_0^\infty J(r')\, \frac{dr'}{r'}\,; \tag{14}$$

using (9) and properties of the elliptic integrals we find $\alpha_0(r) \,=\, \mathcal{O}(r^2)$. This further implies

$$B(r) \,=\, 2\lambda\mu + \beta_0(r)\,, \qquad \beta_0(r) \,:=\, \alpha_0'(r) + \frac{1}{r}\alpha_0(r) \,=\, \mathcal{O}(r)\,. \tag{15}$$

Consider the case $\ell = 0$. We substitute to (7) from (14,15) and employ the rescaled variable $u := r\sqrt{\lambda}$. In this way $H_0^{(\ell)}$ is unitarily equivalent to the operator λA_λ, where $A_\lambda = A_0 + W_\lambda$ on $L^2(\mathbb{R}_+, u\, du)$ with

$$A_0 \,:=\, -\,\frac{d^2}{du^2} \,-\, \frac{1}{u}\frac{d}{du} \,-\, g\mu + \mu^2u^2 \tag{16}$$

and

$$W_\lambda(u) \,:=\, 2\sqrt{\lambda}\mu\, u\alpha_0\left(\frac{u}{\sqrt{\lambda}}\right) \,+\, \lambda\,\alpha_0^2\left(\frac{u}{\sqrt{\lambda}}\right) \,-\, \frac{1}{2}\,g\,\beta_0\left(\frac{u}{\sqrt{\lambda}}\right). \tag{17}$$

The limit $\lambda \to \infty$ changes the spectrum substantially; we have $\sigma_{ess}(A_\lambda) = \sigma_{ess}(\lambda A_\lambda) = [0,\infty)$ for any $\lambda > 0$, while A_0 as the s–wave part of the two-dimensional harmonic oscillator has a purely discrete spectrum. Nevertheless, one can justify the use of the asymptotic perturbation theory for stable (*i.e.*, negative) eigenvalues of A_0; the fact that $W_\lambda \to 0$ pointwise together with the resolvent identity imply $A_\lambda \to A_0$ in the strong resolvent sense as $\lambda \to \infty$ [BEZ]. In that case there is a family of $\nu_n(\lambda) \in \sigma(A_\lambda)$ to any $\nu_n \in \sigma_p(A_0)$ such that $\nu_n(\lambda) \to \nu_n$ [Ka]. The spectrum of A_0 is given explicitly by

$$\nu_n = \mu(4n+2-g), \qquad n = 0, 1, \dots, \tag{18}$$

so ν_0 is stable for $g > 2$ and A_λ has a negative eigenvalue for λ large enough. The analogous argument applies to the case $\ell \neq 0$, where the potential in (16) is replaced by $\mu^2 u^2 + \ell^2 r^{-2} + \mu(2\ell - g)$, and one looks for negative eigenvalues among $\nu_{n,\ell} = \mu(4n + 2(|\ell|+\ell) + 2 - g)$. The critical λ at which the eigenvalue emerges from the continuum is naturally ℓ–dependent. □

V.Z. thanks for the hospitality extended to him at NPI. The research has been partially supported by GACR under the contract 202/96/0218.

References

[AC] Y. Aharonov, A. Casher: Ground state of a spin–1/2 charged particle in a two–dimensional magnetic field, *Phys. Rev.* **A19** (1979), 2641–2642.

[BEZ] F. Bentosela, P. Exner, V.A. Zagrebnov: Electron trapping by a current vortex, *J. Phys.* **A31** (1998), L305–311.

[CC] R.M. Cavalcanti, C.A.A. de Carvalho: Bound states in spin-1/2 charged particle in a magnetic flux tube, *J. Phys.* **A31** (1998), 7061-7063.

[CFC] R.M. Cavalcanti, E.S. Fraga, C.A.A. de Carvalho: Electron localization by a magnetic vortex, *Phys. Rev.* **B56** (1997), 9243–9246.

[EŠSF] P. Exner, P. Šeba, A.F. Sadreev, P. Středa, P. Feher: Strength of topologically induced magnetic moments in a quantum device, *Phys. Rev. Lett.* **80** (1998), 1710–1713.

[Ja] J.D. Jackson: *Classical Electrodynamics*, John Wiley, New York 1962.

[Ka] T. Kato: *Perturbation Theory for Linear Operators*, Springer, Heidelberg 1966.

[Mo] A. Moroz: Single–particle density of states, bound states, phase–shift flip, and a resonance in the presence of an Aharonov–Bohm potential, *Phys. Rev.* **A53** (1996), 669–694.

[Si] B. Simon: The bound state of weakly coupled Schrödinger operators in one and two dimensions, *Ann.Phys.* **97** (1976), 279–288.

[Th] B. Thaller: *The Dirac equation*, Springer, Berlin 1992.

Operator Theory:
Advances and Applications, Vol. 108

On the absolutely continuous energy distribution of a quantum mechanical system in a bounded domain

Johannes F. Brasche
Institute for Applied Mathematics, University of Bonn
Wegelerstr. 10, D-53115 Bonn, Germany

Abstract

Let Ω be a bounded domain in $\mathbf{R}^d$, $d > 1$. It is shown that for every self-adjoint operator M in a separable Hilbert space there exists a self-adjoint realization H of the Laplacian on Ω such that the absolutely continuous part of H is unitarily equivalent to the absolutely continuous part of M. A method to construct H is given.

Introduction

In what follows let Ω be a bounded domain in $\mathbf{R}^d$ and $d > 1$. Consider a quantum mechanical particle which is inside Ω for all times and does not interact with any potential inside Ω. Obviously (up to a constant factor) the corresponding Hamiltonian H must be a self-adjoint realization of the Laplacian on Ω, i. e. H is a self-adjoint operator in $L^2(\Omega)$, its domain $D(H)$ is contained in the Sobolev space of square integrable f such that Δf is also square integrable and

$$Hf = -\Delta f, \quad f \in D(H).$$

Obviously H is a self-adjoint realization of the Laplacian on Ω if and only if it is a self-adjoint extension of the minimal Laplacian on Ω, i. e. the operator $-\Delta_{min}^{\Omega}$ in $L^2(\Omega)$ defined by

$$\begin{aligned} D(-\Delta_{min}^{\Omega}) &:= C_0^\infty(\Omega), \\ -\Delta_{min}^{\Omega} f &:= -\Delta f, \quad f \in C_0^\infty(\Omega). \end{aligned}$$

For a long time one has only known self-adjoint realizations of the Laplacian on a bounded domain which have a discrete spectrum. Simultaneously and independently two different groups found out that there are interesting additional possibilities. B. Simon [1] has given a bounded domain Ω such that the Neumann Laplacian on Ω has non–empty absolutely continuous spectrum. Implications of Simon's result for scattering theory have been discussed by R. Hempel and R. Weder [2].

S. Albeverio, J. F. Brasche, H. Neidhardt and J. Weidmann [3], [4], [5], [6], [7], [8] have introduced new classes of self–adjoint realizations of the Laplacian on a bounded domain Ω with dense point spectrum and non–empty singular continuous and absolutely continuous spectrum.

Let $b_\Omega > 0$ be the largest lower bound of the minimal Laplacian on Ω and

$$J_\Omega := (-\infty, b_\Omega).$$

Let M be any self–adjoint operator in any separable Hilbert space. Let $T \subset \mathbf{R}$ be a set which is "thick" in the sense that it is the topological support of an absolutely continuous measure. In [7] it has been shown that there exists a self–adjoint realization H of the Laplacian on Ω such that H has no singular continuous spectrum below b_Ω, below b_Ω has the same eigenvalues with the same multiplicities as M and satisfies

$$H_{acJ_\Omega} \simeq M_{acJ_\Omega}.$$

Here $\simeq$ means "unitarily equivalent", M_{ac} stands for the absolutely continuous part and

$$M_J := M_{|D(M)\cap \mathrm{ran} 1_J(M)},$$

i. e. M_J equals the restriction of M to the intersection of $D(M)$ and the range of the spectral projector corresponding to M and J. In particular, below b_Ω the absolutely continuous spectrum $\sigma_{ac}(H)$ of H equals the absolutely continuous spectrum of the (arbitrarily chosen) self–adjoint operator M. In [7] also a method to construct an operator H with the mentioned properties has been presented. In [8] it has been shown that there exists a self–adjoint realization H of the Laplacian on Ω such that

$$\sigma_{sc}(H) \cap J_\Omega = T \cap J_\Omega.$$

In addition, as above, below b_Ω the absolutely continuous part and the pure point part of H can be arbitrarily chosen.

In [8] only the existence of an operator H with the mentioned properties has been shown. In general it is an open problem how to construct H. If, however, the set T is "very thick" in the sense that it equals the closure of some open set then one can give an explicit construction [6].

It is also an open problem whether for every perfect set F there exists a self–adjoint realization H of the Laplacian on Ω such that

$$\sigma_{sc}(H) \cap J_\Omega = F \cap J_\Omega.$$

Using essentially results by R. del Rio, S. Jitomirskaja, Y. Last and B. Simon [9] in [8] it has been shown that for every $\alpha \in [0,1]$ there exists a self–adjoint realization H of the Laplacian on Ω such that the set $\sigma_{sc}(H) \cap J_\Omega$ is non–empty and its Hausdorff–dimension equals the given number α.

Above we have always concentrated on the interval J_Ω. In general it is not possible to replace J_Ω by any strictly larger open interval J. For instance as a trivial consequence of the mentioned results we get that for every relatively closed subset C of J_Ω there exists a self–adjoint realization H_C of the Laplacian on Ω such that

$$\sigma(H_C) \cap J_\Omega = C.$$

Obviously the analogous statement is false for every interval J which is a proper superset of J_Ω. For instance there does not exist any self–adjoint realization H of the Laplacian on Ω such that H has no spectrum in J. There are plenty of other rules one can find if one investigates “global spectral properties”, i. e. the spectra on the whole real axis. Some of these rules are far from being trivial and actually the investigation of global spectral properties seems to be extremely difficult. I refer to the work [10] by V. A. Derkach and M. M. Malamud for first steps in this direction.

While it is not true that for every closed subset C of the real axis there exists a self–adjoint realization H_C of the Laplacian on Ω such that

$$\sigma(H_C) = C$$

it is true that self–adjoint realizations can have **every** kind of absolutely continuous spectrum **on the whole real axis**. Actually this statement does not only hold in the (weak) sense that for every set C_{ac} which equals the absolutely continuous spectrum of some self–adjoint operator, or, in other words, the topological support of some absolutely continuous measure there exists a self–adjoint realization $H_{(C_{ac})}$ of the Laplacian on Ω such that

$$\sigma_{ac}(H_{(C_{ac})}) = C_{ac}$$

but even in the strong sense that for every self–adjoint operator M in a separable Hilbert space there exists a self–adjoint realization H such that

$$H_{ac} \simeq M_{ac}.$$

This surprising “global result” on the absolutely continuous part is an easy consequence of considerations in [7]. In what follows I shall give a proof of this result and shall provide a method for the construction of an operator H with the required properties. From the very beginning I shall work within a general framework.

The construction

Let S be a closed symmetric operator in a Hilbert space $\mathcal{H}$. Suppose that S is invertible and its inverse S^{-1} is bounded. Let

$$A := P_{\mathrm{ran}(S)} S^{-1}, \quad B := P_{\mathrm{ran}(S)^\perp} S^{-1}.$$

Here $P_{\mathcal{K}}$ denotes the orthogonal projection from $\mathcal{H}$ to the closed subspace $\mathcal{K}$ of $\mathcal{H}$. A and B can and will be regarded as an operator in the Hilbert space $\mathrm{ran}(S)$ and from the Hilbert space $\mathrm{ran}(S)$ to the Hilbert space $\mathrm{ran}(S)^\perp$, respectively.

It is not difficult to prove that the operator A in $\mathrm{ran}(S)$ is self–adjoint [7]. Suppose now in addition that the absolutely continuous spectrum of A is empty and that there exists an orthogonal projection P in the Hilbert space $\mathrm{ran}(S)^\perp$ such that the range of P is infinite dimensional and the operator PB from $\mathrm{ran}(S)$ to $\mathrm{ran}(P)$ belongs to the trace class.

Let

$$S_0 := S^*_{|\{f+g: f\in D(S), g\in \mathrm{ran}(S)^\perp \ominus \mathrm{ran}(P)\}}.$$

It follows from a general result in [7] and can also easily verified directly that S_0 admits the following decomposition:

$$S_0 = G \oplus 0$$

where "0" denotes the zero–operator in the Hilbert space

$$\mathrm{ran}(P)^\perp = \mathrm{ran}(S)^\perp \ominus \mathrm{ran}(P).$$

Along with S also the operator G in the Hilbert space

$$\mathcal{H}_0 := \mathrm{ran}(S) \oplus \mathrm{ran}(P)$$

is continuously invertible and according to the above decomposition of $\mathcal{H}_0$ the operator G^{-1} is represented by the block matrix

$$G^{-1} = \begin{pmatrix} A \\ PB \end{pmatrix}$$

(cf. [7]).

Now let M be any self–adjoint operator in any separable Hilbert space. Since the space $\mathrm{ran}(P)$ is infinite dimensional we can choose a self–adjoint operator Q in $\mathrm{ran}(P)$ such that

$$Q_{ac}^{-1} \simeq M_{ac}.$$

It is easily verified that the operator in $\mathcal{H}_0$ given by the block matrix

$$\begin{pmatrix} A & (PB)^* \\ PB & Q \end{pmatrix}$$

is invertible and its inverse G_Q is a self–adjoint extension of G. Since PB belongs to the trace class the absolutely continuous part of G_Q^{-1} is unitarily equivalent to the absolutely continuous part of the orthogonal sum $A \oplus Q$. Since the absolutely continuous spectrum of A is empty the absolutely continuous part of $A \oplus Q$ equals Q_{ac}. Thus

$$(G_Q)_{ac} \simeq Q_{ac}^{-1} \simeq M_{ac}.$$

Since G_Q is a self–adjoint extension of G and $G \oplus 0$ is an extension of the given symmetric operator S the orthogonal sum

$$S_Q := G_Q \oplus 0$$

is a self–adjoint extension of S and it satisfies

$$(S_Q)_{ac} \simeq (G_Q)_{ac} \simeq M_{ac}.$$

In order to get now the mentioned result on self–adjoint realizations of the Laplacian on Ω one has only to prove that the closure S_Ω of the minimal Laplacian on Ω has the properties of the symmetric operator S used in the above considerations. This is easily done. Obviously S_Ω is continuously invertible. Since its inverse is compact the operators A_Ω and B_Ω (defined in the obvious way) are also compact. Thus A_Ω has no absolutely continuous spectrum. Along with B_Ω also the operator $|B_\Omega^*|$ is compact. Thus the Hilbert space $\mathrm{ran}(S)^\perp$ has an orthonormal base of eigenvectors e_n of $|B_\Omega^*|$ corresponding to eigenvalues E_n, $n \in \mathbf{N}$, which tend to zero as $n \to \infty$. We can choose an absolutely summable subsequence (E_{n_k}). Then the orthogonal projection P_Ω onto the space spanned by the corresponding eigenvectors e_{n_k} has the required properties, as is easily verified.

References

[1] B. Simon: The Neumann Laplacian of a jelly roll. *Proc. Amer. Math. Soc.* Vol. 114, No. 3, (1993), 783–785.

[2] R. Hempel, R. Weder: On the completeness of wave operators under loss of local compactness. *J. Funct. Analysis* 113 (1993), 391–412.

[3] J. F. Brasche: Generalized Schrödinger operators, an inverse problem in spectral analysis and the Efimov effect, pp. 207–245 in S. Albeverio et al. (eds.): *Stochastic Processes, Physics and Geometry.* World Scientific, Singapore 1989.

[4] J. F. Brasche, H. Neidhardt, J. Weidmann: On the point spectrum of self–adjoint extensions. *Math.Zeitschr.* 214 (1993), 343–355.

[5] J. F. Brasche, H. Neidhardt: On the absolutely continuous spectrum of self–adjoint extensions. *Journ. Funct. Analysis* 131, (1995), 364–385.

[6] J. F. Brasche, H. Neidhardt: On the singular continuous spectrum of self–adjoint extensions. *Math. Zeitschr.* 222 (1996), 533–542.

[7] S. Albeverio, J. F. Brasche, H. Neidhardt: On inverse spectral theory for self–adjoint extensions. *Journ. Funct. Anal.* 154 (1998). 130–173.

[8] J. F. Brasche: On inverse spectral theory for self–adjoint extensions: Nowhere dense singular continuous spectra and Hausdorff dimension of spectra. Preprint of the SFB 237, 1997. Submitted to Journ. Op. Theory.

[9] R. del Rio, S. Jitomirskaya, Y. Last, B. Simon: Operators with singular continuous spectrum IV. Hausdorff dimensions, rank one perturbations, and localization. *J. D'Analyse Mathematique* 69, 153–200 (1996).

[10] V. A. Derkach, M. M. Malamud: Generalized resolvents and the boundary value problems for Hermitean operators with gap. *J. Funct. Analysis* 95, (1991), 1–95.

Operator Theory:
Advances and Applications, Vol. 108

Some Aspects of Generalized Contact Interaction in One-Dimensional Quantum Mechanics

Taksu Cheon[1,2] and T. Shigehara[3]

[1] *Laboratory of Physics, Kochi University of Technology*
Tosa Yamada, Kochi 782-8502, Japan
[2] *Theory Division, High Energy Accelerator Research Organization (KEK)*
Tsukuba, Ibaraki 305-0801, Japan
[3] *Department of Information and Computer Sciences*
Saitama University, Urawa, Saitama 338-8570, Japan

Abstract

A local construction is presented for a one-dimensional contact interaction (ε potential) which induces the discontinuity of the wave function while keeping its derivative continuous. By combining the ε potential and the Dirac's δ function, we construct most general one-dimensional contact interactions allowable under the time reversal symmetry. We present some elementary results for the scattering problem which suggest a dual relation between δ and ε potentials.

I. Introduction

The quantum-mechanical contact interaction is a natural idealization of an obstacle with size much smaller than the wavelength of a particle. This idealization often brings the prized property of solvability to the physical models [1]. In spite of the seeming simplicity of contact interactions, there are several non-trivial aspects which are largely left unexplored. Even in the simplest setting of one dimension, there has been a longstanding problem of approximating the generalized contact interaction with a short-range limit of a local self-adjoint operator. Apart from a rather trivial case of Dirac's δ potential, it has resisted the realization of such representation. Existing approximations have been either non-local or non-Hermitian [2-6], thus have left us with little physical intuition. Also, one could not find much relevance to the experimentally realizable "real-life" quantum mechanics. In this paper, we sketch a scheme for just such local realization of generalized contact interaction. We also show that the object so realized is not only a natural generalization, but also an indispensable element of one-dimensional quantum mechanics.

We consider the one-dimensional Schrödinger equation

$$-\varphi''(x) + V(x)\varphi(x) = k^2\varphi(x), \tag{1}$$

where $V(x)$ belongs to a general class of quantum-mechanical contact interactions; $V(x) = 0$ for $x \neq 0$. The self-adjoint extension theory dictates [1] that if the system has the time reversal symmetry, there exists a three parameter family of solutions (apart from the special exception of "separated boundary" [7]), which is characterized by the connection conditions

$$\begin{pmatrix} \varphi'(0_+) \\ \varphi(0_+) \end{pmatrix} = \mathcal{V} \begin{pmatrix} \varphi'(0_-) \\ \varphi(0_-) \end{pmatrix} \tag{2}$$

with $\mathcal{V} \in SL(2, \mathbf{R})$. Clearly, the connection matrix

$$\mathcal{V}_\delta(v) = \begin{pmatrix} 1 & v \\ 0 & 1 \end{pmatrix} \tag{3}$$

corresponds to the δ potential of strength v;

$$V(x) = v\delta(x). \tag{4}$$

Our main purpose is to show that a potential consisting of several δ functions with disappearing distances can give rise to all possible connection conditions Eq.(2) with appropriate scaling of the coupling strengths. This means that all allowable one-dimensional contact interactions can be realized as the zero-range limit of local self-adjoint (namely, physically realizable) interactions. We start in Sect.II by constructing a zero-range force which leaves the derivative of the wave function $\varphi'(x)$ continuous, but makes $\varphi(x)$ itself discontinuous [2,3];

$$\mathcal{V}_\varepsilon(u) = \begin{pmatrix} 1 & 0 \\ u & 1 \end{pmatrix}. \tag{5}$$

We denote the contact interaction with the connection condition (5) by $\varepsilon(x)$;

$$V(x) = u\varepsilon(x). \tag{6}$$

We then show that the general connection condition of the form Eq.(2) can be constructed from the ε and δ potentials. Some elementary results for the scattering problem are presented in Sect.III, where a dual role of δ and ε potentials is emphasized. This work is summarized in Sect.IV.

II. Local Realization of Discontinuity-Inducing Contact Interactions

We begin with the Schrödinger equation in one-dimensional free space;

$$-\varphi''(x) = k^2\varphi(x), \tag{7}$$

where k is the wave number of a particle. It is convenient to introduce a vector notation for the wave function and its space derivative;

$$\Psi(x) = \begin{pmatrix} \varphi'(x) \\ \varphi(x) \end{pmatrix}. \tag{8}$$

The Schrödinger equation, Eq.(7) now takes the form

$$\Psi'(x) = \mathcal{H}(k)\Psi(x) \tag{9}$$

with

$$\mathcal{H}(k) = \begin{pmatrix} 0 & -k^2 \\ 1 & 0 \end{pmatrix}. \tag{10}$$

The solution of Eq.(9) is given by

$$\Psi(x) = \mathcal{G}(k; x - x_0)\Psi(x_0), \tag{11}$$

where

$$\mathcal{G}(k;x) \equiv e^{\mathcal{H}(k)x} = \begin{pmatrix} \cos kx & -k \sin kx \\ \frac{1}{k}\sin kx & \cos kx \end{pmatrix}. \tag{12}$$

Because of $Tr\mathcal{H}(k) = 0$, we have $\mathcal{G}(k;x) \in SL(2, \mathbf{R})$, namely $\det \mathcal{G}(k;x) = 1$.

We now consider a potential consisting of three nearby δ functions located with equal distances a which we assume to be small [8];

$$V(a;x) = v_1\delta(x+a) + v_0\delta(x) + v_1\delta(x-a). \tag{13}$$

The connection condition between $x = (-a)_-$ and $x = a_+$ is given by

$$\Psi(a_+) = \mathcal{V}_{\varepsilon;a}(k)\Psi((-a)_-) \tag{14}$$

where

$$\mathcal{V}_{\varepsilon;a}(k) = \mathcal{V}_\delta(v_1)\mathcal{G}(k;a)\mathcal{V}_\delta(v_0)\mathcal{G}(k;a)\mathcal{V}_\delta(v_1). \tag{15}$$

For sufficiently small $a \ll 1/k$, we can expand Eq.(12) as

$$\mathcal{G}(k;a) \simeq \begin{pmatrix} 1 & -k^2 a \\ a & 1 \end{pmatrix} \tag{16}$$

Inserting Eqs.(3) and (16) into Eq.(15), we obtain

$$\begin{aligned} [\mathcal{V}_{\varepsilon;a}(k)]_{11} &= [\mathcal{V}_{\varepsilon;a}(k)]_{22} \\ &= 1 + (v_0 + 2v_1 + v_0 v_1 a - k^2 a)a, \end{aligned} \tag{17}$$

$$\begin{aligned} [\mathcal{V}_{\varepsilon;a}(k)]_{12} &= v_0 + 2v_1 + 2v_1 a(v_0 + v_1 - k^2 a) \\ &\quad + v_0 v_1^2 a^2 - 2k^2 a, \end{aligned} \tag{18}$$

$$[\mathcal{V}_{\varepsilon;a}(k)]_{21} = 2a + v_0 a^2. \tag{19}$$

In the limit $a \to +0$ with v_0, v_1 constant, we simply obtain the connection condition for a δ potential of strength $v_0 + 2v_1$. On the other hand, Eq.(19) tells that if the coupling strength v_0 varies as

$$v_0(a) \simeq \frac{u}{a^2} \tag{20}$$

for small a, one obtains $\lim_{a\to+0} [\mathcal{V}_{\varepsilon;a}(k)]_{21} = u$. Furthermore, if we let

$$v_1(a) \simeq \frac{2}{u} - \frac{1}{a}, \tag{21}$$

the diagonal elements (17) converge toward one in the small a limit. We obtain

$$\lim_{a\to+0} \mathcal{V}_{\varepsilon;a}(k) = \mathcal{V}_\varepsilon(u) \tag{22}$$

with Eqs.(20) and (21). Thus our first objective is achieved.

Though both the strengths v_0, v_1 diverge in the small a limit, $V(a;x)$ is nonsingular by construction. Note that $\lim_{a\to+0} [\mathcal{V}_{\varepsilon;a}(k)]_{12} = 0$ is ensured because of $\mathcal{V}_{\varepsilon;a}(k) \in SL(2,\mathbf{R})$.

It appears that our procedure of starting from the singular interaction, eq. (13) is essential in obtaining the ε potential, since a finite "three-peaked" potential, say, will simply converge toward usual δ function in the zero-size limit due to the theorem 3.2.3 of Ref. [1]. Obviously, more rigorous analysis is called for to clarify the exact nature of the convergence of the Hilbert space operator $\mathcal{V}_{\varepsilon;a}$.

By using the ε potential and the δ potential in combination, one can realize the general connection condition

$$\mathcal{V} = \begin{pmatrix} t & v \\ u & s \end{pmatrix}, \tag{23}$$

where $ts - uv = 1$, $t, v, u, s \in \mathbf{R}$. The following matrix decompositions serve for our purpose;

$$\begin{pmatrix} t & v \\ u & s \end{pmatrix} = \begin{cases} \mathcal{V}_\delta(\frac{t-1}{u})\mathcal{V}_\varepsilon(u)\mathcal{V}_\delta(\frac{s-1}{u}) & \text{for } u \neq 0, \\ \mathcal{V}_\varepsilon(\frac{s-1}{v})\mathcal{V}_\delta(v)\mathcal{V}_\varepsilon(\frac{t-1}{v}) & \text{for } v \neq 0. \end{cases} \tag{24}$$

Each matrix on RHS takes a form of either δ or ε potential. This implies that in case of $u \neq 0$, for instance, one can realize a general connection condition in the limit

$$V(x) = \lim_{b\to+0} \left(\frac{s-1}{u}\delta(x+b) + u\varepsilon(x) + \frac{t-1}{u}\delta(x-b) \right). \tag{25}$$

In case of $v = u = 0$, one can use the decomposition

$$\begin{pmatrix} \pm|t| & 0 \\ 0 & \pm|s| \end{pmatrix} = \mathcal{V}_\delta(\rho)\mathcal{V}_\varepsilon(-\frac{1}{\rho})\mathcal{V}_\delta(\rho) \times \mathcal{V}_\delta(\mp\frac{1}{\rho})\mathcal{V}_\varepsilon(\pm\rho)\mathcal{V}_\delta(\mp\frac{1}{\rho}) \tag{26}$$

with $\rho = \sqrt{|t|}$, which is also written in terms of the connection matrices for the δ and ε potentials.

III. Scattering Properties and Delta-Epsilon Duality

In this section, we discuss an aspect of the scattering properties of the one-dimensional contact potential and point out a dual role between the δ and ε potentials.

It is well known that the one-body scattering problem of generalized contact force is solvable in closed form [1]. Here, we consider the scattering of two identical particles interacting through the contact force δ and ε. We now regard the variable x as the relative coordinate, and $\varphi(x)$ as the relative wave function of two identical particles whose statistics is either fermionic or bosonic. One has

$$\varphi(-x) = \pm\varphi(x), \tag{27}$$

where the composite sign takes + for bosons and − for fermions. Important fact to note is that the symmetry (resp. antisymmetry) of $\varphi(x)$ leads to the antisymmetry (resp. symmetry) of it's derivative $\varphi'(x)$. Apart from the contact position $x = 0$, $\varphi(x)$ is given by the free solution consisting of incoming and outgoing waves;

$$\varphi(x) = e^{ikx} + C(k)e^{-ikx} \qquad (x < 0), \tag{28}$$

$$\varphi(x) = \pm e^{-ikx} \pm C(k)e^{ikx} \qquad (x > 0). \tag{29}$$

In this form, the coefficient $C(k)$ represents the scattering matrix. The connection condition at $x = 0$, Eq.(2) now reads

$$\left[\begin{pmatrix} -ik & ik \\ 1 & 1 \end{pmatrix} \mp \mathcal{V} \begin{pmatrix} ik & -ik \\ 1 & 1 \end{pmatrix}\right] \begin{pmatrix} 1 \\ C(k) \end{pmatrix} = 0. \tag{30}$$

We first consider the case for δ potential $\mathcal{V} = \mathcal{V}_\delta(v)$. One obtains

$$C_\delta(k) = 1 \qquad \text{for fermions,} \tag{31}$$

$$C_\delta(k) = \frac{2ik + v}{2ik - v} \qquad \text{for bosons.} \tag{32}$$

The first equation means that the δ function is inoperative as the two-body interaction between identical bosons, which is of course a trivial fact. Next we consider the case of ε potential $\mathcal{V} = \mathcal{V}_\varepsilon(u)$. One has

$$C_\varepsilon(k) = \frac{2ik + 4/u}{2ik - 4/u} \qquad \text{for fermions,} \tag{33}$$

$$C_\varepsilon(k) = 1 \qquad \text{for bosons.} \tag{34}$$

One finds that the role of fermion and boson cases are exchanged: The ε potential as the two-body interaction has no effect on identical bosons, but *does* have an effect on the fermions. Moreover, the scattering amplitude of fermions with $\mathcal{V}_\varepsilon(u)$ is exactly the same as that of bosons with $\mathcal{V}_\delta(v)$ if the two coupling constants are related as

$$vu = 4. \tag{35}$$

Therefore, a two-fermion system with ε potential is dual to a two-boson system with δ potential. As expected, a natural generalization to N-particle systems exists [9].

IV. Conclusion

From the construction Eq. (24), we can easily see the reason why the generalized contact interaction in one-dimension with time reversal symmetry is of three parameter family: There are two kinds of "elementary" contact interactions – δ and ε – and one obtains different physics by different spatial ordering of δ and ε even in the zero-range limit. This construction should be also useful in the actual design of the possible device realizing the generalized contact interaction in mesoscopic scale.

It is clear from the arguments in the last section that the ε interaction is a generic short-range limit of arbitrary two-body interaction between like fermions, just as δ interaction being the generic zero-range force between like bosons. As such, the ε interaction, which has been a rather mysterious object, should be viewed as an integral component of one-dimensional many-body quantum mechanics.

TC is grateful to Prof. Y. Okada and the members of the Theory Group for their hospitality during his stay at KEK, Tsukuba. We also thank Prof. T. Kawai, Prof. I. Tsutsui and Prof. K. Takayanagi for useful discussions. This work has been supported in part by the Grant-in-Aid (No. 10640396) by the Japanese Ministry of Education.

References

[1] S. Albeverio, F. Gesztesy, R. Høegh-Krohn and H. Holden, "Solvable Models in Quantum Mechanics," Springer-Verlag, New York, 1988.
[2] P. Šeba, Czech. J. Phys., B36 (1986) 667.
[3] P. Šeba, Rep. Math. Phys. 24 (1986) 111.
[4] M. Careau, J. Phys. A26 (1993) 427.
[5] P.R. Chernoff and R. Hughes, J. Funct. Anal. 111 (1993) 92.
[6] J.E. Avron, P. Exner, Y. Last, Phys. Rev. Lett. 72 (1994) 896.
[7] S. Albeverio, L. Dabrowski, P. Kurasov, Lett. Math. Phys. 45 (1998) 33.
[8] T. Cheon and T. Shigehara, Phys. Lett. A243 (1998) 111.
[9] T. Cheon and T. Shigehara, LANL preprint quant-ph/9806041.

Operator Theory:
Advances and Applications, Vol. 108

Traces and trace norms for semigroup differences

Michael Demuth
Institute of Mathematics
TU Clausthal
Erzstr. 1
D-38678 Clausthal-Zellerfeld
E-mail: demuth math.tu-clausthal.de

Abstract

If the semigroup difference has a kernel $D_t(x,y)$ their trace class and Hilbert-Schmidt class properties depend on the function $D_t(x) = \int |D_t(x,y)|dy$, called comparison function. The general theory is applied to Feller semigroups.

0. Introduction. Semigroups and sandwiched semigroups can be represented by products of operators. Hence the CDSS-trace class criterion can be applied, if the semigroups are integral operators. If the semigroup difference has the kernel $D_t(x,y)$ the conditions depend on $D_t(x) := \int\limits_E |D_t(x,y)|dy$, which is called the comparison function for the semigroup difference. The difference is trace class if $D_t(.)^{1/2} \in L^1$, it has a finite trace if $D_t(.) \in L^1$ and it is Hilbert-Schmidt if $D_t(.) \in L^2$.

These general conditions are applied to semigroups generated by Feller operators $K_0 \dot{+} V$, where V is a regular potential, and to singularly perturbed K_0 with Dirichlet boundary conditions. In these cases the semigroups are integral operators the kernels of which can be represented in stochastic spectral analysis by the expectation with respect to conditional Feller measures. In the regular case Kato-Feller potentials play a crucial role, for singular perturbations the equilibrium potentials are important.

1. Semigroup differences and spectral theory. Throughout this article we study selfadjoint semibounded operators H_1, H_2 in the Hilbert space $L^2(E,m)$, where E is a second countable, locally compact Hausdorff space and m is a Radon measure. Instead of analyzing H_1, H_2 directly it is often useful to consider bounded functions of them. Here we will study semigroup differences

$$D_t := e^{-tH_2} - e^{-tH_1} \tag{1}$$

and sandwiched semigroup differences of the form

$$e^{-t_1 H_i} D_{t_2} e^{-t_1 H_j} \tag{2}$$

where i, j can be 1 or 2 and t_1, t_2 are in general some positive different real values.

There are obvious motivations to study the difference D_t or the sandwiched difference in (2):

The trace class property of D_t has its own right and is studied in the literature with respect to different aspects. The sandwiched difference is useful for the trace of D_t. For seeing that we notice the standard decomposition of a semigroup difference into one-sided sandwiched differences, i. e.

$$\begin{aligned} D_{2t} &:= e^{-2tH_2} - e^{-2tH_1} \\ &= e^{-tH_2} D_t + D_t e^{-tH_1}. \end{aligned} \tag{3}$$

That implies also that we can investigate D_t by studying products of operators. If now D_{4t} is a trace class operator, its trace is given by

$$\begin{aligned} &\text{trace } [D_{4t}] \\ &= \text{trace } [e^{-tH_2} D_{2t} e^{-tH_2}] + \text{trace } [e^{-tH_1} D_{2t} e^{-tH_1}] \end{aligned} \tag{4}$$

As the last motivation let us mention a modified completeness criterion which is proven in [DC,1] or [DC,2]:

If

$$\begin{aligned} &e^{-tH_2}(e^{-tH_2} - e^{-tH_1})e^{-tH_1} \\ &= e^{-tH_2} D_t \, e^{-tH_1} \end{aligned} \tag{5}$$

is trace class and if

$$D_t = e^{-tH_2} - e^{-tH_1} \tag{6}$$

is a Hilbert-Schmidt operator, then $\{H_2, H_1\}$ form a complete scattering system, i. e. the wave operators exist and the scattering operator is unitary, the absolutely continuous spectra are the same. In view of (6) it is useful to know also when D_t is a Hilbert-Schmidt or a compact operator. By Weyl's Theorem then the essential spectra of H_1 and H_2 are also invariant.

2. CDSS-trace class criterion for semigroup differences. Having in mind the decomposition (3) we have to study the trace class properties of products of integral operators. The CDSS-trace class criterion improves optimally the usual

Hilbert-Schmidt-norm estimates. For the sake of completeness let us repeat this criterion.

THEOREM 1 [CDSS]: Let A and B integral operators in $L^2(E,m)$ with measurable kernels $A(.,.)$, $B(.,.)$ mapping $E \times E \to \mathbb{C}$. Then AB is a trace class operator if

$$\int_E \|A(.,x)\| \; \|B(x,.)\| \; dm(x) = M \tag{7}$$

is finite, and the trace norm of AB can be estimated by

$$\|AB\|_{trace} \leq M. \tag{8}$$

The proof is given in [CDSS].

In order to formulate the main theorem in this note we introduce the comparison function for differences of semigroups.

DEFINITION 2: Let H_1, H_2 be two selfadjoint semibounded operators in $L^2(E,m)$ generating ultracontractive semigroups e^{-tH_j}, $j = 1,2$. Their kernels are denoted by $(e^{-tH_j})(x,y)$. Define the semigroup difference by

$$D_t := e^{-tH_2} - e^{-tH_1}. \tag{9}$$

This is again an integral operator, the kernel is written as $D_t(x,y)$. The crucial function in the conditions of the following theorem is

$$D_t(x) := \int_E |D_t(x,y)| dm(y) \tag{10}$$

We call $D_t(.)$ the **comparison function** for the difference D_t.

THEOREM 3: Let H_1, H_2 be given as above. Assume additionally that the semigroup kernels satisfy

$$\operatorname*{ess-sup}_{x,y \in E} (e^{-tH_j})(x,y) \leq c_j(t)^2 \tag{11}$$

and

$$\operatorname*{ess-sup}_{x \in E} \int_E (e^{-tH_j})(x,y) dm(y) \leq d_j(t)^2 \tag{12}$$

Then we have the following assertions:

(i) The semigroup difference

$$D_{2t} = e^{-2tH_2} - e^{-2tH_1}$$

is a trace class operator if

$$\int_E \sqrt{D_t(x)}dm(x) < \infty. \tag{13}$$

(ii) Let D_{4t} be a trace class operator. Then its trace can be estimated by

$$trace\ [D_{4t}] \leq c \cdot \int_E D_{2t}(x)dm(x). \tag{14}$$

(iii) The semigroup difference D_{2t} is a Hilbert-Schmidt operator if

$$\int_E D_t(x)^2 dm(x) < \infty. \tag{15}$$

And the Hilbert-Schmidt norm can be estimated by this expression.

(iv) The sandwiched difference of the form

$$e^{-tH_i} D_{2t}\ e^{-tH_j}, \quad i,j \in \{1,2\},$$

is trace class, if

$$\int_E D_{2t}(x)dm(x) < \infty \tag{16}$$

and its trace norm can be estimated by this expression.

REMARK on the proof of Theorem 3:

The CDSS-Theorem is used for the standard decompostion

$$D_{2t} = e^{-tH_2} D_t + D_t e^{-tH_1}.$$

Now

$$\|e^{-tH_2}(.,x)\| \leq c_2(t) d_2(t)$$

and

$$\|D_t(x,.)\| \leq [c_1(t)^2 + c_2(t)^2]^{1/2} [\int_E |D_t(x,y)| dy]^{1/2} \leq c \cdot \sqrt{D_t(x)}.$$

That explains the sufficiency of (13). The other conditions in (14), (15), (16) can be verified in a similar way.

3. Applications to Feller semigroups. Theorem 3 will be applied to selfadjoint generators of Feller semigroups. This has the advantage to have explicit stochastic representations for the integral kernels of the semigroups, such that the comparison function $D_t(x)$ becomes explicit.

Let K_0 be a selfadjoint, positive operator, which generates a strongly continuous semigroup e^{-tK_0} in $L^2(E, m)$. Assume that this semigroup has a positive integral kernel $(e^{-tK_0})(x,y)$ satisfying

$$0 \leq (e^{-tK_0})(x,y) \leq c(t) \tag{17}$$

for almost all $x, y \in E$. For a simpler denotation in the following we write

$$(e^{-tK_0})(x,y) =: p_0(t,x,y). \tag{18}$$

In order to find the associated Markov process we assume the following properties of $p_0(t,x,y)$ (see basic assumptions in stochastic spectral analysis [DC,1]).

Ass. 1) $p_0 : (0,\infty) \times E \times E \to [0,\infty)$
is assumed to be continuous in all three variables, to be symmetric in x and y, to satisfy the Chapman-Kolmogorov identity, and to provide

$$\int_E p_0(t,x,y) dm(y) \leq 1. \tag{19}$$

Ass. 2) The Feller property is assumed which says that $e^{-tK_0} C_\infty \subseteq C_\infty$,
i. e. e^{-tK_0} maps the continuous functions vanishing at infinity into

this set. Moreover, for $f \in C_\infty(E)$ we assume

$$\lim_{t \to 0} \int_E p_0(t,x,y) f(y) dm(y) = f(x) \tag{20}$$

for all $x \in E$.

REMARKS:

Assumption 2 is sufficient for the existence of a strong Markov process

$$((\Omega, \mathcal{F}, P_x); (X(t), t \geq 0); (E, \mathcal{E})). \tag{21}$$

$X(.) \in \Omega$ are cadlag trajectories. $\mathcal{F}$ is a σ-field in Ω, $(\Omega, \mathcal{F}, P_x)$ is a measure space, $(E, \mathcal{E})$ is the state space, where $\mathcal{E}$ is the Borel field in E. This process is related to the semigroup by

$$(e^{-tK_0} f)(x) = E_x\{f(X(t))\} \tag{22}$$

where the right hand side can be extended to f for which the integral is finite.

In order to introduce the stochastic representation of the integral kernel for perturbed Feller generators, we define the pinned measure

$$E_x^{y,t}(A) := E_x\{p_0(t-s, X(s), y) 1_A\} \tag{23}$$

where 1_A is the indicator function of the set A, in the field $\mathcal{F}_s, s < t$. The process $s \to p_0(t-s, X(s), y)$ is a martingale on $[0,t)$. It holds

$$E_x^{y,t}(E) = p_0(t,x,y) \tag{24}$$

For more details see [DC,1]. For the purposes of this article it suffices to assume a strong Markov process with the properties mentioned above. That can also be constructed via regular Dirichlet forms following Fukushima [F] or [F0T] (see also Ma, Röckner [MR]).

Next we introduce possible regular perturbations of K_0. A function $V : E \to \mathbb{R}$ is called a Kato-Feller potential if

$$\limsup_{t \downarrow 0} \sup_x \int_0^t ds \int_E dm(y)\ p_0(s,x,y)\ |V(y)| = 0 \tag{25}$$

For such potentials $K_0 \dot{+} V$ (form sum) generates a semigroup in L^2, given by the Feynman-Kac formula. Moreover $e^{-t(K_0 \dot{+} V)}$ is an integral operator with the kernel

$$E_x^{y,t}\{e^{-\int_0^t V(X(u))du}\} =: (e^{-t(K_0 \dot{+} V)})(x,y).$$

For studying singular perturbations we introduce stochastically the following semigroup. Let Γ be a closed subset of E and $\Sigma = E\backslash\Gamma$ be the complement of Γ. Let

$$S_\Gamma := \inf\{s, X(s) \in \Gamma\} \tag{26}$$

the first hitting time of Γ. Then the family of operators

$$E_x\{S_\Gamma > t\}/_{L^2(\Sigma,m)}$$

yields a strongly continuous semigroup in $L^2(\Sigma, m)$. Its generator is denoted by $(K_0)_\Sigma$. One can find conditions for K_0 and Γ such that $(K_0)_\Sigma$ is the Friedrichs extension of K_0 restricted to dom $(K_0) \cap L^2(\Sigma)$ (see [DC,1]). It corresponds to a free Feller operator with Dirichlet boundary conditions on the boundary of Γ. The semigroup $e^{-t(K_0)_\Sigma}$ is an integral operator in $L^2(\Sigma, m)$. Its kernel is

$$E_x^{y,t}\{S_\Gamma > t\} =: (e^{-t(K_0)_\Sigma})(x, y). \tag{27}$$

In a similar way one can combine both kinds of perturbations. Let V be a Kato-Feller potential and Γ be a singularity region described above. Then

$$E_x\{e^{-\int_0^t V(X(u))du} f(X(t)),\ S_\Gamma > t\} =: (e^{-t(K_0\dot{+}V)_\Sigma} f)(x) \tag{28}$$

forms a strongly continuous semigroup in $L^2(\Sigma)$, the generator of which is denoted by $(K_0\dot{+}V)_\Sigma$. Again one can find conditions such that $(K_0\dot{+}V)_\Sigma$ is the Friedrichs extension of the semibounded operator $K_0\dot{+}V$ restricted to
dom $(K_0\dot{+}V) \cap L^2(\Sigma)$ (see [DC,1]).

Its kernel is given by

$$(e^{-t(K_0\dot{+}V)_\Sigma})(x, y) := E_x^{y,t}\{e^{-\int_0^t V(X(u))du},\ S_\Gamma > t\} \tag{29}$$

In order to use the general results of Theorem 3 for these Kato-Feller semigroups we have to ensure the conditions in (11) and (12).

From (17) and (19) it is obvious that $(e^{-tK_0})(x, y)$ satisfies (11) and (12). If we have a Kato-Feller potential V, we obtain (see [DC,1], chapter 2)

$$\begin{aligned}
&(e^{-t(K_0\dot{+}V)})(x, y)\\
&\leq \|e^{-\frac{t}{2}(K_0\dot{+}4V)}\|_{\infty,\infty}^{1/2}\ \|e^{-\frac{t}{2}K_0}\|_{\infty,1}^{1/2} \cdot [(e^{-tK_0})(x, y)]^{1/2}\\
&\leq ce^{bt}c(\tfrac{t}{2})^{1/2}[(e^{-tK_0})(x, y)]^{1/2}
\end{aligned}$$

Hence (11) and (12) hold because they are true for the kernel of the free semigroup.

For $(e^{-t(K_0\dot{+}V)_\Sigma})(x,y)$ the same is then obvious because

$$(e^{-t(K_0\dot{+}V)_\Sigma})(x,y) \leq (e^{-t(K_0\dot{+}V)})(x,y)$$

for $x, y \in \Sigma$.

Therefore we can apply Theorem 3 for such Kato-Feller semigroups.

COROLLARY 4: Let K_0 be a free Kato-Feller operator, let V, W be Kato-Feller potentials, let Γ be a singularity region described above.

Set

$$D_{t,V,W} = e^{-t(K_0\dot{+}V)} - e^{-t(K_0\dot{+}W)}, \tag{30a}$$

$$D_{t,\Sigma} = e^{-tK_0} - J^* e^{-t(K_0)_\Sigma} J, \tag{31a}$$

$$D_{t,V,\Sigma} = e^{-t(K_0\dot{+}V)} - J^* e^{-t(K_0\dot{+}V)_\Sigma} J, \tag{32a}$$

here $Jf = f/_\Sigma$, such that all three differences map $L^2(E,m)$ into $L^2(E,m)$, whereas $e^{-t(K_0)_\Sigma}, e^{-t(K_0\dot{+}V)_\Sigma}$ are operators in $L^2(\Sigma, m)$.

Then the assertions of Theorem 3 are valid if we choose as comparison function

$$D_{t,V,W}(x) = E_x\{|e^{-\int_0^t V(X(s))ds} - e^{-\int_0^t W(X(s))ds}|\} \tag{30b}$$

$$D_{t,\Sigma}(x) = E_x\{S_\Gamma \leq t\} \tag{31b}$$

$$D_{t,V,\Sigma}(x) = E_x\{e^{-\int_0^t V(X(u))du}, S_\Gamma \leq t\}. \tag{32b}$$

4. Discussion. The condition in (31b) is related to the capacity. We have

$$\begin{aligned} &E_x\{S_\Gamma \leq t\} \\ &\leq e^t\, E_x\{e^{-S_\Gamma}, S_\Gamma \leq t\} \\ &\leq e^t\, E_x\{e^{-S_\Gamma}, S_\Gamma < \infty\}\,. \end{aligned}$$

The function $x \to E_x\{e^{-S_\Gamma}, S_\Gamma < \infty\}$ is the 1-equilibrium potential of Γ. Then the integral $\int_E E_x\{e^{-S_\Gamma}, S_\Gamma < \infty\}dm(x)$ is smaller than the capacity of the set Γ. For more details in this direction the interested reader may have a look in the review article [DMN]. In this article we describe also further spectral theoretical results derived from this theory.

In this general framework a large variety of examples for K_0 is contained. K_0 can be a second order differential operator in $\mathbb{R}^d$ given by

$$-\sum_{i,j=1}^{d} \frac{\partial}{\partial x_i} [a_{ij}(x) \frac{\partial}{\partial x_j}]$$

where

$$\frac{\partial^2 a_{ij}(x)}{\partial x_i \partial x_j} \in L^1_{\text{loc}}(\mathbb{R}^d) \cap L^{d/2}(\mathbb{R}^d)$$

and locally Hölder continuous,

$$\sum_j \frac{\partial a_{ij}(x)}{\partial x_j} \in L^d_{\text{loc}}(\mathbb{R}^d)$$

and locally Hölder continuous, and

$$\sum_{i,j} \frac{\partial^2 a_{ij}(x)}{\partial x_i \partial x_j} \leq 0 \text{ for all } x \in \mathbb{R}^d.$$

For details see Kochubei [K]. Notice that $a_{ij}(.)$ are not assumed to be bounded, such that a Gaussian estimate for the kernels is not available.

Jacob [J] proved the Feller property for several pseudo-differential operators. A simple class is

$$L(x, D) := \sum_{j=1}^{n} b_j(x) \Lambda_j^r(D) \varphi(x)$$

with $\varphi \in C_0^\infty(\mathbb{R}^d)$ and

$$(\Lambda_j^r(D)\varphi)(x) = \frac{1}{(2\pi)^{d/2}} \int_{\mathbb{R}^d} e^{ix\xi}(1 + |\xi_i|^2)^{r/2} \hat{\varphi}(\xi) d\xi$$

with $0 < r \leq 2$. Under some conditions for b_j he proved that L has a closed extension K_0 which is selfadjoint and generates a Feller semigroup.

Jacob's class includes the relativistic Hamiltonian of the form $\sqrt{-\Delta + m^2} - m$, studied by Carmona, Masters and Simon [CMS].

Further results in stochastic spectral analysis are related for instance to the existence of bound states [MO], the large coupling behaviour of spectral data for $K_0 + \beta 1_\Gamma$, as $\beta \to \infty$ (see Demuth, Kirsch, McGillivray [DKM]), domain perturbations in scattering theory [DM]. In mathematical scattering theory the two space trace class criterion is due to Pearson. In this context the wave operators satisfy

Pearson's estimate. Improving this estimate the norm of the wave operator for the scattering system $\{(K_0)_{E\setminus\Gamma}, K_0\}$ can be estimated in terms of the capacity of Γ. Finally, the study of the relations between geometrical conditions and spectral theoretical results gave rise to extend the notion of capacities by introducing the capacity of subspaces. That was initiated by Noll [N].

References

[CMS] - Carmona, R., Masters, W.C., Simon, B., Relativistic Schrödinger operators: Asymptotic behaviour of eigenfunctions, J. Funct. Anal. **91** (1990), 117 - 142.

[CDSS] - van Casteren, J., Demuth, M., Stollmann, P., Stolz, G., Trace norm estimates for products of integral operators and diffusion semigroups, Integral Equat. and Op. Theory **23** (1995), 145 - 153.

[DC,1] - Demuth, M., van Casteren, J., Stochastic Spectral Theory for Selfadjoint Feller Operators, To appear in Birkhäuser Verlag, 1998.

[DC,2] - Demuth, M., van Casteren, J., Completeness of scattering systems with obstacles of finite capacity. Preprint TU Clausthal 7 (1994). To appear in proceedings of the international workshop on operator theory and applications, Regensburg, 1995.

[DKM] - Demuth, M., Kirsch, W., McGillivray, I., Schrödinger semigroups - geometric estimates in terms of the occupation time, Comm. Part. Diff. Equat. **20** (1995), 37 - 57.

[DM] - Demuth, M., McGillivray, I., Capacitary estimates for the scattering phase, Preprint, TU Clausthal 5 (1996), to be published in: Journal of Mathematical Analysis and Applications.

[DMN] - Demuth, M., McGillivray, I., Noll, A., Capacity and spectral theory, Advances in Partial Differential Equations, Math. Topics, Vol. 14, p. 12 - 77 (1997).

[F] - Fukushima, M., Dirichlet Forms and Markov Processes, North Holland, Amsterdam, 1980.

[FOT] - Fukushima, M., Oshima, Y., Takeda, M., Dirichlet Forms and Symmetric Markov Processes. Walter de Gruyter, Berlin, 1994.

[J] - Jacob, N., Pseudo-differential Operators and Markov Processes: A Mathematical Survey, Akademieverlag, Berlin, 1996.

[K] - Kochubei, A.N., Parabolic pseudo-differential equations, hyper-singular integrals and Markov processes, Math. USSR Izvestiya **33** (1989), 233 - 259.

[McG,1] - McGillivray, I., Capacitary estimates for Dirichlet eigenvalues. J. Funct. Anal. **139** (1996), 244 - 259.

[McG,2] - McGillivray, I., Capacitary asymptotic expansion of the ground state to second order, Preprint, Univ. Bristol, 1997.

[MO] - McGillivray, I., Ouhabaz, E., Some spectral properties of recurrent semigroups, Arch. Math. **66** (1996), 233 - 242.

[MR] - Ma, Z., Röckner, M., Introduction to the Theory of (non-symmetric) Dirichlet Forms, Universitext, Springer Verlag, Berlin, 1992.

[N] - Noll, A., A generalization of Dynkin's formula and capacitary estimates for semibounded operators. In: Differential Equations, Asymptotic Analysis and Mathematical Physics, Vol. 100 of Mathematical Research, pp. 252 - 259, Akademieverlag, Berlin, 1997.

Operator Theory:
Advances and Applications, Vol. 108

About a resolvent formula

P. Duclos[a], P. Šťovíček[b] and O. Váňa[b]

(a) Centre de Physique Théorique, CNRS, 13288 Marseille–Luminy
and PHYMAT, Université de Toulon et du Var, BP 132,
83957 La Garde Cedex, France

(b) Department of Mathematics, Faculty of Nuclear Science, CTU,
Trojanova 13, 120 00 Prague, Czech Republic

Abstract

A resolvent formula, presented originally by Karner in his Habilitation [3], is discussed. First the formula is considered abstractly and then it is demonstrated on an explicit example – the so-called simplified Fermi accelerator.

1 Introduction

In his habilitation [3] Karner demonstrated on two explicit examples a resolvent formula adjusted to the case when the underlying Hilbert space was written as a tensor product $\mathcal{H}_1 \otimes \mathcal{H}_2$ and the self-adjoint operator in question was of the form $A_1 \otimes \mathbb{I}_2 + \mathbb{I}_1 \otimes A_2$ plus a perturbation generally mixing the two factors $\mathcal{H}_1$ and $\mathcal{H}_2$. Usually the spectral decompositions of A_1 and A_2 are well known. In [3] Karner calls the formula a modified Krein formula. This is not quite exact as the formula depends in fact only on the described particular algebraic structure. Nevertheless it can be combined effectively with the Krein formula. Furthermore, some applications to spectral analysis have been proposed in [3]; however, this program doesn't seem to be completed entirely. Moreover, the formula itself has to be extracted from the text of the Habilitation.

Nevertheless we believe that the Karner's formula deserves a more detailed treatment. One reason is that the particular form of the considered operator, as mentioned above, occurs in various interesting situations. As a prominent example we may mention Floquet Hamiltonians introduced to study time dependent quantum systems [1, 5]. Thus in our short contribution we start from revealing the algebraic structure of Karner's formula. To this end we treat it abstractly and confine ourselves for a while to a finite-dimensional case, for then all the terms occurring in the formula are well-defined. However, this is not so obvious in some concrete examples when the Hilbert space is infinite-dimensional. It seems that one has to consider each case separately in order to verify that the formula actually makes good sense. Here we discuss as an example the so-called simplified Fermi accelerator [4].

2 Karner's formula

Proposition. *Suppose that $\mathcal{T}$ and $\mathcal{H}$ are two finite-dimensional vector spaces, and set $\mathcal{K} := \mathcal{T} \otimes \mathcal{H}$. Furthermore, let $\{Q_k\}_{k=1}^M$, $\{P_j\}_{j=1}^N$ be two complete sets of projectors on the vector space $\mathcal{T}$, i.e., $Q_k Q_{k'} = \delta_{kk'} Q_k$, $\sum_k Q_k = \mathbb{I}_\mathcal{T}$, $P_j P_{j'} = \delta_{jj'} P_j$, $\sum_j P_j = \mathbb{I}_\mathcal{T}$. To a set $\{\lambda_k\}_{k=1}^M$ of complex numbers and to a set $\{H_i\}_{i=0}^N$ of operators on $\mathcal{H}$ we relate the operators*

$$D := \textstyle\sum_{k=1}^M \lambda_k Q_k, \tag{1}$$

$$K_0 := D \otimes \mathbb{I}_\mathcal{H} + \mathbb{I}_\mathcal{T} \otimes H_0, \tag{2}$$

$$K := D \otimes \mathbb{I}_\mathcal{H} + \textstyle\sum_{j=1}^N P_j \otimes H_j, \tag{3}$$

$$\Lambda(z) := \textstyle\sum_{j=1}^N \sum_{k=1}^M P_j Q_k \otimes \Big((H_j + \lambda_k - z)^{-1} - (H_0 + \lambda_k - z)^{-1}\Big). \tag{4}$$

Then it holds

$$(K - z)^{-1} = \Big((K_0 - z)^{-1} + \Lambda(z)\Big)\Big(\mathbb{I} + [D \otimes \mathbb{I}_\mathcal{H}, \Lambda(z)]\Big)^{-1}. \tag{5}$$

Remark. $(K - z)^{-1}$, $(K_0 - z)^{-1}$ and $\Lambda(z)$ are meromorphic functions with values in the space $\mathrm{Lin}(\mathcal{T} \otimes \mathcal{H})$, and all of them converge to 0 as $|z| \to +\infty$. Consequently $\det\Big(\mathbb{I} + [D \otimes \mathbb{I}, \Lambda(z)]\Big)$ is a meromorphic function as well, with a finite number of poles, and converging to 1 as $|z| \to +\infty$. This function has necessarily a finite number of zeroes and the equality (5) makes sense except at a finite number of points $z \in \mathbb{C}$.

Proof. Let us assume that $z \in \mathbb{C}$ is chosen so that all terms in (5) are well defined. Multiplying the relation (5) from the right by the expression $\Big(I + [D \otimes I, \Lambda(z)]\Big)(K_0 - z)$, and from the left by the expression $(K - z)$ one arrives at an equivalent identity, namely

$$K_0 - K = \left(\Lambda(z)(D \otimes I) + \Big(\sum_{j=1}^N P_j \otimes (H_j - z)\Big)\Lambda(z)\right)(K_0 - z). \tag{6}$$

Using the equalities

$$\Lambda(z)(D \otimes \mathbb{I}) = \sum_j \sum_k P_j Q_k \otimes \lambda_k \Big((H_j + \lambda_k - z)^{-1} - (H_0 + \lambda_k - z)^{-1}\Big),$$

$$\Big(\sum_{j=1}^{N} P_j \otimes (H_j - z)\Big)\Lambda(z)$$
$$= \sum_j \sum_k P_j Q_k \otimes (H_j - z)\Big((H_j + \lambda_k - z)^{-1} - (H_0 + \lambda_k - z)^{-1}\Big),$$

one can show that the RHS of (6) equals

$$\begin{aligned}
&\Big(\sum_j \sum_k P_j Q_k \otimes (H_j + \lambda_k - z)\Big((H_j + \lambda_k - z)^{-1} - (H_0 + \lambda_k - z)^{-1}\Big)\Big) \\
&\quad \times (K_0 - z) \\
&= (K_0 - z) - \Big(\sum_j \sum_k P_j Q_k \otimes (H_j + \lambda_k - z)(H_0 + \lambda_k - z)^{-1}\Big)(K_0 - z) \\
&= -\Big(\sum_j \sum_k P_j Q_k \otimes (H_j - H_0)(H_0 + \lambda_k - z)^{-1}\Big)(K_0 - z) \\
&= -\Big(\sum_j P_j \otimes (H_j - H_0)\Big)\Big(\sum_k Q_k \otimes (H_0 + \lambda_k - z)^{-1}\Big)(K_0 - z) \\
&= (K_0 - K)\Big(\sum_k Q_k \otimes (H_0 + \lambda_k - z)\Big)^{-1}(K_0 - z) = K_0 - K\,,
\end{aligned}$$

and this completes the verification. ∎

3 Example: simplified Fermi accelerator

We set $\mathcal{T} = L^2([0,T], dt)$, $\mathcal{H} = L^2([0,1], dx)$, and so $\mathcal{K} = L^2([0,T]\times[0,1], dt\, dx)$. We identify $\mathcal{K}$, as usual, with $L^2([0,T], \mathcal{H}, dt)$. We set further $D = -i\partial_t$ with periodic boundary conditions, and $H_0 = -\partial_x^2$ with Neumann boundary conditions. Both D and H_0 are self-adjoint operators with discrete spectra. The diagonalization (1) of D is now replaced by an infinite sum, with $\lambda_k = k\omega$, $k \in \mathbb{Z}$, where $\omega := 2\pi/T$, and Q_k's are the orthogonal projectors on the eigen-functions $\chi_k(t) := T^{-1/2} \exp(i\, k\omega t)$. Clearly, the operator $K_0 = -i\partial_t \otimes \mathbb{I} - \mathbb{I} \otimes \partial_x^2$ is self-adjoint with a pure point spectrum.

Let us now make a small digression and consider a perturbation H_g of H_0 written in the form sense as $H_g := H_0 + g\tau^*\tau$ where $\tau : \mathcal{H} \to \mathbb{C}$ is the trace operator: $\tau u := u(0)$. Of course, this means nothing but that, in the case of H_g, the boundary condition at the point $x = 0$ reads $(\partial_x f)(0) = g\, f(0)$ while the boundary condition at the point $x = 1$ is still of Neumann type. In fact, H_g is an entire analytic family of type B [2, ch.VII §4] in the variable g.

For a later convenience let us also examine the resolvents of H_0 and H_g. We set $R_0(z) := (H_0 - z)^{-1}$ and $R_g(z) := (H_g - z)^{-1}$. It is easy to calculate the Green function corresponding to H_0 explicitly, namely

$$G_0(x,y) = -\frac{\cos(\sqrt{z}\,x)\cos\left(\sqrt{z}(y-1)\right)\vartheta(y-x) + \{x \leftrightarrow y\}}{\sqrt{z}\,\sin(\sqrt{z})} \tag{7}$$

where $\vartheta(x)$ is the Heaviside step function. Particularly,

$$\tau R_0(z)\tau^* = -\frac{\cot(\sqrt{z})}{\sqrt{z}}. \tag{8}$$

The two resolvents are related by the equality $R(z) = R_0(z) - gR(z)\tau^\star\tau R_0(z)$, and so

$$R_g(z) - R_0(z) = \frac{-g}{1 + g\,\tau R_0(z)\tau^*}\,R_0(z)\tau^*\tau R_0(z). \tag{9}$$

Here $R_0(z)\tau^*\tau R_0(z)$ is a rank-one operator with the norm

$$\begin{aligned} \|R_0(z)\tau^*\tau R_0(z)\|^2 &= \tau R_0(\bar{z})R_0(z)\tau^*\,\|R_0(\bar{z})\tau^*\tau R_0(z)\| \\ &= (\tau R_0(\bar{z})R_0(z)\tau^*)\,(\tau R_0(z)R_0(\bar{z})\tau^*) \\ &= \left(\frac{\operatorname{Im}\tau R_0(z)\tau^*}{\operatorname{Im} z}\right)^2. \end{aligned} \tag{10}$$

Suppose now that $g(t)$ is a T-periodic real function. Following [4] we call the time-dependent quantum system determined by the Hamiltonian $H(t) \equiv H_{g(t)}$ a simplified Fermi accelerator. The corresponding Floquet Hamiltonian $K := -i\partial_t + H(t)$ has the same structure as given in (3) provided one replaces the sum $\sum_j P_j \otimes H_j$ by the direct integral

$$\int_0^{T\oplus} H(t)\,dt. \tag{11}$$

This means that the family of projectors $\{P_j\}_j$ is formally substituted by the spectral decomposition of the multiplication operator $X \in \mathcal{B}(\mathcal{T})$, $(Xf)(t) := t\,f(t)$, which has, however, an absolutely continuous spectrum. Proceeding this way one finds that the definition (4) of the operator $\Lambda(z)$ has to be modified as follows:

$$\Big(\Lambda(z)\psi\Big)(t) := \sum_{k=-\infty}^{+\infty} \chi_k(t)\Big(R_{g(t)}(z-k\omega) - R_0(z-k\omega)\Big)\int_0^T \overline{\chi_k(\tau)}\,\psi(\tau)\,d\tau \tag{12}$$

where $\psi \in L^2([0,T],\mathcal{H},dt)$ and $z \in \mathbb{C}\setminus\mathbb{R}$.

The definition (12) implies that

$$\Big(\Lambda(z)\,\chi_k\otimes f\Big)(t) = \chi_k(t)\Big(R_{g(t)}(z-k\omega)-R_0(z-k\omega)\Big)f, \quad \forall k \in \mathbb{Z} \text{ and } f \in \mathcal{H}, \tag{13}$$

where we have used the convention $(u \otimes f)(t) := u(t)\,f$, with $u \in \mathcal{T}$ and $f \in \mathcal{H}$. So $\Lambda(z)$ is densely defined. To show that $\Lambda(z)$ is even bounded we shall treat the

RHS of (12) perturbatively assuming that z is separated from the real axis, i.e., $|\mathrm{Im}\, z| \geq s_0 > 0$, and $g \in C^0$, with the supremum norm $\|g\|_\infty$ being sufficiently small with respect to s_0. Relying on the estimates

$$|\tau R_0(z)\tau^*| = \left|\frac{\cot(\sqrt{z})}{\sqrt{z}}\right| \leq \frac{2}{|\mathrm{Im}\, z|}\sqrt{1+\frac{|\mathrm{Im}\, z|}{4}} \leq \frac{2}{s_0}\sqrt{1+\frac{s_0}{4}} =: \alpha(s_0) \tag{14}$$

and

$$\|R_0(z)\tau^*\tau R_0(z)\| \leq \left|\frac{\tau R_0(z)\tau^*}{\mathrm{Im}\, z}\right| \leq \frac{\alpha(s_0)}{s_0}\,, \tag{15}$$

we find that if $\|g\|_\infty < 1/\alpha(s_0)$ then

$$\begin{aligned}\Big(\Lambda(z)\psi\Big)(t) \;&=\; \sum_{n=0}^{\infty}(-1)^{n+1}\, g(t)^{n+1} \sum_{k=-\infty}^{+\infty} \chi_k(t)\left(\tau R_0(z-k\omega)\tau^*\right)^n \\ &\quad \times R_0(z-k\omega)\tau^*\tau R_0(z-k\omega)\int_0^T \overline{\chi_k(\tau)}\,\psi(\tau)\,d\tau\end{aligned} \tag{16}$$

and so

$$\begin{aligned}\|\Lambda(z)\psi\| \;&\leq\; \sum_{n=0}^{\infty}\|g\|_\infty^{n+1}\left(\sum_{k=-\infty}^{+\infty}|\tau R_0(z-k\omega)\tau^*|^{2n}\right. \\ &\quad \left.\times\,\|R_0(z-k\omega)\tau^*\tau R_0(z-k\omega)\|^2\left\|\int_0^T\overline{\chi_k(\tau)}\,\psi(\tau)\,d\tau\right\|_{\mathcal{H}}^2\right)^{1/2} \\ &\leq\; \frac{\|g\|_\infty\,\alpha(s_0)}{s_0\big(1-\|g\|_\infty\,\alpha(s_0)\big)}\,\|\psi\|\,.\end{aligned} \tag{17}$$

Next let us consider the commutator $[\,D\otimes\mathbb{I},\,\Lambda(z)\,]$. One deduces from (13) immediately that

$$\Big([\,D\otimes\mathbb{I},\,\Lambda(z)\,]\,\chi_k\otimes f\Big)(t) = -i\,\chi_k(t)\,\partial_t\Big(R_{g(t)}(z-k\omega)-R_0(z-k\omega)\Big)f\,. \tag{18}$$

Assuming that $g\in C^1$ we get a relation similar to (16), namely

$$\begin{aligned}\Big([\,D\otimes\mathbb{I},\,\Lambda(z)\,]\psi\Big)(t) \;&=\; -i\,g'(t)\sum_{n=0}^{\infty}(-1)^{n+1}(n+1)\,g(t)^n\sum_{k=-\infty}^{+\infty}\chi_k(t) \\ &\quad\times\left(\tau R_0(z-k\omega)\tau^*\right)^n R_0(z-k\omega)\tau^*\tau R_0(z-k\omega) \\ &\quad\times\int_0^T\overline{\chi_k(\tau)}\,\psi(\tau)\,d\tau\,,\end{aligned} \tag{19}$$

and consequently the estimate

$$\|[\,D\otimes\mathbb{I},\,\Lambda(z)\,]\psi\| \leq \frac{\|g'\|_\infty\,\alpha(s_0)}{s_0\big(1-\|g\|_\infty\,\alpha(s_0)\big)^2}\,\|\psi\|\,. \tag{20}$$

Since, apart of the problems with the precise definition of the operator $\Lambda(z)$, the algebraic structure remains the same as in the finite-dimensional case we conclude that the formula (5) extends, as it is, also to our example of simplified Fermi accelerator provided $|\mathrm{Im}\, z| \geq s_0$, $\|g\|_\infty\, \alpha(s_0) < 1$ and

$$\frac{\|g'\|_\infty\, \alpha(s_0)}{s_0\big(1 - \|g\|_\infty\, \alpha(s_0)\big)^2} < 1\,. \tag{21}$$

Of course, from the point of view of spectral analysis, a truly interesting result would involve the limit $|\mathrm{Im}\, z| \downarrow 0$. There is no doubt that this goal requires a much more subtle analysis than that based on the elementary estimate (14), and this program goes well beyond the scope of this short contribution. We hope anyway to have shown that the formula (5) may be given a good sense.

Acknowledgements. P. S. wishes to acknowledge gratefully partial support from the grant 202/96/0218 of Czech Grant Agency.

References

[1] Howland J. S.: *Scattering theory for Hamiltonians periodic in time*, Indiana J. Math. **28**, 471-494 (1979)

[2] Kato T.: *Perturbation theory of linear operators*, Springer-Verlag, New York, 1966

[3] Karner B.: *De l'application standard et l'application de Fermi en méchaniques quantique et classique*, habilitation, Université de Toulon et du Var (1993)

[4] Karner G.: *The simplified Fermi accelerator in classical and quantum mechanics*, J. Stat. Phys. **77**, 867-879 (1994)

[5] Yajima K.: *Scattering Theory for Schrödinger Equations with Potential Periodic in Time*, J. Math. Soc. Japan **29**, 729-743 (1977)

Operator Theory:
Advances and Applications, Vol. 108

The determinant anomaly in low-dimensional quantum systems

Emilio Elizalde

Consejo Superior de Investigaciones Científicas
Institut d'Estudis Espacials de Catalunya
Edifici Nexus 201, Gran Capità 2-4, 08034 Barcelona, Spain
and
Departament ECM and IFAE, Facultat de Física, Universitat de Barcelona
Diagonal 647, 08028 Barcelona, Spain
e-mail: elizalde@ieec.fcr.es, eli@zeta.ecm.ub.es

Abstract

The determinant of a product of two elliptic ΨDOs is not always equal to the product of the corresponding determinants. This anomaly is already present in the very simple situations of two commuting Dirac-like or Laplacian operators in a low-dimensional space ($d = 1, 2, \ldots$). It cannot always be absorbed by renormalization, being in such cases physically relevant.

1 Introduction

Pseudodifferential operators (ΨDO) are basic entities in the formulation of contemporary physics. In particular, they constitute the basic starting point of any rigorous formulation of Quantum Field Theory. At the level of Quantum Mechanics (where it was originally introduced by Feynman), the path-integral approach is just an alternative formulation of the theory. In QFT, it is much more than this, being in many ocassions *the* actual formulation of QFT. In short, consider the Gaussian integration:

$$\int [d\Phi] \ \exp\left\{ - \int d^D x \left[\Phi^\dagger(x) (\quad) \Phi(x) + \cdots \right] \right\} \quad \longrightarrow \quad \det(\quad)^{\pm}. \tag{1}$$

Assume that the operator matrix has the following simple structure (being each A_i an operator on its own):

$$\begin{pmatrix} A_1 & A_2 \\ A_3 & A_4 \end{pmatrix} \quad \longrightarrow \quad \begin{pmatrix} A & \\ & B \end{pmatrix}, \tag{2}$$

where the last expression is the result of diagonalizing the operator matrix. The question is, what is the determinant of the operator matrix

$$\det(AB) \qquad \text{or} \qquad \det A \cdot \det B \qquad ?$$

This has been much discussed recently [1]. There is agreement in that:

- In a situation where a superselection rule exists, AB has no sense (much less its determinant), and then the determinant must be $\det A \cdot \det B$.
- If the diagonal form is obtained after a change of basis (diagonalization process), then the quantity that is preserved by such transformations is the value of $\det(AB)$ and *not* the product of the individual determinants (for counterexamples supporting this point, see [2]).

It turns out that these two calculations produce different results, in general. The log of the difference has been termed the noncommutative anomaly.

2 Determinant of a ΨDO

Let A be a ΨDO operator with a spectral decomposition: $\{\varphi_i, \lambda_i\}$, where $i \in I$, *any* set of indices. The definition of determinant starts by making sense of the product $\prod_{i\in I} \lambda_i$, which can be easily transformed into a sum: $\ln \prod_{i\in I} \lambda_i = \sum_{i\in I} \ln \lambda_i$.

Recall now the conditions for the existence of $\zeta_A(s)$ [3]. Let A a positive-definite elliptic ΨDO of positive order $m \in \mathbf{R}$, acting on the space of smooth sections of E, an n-dimensional vector bundle over M, a closed n-dimensional manifold. The *zeta function* ζ_A is defined as

$$\zeta_A(s) = \text{tr } A^{-s} = \sum_j \lambda_j^{-s}, \qquad \text{Re } s > \frac{n}{m} \equiv s_0. \tag{3}$$

where $s_0 = \dim M/\text{ord}\, A$ is called the *abscissa of convergence* of $\zeta_A(s)$. Under these conditions, it can be proven that $\zeta_A(s)$ has a meromorphic continuation to the whole complex plane $\mathbf{C}$ (regular at $s = 0$), provided that the principal symbol of A (that is $a_m(x,\xi)$) admits a *spectral cut*: $L_\theta = \{\lambda \in \mathbf{C}; \text{Arg}\,\lambda = \theta, \theta_1 < \theta < \theta_2\}$, $\text{Spec}\, A \cap L_\theta = \emptyset$ (Agmon-Nirenberg condition). The definition of $\zeta_A(s)$ depends on the position of the cut L_θ. The only possible singularities of $\zeta_A(s)$ are *simple poles* at $s_k = (n-k)/m$, $k = 0, 1, 2, \dots, n-1, n+1, \dots$. M. Kontsevich and S. Vishik have managed to extend this definition to the case when $m \in \mathbf{C}$ (no spectral cut exists) [4].

From the definition of the zeta function of A: $\zeta_A(s) = \sum_{i\in I} \lambda_i^{-s}$, by taking the derivative at $s = 0$: $\zeta_A'(0) = -\sum_{i\in I} \ln \lambda_i$, we arrive to the following definition of determinant of A (Ray & Singer, '67): $\det_\zeta A = \exp[-\zeta_A'(0)]$. When A has zero-modes, the zeta function $\zeta_A(s)$ is just defined as that of A restricted to the orthogonal complement of the kernel ker A, and the definition of determinant follows then from taking the derivative of *this* zeta function.

An older definition (due to Weierstrass) is obtained by subtracting in the series above the leading behavior of λ_i as a function of i, as $i \to \infty$, until the series $\sum_{i\in I} \ln \lambda_i$ is made to converge. The shortcoming is here (for physical applications) that these additional terms turn out to be *non-local* and thus are non-admissible in any renormalization process.

3 The noncommutative anomaly

Given A, B, and AB ΨDOs, even if ζ_A, ζ_B, and ζ_{AB} exist, it turns out that, in general [5] $\det_\zeta(AB) \neq \det_\zeta A \det_\zeta B$. The multiplicative (or noncommutative) anomaly is defined as:

$$\delta(A,B) = \ln\left[\frac{\det_\zeta(AB)}{\det_\zeta A \ \det_\zeta B}\right] = -\zeta'_{AB}(0) + \zeta'_A(0) + \zeta'_B(0). \tag{4}$$

A formula due to Wodzicki for the multiplicative anomaly reads [6]:

$$\delta(A,B) = \frac{\text{res}\left\{[\ln \sigma(A,B)]^2\right\}}{2\,\text{ord}\,A\,\text{ord}\,B(\text{ord}\,A + \text{ord}\,B)}, \tag{5}$$

where $\sigma(A,B) = A^{\text{ord}\,B} B^{-\text{ord}\,A}$. Here res is the Wodzicki (or noncommutative) residue, i.e., the *only* trace at all one can define in the algebra of ΨDOs (up to a multiplicative constant): res $A = 2\ \text{Res}_{s=0}\ \text{tr}\,(A\Delta^{-s})$, with Δ the Laplacian. It satisfies the trace condition: res (AB) = res (BA). A very important property is that it can be expressed as an integral (local form): res $A = \int_{S^*M} \text{tr}\ a_n(x,\xi)\, d\xi$, being $S^*M \subset T^*M$ the co-sphere bundle on M and $a_n(x,\xi)$ the monomial of degree n on the variable ξ^{-1}, in the expansion of the symbol $a(x,\xi)$ of the ΨDO (some authors put a coefficient in front of the integral: Adler-Manin residue).

4 Examples

4.1 Dirac-like operator in one dimension

Consider the square root of the harmonic oscillator obtained by Delbourgo. This example has potentially some interesting physical applications, for it is well known that a fermion in an external constant electromagnetic field has a similar spectrum (Landau spectrum). Exactly in the same way as when going from the Klein-Gordon to the Dirac equation and at the same price of doubling the number of components (e.g., introducing spin), Delbourgo has constructed a model for which there exists a square root of its Hamiltonian, which is very close to the one for the harmonic oscillator. The main difference lies in the introduction now of the parity operator, Q. Whereas creation and destruction operators for the harmonic oscillator, $a^\pm = P \pm iX$, are non-hermitian, the combinations $D^\pm = P \pm iQX$ are hermitian and

$$H^\pm \equiv (D^\pm)^2 = P^2 + X^2 \mp Q = 2H_{\text{OSC}} \mp Q. \tag{6}$$

Notice that the parity term commutes with $H_{\rm OSC}$. Doubling the components (σ_i are the Pauli matrices) $P \to -i\sigma_1 \frac{\partial}{\partial x}$, $X \to \sigma_1 x$, $Q \to \sigma_2$, the operators $D^\pm$ are represented by $D^\pm \to -i\sigma_1 \frac{\partial}{\partial x} \pm \sigma_3 x$.

In what follows, we will only consider the operator $D \equiv D^+$. It has for eigenfunctions and eigenvalues, respectively,

$$\psi_n^\pm(x) = \frac{-ie^{-x^2/2}}{\sqrt{2^{n+1}(n-1)!\sqrt{\pi}}} \begin{pmatrix} -i\left[H_{n-1}(x) \pm H_n(x)/\sqrt{2n}\right] \\ \left[H_{n-1}(x) \mp H_n(x)/\sqrt{2n}\right] \end{pmatrix},$$

$$\lambda_n = \pm\sqrt{2n}, \ n \geq 1, \qquad \psi_0(x) = \frac{e^{-x^2/2}}{\sqrt{2\sqrt{\pi}}} \begin{pmatrix} 1 \\ i \end{pmatrix}, \qquad \lambda_0 = 0, \tag{7}$$

where the $H_n(x)$ are Hermite polynomials. Observe that the operator D has a zero mode; thus, according to the precise definition of determinant in such case (see above), when we write $\zeta_D(s)$ it will correspond to the zeta function of the restriction of D to the complement of its kernel.

The two operators we shall consider for the calculation of the anomaly are D and $D_V = D+V$, V being a real, constant potential with $|V| < \sqrt{2}$, that goes multiplied with the identity matrix in the two (spinorial) dimensions (omitted here). Notice that D and D_V are hermitian, commuting operators. The corresponding zeta functions are:

$$\begin{aligned} \zeta_D(s) &= \sum_i{}' \lambda_i^{-s} = \sum_{n=1}^\infty [1+(-1)^{-s}]\left(\sqrt{2n}\right)^{-s} = [1+(-1)^{-s}]2^{-s/2}\zeta_R(s/2), \\ \zeta_{D_V}(s) &= \sum_i (\lambda_i + V_i)^{-s}, \qquad\qquad \zeta_{DD_V}(s) = \sum_i{}' \left(\lambda_i^2 + V\lambda_i\right)^{-s}, \end{aligned} \tag{8}$$

where the prime means here that the zero-mode is excluded from the sum, and $\zeta_R(s)$ is the usual Riemann zeta function, which has a simple pole at $s = 1$. Binomial expansion yields

$$\begin{aligned} \zeta_{D_V}(s) &= [1+(-1)^{-s}]2^{-s/2}\zeta_R(s/2) + \\ &\quad \sum_{k=1}^\infty \frac{(-V)^k\Gamma(k+s)}{2^{(s+k)/2}\,k!\,\Gamma(s)}[1+(-1)^{-s-k)}]\zeta_R(s/2+k/2), \\ \zeta_{DD_V}(s) &= [1+(-1)^{-2s}]2^{-s}\zeta_R(s) + \\ &\quad \sum_{k=1}^\infty \frac{(-V)^k\Gamma(k+s)}{2^{s+k/2}\,k!\,\Gamma(s)}[1+(-1)^{-2s-k}]\zeta_R(s+k/2). \end{aligned}$$

Form these expressions, in spite of the manifold being non-compact, we get

$$\delta(D, D_V) = \frac{V^2}{2} - \ln V, \tag{9}$$

what is in perfect accordance with the Wodzicki formula. The logarithmic term is due to the presence of the zero mode.

We have here the first and most simple example of the presence of a non-trivial anomaly for operators of degree one in a space of dimension one (spinorial, however).

4.2 Harmonic oscillators in d dimensions

Let us recall the case of the harmonic oscillators in d dimensions, with angular frequencies $(\omega_1, \ldots, \omega_d)$. The eigenvalues read

$$\lambda_{\bar{n}} = \bar{n}\cdot\bar{\omega} + b\,, \quad \bar{n} \equiv (n_1, ..., n_d)\,, \quad \bar{\omega} \equiv (\omega_1, ..., \omega_D)\,, \quad b = \frac{1}{2}\sum_{k=1}^{d} \omega_k \tag{10}$$

and the related zeta function is the Barnes one $\zeta_d(s, b|\bar{\omega})$, whose poles are to be found at the points $s = k$ $(k = d, d-1, \ldots, 1)$. Their corresponding residua can be expressed in terms of generalized Bernoulli polynomials $B^{(d)}_{d-k}(b|\bar{\omega})$, defined by

$$\frac{t^d e^{-at}}{\prod_{i=1}^d (1 - e^{-b_i t})} = \frac{1}{\prod_{i=1}^d b_i}\sum_{n=0}^{\infty} B^{(d)}_n(a|b_i)\frac{(-t)^n}{n!}. \tag{11}$$

The residua of the Barnes zeta function are

$$\mathrm{Res}\,\zeta_d(k, b|\bar{\omega}) = \frac{(-1)^{d+k}}{(k-1)!(d-k)!\prod_{j=1}^d \omega_j} B^{(d)}_{d-k}(b|\bar{\omega}), \qquad k = d, d-1, \ldots. \tag{12}$$

Now, if V is a constant potential, we obtain

$$\delta(H, H_V) = \frac{(-1)^d}{2\prod_{j=1}^d \omega_j}\sum_{k=1}^{[d/2]} \frac{[\gamma + \psi(d-2k)]\, B^{(d)}_{2k}(b|\bar{\omega})}{(2k)!\,(d-2k)!} V^{2k}. \tag{13}$$

Here the generalized Bernoulli polynomials of odd order vanish. Even if the manifold is not compact, we confirm the validity of the Wodzicki formula. On the other hand, the remaining generalized Bernoulli polynomials are *never* zero, in fact

$$\begin{aligned}
B^{(d)}_0(b|\bar{\omega}) &= 1, \qquad B^{(d)}_2(b|\bar{\omega}) = -\frac{1}{12}\sum_{i=1}^{d}\omega_i^2, \\
B^{(d)}_4(b|\bar{\omega}) &= \frac{1}{24}\left[\frac{7}{10}\sum_{i=1}^{d}\omega_i^4 + \sum_{i<j}\omega_i^2\omega_j^2\right], \\
B^{(d)}_6(b|\bar{\omega}) &= -\frac{5}{96}\left[\frac{31}{70}\sum_{i=1}^{d}\omega_i^6 + \frac{7}{10}\sum_{i\neq j}\omega_i^4\omega_j^2 + \sum_{i<j<k}\omega_i^2\omega_j^2\omega_k^2\right], \quad \ldots
\end{aligned} \tag{14}$$

As a consequence, the anomaly does not vanish in any case, not for d odd or $d = 2$, whatever the frequencies ω_i be. Moreover, only even powers of the potential V appear.

It can be easily shown, that in the more elaborated case of a massive (free) Bose gas at finite temperature, with a chemical potential, the multiplicative anomaly term cannot be reabsorbed by a renormalization of the physical constants [2].

Acknowledgements. It is a pleasure to thank G. Cognola, A. Filippi, L. Vanzo and S. Zerbini for fruitful collaboration. This work has been financed by DG-ICYT (Spain), project PB96-0925, and by CIRIT (Generalitat de Catalunya), grant 1997SGR-00147.

References

[1] N. Evans, *Regularization schemes and the multiplicative anomaly* hep–th/9803184 (1998); J.S. Dowker, *On the relevance on the multiplicative anomaly* hep–th/9803200 (1998);E. Elizalde, A. Filippi, L. Vanzo and S. Zerbini, *Is the multiplicative anomaly dependent on the regularization?* hep–th/9804071 (1998); E. Elizalde, A. Filippi, L. Vanzo and S. Zerbini, *Is the multiplicative anomaly relevant?* hep–th/9804072 (1998).

[2] E. Elizalde, G. Cognola and S. Zerbini, *Applications in physics of the multiplicative anomaly formula involving some basic differential operators*, hep-th/9804118 (1998), Nucl. Phys. B, to appear; J.J. McKenzie-Smith and D.J. Toms, *ζ–function regularization, the multiplicative anomaly, and finite temperature field theory*, hep–th/9805184 (1998).

[3] D.B. Ray and I.M. Singer. Advances in Math. **7**, 145 (1971); E. Elizalde, *Ten Physical applications of spectral zeta functions* (Springer, Berlin, 1995).

[4] M. Kontsevich and S. Vishik. *Functional Analysis on the Eve of the 21st Century.* volume 1, 173–197, (1993).

[5] E. Elizalde, J. Phys. **A** (1997); E. Elizalde, L. Vanzo and S. Zerbini, hep-th/9701060, Commun. Math. Phys., **194**, 613 (1998); E. Elizalde, hep-th/9707258, Commun. Math. Phys., to appear.

[6] M. Wodzicki. *Non-commutative Residue Chapter I.* In *Lecture notes in Mathematics.* Yu.I. Manin, editor, volume 1289, 320. Springer-Verlag, (1987).

Operator Theory:
Advances and Applications, Vol. 108

Linear Boltzmann Equation as the Weak Coupling Limit of the Random Schrödinger Equation

László Erdős *
Courant Institute, New York University
251 Mercer Str. New York, NY-10012
erdos@cims.nyu.edu

Abstract

We study the long time evolution of a quantum particle in a Gaussian random environment. We show that in the weak coupling limit the Wigner distribution of the wave function converges to the solution of a linear Boltzmann equation globally in time. The Boltzmann collision kernel is given by the Born approximation of the quantum scattering cross section.

1 Statement of the result

Let $V(x) = V_\omega(x)$ be a Gaussian field in $\mathbf{R}^d$, $d \geq 3$, with $\mathbf{E}V(x) = 0$ and covariance $G(x-y) := \mathbf{E}V(x)V(y)$. We assume that G is spherically symmetric, smooth with compact support. Let $\psi_t = \psi_{t,\omega}^\lambda \in L^2(\mathbf{R}^d)$ solve the random Schrödinger equation

$$i\partial_t \psi_{t,\omega}^\lambda(x) = \Big(-\frac{1}{2}\Delta_x + \lambda V_\omega(x) \Big) \psi_{t,\omega}^\lambda(x) \tag{1}$$

with a coupling constant $0 < \lambda \ll 1$. Here x, t are atomic (microscopic) space and time coordinates. Let ε be the scaling parameter which separates the microscopic and macroscopic scales, and let

$$(X, T) := (x\varepsilon, t\varepsilon)$$

be the macroscopic coordinates. In physical reality $\varepsilon \sim \frac{1\mathring{A}}{1cm} = 10^{-8}$, and in this model we take the idealized $\varepsilon \to 0$ limit. Note that the velocity is not rescaled, $V = v$.

*Affiliation from Sep 1, 1998: School of Mathematics, Georgiatech, ATLANTA, GA-30332-0160, USA

Our goal is to solve (1) up to any fixed *macroscopic* time $T > 0$, i.e. up to a *microscopic* time $t = T\varepsilon^{-1}$. The first scale for the coupling λ yielding a nontrivial (non-free) evolution on this time scale is $\lambda = \sqrt{\varepsilon}$. This is called the *weak coupling limit (WCL)*.

A similar model, the *low density* (or Boltzmann-Grad) limit (LDL), has been discussed in **[EY1]**. In this model, the effect of the potential is suppressed by considering rarified impurities. The potential is $V_\omega(x) = \sum_\alpha V_0(x - x_\alpha(\omega))$, where $\{x_\alpha(\omega)\}$ is a random point process (say Poisson) with a density $\varrho \ll 1$, which depends on ε. Here the first nontrivial regime is obtained by choosing $\varrho \sim \varepsilon$.

To formulate our result we recall that the Wigner function W_ψ of a wave function $\psi \in L^2(\mathbf{R}^d)$ is defined as

$$W_\psi(x, v) := \frac{1}{(2\pi)^d} \int \overline{\psi\Big(x + \frac{z}{2}\Big)} \psi\Big(x - \frac{z}{2}\Big) e^{iv\cdot z} \mathrm{d}z.$$

and let

$$W_\psi^\varepsilon(X, V) := \varepsilon^{-d} W_\psi\Big(\frac{X}{\varepsilon}, V\Big)$$

be the rescaled Wigner function. Typically the Wigner function has no definite sign, but its proper mollification, the Husimi function, does. We shall prove that the weak limit of the rescaled Wigner function is nonnegative and thus can be considered as the phase space density.

Finally we recall that the linear Boltzmann equation with collision kernel $\sigma(U, V)$ is

$$\partial_T F_T(X, V) + V \cdot \nabla_X F_T(X, V) = \int \Big[\sigma(U, V) F_T(X, U) - \sigma(V, U) F_T(X, V)\Big] \mathrm{d}U$$

$$= \int \sigma(U, V) F_T(X, U) \mathrm{d}U - \sigma_0(V)\, F_T(X, V) \tag{2}$$

with total cross section $\sigma_0(V) := \int \sigma(V, U) \mathrm{d}U$. In case of elastic collision, $\sigma(U, V)$ is supported on the set $\{(U, V) \,:\, U^2 = V^2\}$ (onshell condition). If, in addition, the collision mechanism is spherically symmetric, then $\sigma_0(V)$ depends only on $|V|$ and $\sigma(U, V) = \widetilde{\sigma}(U - V)\delta(U^2 - V^2)$.

Theorem 1.1 [EY2] *Let $\lambda = \sqrt{\varepsilon}$ and let $\psi_{t,\omega}^\lambda$ solve the Schrödinger equation (1) with initial condition*

$$\psi_0^\varepsilon(x) := \varepsilon^{3/2} h(\varepsilon x) \exp\Big(\frac{iS(x\varepsilon)}{\varepsilon}\Big). \tag{3}$$

Assume that $h(X), S(Y)$ are in the Schwarz class such that S has only isolated nondegenerate critical points. In particular

$$W_{\psi_0^\varepsilon}(X, V) \to |h(X)|^2 \delta(V - \nabla S(X)) := F_0(X, V) \tag{4}$$

weakly in H_{-2} *on* $\mathbf{R}^{2d}$ *as* $\varepsilon \to 0$. *Then for any* $T > 0$

$$\mathbf{E}W^{\varepsilon}_{\psi^{\varepsilon}_{T/\varepsilon,\omega}}(X,V) \to F_T(X,V)$$

weakly in H_{-2} *as* $\varepsilon \to 0$, *and* F_T *satisfies the linear Boltzmann equation (2) with initial condition* $F_0(X,V)$ *and collision kernel* $\sigma(U,V) = 4\pi\widehat{G}(U-V)\delta(U^2 - V^2)$.

Remark 1. The analogous theorem for the low density limit (LDL) was formulated in [**EY1**]. In that model the Boltzmann collision kernel $\sigma(U,V)$ is the quantum scattering cross section, hence the full quantum collision mechanism is present even after the scaling limit. In the weak coupling limit only the first Born approximation of the quantum scattering cross section appears in the limiting Boltzmann equation.

Remark 2. Certain version of Theorem 1.1 was proved by H. Spohn [**S**] for sufficiently small T. His result was extended to higher order correlation functions by Ho, Landau and Wilkins [**HLW**] under the same assumption.

This paper is a summary of our joint work with H.-T. Yau [**EY2**].

2 Outline of the proof

Let $H = -\frac{1}{2}\Delta + \lambda V_\omega$ and we usually omit ω from the notation. We shall express the time evolution $\psi(t) := e^{-itH}\psi$ using the Duhamel formula, where $t = T\varepsilon^{-1}$ and T is fixed. For any thresholds M and $N(\varepsilon)$ we have

$$\psi(t) = \psi_M^{main}(t) + \psi_M^{error}(t)$$

$$\psi_M^{main}(t) := \sum_{n=0}^{M-1} \psi_n(t) \qquad \text{and} \qquad \psi_M^{error}(t) := \sum_{n=M}^{N(\varepsilon)} \psi_n(t) + \Psi_{N(\varepsilon)}(t),$$

where in momentum space

$$\widehat{\psi}_n(t,p_0) := \int \Big(\prod_{j=1}^{n} \mathrm{d}p_n\Big) K(t,p_0,p_1,\dots p_n)\Big(\prod_{j=0}^{n-1} \lambda\widehat{V}(p_j - p_{j+1})\Big)\widehat{\psi}(p_n), \tag{5}$$

$$\Psi_{N(\varepsilon)}(t) := (-i)\int_0^t \mathrm{d}s\, e^{-i(t-s)H}\lambda V \psi_{N(\varepsilon)-1}(s),$$

and the kernel K is defined as

$$K(t,\mathbf{p}_n) = (-i)^n \int_0^t \cdots \int_0^t \Big(\prod_{j=0}^{n} \mathrm{d}s_j\Big)\delta\Big(t - \sum_{j=0}^{n} s_j\Big)\prod_{j=0}^{n} e^{-is_j p_j^2/2} \tag{6}$$

with $\mathbf{p}_n := (p_0, p_1, \dots p_n)$.

The proof consists of three steps. Step 1 is an apriori estimate for $\mathbf{E}\|\psi_n\|^2$ (Lemma 2.1) which, among other purposes, is used to show that

$$\lim_{M\to\infty}\lim_{\varepsilon\to 0}\mathbf{E}\Big\|\sum_{n=M}^{N(\varepsilon)}\psi_n(t)\Big\|^2 = 0. \tag{7}$$

In Step 2 (Lemma 2.2) we use a *partial time integration* scheme to show that

$$\lim_{\varepsilon\to 0}\mathbf{E}\|\Psi_{N(\varepsilon)}(t)\|^2 = 0. \tag{8}$$

To achieve both (7) and (8) requires a careful choice of the threshold $N(\varepsilon)$. Step 1 and 2 ensure that ψ_M^{error} is negligible. Finally, in Step 3 we compute the Wigner transform of ψ_M^{main} and we let $M\to\infty$ at the end.

2.1 Step 1. L^2-estimate; the proof of (7)

Lemma 2.1 *There is a constant C depending only on $\|\psi\|_{H^k}$ and on G such that*

$$\mathbf{E}\|\psi_n(t)\|^2 \le \frac{(C\lambda^2 t)^n}{(n!)^{1/2}}\left[1+\frac{(n!)^{3/2}(\log t)^{n+3}}{t^{1/2}}\right]. \tag{9}$$

Here H^k denotes the Sobolev space and k is large enough.

Remark. Recalling that $t=T/\varepsilon$, $\lambda=\sqrt{\varepsilon}$, we see that $\mathbf{E}\|\psi_n(t)\|^2\le 2(CT)^n(n!)^{-1/2}$ if $n\le N(\varepsilon)$ and

$$\varepsilon^{1/4}|\log\varepsilon|^{N(\varepsilon)+3}\to 0 \qquad \text{and} \qquad \varepsilon^{1/4}[N(\varepsilon)!]^4\to 0. \tag{10}$$

For the sketch of the proof, let $R(p):=\widehat{G}(p)^{1/2}$, where $\widehat{G}$ is the Fourier transform of G, hence it is nonnegative and symmetric. The two point correlations of the random field $\widehat{V}$ are

$$\mathbf{E}\widehat{V}(p)\widehat{V}(q)=\delta(p+q)R(p)R(q) \qquad \text{and} \qquad \mathbf{E}\overline{\widehat{V}(p)}V(q)=\delta(p-q)R(p)R(q). \tag{11}$$

After substituting (5) into $\mathbf{E}\|\psi_n(t)\|^2 = \mathbf{E}\int\overline{\widehat{\psi}_n(t,p_0)}\widehat{\psi}_n(t,p_0')\mathrm{d}p_0$, higher order correlation functions appear. We use Wick's theorem to express them as a sums of products of two point correlation functions, where the summands are labelled by pairings $\pi\in\Pi_n$, and Π_n denotes the set of all pairings of a set of $2n$ elements. Hence we have

$$\mathbf{E}\prod_{j=0}^{n-1}\overline{\widehat{V}(p_j-p_{j+1})\widehat{V}(p_j'-p_{j+1}')} = \sum_{\pi\in\Pi_n}\Big(\prod_{j=0}^{n-1}R(p_j-p_{j+1})R(p_j'-p_{j+1}')\Big)\Delta_\pi(\mathbf{p}_n,\mathbf{p}_n'),$$

where $\Delta_\pi(\mathbf{p}_n, \mathbf{p}'_n)$ denotes the product of the n delta functions from the two point correlation functions. As a result, we can express

$$\mathbf{E}\|\psi_n(t)\|^2 = \sum_{\pi \in \Pi_n} C_\pi \tag{12}$$

with

$$C_\pi := \lambda^{2n} \int \mathrm{d}\mathbf{p}_n \mathrm{d}\mathbf{p}'_n \delta(p_0 - p'_0) \overline{K(t, \mathbf{p}_n)} K(t, \mathbf{p}'_n) H(\mathbf{p}_n, \mathbf{p}'_n) \Delta_\pi(\mathbf{p}_n, \mathbf{p}'_n), \tag{13}$$

where

$$H(\mathbf{p}_n, \mathbf{p}'_n) := \Big(\prod_{j=0}^{n-1} R(p_j - p_{j+1}) R(p'_j - p'_{j+1}) \Big) \overline{\widehat{\psi}(p_n)} \widehat{\psi}(p'_n).$$

To control C_π, we need the following basic estimate on the kernel K :

$$\int |K(t, \mathbf{p}_n)|^2 F(\mathbf{p}_n) \mathrm{d}\mathbf{p}_n \le \frac{(C_a t)^n}{[n!]^a} \|F\|_{\mathcal{S}} \tag{14}$$

for any $0 < a < 1$, where $\mathcal{S}$ refers to the Schwarz class and C_a is a universal constant depending only on a. In fact the same estimate is true under much weaker conditions on F. Recall that K was an n-fold time integral under the constraint that $\sum_{j=1}^n s_j \le t$, hence $|K(t, \mathbf{p}_n)| \le t^n/n!$ by a trivial estimate. What (14) claims is that essentially the same is true for $|K|^2$ in integral sense. The proof of (14) heavily uses the cancellation effects of the phases in K via the following representation

$$K(t, \mathbf{p}_n) = i e^{t\eta/2} \int_{-\infty}^{\infty} \mathrm{d}\alpha \, e^{-i\alpha t} \prod_{j=0}^{n} \frac{1}{\alpha - p_j^2/2 + i\eta} \tag{15}$$

for any $\eta > 0$, but we mostly use it for $\eta = t^{-1}$. For large t it is useful to look at these denominators as approximate delta functions $\delta(\alpha - p_j^2/2)$, hence $K(t, \mathbf{p}_n)$ has a singularity at $2\alpha \sim p_0^2 \sim p_1^2 \sim \ldots p_n^2$, which will be the source of the onshell condition in the Boltzmann collision kernel.

Therefore, if $\Delta_\pi(\mathbf{p}_n, \mathbf{p}'_n)$ has a structure such that

$$\mathbf{p}'_n \mapsto \int \Delta_\pi(\mathbf{p}_n, \mathbf{p}'_n) \delta(p_0 - p'_0) H(\mathbf{p}_n, \mathbf{p}'_n) \mathrm{d}\mathbf{p}_n$$

is in Schwarz class as a function of $\mathbf{p}'_n$ and the same is true for the $\int \ldots \mathrm{d}\mathbf{p}'_n$ integral viewed as a function of $\mathbf{p}_n$, then by a simple Schwarz inequality

$$|\overline{K(t, \mathbf{p}_n)} K(t, \mathbf{p}'_n)| \le |K(t, \mathbf{p}_n)|^2 + |K(t, \mathbf{p}'_n)|^2$$

we obtain

$$|C_\pi| \le \frac{(C\lambda^2 t)^n}{(n!)^{1/2}} = \frac{(CT)^n}{(n!)^{1/2}} \tag{16}$$

with a constant independent of n and t.

Typically the structure of $\Delta_\pi(\mathbf{p}_n, \mathbf{p}'_n)$ is more complicated causing extra logarithmic factors, and the Schwarz space has to be replaced by a weaker weighted Sobolev space but for these technicalities we refer to the complete proof [**EY2**].

However, (16) is still not enough to prove (9), since the total number of pairings is $C^n n!$. Most pairings do not correspond to classical trajectories, and there is a stronger estimate for them.

A pairing π is called *crossing* if $\Delta_\pi(\mathbf{p}_n, \mathbf{p}'_n)$ cannot be written as a product of deltafunctions, each depending only on two variables. A non-crossing pairing is called *nested* if $\Delta_\pi(\mathbf{p}_n, \mathbf{p}'_n)$ contains a factor of the form $\delta(p_i - p_{i+1} + p_j - p_{j+1})\delta(p_k - p_{k+2})$ with $i < k < j$ or a similar factor with the primed variables. Finally a non-crossing, non-nested pairing is called *simple*. The names come from a graphical representation of the pairings; for the details and the precise definitions, see [**EY2**]. We show that

$$|C_\pi| \leq \frac{(C\lambda^2 t)^n (\log t)^{n+3}}{(n!)^{1/2} t} \tag{17}$$

for any crossing or nested pairing π, hence we gain an extra factor of $t^{-1} \sim \varepsilon$ compared to (16). The extra logarithmic factors do not matter by (10) and $n \leq N(\varepsilon)$.

Finally a combinatorial lemma shows that the number of simple pairings is only C^n, hence one can use (16) to control them, which completes the proof of Lemma 2.1.

2.2 Step 2. Partial time integration: proof of (8)

Let the integer $\kappa(\varepsilon)$ be chosen later and let $\theta_j := jt/\kappa(\varepsilon)$ for $j = 0, 1, \dots \kappa(\varepsilon)$. Write

$$\Psi_{N(\varepsilon)}(t) = (-i) \sum_{j=0}^{\kappa(\varepsilon)-1} e^{-i(t-\theta_{j+1}H)} \int_{\theta_j}^{\theta_{j+1}} e^{-i(\theta_{j+1}-\sigma)H} \lambda V \psi_{N(\varepsilon)-1}(\sigma) \mathrm{d}\sigma.$$

Let

$$[\mathcal{D}_t(m)\psi]^\wedge(p_0) := \int \mathrm{d}p_1 \dots \mathrm{d}p_m K(t, p_0, \dots p_{m-1}) \Big(\prod_{j=0}^{m-1} \lambda \widehat{V}(p_j - p_{j+1}) \Big) \widehat{\psi}(p_m),$$

and define for $n \geq m$

$$\psi_{n,m,\theta}(s) := \int_\theta^s \mathrm{d}\sigma \mathcal{D}_{s-\sigma}(n-m) \psi_m(\sigma),$$

i.e. this is the wavefunction up to time s with m collisions up to time θ and altogether n collisions. For $n > m$ we also define

$$\widehat{\psi}^*_{n,m,\theta}(s, p_0) := \int \mathrm{d}p \, \lambda \widehat{V}(p_0 - p) \widehat{\psi}_{n-1,m,\theta}(s, p).$$

We expand the unitary group $e^{-i(\theta_{j+1}-\sigma)H}$ further, by using the Duhamel formula, to have

$$\Psi_{N(\varepsilon)} = U_1 + U_2,$$

where

$$U_1(t) = \sum_{N(\varepsilon)\leq n<4N(\varepsilon)} \sum_{j=0}^{\kappa(\varepsilon)-1} e^{-i(t-\theta_{j+1})H}\psi_{n,4N(\varepsilon),\theta_j}(\theta_{j+1}),$$

$$U_2(t) = \sum_{j=0}^{\kappa(\varepsilon)-1} e^{-i(t-\theta_{j+1})H}\int_{\theta_j}^{\theta_{j+1}} e^{-i(\theta_{j+1}-\sigma)H}\psi^*_{4N(\varepsilon),N(\varepsilon),\theta_j}(\sigma)\mathrm{d}\sigma.$$

By unitarity and the triangle inequality

$$\| U_1 \|_2^2 \leq CN(\varepsilon)\kappa(\varepsilon) \sum_{N(\varepsilon)\leq n<4N(\varepsilon)} \sum_{j=0}^{\kappa(\varepsilon)-1} \| \psi_{n,N(\varepsilon),\theta_j}(\theta_{j+1}) \|_2^2$$

and

$$\begin{aligned} \| U_2(t) \|_2^2 &\leq t\sum_{j=0}^{\kappa(\varepsilon)-1}\int_{\theta_j}^{\theta_{j+1}} \| \psi^*_{4N(\varepsilon),N(\varepsilon),\theta_j}(s) \|_2^2\mathrm{d}s \\ &\leq t^2 \sup_j \sup_{\theta_j\leq s\leq\theta_{j+1}} \| \psi^*_{4N(\varepsilon),N(\varepsilon),\theta_j}(s) \|_2^2. \end{aligned}$$

Lemma 2.2 *For* $0 \leq t-\theta \leq t/\kappa(\varepsilon)$, $\kappa(\varepsilon) \geq 2$ *and* $N(\varepsilon) \leq n$,

$$\mathbf{E}\|\psi_{n,N(\varepsilon),\theta}(t)\|^2 \leq \frac{(C\lambda^2 t)^n}{(n!)^{1/2}}\Big[(\log t)^4 + \frac{(n!)^{3/2}(\log t)^{n+5}}{t^{1/2}}\Big],$$

while for $4N(\varepsilon) \leq n$

$$\mathbf{E}\|\psi^*_{n,N(\varepsilon),\theta}(t)\|^2 \leq n!(C\lambda^2 t)^n\Big(\frac{1}{\kappa(\varepsilon)}\Big)^{N(\varepsilon)}(\log t)^{n+4}.$$

Remark. In the last estimate one could get an extra $\frac{1}{t}$ due to the definition of ψ^* starting with a potential, but we do not need it. The true exponent of $\frac{1}{\kappa(\varepsilon)}$ should be $n - N(\varepsilon)$ instead of $N(\varepsilon)$, corresponding to squeezing $n - N(\varepsilon)$ collisions in a short time interval of size $t/\kappa(\varepsilon)$, but we do not need this stronger result.

The proof of Lemma 2.2 is similar to that of Lemma 2.1 and we refer to [**EY2**] for details.

Given Lemma 2.2 and recalling that $t = T\varepsilon^{-1}$, $\lambda = \varepsilon^{1/2}$, we see that (8) follows if the following requirements are satisfied in addition to (10)

$$N(\varepsilon) \to \infty\,, \qquad \frac{\kappa(\varepsilon)^2 N(\varepsilon)^2|\log\varepsilon|^4}{[N(\varepsilon)!]^{1/2}} \to 0$$

$$\varepsilon^{1/8}|CT\log\varepsilon|^{4N(\varepsilon)}\to 0\,,\qquad\text{and}\qquad \frac{|CT\log\varepsilon|^{4N(\varepsilon)+4}[4N(\varepsilon)]!}{\varepsilon^2\kappa(\varepsilon)^{N(\varepsilon)}}\to 0.$$

Choosing

$$N(\varepsilon):=\frac{\gamma|\log\varepsilon|}{\log|\log\varepsilon|}\,,\qquad \kappa(\varepsilon):=\Big[|\log\varepsilon|^{1/\gamma^2}\Big]$$

with some $0<\gamma\ll 1$ sufficiently small, all these requirements are satisfied.

2.3 Step 3. Computation of the Wigner transform

Since $\widehat{W}_\psi(\xi,v)=\overline{\widehat{\psi}(v-\xi/2)}\widehat{\psi}(v+\xi/2)$ (hat denotes Fourier transform in the first variable), after rescaling and testing it against a smooth function J we obtain

$$\mathbf{E}\int\overline{J(X,V)}W^{\varepsilon}_{\psi_{T/\varepsilon}}(X,V)\mathrm{d}X\mathrm{d}V=\int\mathrm{d}\xi\mathrm{d}v\;\varepsilon^{-d}\overline{\widehat{J}(\xi\varepsilon^{-1},v)}\mathbf{E}\overline{\widehat{\psi}_t}\Big(v-\frac{\xi}{2}\Big)\widehat{\psi}_t\Big(v+\frac{\xi}{2}\Big). \tag{18}$$

Since $\mathbf{E}\|\psi_M^{error}\|^2$ is negligible, we can replace ψ_t by $\psi_M^{main}(t)$. Similarly to the L^2-norm computation

$$\int\mathrm{d}\xi\mathrm{d}v\;\varepsilon^{-d}\overline{\widehat{J}(\xi\varepsilon^{-1},v)}\mathbf{E}\overline{\widehat{\psi}_M^{main}}\Big(t,v-\frac{\xi}{2}\Big)\widehat{\psi}_M^{main}\Big(t,v+\frac{\xi}{2}\Big)=\sum_{\substack{n,n'\\ n+n'\le M}}\;\sum_{\pi\in\Pi_{n,n'}}C_\pi^J \tag{19}$$

with

$$C_\pi^J=\lambda^{n+n'}\int\mathrm{d}\xi\mathrm{d}v\;\varepsilon^{-d}\overline{\widehat{J}(\xi\varepsilon^{-1},v)}\int\mathrm{d}\mathbf{p}_n\mathrm{d}\mathbf{p}'_{n'}\delta(p_0-v+\xi/2)\delta(p'_0-v-\xi/2)$$

$$\times\overline{K(t,\mathbf{p}_n)}K(t,\mathbf{p}'_{n'})\Big(\prod_{j=0}^{n-1}R(p_j-p_{j+1})\prod_{j'=0}^{n'-1}R(p'_{j'}-p'_{j'+1})\Big)\Delta_\pi(\mathbf{p}_n,\mathbf{p}'_{n'})\overline{\widehat{\psi}(p_n)}\widehat{\psi}(p'_{n'}),$$

and $\Pi_{n,n'}$ is the set of pairings with n' potentials V and n conjugate potentials $\overline{V}$.

Similarly to (17), one can show that for crossing and nested pairings C_π^J is negligible.

The structure of a simple pairing can explicitly be given. Those momenta which $\Delta_\pi(\mathbf{p}_n,\mathbf{p}'_{n'})$ does not depend on, are called *internal*, the rest are *external*. The internal momenta can easily be integrated out. The delta functions in a simple pairing identify certain external momenta. Hence it is convenient to rename the momenta variables according to this identification. The details are notationally somewhat tedious (see [**EY2**]), but the bottomline is the following formula, where also (15) was used

$$C_\pi^J=e^{t\eta}\lambda^{n+n'}\int\mathrm{d}\xi\mathrm{d}\mathbf{v}_m\;\varepsilon^{-d}\overline{\widehat{J}(\xi\varepsilon^{-1},v_0)}\int_{-\infty}^{\infty}\mathrm{d}\alpha\mathrm{d}\beta\,e^{i(\alpha-\beta)t}$$

$$\times \prod_{j=0}^{m} \left[\frac{1}{\alpha - \frac{1}{2}(v_j - \frac{\xi}{2})^2 - i\eta} \right]^{k_j+1} \left[\frac{1}{\beta - \frac{1}{2}(v_j + \frac{\xi}{2})^2 + i\eta} \right]^{k_j'+1}$$

$$\times \left[\prod_{j=0}^{m-1} R^2(v_j - v_{j+1}) \left[\overline{\Theta}_{\alpha,\eta}\Big(v_j - \frac{\xi}{2}\Big)\right]^{k_j} \left[\Theta_{\beta,\eta}\Big(v_j + \frac{\xi}{2}\Big)\right]^{k_j'} \right] \overline{\widehat{\psi}\Big(v_m - \frac{\xi}{2}\Big)} \widehat{\psi}\Big(v_m + \frac{\xi}{2}\Big) \tag{20}$$

with

$$\Theta_{\alpha,\eta}(r) := \int \frac{R^2(r-q)}{\alpha - q^2/2 + i\eta} \mathrm{d}q$$

and η is chosen t^{-1}. Here m is the number of *different* external momenta while k_j and k_j' count the number of internal p and p' momenta, respectively. These numbers depend on π; $m = m(\pi)$ etc. We have $m + 2\sum_{j=0}^{m} k_j = n$ and $m + 2\sum_{j=0}^{m} k_j' = n'$.

Recalling the remark after (15), we show that $\Theta_{\alpha,\eta}(r)$ can be replaced by $\Theta(r) := \lim_{\nu\to 0+0} \Theta_{r^2/2,\nu}(r)$ and similarly for $\Theta_{\beta,\eta}(r)$. Moreover, $\widehat{J}(\xi\varepsilon^{-1}, v_0)$ forces $|\xi| \sim \varepsilon$, hence $\Theta(v_j \pm \xi/2)$ can be replaced by $\Theta(v_j)$, and again by the effective onshell condition and $\Theta(v_j) \sim \Theta(v_0)$. In the last step we used that R, hence Θ are radially symmetric. For the details of the validity of these replacements, see **[EY2]**.

Now we sum up C_π^J for all $\pi's$ with a fixed m, in other words we sum up all the k_j's and (k_j')'s. Though the sum in (19) allows only summation under the constraint $2\sum_{j=0}^{m} k_j \leq M - m$, the extension of this sum to infinity yields a small error. The result is that for any m, up to a negligible error,

$$\sum_{\pi : m(\pi) = m} C_\pi^J = \lambda^{2m} \int \mathrm{d}\xi \mathrm{d}\mathbf{v}_m \, \varepsilon^{-d} \overline{\widehat{J}(\xi\varepsilon^{-1}, v_0)} e^{it\lambda^2[\overline{\Theta}(v_0) - \Theta(v_0)]} \tag{21}$$

$$\times \overline{K(t, \mathbf{v}_m - \xi/2)} K(t, \mathbf{v}_m + \xi/2) \prod_{j=0}^{m-1} R^2(v_j - v_{j+1}) \widehat{W}_{\psi_0}(\xi, v_m).$$

Notice that the effect of summing up all internal momenta integrations (all the Θ's) culminates in a simple factor

$$e^{it\lambda^2[\overline{\Theta}(v_0) - \Theta(v_0)]} = e^{2T \mathrm{Im}\Theta(v_0)}$$

and $\mathrm{Im}\Theta(v_0) = -2\pi \int R^2(r-q)\delta(r^2 - q^2)\mathrm{d}q$.

By the definition of the K-kernel (6) and a change of variables we have

$$\overline{K\Big(t, \mathbf{v}_m - \frac{\xi}{2}\Big)} K\Big(t, \mathbf{v}_m + \frac{\xi}{2}\Big) = 2^m \left[\prod_{j=0}^{m} \int_0^t \mathrm{d}a_j \right] \delta\Big(t - \sum_{j=0}^{m} a_j\Big) \exp\Big(-i\sum_{j=0}^{m} a_j \xi\cdot v_j\Big)$$

$$\times \left[\prod_{j=1}^{m} \int_{-a_j}^{a_j} \mathrm{d}b_j \right] \chi\Big(-a_0 \leq \sum_{j=1}^{m} b_j \leq a_0\Big) \exp\left[i \sum_{j=1}^{m} b_j(v_j^2 - v_0^2) \right].$$

By a change of variables $t = T\varepsilon^{-1}$, $a = \tau\varepsilon^{-1}$, we see that, as $\varepsilon \to 0$,

$$\int_{-\tau_j\varepsilon^{-1}}^{\tau_j\varepsilon^{-1}} e^{ib_j(v_j^2-v_0^2)}\mathrm{d}b_j \to 2\pi\delta(v_0^2 - v_j^2),$$

giving the onshell condition. Finally, performing an inverse Fourier transform in ξ, and using $\lambda = \varepsilon^{1/2}$ we obtain in the limit ($V = V_0$)

$$\sum_{\pi:m(\pi)=m} C_\pi^J = \int \mathrm{d}X\mathrm{d}V\overline{J(X,V)}e^{2T\mathrm{Im}\Theta(V)}\left(\prod_{j=0}^{m}\int_0^T \mathrm{d}\tau_j\right)\delta\Big(T - \sum_{j=0}^{m}\tau_j\Big)$$

$$\times \int \mathrm{d}V_1\ldots\mathrm{d}V_m\sigma(V,V_1)\sigma(V_1,V_2)\ldots\sigma(V_{m-1},V_m)F_0\Big(X - \sum_{j=0}^{m}\tau_jV_j,V_m\Big). \quad (22)$$

Here we recall the collision kernel $\sigma(U,V) = 4\pi R^2(U-V)\delta(U^2 - V^2)$ and the initial condition $\psi = \psi_0^\varepsilon$ with its Wigner transform from (3) and (4). Notice that (22) is the mth term in the Dyson series of the Boltzmann equation (2) (tested with J), as $2\mathrm{Im}(\Theta)(V) = -\sigma_0(V)$. This completes the proof of our theorem.

References

[EY1] L. Erdős and H.-T. Yau: *Linear Boltzmann equation as scaling limit of quantum Lorentz gas*, Advances in Differential Equations and Mathematical Physics. Contemporary Mathematics **217**, 137-155 (1998).

[EY2] L. Erdős and H.-T. Yau: *Linear Boltzmann equation as a weak coupling limit.* Submitted to Comm. Pure Appl. Math.

[HLW] T. G. Ho, L. J. Landau and A. J. Wilkins: *On the weak coupling limit for a Fermi gas in a random potential*, Rev. Math. Phys. **5**, 209-298 (1992).

[S] H. Spohn: *Derivation of the transport equation for electrons moving through random impurities*, J. Stat. Phys. **17**, 385 (1977).

Operator Theory:
Advances and Applications, Vol. 108

Coexistence of different spectral types for almost periodic Schrödinger equations in dimension one

Alexander Fedotov[a], Frédéric Klopp[b]

a) Department of Mathematical Physics, St Petersburg State University, 1, Ulianovskaja, 198904 St Petersburg-Petrodvoretz, Russia
fedotov@mph.niif.spb.su
b) Département de Mathématique, Institut Galilée, U.R.A 7539 C.N.R.S, Université de Paris-Nord, Avenue J.-B. Clément, F-93430 Villetaneuse, France
klopp@math.univ-paris13.fr

Abstract
This paper is an enhanced version of a talk given at the conference QMath 7 that was held in Prague (Czech Republic), 22-26 June 1998. It is devoted to the study of almost periodic Schrödinger equations in dimension 1. The main result is the proof of the coexistence of absolutely continuous and singular spectrum for such equations. We also give a criterion for the existence of absolutely continuous and singular spectrum.

The results presented here are taken from [7] (see also [8]). In [7], we deal with a more general class of equations, but to simplify and shorten the assumptions and statements we will restrict ourselves to the following family of almost periodic Schrödinger equations

$$H_{\phi,\varepsilon,\alpha}\psi = -\frac{d^2}{dx^2}\psi(x) + (V(x-\phi) + \alpha\cos(\varepsilon x))\psi(x) = E\psi(x), \quad x \in \mathbb{R} \tag{1}$$

where

- V is L^2_{loc}, real valued and 1-periodic i.e. $V(x+1) = V(x)$ for $x \in \mathbb{R}$,
- ε is a positive number,
- $\alpha > 0$ is a coupling constant,
- ϕ is a real parameter indexing the family of equations.

To be able to control the spectral behavior of $H_{\phi,\varepsilon,\alpha}$, we will need ε to be in a Diophantine class that we define now: for $\aleph > 0$, we introduce

$$\mathcal{D}_\aleph := \left\{\varepsilon \in \mathbb{R};\ \forall q \in \mathbb{Z}\setminus\{0\},\ \inf_{p\in\mathbb{N}}\left|\frac{2\pi}{\varepsilon} - \frac{p}{q}\right| \geq \frac{\aleph}{|q|^3}\right\}.$$

It is well known that $\cup_{\aleph>0}\mathcal{D}_\aleph$ is a set of full measure in $\mathbb{R}$. Moreover, for any $0<\aleph<4/9$, 0 is an accumulation point of $\mathcal{D}_\aleph$. We will study the family (1) in the adiabatic limit i.e. when $\varepsilon\to 0$ within a fixed $\mathcal{D}_\aleph$.

1 The periodic equation and the complex momentum

Consider the periodic Schrödinger equation

$$-\frac{d^2}{dx^2}\psi(x)+V(x)=E\psi(x),\quad x\in\mathbb{R} \tag{2}$$

The spectrum of equation (2) on $L^2(\mathbb{R})$ consists of intervals $[E_1,E_2]$, $[E_3,E_4]$, ..., $[E_{2n+1},E_{2n+2}]$, ..., of the real axis such that

$$E_1<E_2\leq E_3<E_4\dots E_{2n}\leq E_{2n+1}<E_{2n+2}\leq\cdots$$
$$E_n\to+\infty,\quad n\to+\infty.$$

In the sequel we will assume that $E_2<E_3$ i.e. that the first gap of (2) is open. Denote the quasi-momentum associated to (2) by $k(E)$, and its dispersion law by $\mathcal{E}(\kappa)$. Then the points $(E_n)_{n\in\mathbb{N}}$ are branching points of k; they are of square root type (see [14, 8] section 2 and references therein).
The *complex momentum* associated to the family of equations (1) is defined by

$$\kappa_\alpha(\varphi)=k(E-\alpha\cos(\varphi)).$$

The function κ_α is multi-valued and its branching points are given by the equation

$$E-\alpha\cos(\varphi)=E_n,\ n\in\mathbb{N}.$$

These branching points form a 2π-periodic lattice. In Fig. 1, we show how these points move when E is increasing. We represented 3 different cases:

1. when $E<E_1-\alpha$.
2. when $E_1-\alpha<E<\inf(E_2-\alpha,E_1+\alpha)$.
3. when $\sup(E_1+\alpha,E_2-\alpha)<E<\inf(E_3-\alpha,E_2+\alpha)$.

The arrows indicate the direction in which the branching points are moving as E is increasing.

2 The actions and the phase

Obviously, the spectrum of $H_{\phi,\varepsilon,\alpha}$ is contained in $[E_1-\alpha,+\infty)$. We will study it near its infimum. More precisely, pick $\delta>0$ and define the interval J_δ to be

$$J_\delta=[E_1-\alpha,\inf(E_1+\alpha,E_2-\alpha)-\delta].$$

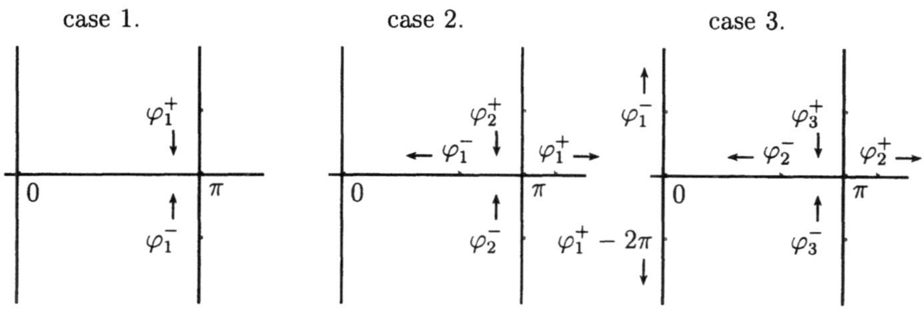

Figure 1: The branching points

We will study the spectrum of the family of equations (1) in J_δ. For $E \in J_\delta$, the branching points of κ_α are located like in case 2 of Fig. 1.

The main objects needed to describe the spectrum of a typical equation of the family (1) in J_δ will be the actions and phase that we define now. We define the *action integrals* S and S_1 and the *phase* Φ as

$$S(\alpha,E) = -i\int_{\varphi_1^+-2\pi}^{\varphi_1^-} \kappa_\alpha d\varphi,\ S_1(\alpha,E) = i\int_{\varphi_2^-}^{\varphi_2^+} (\kappa_\alpha - \pi)d\varphi,\ \Phi(\alpha,E) = \int_{\varphi_1^-}^{\varphi_1^+} \kappa_\alpha d\varphi.$$

These integrals depend on the choice of the branch of the multi-valued function κ_α. We choose this branch (see [8]) so that, for $E \in J_\delta$, one has

$$S(\alpha,E) > 0,\quad S_1(\alpha,E) > 0,\quad \Phi(\alpha,E) \geq 0 \text{ and } \Phi'(\alpha,E) > 0.$$

Here Φ' denotes the derivative of Φ with respect to E. The actions S and S_1 and the phase Φ admit a natural geometric interpretation (cf [7, 8]).

3 The location of the spectrum

We first state a result on the location of the spectrum in J_δ. Therefore, we notice that the function $\Phi(\alpha, E)$ is monotonically increasing on J_δ and that its derivative does not vanish on J_δ. Moreover $\Phi(\alpha, E_1 - \alpha) = 0$. In J_δ, consider the points $E^{(l)}$, $l \in \mathbb{N}$, defined by

$$\frac{1}{\varepsilon}\Phi(\alpha, E^{(l)}) = \pi/2 + \pi l, l \in \mathbb{N}.$$

The number of these points is finite; we denote it by L. For sufficiently small ε, the distances between these points satisfy the inequalities

$$c_1\varepsilon \leq E^{(l)} - E^{(l-1)} \leq c_2\varepsilon,\ l = 1, \ldots, L$$

where c_1 and c_2 are two positive constant independent of ε. We then prove

Theorem 1. *There exists a collection of intervals* $(I_l)_{1\leq l\leq L}$ *such that* $I_l \subset J_\delta$ *and such that the spectrum of equation* (1) *in* J_δ *is contained in* $\cup_{1\leq l\leq L} I_l$ *where*

- *the interval* I_l *lies in an* $o(\varepsilon)$*-vicinity of* $E^{(l)}$,
- *the measure of* I_l *is given by*

$$|I_l| = 2\frac{\varepsilon\left(e^{-\frac{1}{\varepsilon}S(\alpha,E)} + e^{-\frac{1}{\varepsilon}S_1(\alpha,E)}\right)}{F(1/2+l)\Phi'(\alpha,E^{(l)})}(1+o(1))$$

$$\text{where } F(\zeta) = \frac{1}{\sqrt{\pi}}\Gamma(\zeta+1/2)e^{-\zeta\ln\zeta+\zeta},\ \zeta>0.$$

Moreover, if $dN_\varepsilon(E)$ *denotes the density of states measure of* $H_{\phi,\varepsilon,\alpha}$ *at energy* E, *then, one has*

$$\int_{I_l} dN_\varepsilon(E) = \frac{1}{2\pi}\varepsilon. \tag{3}$$

Note that the intervals I_l are exponentially small and separated by a distance of order $O(\varepsilon)$. Theorem 1 can be proved without number theoretical assumptions on ε.

4 The nature of the spectrum

We fix α. The function that will tell us the nature of the spectrum of (1) is

$$\Delta S(\alpha,E) := S_1(\alpha,E) - S(\alpha,E).$$

For $\delta \geq 0$, define the intervals

$$J_\delta^- = \{E \in J_\delta;\ \Delta S(\alpha,E) < -\delta\} \text{ and } J_\delta^+ = \{E\in J_\delta;\ \Delta S(\alpha,E) > \delta\}.$$

If $J_\delta^+ \neq \emptyset$ (resp. $J_\delta^- \neq \emptyset$), then, for sufficiently small ε, the number of intervals I_l lying in J_δ^+ (resp. in J_δ^-) is of order $O(1/\varepsilon)$. We prove

Theorem 2. *Fix* $0 < \aleph < 4/9$ *and* $\delta > 0$. *Then, there exists* $\varepsilon_0 = \varepsilon_0(\aleph,\delta)$ *such that, for* $\varepsilon \in (0,\varepsilon_0)\cap\mathcal{D}_\aleph$, *we have*

- *each of the intervals* I_l *contained in* J_δ^- *contains only singular spectrum (i.e. singular continuous or pure point spectrum),*
- *each of the intervals* I_l *contained in* J_δ^+ *contains absolutely continuous spectrum. Moreover, for* $I_l \subset J_\delta^+$, *one has*

$$|I_l \cap \sigma_{ac}(H_{\phi,\varepsilon,\alpha})|/|I_l| = 1 + o(1) \text{ as } \varepsilon \to 0,\ \varepsilon\in\mathcal{D}_\aleph.$$

For fixed α, it is natural to define the *asymptotic mobility edges* as the solutions of the equations $\Delta S(\alpha, E) = 0$. According to Theorem 2, asymptotically, these are the points where a transition from absolutely continuous to singular spectrum (or vice versa) occurs.
We also note that our result does not tell us what happens at the transitions.

5 The phase diagram

To complete our understanding of the spectrum, we now need to determine whether transitions actually occur i.e. to study the locus of the asymptotic mobility edges. Define

$$\mathcal{S} = \{(\alpha, E);\;\; E_1 - \alpha \leq E \leq \inf(E_2 - \alpha, E_1 + \alpha)\}.$$

and

$$\mathcal{S}_\pm = \{(\alpha, E) \in \mathcal{S};\; \Delta S(\alpha, E) \gtrless 0\} \text{ and } \mathcal{S}_0 = \{(\alpha, E) \in \mathcal{S};\; \Delta S(\alpha, E) = 0\}.$$

$\mathcal{S}_+$, $\mathcal{S}_0$ and $\mathcal{S}_-$ form a partition of $\mathcal{S}$. Roughly, Theorem 2 tells us that $\mathcal{S}_+$ (resp. $\mathcal{S}_-$) is the set of (α, E) where we have absolutely continuous (resp. singular) spectrum. We prove

Theorem 3. *Under the assumptions made above, we know that*

- $\mathcal{S}_0$ *is a non-empty piecewise analytic curve.*
- $\mathcal{S}_+$ *is connected and there exists* $\alpha_0 > 0$ *such that* $\mathcal{S} \cap \{\alpha < \alpha_0\} \subset \mathcal{S}_+$.
- $\mathcal{S}_-$ *is connected and there exists* $\alpha_1 > \alpha^* > 0$ *such that* $\mathcal{S} \cap \{\alpha > \alpha_1\} \subset \mathcal{S}_-$.

To illustrate this, we compute (numerically) the action integrals S and S_1 in the case when V is a 1-periodic one-gap potential associated to the periodic spectrum $[E_1, E_2] \cup [E_3, +\infty)$ where $E_1 = 0$, $E_2 = 8$. The value of E_3 is computed so that the potential do have period 1. Such a potential is unique and the quasi-momentum k associated to it can be explicitly expressed in terms of elliptic integrals (see e.g. [14] and the references therein).
In this case, the *phase diagram* (i.e. the curve $\mathcal{S}_0$ in (α, E)) is given in Fig. 2. We see that, for any $\delta > 0$, for ε small enough,

- for $\alpha < \alpha_0$: we have no transition and most of the spectrum is ac (i.e absolutely continuous). This is the case of "a small perturbation" of a periodic Schrödinger operator.
- for $\alpha_0 < \alpha < \alpha^*$: there exists a single mobility edge $E_{\text{mob}}(\alpha)$ such that
  - for $E < E_{\text{mob}} - \delta$, the spectrum is singular.
  - for $E > E_{\text{mob}} + \delta$, most of the spectrum is ac.

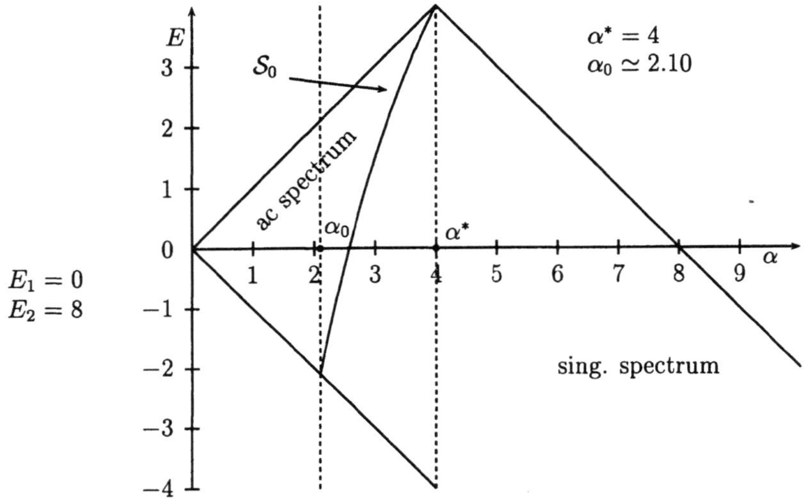

Figure 2: An example of phase diagram

- for $\alpha > \alpha^*$: there is no transition and the spectrum is purely singular. We are in a large perturbation regime.

The theory developed in [7] allows more general potentials than $V(x) + \alpha\cos(\varepsilon x)$ i.e. one can replace cos by a more general 2π-periodic function. In this case the phase diagram can be quite different from the one shown in Fig. 2; e.g. for a given α, one may have two or more mobility edges (see [7]).

6 A word about the proofs

The method used to prove the results described above is to reduce the family of equations (1) to an effective Hamiltonian. This is done locally in energy. Therefore we introduce a notion of *monodromy matrix* suitable for (1) (see [8, 7]). The monodromy matrix is a very general idea. For periodic Schrödinger equations, the notion of a monodromy matrix is classical (see, for example, [16]). For difference equations with periodic coefficients, the monodromy matrix was introduced in [2]. For almost periodic Schrödiger equations, it was introduced and its role was understood in [8].
For any fixed ϕ, let $\psi_{1,2}(x,\,\phi)$ be two linearly independent solutions of equation (1). We say that they form a *consistent basis* if their Wronskian is independent of ϕ and these solutions are 1-periodic in ϕ i.e. $\forall x, \phi,\ \psi_{1,2}(x,\,\phi+1) = \psi_{1,2}(x,\,\phi)$.
Let $\psi_{1,2}$ be a consistent basis. As the cosine is 2π-periodic, the functions $\psi_{1,2}(x + 2\pi/\varepsilon, \phi + 2\pi/\varepsilon)$ are linearly independent solutions of equation (1). Therefore, one

can write

$$\Psi(x+2\pi/\varepsilon,\phi+2\pi/\varepsilon)=M(\phi)\Psi(x,\phi), \tag{4}$$

where Ψ is the transposed of the vector $(\psi_1(x,\phi),\psi_2(x,\phi))$ and $M(\phi)$ is a matrix with coefficients independent of x. $M(\phi)$ is the *monodromy matrix* associated to the consistent basis $\psi_{1,2}$. As $\psi_{1,2}$ is consistent, M is 1-periodic in ϕ and $\forall\phi$, $\det M(\phi)\equiv 1$.
Set $h=\frac{2\pi}{\varepsilon}\mathrm{mod}1$ so that $0<h<1$. The spectral analysis of (1) can be reduced to the investigation of the solutions of the *monodromy equation*

$$\chi\left(\phi+h\right)=M\left(\phi\right)\chi\left(\phi\right)\qquad\forall\phi\in\mathbb{R}. \tag{5}$$

Indeed, it is well known that the spectral properties of one dimensional Schrödinger equations can be completely described in terms of the behavior of its solutions as $x\to\pm\infty$ (see e.g. [9]). It appears that the behavior of solutions of (1) for $x\to\pm\infty$ mimics the behavior of solutions of the monodromy equation for $\phi\to\mp\infty$. More precisely, consider the restriction $\chi_n=\chi(\phi+nh)$ of a solution χ of (5) to the lattice $\phi+h\mathbb{Z}$. It satisfies the equation

$$\chi_{n+1}=M(\phi+nh)\chi_n,\quad\forall n. \tag{6}$$

One proves

Proposition 1. *Let $\psi_{1,2}$ be consistent basis solutions of* (1), *and let M be the corresponding monodromy matrix. Fix $\phi=\phi_0\in\mathbb{R}$. Then, there exists $C>1$ such that, for any vector solution $(F_n)_{n\in\mathbb{Z}}$ of the equation* (6), *there exists a unique solution f of* (1) *such that*

$$\frac{1}{C}\|F_{-n}\|\leq\left\|\begin{pmatrix} f(x+2\pi n/\varepsilon)\\ f'(x+2\pi n/\varepsilon)\end{pmatrix}\right\|\leq C\|F_{-n}\|,\quad\forall x\in\left[0,\frac{2\pi}{\varepsilon}\right],\ \forall n\in\mathbb{Z}. \tag{7}$$

This proposition reduces the spectral analysis of equations (1) to the analysis of the solutions of the finite difference equations (6). This is the effective Hamiltonian referred to in the beginning of this section.

This now leaves us with 2 problems. First we have to find a consistent basis for which we can compute the monodromy matrix (at least to some degree of precision). And then we have to analyze the thus obtained matrix.

To solve the first problem, we use a WKB method for quasiperiodic ordinary differential equations in the adiabatic regime $\varepsilon\to 0$. This method was developed in [8]. The form of the monodromy matrix (i.e. which of the coefficients are of leading order when $\varepsilon\to 0$) that one obtains from this computation essentially depends on the picture of the branching points obtained in Fig. 1 (as this picture is related to the geometry of the canonical domains used in the WKB method). In the energy region we study (i.e. in case 2 in Fig. 1), the asymptotics of the leading coefficients of the monodromy matrix are given in terms of the actions S, S_1 and the phase Φ.

The first order vector valued finite difference equation (6) can be rewritten as a second order scalar finite difference equation. It turns out that, when the branching points are in the configuration of case 2 in Fig. 1, this equation is nothing but a small perturbation of Harper's equation

$$(u_{n+1} + u_{n-1}) + 2\lambda(\alpha, E)\cos(nh + \phi)u_n = F(\alpha, E)u_n. \tag{8}$$

Harper's equation and its perturbations have been the subject of many studies during the last 20 years (for more references, see e.g. [11, 12, 13, 2, 15, 3, 10, 4]). Roughly, it is known that, if $\lambda(\alpha, E)$ is large then the spectrum of (8) is pure point, and in $\lambda(\alpha, E)$ is small then it is absolutely continuous.

In our case, the parameter $\lambda(\alpha, E)$ is given quite explicitly in terms of the action difference ΔS: one has

$$\lambda(\alpha, E) \sim e^{-\Delta S(\alpha,E)/\varepsilon}.$$

This explains roughly the results of Theorem 2. To get a rigorous proof, we need to take into account the fact that (6) is a small perturbation of (8) and the spectrum is very sensitive to these perturbations. Hence, to conclude our study, we use results on more general quasi-periodic finite difference equations (see [1, 5, 6]). To complete this section, let us add that the fact that both the effective energy F and the coupling constant in (8) depend on the spectral parameter E of the initial equation leads to another technical problem that, in particular, prevents us from showing that the singular spectrum is pure point.

References

[1] J. Bellissard, R. Lima, and D. Testard. Metal-insulator transition for the Almost Mathieu model. *Communications in Mathematical Physics*, 88:207–234, 1983.

[2] V. Buslaev and A. Fedotov. Monodromization and Harper equation. In *Séminaires d'équations aux dérivées partielles*, volume XXI, Palaiseau, 1994. Ecole Polytechnique.

[3] R. Carmona and J. Lacroix. *Spectral Theory of Random Schrödinger Operators.* Birkhäuser, Basel, 1990.

[4] H.L. Cycon, R.G. Froese, W. Kirsch, and B. Simon. *Schrödinger Operators.* Springer Verlag, Berlin, 1987.

[5] L. H. Eliasson. Discrete one-dimensional quasi-periodic Schrödinger operators with pure point spectrum. *Acta Mathematica*, 179:153–196, 1997.

[6] L. H. Eliasson. Reducibility and point spectrum for linear quasi-periodic skew products. In *Proceedings of the ICM 1998,Berlin*, volume II, pages 779–787, 1998.

[7] A. Fedotov and F. Klopp. Anderson transitions for almost periodic Schrödinger operators in dimension 1. in progress.

[8] A. Fedotov and F. Klopp. The monodromy matrix for a family of almost periodic equations in the adiabatic case. Preprint, Fields Institute, Toronto, 1997.

[9] D. Gilbert and D. Pearson. On subordinacy and analysis of the spectrum of one-dimensional Schrödinger operators. *Journal of Mathematical Analysis and its Applications*, 128:30–56, 1987.

[10] B. Helffer and J. Sjöstrand. Analyse semi-classique pour l'équation de Harper. *Mémoires de la Société Mathématique de France*, 34, 1988.

[11] T. Janssen. Aperiodic Schrödinger operators. In R. Moody, editor, *The Mathematics of Long-Range Aperiodic Order*, pages 269–306. Kluwer, 1997.

[12] S. Jitomirskaya. Almost everything about the almost Mathieu operator. II. In *XIth International Congress of Mathematical Physics (Paris, 1994)*, pages 373–382, Cambridge, 1995. Internat. Press.

[13] Y. Last. Almost everything about the almost Mathieu operator. I. In *XIth International Congress of Mathematical Physics (Paris, 1994)*, pages 366–372, Cambridge, 1995. Internat. Press.

[14] H. McKean and P. van Moerbeke. The spectrum of Hill's equation. *Inventiones Mathematicae*, 30:217–274, 1975.

[15] L. Pastur and A. Figotin. *Spectra of Random and Almost-Periodic Operators.* Springer Verlag, Berlin, 1992.

[16] E.C. Titschmarch. *Eigenfunction expansions associated with second-order differential equations. Part II.* Clarendon Press, Oxford, 1958.

Operator Theory:
Advances and Applications, Vol. 108

Dynamical Localization for Random Schrödinger Operators and an Application to the Almost Mathieu Operator

François Germinet *
UFR de Mathématiques and LPTMC
Université Paris VII - Denis Diderot
75251 Paris Cedex 05 France

1 Introduction and main result

We show for a large class of random Schrödinger operators H_ω on $\ell^2(\mathbb{Z}^\nu)$ and on $L^2(\mathbb{R}^\nu)$ that **dynamical localization** holds, *i.e.* that, with probability one, for a suitable energy interval I and $q > 0$,

$$\sup_t r^q_{\psi,I}(t) \equiv \sup_t \langle P_I(H_\omega)\psi_t,\ |X|^q P_I(H_\omega)\psi_t\rangle < \infty.$$

Here ψ is a function of sufficiently rapid decrease, $\psi_t = e^{-iH_\omega t}\psi$, $P_I(H_\omega)$ is the spectral projector of H_ω corresponding to the interval I and X is the usual position operator. The result covers all random Schrödinger operators for which exponential localization has been proved using a multi-scale argument, including, for instance, the one dimensional and discrete Anderson model with Bernoulli potential and also random Landau Hamiltonians (continuous case). In section 3 we propose a second way, more direct, to reach dynamical localization, and give an application of this result through the Almost Mathieu model (Theorem 3.1).

The main result of this talk is given in Theorem 1.1 below, and has been obtained in collaboration with S. De Bièvre [9]. This theorem is an illustration, with a precise model studied by Combes and Hislop [3], of a more general theorem contained in [9], that provides dynamical localization for random Hamiltonians.

So we consider here the following family of random Hamiltonians defined on $L^2(\mathbb{R}^\nu)$ ($\nu \geq 1$):

$$H_\omega = -\Delta + \lambda V_\omega,$$

*Current address: UFR de Mathématiques and URA, USTL, 59655 V. d'Ascq, France. E-mail: germinet@gat.univ-lille1.fr

where

$$V_\omega = \sum_{i \in \mathbb{Z}^\nu} \lambda_i(\omega) u(x - i),$$

with $\omega \in \Omega$, Ω being the probability space with measure $\mathrm{d}\mathbb{P}$. Suppose moreover that
(i) the variables $\lambda_i(\omega)$, $i \in \mathbb{Z}^\nu$ are independent and identically distributed, with common distribution $\mathrm{d}\mu(v) = g(v)\mathrm{d}v$, and $g \in \mathrm{L}^\infty(\mathbb{R}^\nu)$.
(ii) the function u belongs to $C_0^2(\mathbb{R}^\nu)$, has compact support, and satisfies

$$u(x) > \chi_{3/2}(x) \geq 0.$$

Then one has the following result:

Theorem 1.1. *Fix $E_0 > 0$ and $\gamma_0 > 0$. Then, for λ large enough and for $\mathbb{P}$-almost every ω:*
(i) for all $E \in I = [0, E_0]$, there exist "centers of localization" $x_{E,\omega}$, associated to the eigenfunctions $\varphi_{E,\omega}$ with energy E so that $\forall \gamma \in]0, \gamma_0[$, one has, for all $x \in \mathbb{R}^\nu$,

$$|\varphi_{E,\omega}(x)| \leq C_E e^{-\gamma|x - x_{E,\omega}|}, \tag{1.1}$$

where

$$C_E = C(E_0, \lambda, \omega, \gamma) e^{\sqrt{|x_{E,\omega}|}}, \tag{1.2}$$

*(ii) (*Dynamical Localization*) if $q > 0$ and $\psi \in \mathrm{L}^2(\mathbb{R}^\nu)$ decays exponentially, then*

$$\sup_t r^q_{\psi,I}(t) \leq C_{\psi,\omega}. \tag{1.3}$$

Remarks:
a) A discrete version of this theorem also holds; it has been supplied, with a measure μ with bounded density, by Del Rio, Jitomirskaya, Last and Simon [5], using some results of Aizenman [1]. But the dynamical localization also holds for a larger class of probability measure μ [9], including the Bernoulli case ($\nu = 1$). An application of this last case has been given for the random dimer model in [4].
b) The result still holds if one adds a constant magnetic field $B = \mathrm{rot} A$, and thus considers $H_\omega = (P - A)^2 + \lambda V_\omega$.
c) There exists a recent and related result of Barbaroux, Fischer, Müller [2]. Taking ψ with compact support, the dynamical localization of the random Hamiltonian H_ω is proven, in the strong localized regime and in the following sense:

$$\int \mathrm{d}\mathbb{P}(\omega) \sup_{T>1} \frac{1}{T} \int_0^T r^2_{\psi,\omega}(t)\mathrm{d}t < C_\psi.$$

We want to emphasize that the important and interesting point of this result, in comparison of Theorem 1.1, is that it is uniform in ω. Indeed we stress that in line (1.3) the constant $C_{\psi,\omega}$ does depend on ω. But on the other hand, in order to yield this new result, it is necessary to average in time, to look only at the second moment ($q = 2$), and to consider initial conditions that no longer decay exponentially as in Theorem 1.1, but with compact support. Moreover, we point out that no information on the eigenfunctions (part (i) of Theorem 1.1) is obtained in this way.

2 Some ideas of the proof

The decisive role played by the multi-scale analysis [7] for such models is now well known. More precisely, a Wegner estimate plus the multi-scale analysis developed by Von Dreifus and Klein in [6] is sufficient, in the strong localized regime, to entail a pure point spectrum with the exponential decay of the eigenfunctions, *i.e.*: for some constants C_E

$$|\varphi_{E,\omega}(x)| \leq C_E e^{-\gamma|x-x_{E,\omega}|}. \tag{2.4}$$

But it is now established that exponential localization does not systematically entail dynamical localization. The authors of [5] exhibited a one-dimensional discrete Hamiltonian H, which has pure point spectrum and exponentially decaying eigenfunctions, but such that for all $\varepsilon > 0$ there exist a time sequence $(t_n)_{n\geq 0}$ and a constant C_ε such that $r^2_{\delta_0}(t_n) > C_\varepsilon t_n^{2-\varepsilon}$.

It is easy to understand that this surprising behaviour of the quantity $\langle\psi_t, |X|^2\psi_t\rangle$ comes from the lack of control, in bound (2.4), on the constants C_E and on the way the centers of localization x_E go to infinity (with E). It turns out that a sufficient condition to overcome this difficulty, and hence to prove dynamical localization, is to suppose that $C_E = C_\varepsilon e^{\varepsilon|x_E|}$, for all $\varepsilon > 0$. This condition has been introduced in [5] and is called SULE (Semi-Uniform Localized Eigenfunctions). Hence, going back to Theorem 1.1, it means that one has $(i) \Longrightarrow (ii)$ (for the proof, see [5] and [9] for the respectively discrete and continuous cases).

To prove (i), and then the theorem, we go back to Von Dreifus and Klein's proof [6] of exponential localization and proceed as follows:
a) the Borel-Cantelli Lemma applied to the results of the multi-scale analysis provides the exponential decay of any polynomially bounded eigenfunction $\varphi_{E,\omega}$,
b) since one thus knows, now, that $\varphi_{E,\omega}$ is L^2, one can define a center $x_{E,\omega}$, where, roughly speaking, the eigenfunction $\varphi_{E,\omega}$ is maximal,
c) and finally, applying again the Borel-Cantelli Lemma to the results of the multi-scale analysis but in a more technical way than the first time and taking into account the existence of x_E, one is in position to obtain the decisive control (1.2), and thus the first part of the theorem.

3 Another way to obtain dynamical localization

If now one is more interested in the dynamical side of Theorem 1.1 (point (ii)), rather than in the spectral properties of the Hamiltonian (point (i)), one can consider a second criterion (formulated here in the discrete case), that we shall call WULE (Weak Uniform Localized Eigenfunctions) since it is easy to check that SULE $\Longrightarrow$ WULE, and which says that for some constants $\gamma > 0$ and $C(l) > 0$, $l \in \mathbb{Z}^\nu$, one has : $\forall E \in I$, $\forall x, l \in \mathbb{Z}^\nu$,

$$|\varphi_E(x)\varphi_E(l)| \leq C(l)e^{-\gamma|x-l|}. \tag{3.5}$$

Let's point out that this weaker condition (WULE) still implies dynamical localization, taking an initial condition with compact support, and this in the discrete as well as in the continuous case [8].

Remarks concerning bound (3.5):

a) notice that the constants $C(l)$ are E-independent,

b) the explicit control of the constant C_E in terms of centers of localization x_E is not required here, as it was in SULE. And, as a consequence, it avoids the technicalities related to this part of the proof of dynamical localization, specially for the Almost Mathieu case (see Theorem 3.1 below).

c) WULE is closely related to a natural and well known condition on the two-point function, which is:

$$\langle \delta_x, e^{-iHt}\delta_l \rangle \leq \tilde{C}(l)e^{-\gamma|x-l|}.$$

This condition is in fact the key point that leads to dynamical localization.

Although SULE implies WULE, it is important to note that a direct proof of WULE is given in [8], avoiding the use of the centers of localization x_E. This proof applies equally well to the random models mentioned previously or the Almost Mathieu model described below.

To finish, and as an application, let's consider the quasi-periodic Hamiltonian defined on $\ell^2(\mathbb{Z})$ by:

$$H_\theta = -\Delta + \lambda\cos\left(2\pi(\alpha x + \theta)\right),$$

with $\theta \in [0,1]$, and $\alpha \in \mathbb{R}\backslash\mathbb{Q}$ with good Diophantine properties, *i.e.* for $r > 2$ and some constant $C > 0$, p_n/q_n being the nth continuous fraction approximant of ω, one has $|\omega - p_n/q_n| > C/q_n^r$.

Theorem 3.1. *Let $\lambda \geq 15$. Then there exists a set Θ (known explicitly) with full Lebesgue measure in $[0,1]$, such that for all $\theta \in \Theta$, $q > 0$, and for all $\psi \in \ell^2(\mathbb{Z})$ with compact support, there exists some constant $C(\psi,\theta)$ so that*

$$\sup_t r_\psi^q(t) < C_{\psi,\theta}.$$

In a few words, the proof of Theorem 3.1 given in [8] is divided in two parts: in order to yield WULE, one uses results of Jitomirskaya [10], where the multiscale analysis of [6] is adapted to this particular model, and also some ideas of Jona-Lasinio, Martinelli and Scoppola [12]. Then, since we showed that WULE implies dynamical localization, the theorem follows.

Finally, it must be mentioned that this result (and even SULE) has been obtained, under the same hypotheses (in fact those contained in [10]), by Jitomirskaya and Last in a preprint [11].

References

[1] M. Aizenman, *Localization at weak disorder: some elementary bounds*, Rev. Math. Phys. **6**, 1163-1182 (1994).

[2] J.M. Barbaroux, W. Fisher, P. Müller, *A Criterion for Dynamical Localization in Random Schrödinger Models*, Preprint (1998).

[3] J.M. Combes, P.D. Hislop, *Localization for some continuous, random Hamiltonian in d-dimension*, J. Funct. Anal. **124**, 14 9-180 (1994).

[4] S. De Bièvre, F. Germinet, *Dynamical Localization for the random dimer model*, preprint (1998).

[5] R. Del Rio, S. Jitomirskaya, Y. Last, B. Simon, *Operators with singular continuous spectrum IV: Hausdorff dimensions, rank one pertubations and localization*, J. d'Analyse Math. **69**, 153-200 (1996).

[6] A. von Dreifus, A. Klein, *A new proof of localization in the Anderson tight binding model*, Commun. Math. Phys. **124**, 285-299 (1989).

[7] J. Fröhlich, T. Spencer, *Absence of diffusion with Anderson tight binding model for large disorder or low energy*, Commun. Math. Phys. **88**, 151-184 (1983).

[8] F. Germinet, *Dynamical Localization II with an Application to the Almost Mathieu Operator*, submitted to J. Stat. Phys. (1998).

[9] F. Germinet, S. De Bièvre, *Dynamical Localization for Discrete and Continuous Random Schrödinger Operators*, Commun. Math. Phys. **194**, 323-341 (1998).

[10] S. Jitomirskaya, *Anderson localization for the Almost Mathieu Equation: A Non pertubative Proof*, Commun. Math. Phys. **165**, 49-57 (1994).

[11] S. Jitomirskaya, Y. Last *Anderson Localization for the Almost Mathieu Equation, III. Semi-Uniform Localization, Continuity of Gaps, and Measure of the spectrum*, preprint (1997).

[12] G. Jona-Lasinio, F. Martinelli, E. Scoppola, *Mutiple tunnelings in d-dimensions: a quantum particle in a hierarchical potential*, Ann. Inst. Henri Poincaré, **vol. 42**, 73-108 (1985).

Operator Theory:
Advances and Applications, Vol. 108

On fractal structure of the spectrum for periodic point perturbations of the Schrödinger operator with a uniform magnetic field

V. A. Geyler, K. V. Pankrashkin

Department of Mathematics, Mordovian State University
430000 Saransk, Russia

Introduction

The band structure of the spectrum for a periodic Schrödinger operator $H = -\Delta + V$ in $L^2(\mathbf{R}^\nu)$ with a sufficiently nice potential V is a common thing in both the theoretical and mathematical physics [1]. On the other hand, using the direct integral decomposition over the torus $\mathbf{T}^\nu$ one can give simple examples of periodic self-adjoint operators with a Cantor spectrum [2], but these operators are not local and therefore are not Schrödinger operators. (Recall that a closed operator A in $L^2(\mathbf{R}^\nu)$ with domain $\mathbf{D}(A)$ is said to be *local* if for every $\varphi \in \mathbf{D}(A)$ the relation $\operatorname{supp} A\varphi \subset \operatorname{supp}\varphi$ holds.) A curious phenomenon of a hidden fractal structure of the spectrum for periodic graph superlattices has been discovered in [3], this effect being no doubt of physical interest. In this connection, it is intriguing that there are local self-adjoint operators, namely the Schrödinger operators H with some periodic point potential, which have a Cantor spectrum on the semi-axis $(-\infty, 0)$ [4].

Now let $H = H^0 + V$, where V is a periodic potential and H^0 is the Schrödinger operator in $L^2(\mathbf{R}^\nu)$ ($\nu = 2$ or 3) with a constant and uniform magnetic field $\mathbf{B}$, i.e., $H^0 = (i\nabla + \mathbf{A})^2$ where $\mathbf{A}$ is a vector potential of the field $\mathbf{B}$. Then the spectral properties of H depend drastically on the field $\mathbf{B}$. Namely, if $\mathbf{B}$ is a so-called rational magnetic field, then H has a band spectrum, otherwise at least in the case $\nu = 2$ the spectrum of H has components that are Cantor sets [5]. This results are valid for sufficiently nice potentials V again. The aim of this paper is to show that for some periodic point perturbations H of H^0 the spectrum of H contains a Cantor set lying on the semiaxis $(-\infty, a)$ where a is the ground energy of H^0; we note that these operators H are also local.

A remark of Yu. G. Shondin [6] establishes the reasons for such (somewhat unexpected) results: Yu. G. Shondin has observed that the usual definition of local operators is in some sense inadequate for point perturbations of Schrödinger operators. He has proposed to employ in the context of point perturbations the notion of a *form local* operator instead of the notion of a local operator. (Following Yu. G. Shondin a closed operator A in $L^2(\mathbf{R}^\nu)$ with form domain $\mathbf{Q}(A)$ is said to be *form local* if for every φ and ψ from $\mathbf{Q}(A)$ the relation $\operatorname{supp}\varphi \cap \operatorname{supp}\psi = \emptyset$ implies $\langle A\varphi \,|\, \psi \rangle = 0$). It follows from Theorem 2 of this paper and from the results of [7], [8] that a form local periodic point perturbation of H^0 has a band spectrum if the field $\mathbf{B}$ is rational.

Cantor spectrum for point perturbations of H^0

First we recall the notion of a periodic point perturbation of H^0 (see [7] and [8] for details). Let Λ be a ν-dimensional lattice in $\mathbf{R}^\nu$ ($\nu = 2$ or 3) and C be an elementary cell of Λ spanned by some generators $\mathbf{a}_1, \ldots, \mathbf{a}_\nu$ of this lattice. We choose a finite subset $\mathrm{K} \subset C$ such that $0 \in \mathrm{K}$ and denote by A the set $A = \{\kappa + \lambda : \kappa \in \mathrm{K},\, \lambda \in \Lambda\}$. To define a point perturbation of H^0 supported by A we use the so-called "restriction – extension" procedure [9], [10]. Namely, let $\mathbf{D}(S) = \{f \in \mathbf{D}(H^0) : f(a) = 0 \quad \forall\, a \in A\}$; denote by S the restriction of H^0 to the domain $\mathbf{D}(S)$. Obviously, S is a symmetric operator. The point perturbation of H^0 supported by A is, by definition, any self-adjoint extension H of S such that $\mathbf{D}(H) \cap \mathbf{D}(H^0) = \mathbf{D}(S)$. It is significant that every point perturbation H of H^0 is a local operator (see [10], Lemma C.2).

The Krein resolvent formula [11] supplies a convenient tool to describe all such extensions. Using this formula, we are able to obtain the Green function $G(x, y; \zeta)$ of H in the following form

$$G(x,y;\zeta) = G^0(x,y;\zeta) - \sum_{a,b \in A} [Q(\zeta) - T]^{-1}_{ab}\, G^0(x,a;\zeta) G^0(b,y;\zeta). \tag{1}$$

Here $\zeta \in \mathbf{C} \setminus (\sigma(H) \cup \sigma(H^0))$ is the spectral parameter, $Q(\zeta)$ is a bounded operator in $l^2(A)$ for every $\zeta \in \mathbf{C} \setminus \sigma(H^0)$ (an explicit form of the matrix of $Q(\zeta)$ in the standard basis of $l^2(A)$ will be given below), and T is a self-adjoint operator in $l^2(A)$. Conversely, for every self-adjoint (not necessary bounded) operator T in $l^2(A)$ the equation (1) defines the Green function of a periodic point perturbation of H^0 supported by A, we shall denote this perturbation by H_T. The matrix $Q_{ab}(\zeta)$ has the following form:

$$Q_{ab}(\zeta) := \begin{cases} G^0(a,b;\zeta), & \text{if } a \neq b; \\ -\frac{1}{4\pi}\left[\psi\left(\frac{1}{2} - \frac{\zeta}{2|B|}\right) + \ln(2\pi) + 2C_E\right], & \text{if } a = b \text{ and } \nu = 2; \\ \frac{|B|^{1/2}}{4\sqrt{2\pi}} \mathrm{Z}\left(\frac{1}{2}, \frac{1}{2} - \frac{\zeta}{2|B|}\right), & \text{if } a = b \text{ and } \nu = 3. \end{cases} \tag{2}$$

Here $\psi(z)$ is the logarithmic derivative of the Euler Γ-function, C_E is the Euler constant, and $\mathrm{Z}(z, u)$ is the Hurwitz ζ-function.

We shall consider only the Λ-periodic operators H_T, more precisely, the operators H_T that are invariant with respect to the group $W(\Lambda)$ of magnetic translations along vectors of Λ [7], [8]. For this purpose we shall suppose that T satisfies the following condition:

(P) *The operator T is invariant with respect to the standard unitary representation of the group $W(\Lambda)$ in the space $l^2(A)$ [7], [8], i.e., $(T+i)^{-1}$ commutes with every operator of this representation.*

In addition, to avoid some pathologies we restrict ourselves to the operator T obeying the condition:

(F) *The domain $\mathbf{D}(T)$ contains all sequences with finite supports.*

It is well known that $\inf \sigma(H^0) = |B|$, and formula (1) shows that $(-\infty, |B|) \cap \sigma(H)$ consists of those $\zeta \in (-\infty, |B|)$ for which the operator $Q(\zeta) - T$ is not invertible in $l^2(A)$. Now we can state the main result of this section.

Theorem 1. *Suppose that the magnetic field $\mathbf{B}$ is rational (this means that all the numbers $\mathbf{B}(\mathbf{a}_i \times \mathbf{a}_j)/2\pi$ are rational). Let X be a Cantor subset of $(-\infty, |B|)$. Then there are bounded self-adjoint operators T and U in $l^2(A)$ such that*
1) $\sigma_c(H_T) \cap (-\infty, |B|) = X$ and $\sigma_{pp}(H_T) \cap (-\infty, |B|) = \emptyset$;
2) $\sigma_c(H_U) \cap (-\infty, |B|) = \emptyset$ and the closure of $\sigma_{pp}(H_U) \cap (-\infty, |B|)$ is equal to X.

Proof. Replacing the lattice Λ by some sublattice, we can suppose that all the numbers $\mathbf{B}(\mathbf{a}_i \times \mathbf{a}_j)/2\pi$ are integer. In this case there is an isomorphism J of the space $l^2(A)$ on the space

$$L^2(\mathbf{T}^\nu) \otimes l^2(\mathrm{K}) = \int_{\mathbf{T}^\nu}^{\oplus} l^2(\mathrm{K})\, d\lambda, \tag{3}$$

(here λ is the Lebesgue measure on the standard torus $\mathbf{T}^\nu$) such that the following assertion is true (see [7], [8]): *Let S be a bounded or self-adjoint operator in $l^2(A)$; then S is invariant with respect to the standard unitary representation of $W(\Lambda)$ if and only if the operator $\widetilde{S} = JSJ^{-1}$ acts fiberwise in the direct integral* (3). We shall denote by $\widetilde{S}(p)$ the fiber of $\widetilde{S}$ over a point $p \in \mathbf{T}^\nu$. The fiber of $Q(\zeta)$ over p will be denoted by $\widetilde{Q}(p; \zeta)$.

Fix a homeomorphism h of X on the standard Cantor discontinuum $\mathcal{C} = \{0, 1\}^{\mathbf{N}}$ and let β be the image of the standard Bernoulli measure on $\mathcal{C}$ under h. Then there exist Borel subsets $M \subset \mathbf{T}^\nu$, $Y \subset X$ and a Borel isomorphism $f : M \to Y$ such that $\beta(X \setminus Y) = \lambda(\mathbf{T}^\nu \setminus M) = 0$ and $\beta(f(Z)) = \lambda(Z)$ for any Borel subset $Z \subset Y$ [12]. Let $\Phi(p) = \widetilde{Q}(p; f(p))$, then by virtue of equation (2) Φ is a bounded Borel function on M with values in the finite-dimensional space $l^2(\mathrm{K})$. Hence there exists a bounded self-adjoint operator T in $l^2(A)$ obeying the condition (P) and such that $\widetilde{T}(p) = \Phi(p)$ a.e. It is known that $\zeta \mapsto Q(\zeta)$ is an analytical operator-valued function on $\mathbf{C} \setminus \sigma(H^0)$ and $dQ(\zeta)/d\zeta$ is a positive definite operator at points $\zeta \in (-\infty, |B|)$ [11]. Therefore, if $p \in M$, then the relation $0 \in \sigma(\widetilde{Q}(p;x)) \cap (-\infty, |B|)$ is equivalent to the relation $x = f(p)$. Hence, by virtue of Theorem XII.85 from [1], $x \in \sigma(H_T) \cap (-\infty, |B|)$ if and only if $\lambda\{p \in M : f(p) \in (x-\varepsilon, x+\varepsilon)\} > 0$ for every $\varepsilon > 0$. On the other hand, $\lambda\{p \in M : f(p) \in (x-\varepsilon, x+\varepsilon)\} = \beta(Y \cap (x-\varepsilon, x+\varepsilon))$, hence $\sigma(H_T) \cap (-\infty, |B|) = X$. It remains to prove that $\sigma_{pp}(H_T) \cap X = \emptyset$. Let $x \in X$; according to Theorem XII.85 in [1] it is sufficient to prove that $\lambda\{p \in M : 0 \in \sigma(\widetilde{Q}(p;x) - \widetilde{T}(p))\} = 0$. But $\{p \in M : 0 \in \sigma(\widetilde{Q}(p;x) - \widetilde{T}(p))\} = f^{-1}(x)$ and the first part of the theorem is proved.

Let now $\{x_n\}$ be a countable dense subset of X. Take a sequence (Y_n) of nonvoid mutually disjoint open subsets of $\mathbf{T}^\nu$ in such a way that $\lambda(\mathbf{T}^\nu \setminus \bigcup_n Y_n) = 0$ and define a function $\Psi : \mathbf{T}^\nu \to X$ by the rule

$$\Psi(p) = \widetilde{Q}(p;x_n), \quad \text{if } p \in Y_n.$$

Arguing as above we verify that there is an operator U in $l^2(A)$ obeying the condition $\widetilde{U}(p) = \Psi(p)$ a.e. and this operator satisfies the condition in 2). ∎

Remark. If $\lambda(X) = 0$ then, in fact, $X = \sigma_{sc}(H_T) \cap (-\infty, |B|)$.

Form local point perturbations

If T is a form local operator in $l^2(A)$, then it follows from the conditions (P) and (F) that T is a bounded operator with a diagonal matrix (with respect to the standard basis of $l^2(A)$). Therefore, the operator H_T has a band spectrum [7], [8]. In this section we prove that under a natural additional condition the operator H_T is form local simultaneously with T. Namely, we assume in this section that the operator T is *semibounded from below*. Since $\inf Q(\zeta) \to -\infty$ as $\zeta \to -\infty$ [7], [8] then there is $E_0 < |B|$ such that the operator $T - Q(\mu)$ is positive definite provided $\mu \leq E_0$. From here on we fix a number $\mu < E_0$.

Now for every $\zeta \in \mathbf{C} \setminus \sigma(H^0)$ we define an operator $\Gamma_\zeta : l^2(A) \to L^2(\mathbf{R}^\nu)$ by the rule $\Gamma_\zeta((\xi_a)_{a\in A}) = \sum_{a\in A} \xi_a G^0(\cdot, a; \zeta)$. The mapping Γ_ζ is well defined and is a linear topological isomorphism onto its image [14].

Lemma 1. *The subspaces* $\mathbf{Q}(H^0)$ *and* $\operatorname{Ran}\Gamma_\zeta$ *are linearly independent.*

Proof. Let $f \in \mathbf{Q}(H^0) \cap \operatorname{Ran}\Gamma_\zeta$ and $f \neq 0$. Then $f = \sum_{a\in A} \xi_a G^0(\cdot, a; \zeta)$ where $\xi_b \neq 0$ for some $b \in A$. Hence

$$\xi_b G^0(\cdot, b; \zeta) = f - \sum_{a \neq b} \xi_a G^0(\cdot, a; \zeta). \tag{4}$$

The function at the right-hand side of Eq. (4) belongs to the Sobolev space $H^1_{loc}(\{b\} \cup (\mathbf{R}^\nu \setminus A))$ but the function $G^0(\cdot, b; \zeta)$ does not. This contradiction concludes the proof. ∎

The same arguments prove the following lemma.

Lemma 2. *Let* $f = f_0 + \Gamma_\zeta \varphi$ *where* $f_0 \in \mathbf{Q}(H^0)$, $\varphi \in l^2(A)$. *Then* $\operatorname{supp} \varphi \subset \operatorname{supp} f$. ∎

Now set $Q_T = \mathbf{Q}(H^0) + \Gamma_\mu \mathbf{Q}(T)$; by virtue of Lemma 1 there is a unique operator $P : Q_T \to \mathbf{Q}(T)$ with the property $f - \Gamma_\mu Pf \in \mathbf{Q}(H^0)\ \forall f \in Q_T$. It is easy to show that Q_T and P are independent of μ. Below we shall denote by $q^0(\mu)$, $q_T(\mu)$, and $t(\mu)$ the quadratic forms of the operators $H^0 - \mu$, $H_T - \mu$, and $T - Q(\mu)$, respectively.

Proposition. *The form domain* $\mathbf{Q}(H_T)$ *coincides with* Q_T *and for every* $f \in Q_T$ *we have*

$$q_T(\mu)[f,f] = q^0(\mu)[f - \Gamma_\mu Pf, f - \Gamma_\mu Pf] + t(\mu)[Pf, Pf].$$

Proof is similar to that for the case $B = 0$ (i.e., $H^0 = -\Delta$) given in [13] and is thus omitted. ∎

Lemma 3. *Let* $\varphi \in \mathbf{Q}(T)$. *Take an open subset* $\Omega \subset \mathbf{R}^\nu$ *obeying the conditions:* $\operatorname{supp}\varphi \subset \Omega$ *and* $\operatorname{dist}(\operatorname{supp}\varphi, \mathbf{R}^\nu \setminus \Omega) > 0$. *Then there exists* $f \in \mathbf{Q}(H^0)$ *such that* $\varphi = Pf$ *and* $\operatorname{supp} f \subset \Omega$.

Proof. Let numbers δ and d be chosen such that $0 < \delta < d < \operatorname{dist}(\operatorname{supp}\varphi, \mathbf{R}^\nu \setminus \Omega)$. Fix a function $e \in C_0^\infty(\mathbf{R}^\nu)$ with the properties: 1) $0 \le e(x) \le 1$; 2) $e(x) = 1$, if $|x| \le \delta$; 3) $e(x) = 0$, if $|x| > d$. Now denote $\psi = 1 - e$, $\psi_a = \psi(\cdot - a)$, $g_a = \psi_a G^0(\cdot, a; \mu)$ $(a \in A)$. Then the following equation is valid in the sense of distributions:

$$((i\nabla + \mathbf{A})^2 - \mu)g_a = (-\Delta\psi_a)G^0(\cdot, a; \mu) + 2i(\nabla\psi_a)(i\nabla + \mathbf{A})G^0(\cdot, a; \mu). \tag{5}$$

Since the right-hand side of Eq.(5) belongs to $L^2(\mathbf{R}^\nu)$, it follows that $g_a \in \mathbf{D}(H^0)$. Let $\varphi = (\xi_a)_{a\in A}$, denote $g = \sum_{a\in A} \xi_a g_a$. Using the closed graph theorem and the equalities

$$|G^0(x + a, y + a; \mu)| = |G^0(x, y; \mu)|,$$
$$|(i\nabla + \mathbf{A}(x + a))G^0(x + a, y + a; \mu)| = |(i\nabla + \mathbf{A}(x))G^0(x, y; \mu)|,$$

where x, y, $a \in \mathbf{R}^\nu$, we get that $g \in \mathbf{D}(H^0)$. Therefore, the function $f = \Gamma_\mu\varphi - g$ is what we need. ∎

Lemma 4. *Let* $f, g \in \mathbf{Q}(H_T)$ *and* $\operatorname{supp} f \cap \operatorname{supp} g = \emptyset$. *Then* $q^0(\mu)[f - \Gamma_\mu Pf, g - \Gamma_\mu Pg] = \langle Q(\mu)Pf | Pg \rangle$.

Proof is by direct calculation (cf. the proof of Lemma 1 from [6]. ∎

The main result of this section is the following theorem.

Theorem 2. *Let the operator* T *is semibounded from below. Then the operator* H_T *is form local if and only if the operator* T *is form local.*

Proof. Let T be a form local operator. Choose f, $g \in \mathbf{Q}(H_T)$ such that $\operatorname{supp} f \cap \operatorname{supp} g = \emptyset$. By virtue of Lemma 4

$$q_T(\mu)[f, g] = q^0(\mu)[f - \Gamma_\mu Pf,\ g - \Gamma_\mu Pg] + t(\mu)[Pf,\ Pg] = \langle TPf\,|Pg\rangle.$$

On the other hand, Lemma 2 shows that $\operatorname{supp} Pf \cap \operatorname{supp} Pg = \emptyset$, hence $\langle TPf\,|Pg\rangle = 0$.

Let now H_T be a form local operator. Choose $\varphi, \psi \in \mathbf{Q}(T)$ such that $\operatorname{supp}\varphi \cap \operatorname{supp}\psi = \emptyset$. Using Lemma 3 we find $f, g \in \mathbf{Q}(H^0)$ such that $\varphi = Pf$, $\psi = Pg$, and $\operatorname{supp} f \cap \operatorname{supp} g = \emptyset$. According to Lemma 4 we get

$$\begin{aligned} \langle T\varphi | \psi \rangle &= \langle TPf \,|Pg\rangle = q^0(\mu)[f - \Gamma_\mu Pf,\, g - \Gamma_\mu Pg] + t(\mu)[Pf,\, Pg] \\ &= q_T(\mu)[f, g] = 0, \end{aligned}$$

and the proof is completed. ∎

Acknowledgements

The authors are grateful to S. Albeverio, V. A. Margulis, I. Yu. Popov, and Yu. G. Shondin for valuable discussions and remarks and to the referee for very useful suggestions. The work is partially supported by RFFI, the Russian Ministry of Education, and by the Program "University of Russia".

References

[1] Reed, M., Simon, B. *Methods of Modern Mathematical Physics.* Vol IV, Academic Press, 3rd ed., New York – San Francisko – London (1987).

[2] Skriganov, M. M. *The Multidimensional Schrödinger operator with a periodic potential* (Russian). Izv. Akad. Nauk SSSR. Ser. Mat., **47** (1983), 659-687; Engl. transl.: Math. USSR Izv., **22** (1984), 619-645.

[3] Exner, P., Gawlista, R. *Band spectra of rectangular graph superlattices.* Phys. Rev. B. **53** (1996), 4275-4286.

[4] Albeverio, S., Geyler, V. A. *The band structure of the general periodic Schrödinger operator with point interactions.* Inst. Mathematik, Ruhr-Univ. Bochum, SFB 237-Preprint (1998).

[5] Helffer, B., Sjöstrand, J. *Semi-classical analysis for Harper's equation.* III: *Cantor structure of the spectrum.* Bull. Soc. Math. France. **117** Suppl. (Mémoir) No. 39 (1989), 1-124.

[6] Shondin, Yu. G. *Semibounded local Hamiltonians for perturbation of the Laplacian supported by curves with angle points in* $\mathbf{R}^4$ (Russian). Teoret. i Mat. Fiz. **106** (1996), 179-199.

[7] Geyler, V. A. *The two-dimensional Schrödinger operators with a uniform magnetic field and its perturbation by periodic zero-range potentials* (Russian). Algebra i Anal. **3**, No. 3 (1991), 1-48; Engl. transl.: St.-Petersburg Math. J. **3** (1992), 489-532.

[8] Geyler, V. A., Demidov, V. V. *Spectrum of three-dimensional Landau operator perturbed by a periodic point potential* (Russian). Teoret. i Mat. Fiz. **103** (1995), 283-294; Engl. transl.: Theor. Math. Phys. **103** (1995), 561-569.

[9] Pavlov, B. S.: *The theory of extensions and explicitly solvable models* (Russian). Uspekhi Mat. Nauk **42**, No. 6 (1987), 99-131; Engl. transl.: Russ. Math. Surv. **42**, No. 6 (1987), 127-168.

[10] Albeverio, S., Gesztesy, F., Høegh-Krohn, R., Holden, H. *Solvable models in quantum mechanics*, Springer, Berlin (1988).

[11] Krein, M. G., Langer, H. K. *Defect subspaces and generalized resolvents of an Hermitian operators in the space* Π_κ (Russian). Funk. Anal. i Prilozhen. **5**, No. 2 (1971), 59-71; Engl. transl.: Funct. Anal. and its Appl. **5** (1971), 217-228

[12] Rokhlin, V. A. *Main notions of measure theory.* Math. Sb. **67** (1949), 107-150.

[13] Teta, A. *Quadratic forms for singular perturbations of the Laplacian.* Publ. Res. Inst. Math. Sci. **26** (1990), 803-817.

[14] Geyler, V. A., Margulis, V. A., Chuchaev, I. I. *Zero-range potentials and Carleman operators* (Russian). Sibir. Mat. Zhurn. **36** (1995), 828-841; Engl. transl.: Siberian Math. J. **36** (1995), 714-726.

Operator Theory:
Advances and Applications, Vol. 108

A Weyl-Berry formula for the scattering operator associated to self-similar potentials on the line

Charles-Antoine GUERIN
Centre de Physique Théorique,
CNRS-Luminy case 907,
F-13288 Marseille cedex 09
and
Chalmers University of Technology,
Department of Mathematics,
S-41296 Göteborg.

Matthias HOLSCHNEIDER
Centre de Physique Théorique,
CNRS-Luminy case 907,
F-13288 Marseille cedex 09.

Abstract

We consider the one-dimensional potential scattering problem on the line when the potential is given by a self-similar fractal measure. We show that the scattering operator admits a Weyl-Berry like expansion in terms of operators, where the second order term has a non-trivial scaling which reflects the fractal nature of the potential.

1 Introduction

The aim of this paper is to study the one-dimensional potential scattering problem when the interaction term is a self-similar fractal measure. We want to show how the self-similarity of the potential can be retrieved in the scattering operator. First, we will recall some basic facts on the scattering theory for singular Schrödinger operators on the line. Then we introduce the class of self-similar measures and the associated concept of renormalizable measures and operators. Then we establish a "high-frequency" expansion (in a certain meaning) for the scattering operator, where the second order term mirrors the self-similar nature of the potential.

2 Scattering formalism for measures

We equip $L^2(\mathbb{R})$ with the usual scalar product $< f, g >= \int \overline{f} g$ and adopt the notation $||f||$ for the induced norm. Let H_0 be the self-adjoint realization of the negative Laplacian $-\Delta$ on $L^2(\mathbb{R})$, with domain $D(H_0) = W^{2,2}(\mathbb{R})$, where $W^{p,q}(\mathbb{R})$ is the Sobolev space of functions whose q^{th} derivative is in $L^p(\mathbb{R})$. Let V be a finite real Borel measure on $\mathbb{R}$. The operator sum $H = H_0 + V$ is in general not defined on $D(H_0)$, but there exists a unique self-adjoint operator H with domain $D(H) \subset W^{2,1}(\mathbb{R})$ corresponding to this formal expression in the sense of quadratic forms (see e.g [Bra85]). The action of this operator is given by

$$< f, Hg >= \int dx \overline{f'(x)} g'(x) + \int \overline{f(x)} g(x) V(dx)$$

for all $f \in W^{2,1}(\mathbb{R})$ and $g \in D(H)$. It has been shown in [GH98] that an explicit scattering theory can be constructed for such operators. We briefly summarize the results that are needed for the sequel. We assume that the measure V satisfies the typical short-range condition $\int (1 + x^2) \, |V| \, (dx) < \infty$. Then for every $k \in \mathbb{R}$, there exists a unique continuous bounded function $\varphi_k^{\pm}$ satisfying the Helmholtz equation in the sense of distributions,

$$(\Delta + k^2)\varphi_k^{\pm} = V\varphi_k^{\pm} \text{ in } \mathcal{D}'(\mathbb{R}), \tag{2.1}$$

together with the radiation condition:

$$\varphi_k^{\pm}(x) \to \begin{cases} t_k e^{\pm ikx}, & x \to \pm\infty, \\ e^{\pm ikx} + r_k^{\pm} e^{\mp ikx}, & x \to \mp\infty. \end{cases} \tag{2.2}$$

This asymptotics becomes exact if the measure V has compact support. By analogy with the regular case, we call $\varphi_k^{\pm}$ the *scattering wave functions*, the coefficients t_k, r_k^+, r_k^- the *transmission* and *reflection coefficients from the right and left*, respectively, and the matrix $S = \begin{pmatrix} t_k & r_k^- \\ r_k^+ & t_k \end{pmatrix}$ the *scattering matrix*. The transmission and reflection coefficients satisfy a couple of relations which express the unitarity of the scattering matrix,

$$r_k^+ \overline{t_k} + t_k \overline{r_k^-} = 0, \ |t_k|^2 + |r_k^{\pm}|^2 = 1, \tag{2.3}$$

and, together with the scattering wave functions, obey symmetries which account for the time-reversal invariance property,

$$\overline{\varphi_k^{\pm}} = \varphi_{-k}^{\pm}, \ \overline{t_k} = t_{-k}, \ \overline{r_k^{\pm}} = r_{-k}^{\pm}. \tag{2.4}$$

At high-frequency, the transmission and reflection can be well estimated by the Born approximation. It has been shown in [GH96] that

$$r_k^\pm = \pm\frac{\sqrt{2\pi}}{2ik}\widehat{V}(\mp 2k) + O(k^{-2}),\ t_k = 1 + \frac{\sqrt{2\pi}}{2ik}\widehat{V}(0) + O(k^{-2}). \tag{2.5}$$

where $\widehat{V}$ is the Fourier transform of the measure V,

$$\widehat{V}(\xi) = \frac{1}{\sqrt{2\pi}}\int e^{-i\xi x}V(dx).$$

We will adopt the same notation $\widehat{f}$ for the (unitary) Fourier transform of a function f in $L^2(\mathbb{R})$. Thanks to the analytic properties of the two families of functions $\varphi_k^\pm$ in the complex $k-$plane, the spectral and scattering properties of H have been investigated. The positive spectrum is purely absolutely continuous and the negative spectrum consists in a finite number of eigenvalues. The wave operators $W_\pm = \text{s.lim}_{t\to\pm\infty} e^{itH}e^{-itH_0}$ exist and are complete. The scattering operator $\mathcal{S} = W_+^* W_-$ is given by (for details, see [Gué98])

$$(\mathcal{S}f)(x) = \int_{-\infty}^{+\infty} dk\ e^{ikx}t_k\widehat{f}(k) + \int_0^{+\infty} e^{-ikx}r_k^+\widehat{f}(-k) + \int_{-\infty}^0 dk\ e^{-ikx}r_k^-\widehat{f}(-k). \tag{2.6}$$

Thanks to this explicit formula, we will relate the properties of the scattering operator to the fractal characteristics of the potential, which appears here via the Born approximation.

3 Self-similar measures

The family of self-similar measures provides simple examples of "fractal" measures which are easy to construct and whose fractal dimensions are known. A self-similar measure on $\mathbb{R}$ was defined by Hutchinson ([Hut81]) to be a probability measure V satisfying a scale equation (in the sense of sets)

$$V = \sum_{i=1}^N p_i\ V\circ S_i^{-1}, \tag{3.1}$$

where the p_i are a set of weights ($0 < p_i < 1$, $\sum_{i=1}^N p_i = 1$) and the S_i a set of similitudes $S_i(x) = l_i x + b_i$, with $0 < l_i < 1$. Thus, the measure V may be written as a superposition of dilated and translated copies of itself (hence the term 'self-similar'). The support of V is the unique non empty compact K globally invariant under these contractions, i.e. $K = \bigcup_{i=1}^N S_i(K)$. This support may be of zero Lebesgue measure. This occurs when the so-called ***strong open set condition*** ([Str93]) is satisfied, that is there exists a closed bounded set U such that $S_i(U)\cap S_j(U) = \emptyset$ for $i\neq j$. It can be shown that the set U at least contains K. When there are only two similitudes satisfying the strong open set condition with same contraction ratio $l_1 = l_2 = 1/3$ and same weights $p_1 = p_2 = 1/2$, V is the well-known triadic Cantor measure.

Self-similar measures do not exhibit simple scaling invariance such as power-laws for the mass of balls. However, they possess scaling properties which can be expressed in term of a renormalisation procedure. The following definition was introduced in ([Hol96]):

Definition 3.1 *We say a tempered distribution $\eta \in \mathcal{S}'(\mathbb{R})$ has a large scale renormalisation of type (α, β) if there exists $\beta \in \mathbb{C}$, $\alpha > 1$ and a non-zero tempered distribution $\eta^* \in \mathcal{S}'(\mathbb{R})$ such that*

$$w^*. \lim_{n\to\infty} \beta^n \eta(\alpha^n \cdot) = \eta^*,$$

the convergence taking place in the weak-$*$ topology of $\mathcal{S}'(\mathbb{R})$. Conversely, we say a distribution η has a small scale renormalisation of type (α, β) around x_0 if for $\beta \in \mathbb{C}$, $\alpha < 1$ and $\eta^* \neq 0 \in \mathcal{S}'$

$$w^*. \lim_{n\to\infty} \beta^n \eta(\alpha^n (x - x_0) + x_0) = \eta^*.$$

In the case of self-similar measures, we have ([GH96])

Theorem 3.2 *Let V be a self-similar measure satisfying the scale equation (3.1) and the strong open set condition. Then V has a small scale renormalisation of type $(l_i, l_i/p_i)$ around each fixed point $x_i = b_i/(1 - l_i)$ of the similitudes S_i. Consequently, $e^{ix_i\xi}\widehat{V}(\xi)$ has a large scale renormalization of type $(1/l_i, 1/p_i)$.*

When the similitudes are equicontractive (i.e. $l_i = l$ for all i), the convolution $V * \tilde{V}$, where $\tilde{V}(x) = V(-x)$, is still a self-similar measure for some other sets of weights and similitudes. If moreover $V * \tilde{V}$ also satisfies the strong open set condition, we may apply the previous theorem to it, and an explicit computation shows that $\left|\widehat{V}\right|^2 = (V * \tilde{V})^\wedge$ has a large scale renormalization of type $(1/l, 1/\sum_{i=1}^N p_i^2)$. This shall be useful in the following.

4 Scattering operator for self-similar potentials

We now suppose that the potential is given by a self-similar measure. We will show that the self-similarity of the potential extend in a certain way to the scattering operator. Define the operator $\mathcal{T}$ by $\mathcal{S} = 1 + \mathcal{T}$. At high frequency, it is intuitively clear that the contribution of the $\mathcal{T}$ operator will become negligible because incoming states with high energy become less sensitive to the potential barrier. A manner to make this idea precise is the following. For any operator A on $L^2(\mathbb{R})$, let us denote $A_a = D_a^* A D_a$ its "rescaled" version, where $(D_a f)(x) = a^{-1/2} f(a^{-1}x)$ is the unitary dilation operator. Then

Definition 4.1 *We say that a bounded operator A on $L^2(\mathbb{R})$ is* vanishing at small scales *if $A_a \to 0$ strongly as $a \to 0$.*

A natural question might be at which "speed" such operators vanish and whether there is a renormalization which make them converge. This leads to introduce another

Definition 4.2 *A bounded operator A on $L^2(\mathbb{R})$* vanishes at speed $\alpha > 0$ *if*

$$a^{-\alpha} \|A_a\| \sim 1, \ a \to 0.$$

A bounded operator A on $L^2(\mathbb{R})$ vanishes faster than $\alpha > 0$ *if $a^{-\alpha} A_a \to 0$ strongly as $a \to 0$. A bounded operator A on $L^2(\mathbb{R})$ is said to be renormalizable of type (α, β) if there exists $0 < \alpha < 1$, $\beta > 0$ and a positive quadratic form q with dense domain $D(q)$ such that for all $f \in D(q)$,*

$$\lim_{n\to\infty} \|\beta^n A_{\alpha^n} f\|^2 = q(f, f).$$

Remark: To prove that an operator is vanishing at small scales [resp. vanishing at speed α or faster that α], it is enough to verify that $\|AD_a f\| \to 0$ [resp. $a^{-\alpha}\|AD_a f\| \sim 1$ or $a^{-\alpha}\|AD_a f\| \to 0$] for a dense set of functions f. It can then be easily verified that a renormalizable operator of type (α, β) is vanishing at speed $-\log\beta/\log\alpha$.

A natural hierarchy among the vanishing operators is obtained by ordering them according to their speed exponent α. By analogy with the scalar case (see e.g [Lap91]), we say an operator A has a *Weyl-Berry* expansion, at small scale if it can be written:

$$A = A_1 + A_2 + \ldots + A_n,$$

where the A_i, $i = 1\ldots, n-1$ are vanishing operators with speed $\alpha_1 < \alpha_2 < \ldots < \alpha_{n-1}$ and A_n vanishes faster than α_{n-1}. We are now ready to prove the main result of this paper:

Theorem 4.3 *Let V be a self-similar potential on $\mathbb{R}$ satisfying the scale equation (3.1) with equicontractive similitudes such that the strong open set condition is verified for V as well as for its convolution $V * \tilde{V}$. Then we have the following Weyl-Berry expansion for the scattering operator*

$$\mathcal{S} = 1 + \mathcal{T}_1 + \mathcal{T}_2 + \mathcal{T}_3,$$

where

- *$\mathcal{T}_1$ is vanishing at speed* 1

- $\mathcal{T}_2$ *is renormalizable of type* $(l^{-1}(\sum p_i^2)^{-1/2}, l)$ *(hence vanishing with speed* $1+\log(\sum p_i^2)/\log l$*.)*
- $\mathcal{T}_3$ *vanishes faster than* $1+\log(\sum p_i^2)/\log l$.

Proof. Let $\chi \in C_0^\infty(\mathbb{R})$ be some function such that $\chi(k) = 1$ if $|k| \leq 1$ and $\chi(k) = 0$ if $|k| > 2$, and denote $\chi^c = 1 - \chi$. Define the bounded operators $\mathcal{T}_i$ by

$$\begin{aligned}
\mathcal{T}_1 f(x) &= \gamma \int_{-\infty}^{+\infty} dk \chi^c(k) k^{-1} \hat{f}(k) e^{ikx}, \\
\mathcal{T}_2 f(x) &= \gamma \int_0^{\infty} dk \chi^c(k) \hat{V}(-2k) k^{-1} \hat{f}(-k) e^{-ikx} \\
&\quad -\gamma \int_{-\infty}^{0} dk \chi^c(k) k^{-1} \hat{V}(2k)\, \hat{f}(-k) e^{-ikx}, \\
\mathcal{T}_3 f(x) &= \gamma \int_{-\infty}^{+\infty} dk [(t_k - 1)\chi(k) + \tau(k)\chi^c(k)] \hat{f}(k) e^{ikx} \\
&\quad +\gamma \int_0^{\infty} dk [r_k^+ \chi(k) + \rho^+(k)\chi^c(k)] \hat{f}(-k) e^{-ikx} \\
&\quad -\gamma \int_{-\infty}^{0} dk [r_k^- \chi(k) + \rho^-(k)\chi^c(k)] \hat{f}(-k) e^{-ikx},
\end{aligned}$$

where $\gamma = \sqrt{2\pi}/2i$ and the ρ_i are the remainders of the Born approximation:

$$\tau(k) = t_k - 1 - \gamma/k, \ \rho^\pm(k) = r_k^\pm \mp \gamma \hat{V}(\mp 2k)/k$$

which satisfy $|\tau(k)| \leq O(k^{-2})$ and $|\rho^\pm(k)| \leq O(k^{-2})$. Then obviously $\mathcal{S} = 1 + \mathcal{T}_1 + \mathcal{T}_2 + \mathcal{T}_3$. Let us verify that the operators $\mathcal{T}_i$ satisfy the requested conditions. For all $f \in L^2(\mathbb{R})$ we have

$$\begin{aligned}
\|\mathcal{T}_1 f\|^2 &= \gamma^2 \int_{-\infty}^{+\infty} dk\ k^{-2} |\chi^c(k)|^2 \left|\hat{f}(k)\right|^2, \\
\|\mathcal{T}_2 f\|^2 &= \gamma^2 \int_{-\infty}^{+\infty} dk\ k^{-2} |\chi^c(k)|^2 \left|\hat{V}(2k)\right|^2 \left|\hat{f}(k)\right|^2, \\
\|\mathcal{T}_3 f\|^3 &\leq O(1) \int_{-\infty}^{+\infty} dk \left(|\chi(k)|^2 + k^{-4} |\chi^c(k)|^2\right) \left|\hat{f}(k)\right|^2.
\end{aligned}$$

Now take some function $f \in L^2(\mathbb{R})$ such that $\hat{f} \in C_0^\infty(\mathbb{R}\backslash\{0\})$ (such functions form a dense set). Then for a small enough we have

$$\begin{aligned}
\|\mathcal{T}_1 D_a f\|^2 &= \gamma^2\, a \int_{-\infty}^{+\infty} dk\ k^{-2} \left|\hat{f}(ak)\right|^2, \\
\|\mathcal{T}_2 D_a f\|^2 &= \gamma^2\, a \int_{-\infty}^{+\infty} dk\ k^{-2} \left|\hat{V}(2k)\right|^2 \left|\hat{f}(ak)\right|^2, \\
\|\mathcal{T}_3 D_a f\|^3 &\leq O(a) \int_{-\infty}^{+\infty} dk\ k^{-4} \left|\hat{f}(ak)\right|^2.
\end{aligned}$$

It follows that $\|\mathcal{T}_1 D_a f\| \sim a$ and $\|\mathcal{T}_3 D_a f\| \leq O(a^2)$, $a \to 0$. Thus $\mathcal{T}_1$ is vanishing at speed 1 and $\mathcal{T}_3$ vanishes faster than 2. On the other hand, the distribution $\left|\widehat{V}(2k)\right|^2$ has a large scale renormalization with

$$w^*. \lim_{n\to\infty} (\sum_{i=1}^{N} p_i^2)^{-n} \left|\widehat{V}(2l^{-n}k\right|^2 = \eta$$

for some positive tempered distribution η. It follows that

$$\lim_{n\to\infty} l^{-n} (\sum_{i=1}^{N} p_i^2)^{-n} \|\mathcal{T}_2 D_{l^n} f\|^2 = \gamma^2 \eta \left(\left|\widehat{f}\right|^2\right) =: q(f,f),$$

and thus $\mathcal{T}_2$ is renormalizable of type $(l^{-1}(\sum p_i^2)^{-1/2}, l)$, hence vanishing with speed $1 + \log(\sum p_i^2)/\log l > 1$. It remains to show that $\mathcal{T}_3$ vanishes faster than $\mathcal{T}_2$. Now $\sum_{i=1}^{N} p_i^2 \geq \sum_{i=1}^{N} N^{-2} = N^{-1}$ whereas the strong open set condition imposes $l < N^{-1}$. Thus $1 + \log(\sum p_i^2)/\log l < 2$ and this concludes the proof. □

Remark: The index $\log(\sum p_i^2)/\log l$ ruling the speed of decay of the operator $\mathcal{T}_2$ is in fact a fractal dimension of the measure V, namely its so-called *correlation dimension* (see [GH96]).

References

[Bra85] J. F. Brasche. Perturbation of schrödinger hamiltonian by measures. *J. Math. Phys.*, 26(4):621–626, 1985.

[GH96] C.A. Guérin and M. Holschneider. Scattering on fractal measures. *J. Phys. A: Math. Gen.*, 29:7651–7667, 1996.

[GH98] C.A. Guérin and M. Holschneider. Time-dependent scattering on fractal measures. *Journ. Math. Phys.*, 39(8):4165–4194, august 1998.

[Gué98] C. A. Guérin. *Diffusion sur les mesures fractales.* PhD thesis, Université de Provence, Marseille, 1998.

[Hol96] M. Holschneider. Large scale renormalisation of fourier transforms of self-similar measures and self-similarity of riesz measures. *J. of Math. Anal and Appl.*, 200:307–314, 1996.

[Hut81] J.E. Hutchinson. Fractals and self similarity. *Ind. Univ. Math. Jour.*, 30(5):713–746, 1981.

[Lap91] M.L. Lapidus. Fractal drum, inverse spectral problem for elliptic operators and a partial resolution of the Weyl-Berry conjectue. *Trans. Amer. Math. Soc.*, 325(2):465–529, 1991.

[Str93] R.S. Strichartz. Self-similar measures and their fourier transform 2. *Trans. Am. Math. Soc.*, 336(1):335–361, March 1993.

Operator Theory:
Advances and Applications, Vol. 108

Localization and Lifshitz tails for Random Quantum Waveguides

Frank Kleespies
Fachbereich Mathematik
Johann Wolfgang Goethe-Universität
D-60054 Frankfurt am Main, Germany
kleespie@math.uni-frankfurt.de

1 Introduction

The aim of the theory of quantum waveguides is the description of motion of electrons in two- or three-dimensional small channels, tubes or layers of crystalline matter of high purity – for example, one or several films or lines of a semiconducting substance epitaxed on a wafer. The mesoscopic lengthscale of such structures allows an experimentator to shape them as requested, yet on the other hand, structures on this scale are open for quantum effects. In fact, it has been shown in experiments that changes in the shape of such structures have an influence on their conductivity properties, see, for example, the physical literature cited in [DE], whose extensive bibliography provides a good overview of work on the subject in general at the same time. Mathematically speaking, we are interested in the impact of a changing geometry on the spectral properties of a Schrödinger operator in this geometry, a challenging and attractive task. Various models constituted by (minus) the Laplacian operator on a domain of $\mathbb{R}^2$ or $\mathbb{R}^3$ with Dirichlet boundary conditions have been investigated in this context, see [BGRS, DE, ES, ESTV, EV]. One has considered the bending of a strip as in Figure 1, opening a window be-

Figure 1: bending a strip

tween two adjacent strips as in Figure 2 or bulging a strip as in Figure 3 and

Figure 2: opening a window

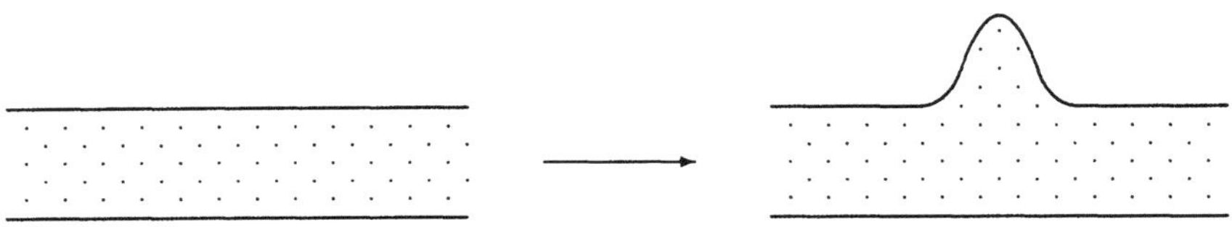

Figure 3: bulging a strip

analogous situations in three dimensions. In all cases, the existence of a number of bound states with eigenvalues below the essential spectrum emerged from it during the deformation could be proven. Thus, the conjecture put up in [DE], that a random deformation, e. g. a random denting or bulging, of a quantum waveguide strip on its full infinite length will cause infinitely many bound states or dense pure point spectrum at the bottom of the spectrum, is well motivated. In a joint paper ([KS]), P. Stollmann and the author were able to deal with such a model of a quantum waveguide irregularly deformed over its full length; one of the results is the verification of the mentioned conjecture.

2 The model

Our random quantum waveguide model consists of a collection of randomly dented versions of a parallel strip $\mathbb{R} \times (0, d_{max}) = D_{max}$; of course, this can be looked at as bulged versions of a smaller strip or a mixture of both. More precisely, let $d_{max} > 0$, $0 < d < d_{max}$, and consider $\Omega = [0, d]^{\mathbb{Z}}$. The i-th coordinate $\omega(i)$ of $\omega \in \Omega$ gives the deviation of the width of the random strip from d_{max}, i. e.

$$d_i(\omega) := d_{max} - \omega(i),$$

which lies between $d_{min} = d_{max} - d$ and d_{max}. Define $\gamma(\omega) : \mathbb{R} \to [d_{min}, d_{max}]$ as the polygon in $\mathbb{R}^2$ joining the points $\{(i, d_i(\omega))\}_{i \in \mathbb{Z}}$ and

$$D(\omega) = \{(x_1, x_2) \in \mathbb{R}^2 \mid 0 < x_2 < \gamma(\omega)(x_1)\}.$$

Figure 4 will help in visualizing this domain. We fix a probability measure μ on

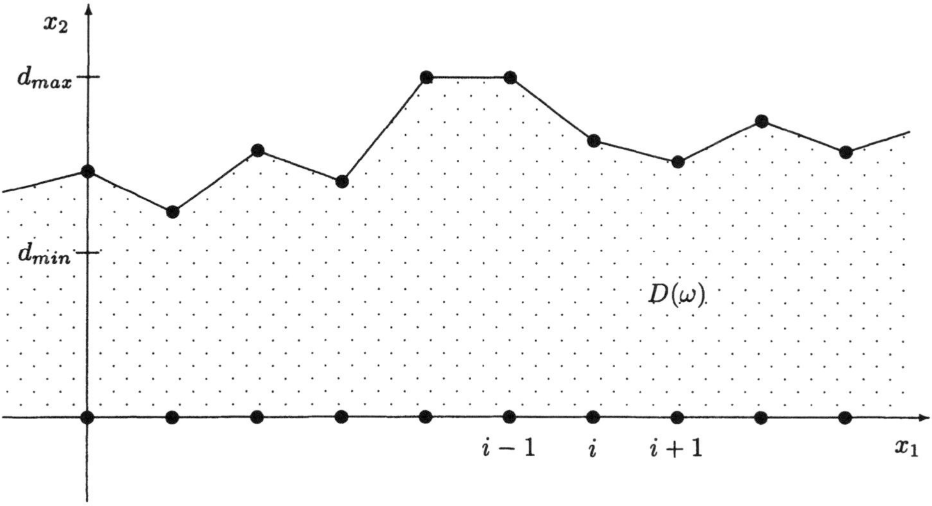

Figure 4: A random quantum waveguide

$[0, d]$ with $0 \in \text{supp}\mu \neq \{0\}$ and introduce $\mathbb{P} = \mu^{\mathbb{Z}}$, a probability measure on Ω. Consider $H(\omega) = -\Delta_{D(\omega)}$, the Laplacian on $D(\omega)$ with Dirichlet boundary conditions, which is a self-adjoint operator in $L^2(D(\omega))$. Ergodicity of the family $(H(\omega))_{\omega \in \Omega}$ implies that the spectra and spectral parts are deterministic in the sense that they coincide on a set of full measure.

There are two technical assumptions needed in the proofs; to prove Lifshitz asymptotics, we need:

(M1) There exist $a, \delta > 0$ such that $\mu[0, \epsilon] \geq a \cdot \epsilon^{\delta}$.

Lifshitz asymptotics deals with the integrated density of states

$$N(E) = \lim_{\Lambda \to \mathbb{R}} \frac{1}{|\Lambda|} N(H_\Lambda(\omega), E),$$

where $H_\Lambda(\omega)$ denotes $H(\omega)$ restricted to $(\Lambda \times \mathbb{R}) \cap D(\omega)$, Λ an open interval, and $N(T, E)$ is the number of eigenvalues of the operator T which are $\leq E$, counted with multiplicity; Lifshitzian behavior of this quantity means that it decays exponentially near the boundary of a spectral gap. Assumption (M1) can indeed guarantee that the strip is dented often enough and, consequently, the eigenvalues of the operators $H_\Lambda(\omega)$ are shifted away from the bottom energy with sufficient probability in order to allow a proof of such a phenomenon near the infimum of the spectrum.

In the proof of localization (more precisely for the Wegner estimate) we use

(M2) μ is Hölder continuous, i. e. there exist $b, \alpha > 0$ such that $\mu(I) \leq b|I|^\alpha$ for every interval $I \subset [0, d]$; here $|I|$ denotes the length of I.

One would very much like to get rid of this assumption and, in particular, admit probability measures with pure point parts.

3 The results

Our first result contributes to one of the central topics in disordered systems, the occurrence of pure point spectrum with exponentially decreasing eigenfunctions, usually called localization. According to the general philosophy this should happen near so-called fluctuation boundaries. Our model has $E_0 = \frac{\pi^2}{d_{max}^2}$, the infimum of the spectrum, as fluctuation boundary, and we can in fact prove:

Theorem 1 (localization) *Assume* (M2). *Then there exists $\delta > 0$ such that $\mathbb{P}$-a.s. the spectrum of $H(\omega)$ is pure point in $[E_0, E_0+\delta]$ with exponentially decreasing eigenfunctions.*

By exponential decay of eigenfunctions we mean exponential decay in x_1-direction.

Roughly speaking the second result says that the number of electrons per unit volume (which is given by the integrated density of states) decreases very rapidly, as the bottom energy is approached:

Theorem 2 (Lifshitz tails) *Assume* (M1). *Then the integrated density of states $N(t)$ for $H(\omega)$ satisfies*

$$\lim_{t \searrow 0} \frac{\log(-\log N(E_0 + t))}{\log t} = -\frac{1}{2}.$$

For the proof of these theorems, which can be found in detail in [KS], we use a quite abstract form of multi-scale analysis based on the von Dreifus-Klein method ([vDK]) and taken from [SLN]; as input, we have to supply two kinds of estimates on eigenvalues of the operators $H_\Lambda(\omega)$: a Wegner estimate, which states that eigenvalues spread when denting the strip, and an initial length scale estimate, which guarantees that eigenvalues are away from the bottom energy with overwhelming probability. To get these estimates, domain perturbations of quadratic forms have to be studied extensively.

4 Modifications and extensions

Of course, the presented model admits several modifications and extensions: it is possible to interpolate the points $\{(i, d_i(\omega))\}_{i\in\mathbb{Z}}$ in a different manner, for example smoothly. Moreover, the given strip can be bulged or dented on both sides. Finally, random quantum waveguides in three or more dimensions are possible: one might rotate the described configuration about the x_1-axis or, in order to end up with a model in the form of a layer, triangulate $\mathbb{R}^2$ by equilateral triangles, prescribe the height of the layer at the vertices v_i of the triangulation randomly by $d_{v_i}(\omega)$ and interpolate.

References

[BGRS] W. Bulla, F. Gesztesy, W. Renger, B. Simon: Weakly Coupled Bound States in Quantum Waveguides, *Proc. Amer. Math. Soc.* **127**, 1487 - 1495 (1997)

[vDK] H. von Dreifus, A. Klein: A new proof of localization in the Anderson tight binding model, *Commun. Math. Phys.* **124**, 285 - 299 (1989)

[DE] P. Duclos, P. Exner: Curvature-induced Bound States in Quantum Waveguides in two and three Dimensions, *Rev. Math. Phys.* **7**, 73 - 102 (1995)

[ES] P. Exner, P. Šeba: Bound States in Curved Quantum Waveguides, *J. Math. Phys.* **30**, 2574 - 2580 (1989)

[ESTV] P. Exner, P. Šeba, M. Tater, D. Vaněk: Bound States and Scattering in Quantum Waveguides Coupled through a Boundary Window, *J. Math. Phys.* **37**, 4867 - 4887 (1996)

[EV] P. Exner, S. A. Vugalter: Asymptotic Estimates for Bound States in Quantum Waveguides Coupled Laterally through a Narrow Window, *Ann. Inst. H. Poincaré Phys. Théor.* **65**, 109 - 123 (1996)

[KS] F. Kleespies, P. Stollmann: Lifshitz Asymptotics and Localization for Random Quantum Waveguides, Preprint (1998)

[SLN] P. Stollmann: Caught by Disorder: Lectures on Bound States in Random Media, in preparation

Operator Theory:
Advances and Applications, Vol. 108

Birman-Schwinger Analysis for Bound States in a Pair of Parallel Quantum Waveguides with a Semitransparent Boundary

David Krejčiřík
Faculty of Mathematics and Physics, Charles University,
V Holešovičkách 2, 180 00 Prague,
and
Nuclear Physics Institute, Academy of Sciences, Řež,
Czech Republic

A recent paper [4] presented a detailed analysis of a waveguide consisting of a pair of parallel straight Dirichlet strips in the plane coupled laterally through a window in the common boundary. Here we want to introduce a waveguide with a similar geometry. We consider the two infinite strips but they are now separated mutually by an arbitrary semitransparent barrier. We suppose that the barrier is supported by the common line only and that the transparentness is modulated by a function $\alpha = \alpha(x)$ (see Figure 1 for specification of the coordinate system). Under these presumptions we can identify the Hamiltonian H_α of such system with the formal expression $-\Delta + \alpha(x)\delta(y)$.

The motivation for this problem is the following. While physically closely related (the waveguide with a window can be regarded as a limit case for $\alpha = \infty$ everywhere except in the window's interval), the two systems differ mathematically. In our case of a semitransparent boundary, the perturbed and unperturbed Hamiltonians have the same form domain. This makes it possible to employ other mathematical methods; in the first place the Birman-Schwinger method to analyze the discrete spectrum.

By means of the technique we will find the perturbative expansion (13) for the ground-state energy when the perturbation of the asymptotic value α_0 is small in the *supremum* norm. Such bound state exists if and only if the effective coupling function $\alpha - \alpha_0$ is not repulsive in the mean (see Theorem 4.2).

However, we could find an analogous expansion in the case that the perturbation is small in the L^1-norm, too. This becomes interesting if we want to investigate the threshold behaviour of the ground-state energy for large α_0 because we should approach the results of the article studying the waveguide with a window (*cf.* Sec. VI.1 of [4]).

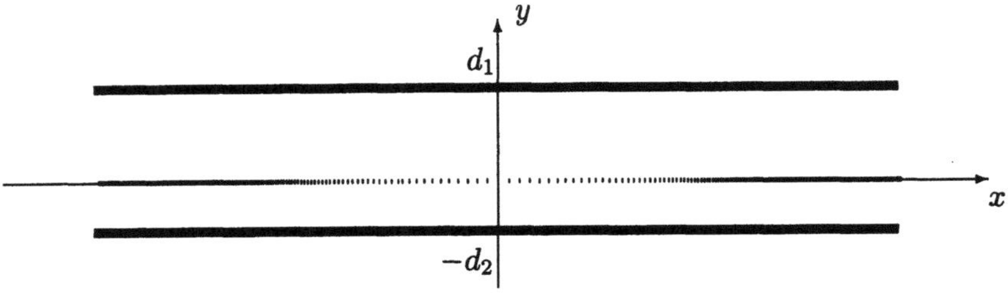

Figure 1: Double waveguide with a δ-barrier

1 Introduction

1.1 The Hamiltonian

In order to introduce the Hamiltonian describing the singular interaction in a mathematically rigorous method, we define it by means of the boundary conditions. Denote $\mathcal{O} := \mathcal{O}_2 \cup \mathcal{O}_1 = (-d_2, 0) \cup (0, d_1)$, $\Omega := \mathbb{R} \times \mathcal{O}$ and put $\hbar^2/2m = 1$. Then we have $H_\alpha := -\Delta$ with the domain

$$D(H_\alpha) := \left\{ \psi \in W_2^2(\Omega) | \; \forall x \in \mathbb{R} : \begin{array}{c} \psi(x, -d_2) = \psi(x, d_1) = 0 \\ \psi(x, 0+) = \psi(x, 0-) =: \psi(x, 0) \\ \psi_y(x, 0+) - \psi_y(x, 0-) = \alpha(x)\psi(x, 0) \end{array} \right\}. \quad (1)$$

The associated quadratic form is obtained by a simple integration of the Schrödinger equation by parts:

$$\begin{aligned} q_\alpha[\psi] &:= \int_{\mathbb{R} \times \bar{\mathcal{O}}} |\nabla\psi|^2(x, y) dx dy + \int_{\mathbb{R}} \alpha(x) |\psi(x, 0)|^2 dx \qquad (2) \\ D(q_\alpha) &:= \{\psi \in W_2^1 (\mathbb{R} \times (-d_2, d_1)) \mid \forall x \in \mathbb{R} : \psi(x, -d_2) = \psi(x, d_1) = 0\}. \end{aligned}$$

It is obviously symmetric; one can check that it is closed and thus really associated with H_α.

1.2 The Unperturbed System

If $\alpha(x) = \alpha$ is a constant function, we can solve the Schrödinger equation by separation of variables. In this situation we can find the spectral condition $-\alpha = k(\cot kd_1 + \cot kd_2)$, which gives arise to the increasing sequence of eigenvalues of the transverse operator; we denote it by $\{\nu_n(\alpha)\}_{n=1}^\infty$. The corresponding eigenfunctions will be:

$$\chi_n(y; \alpha) = \left\{ \begin{array}{ll} N_n \sin \sqrt{\nu_n} d_1 \sin \sqrt{\nu_n}(y + d_2) & \text{if} \quad y \in \mathcal{O}_2 \\ -N_n \sin \sqrt{\nu_n} d_2 \sin \sqrt{\nu_n}(y - d_1) & \text{if} \quad y \in \mathcal{O}_1 . \end{array} \right. \quad (3)$$

N_n is the appropriate normalization factor chosen in such a way that χ_n would be a unit vector in $L^2(-d_2, d_1)$. Furthermore, the Green's function of the Hamiltonian can be written down explicitly:

$$G_\alpha(x,y,x',y';k) = \sum_{n=1}^{\infty} \frac{i}{2k_n} e^{ik_n|x-x'|}\, \chi_n(y;\alpha)\, \bar{\chi}_n(y';\alpha)\,, \tag{4}$$

where $k_n := \sqrt{k^2 - \nu_n(\alpha)}$.

2 The Existence of a Bound State

One of the most interesting situations is when $\lim_{|x|\to\infty} \alpha(x) = \alpha_0$, so the function α approaches an unperturbed value α_0 at large x-distances. Then by a simple bracketing argument (as in [4], Sec. II) employing the *minimax principle* we can localize the essential spectrum, $\sigma_{ess}(H_\alpha) = [\nu_1(\alpha_0), \infty)$.

In [4] it is proved that an arbitrary "window" leads to the existence of a bound state. For our case one can show that it is sufficient to make the effective coupling function smaller in the mean in order to create a locally stronger tunneling between the both channels.

This general existence result is given by the following theorem. Its variational proof is based on the idea of [5], Sec. II. (see also Thm.2.1. of [3] and [4], Sec. III).

Theorem 2.1 *Assume*

(i) $\alpha - \alpha_0 \in L^1_{loc}(\mathbb{R})$

(ii) $\alpha(x) - \alpha_0 = \mathcal{O}(|x|^{-1-\varepsilon})$ *for some* $\varepsilon > 0$ *as* $|x| \to \infty$.

If $\int_{\mathbb{R}} (\alpha(x) - \alpha_0) dx < 0$, *then* H_α *has at least one isolated eigenvalue below its essential spectrum.*

3 The Resolvent

To learn more about the discrete spectrum we would like to know the resolvent of our Hamiltonian. In [2] the resolvent of a Hamiltonian with a singular interaction was found on condition that the configuration space is all Euclidean space. The procedure can be modified for our case when the configuration space is the Dirichlet strip and the δ-interaction is supported by a line.

As a result we get an analogous resolvent formula

$$R^k(\alpha) = R^k(\alpha_0) - R^k_{m,dx}(\alpha_0)\left[I + (\alpha - \alpha_0) R^k_{m,m}(\alpha_0)\right]^{-1} (\alpha - \alpha_0) R^k_{dx,m}(\alpha_0) \tag{5}$$

for any $k \in \mathbb{C}^+ := \{k \in \mathbb{C} \mid \operatorname{Im} k > 0\}$. $R^k_{\mu,\nu}(\alpha_0)$ is the integral operator from $L^2(\Omega, d\mu)$ to $L^2(\Omega, d\nu)$ with the kernel $G_{\alpha_0}(\cdot,\cdot;k)$. In particular, we have

$$(R^k_{\mu,\nu}(\alpha_0)\,\psi)(x) = \int_\Omega G_{\alpha_0}(x,y;k)\psi(y)d\mu(y) \tag{6}$$

for all $\psi \in D(R^k_{\mu,\nu}) \subset L^2(\Omega, d\mu)$. (There is no ν in the definition; we compute the right-hand-side and interpret it as values of a function in $L^2(\Omega, d\nu)$.) μ, ν are positive Radon measures without a discrete component; especially, dx is the standard Lebesgue measure and m denotes the δ-measure (formally, $dm(x,y) = dx\,\delta(y)dy$).

The formula (5) can be used as a basis of the Birman-Schwinger analysis in analogy with [2]. However, one can derive the symmetrized expression:

$$\begin{aligned} R^k(\alpha) &= R^k(\alpha_0) - R^k_{m,dx}(\alpha_0)\,(\alpha-\alpha_0)^{\frac{1}{2}}. \\ &\quad . \left[I + |\alpha-\alpha_0|^{\frac{1}{2}}\,R^k_{m,m}(\alpha_0)\,(\alpha-\alpha_0)^{\frac{1}{2}}\right]^{-1} |\alpha-\alpha_0|^{\frac{1}{2}} R^k_{dx,m}(\alpha_0)\,, \end{aligned} \tag{7}$$

which gives better results. We have introduced here the common notation $(\alpha-\alpha_0)^{\frac{1}{2}} := |\alpha-\alpha_0|^{\frac{1}{2}}\,\mathrm{sgn}(\alpha-\alpha_0)$.

For the integral operator $K^k_\alpha := |\alpha-\alpha_0|^{\frac{1}{2}}\,R^k_{m,m}(\alpha_0)\,(\alpha-\alpha_0)^{\frac{1}{2}}$ we can derive the following assertion which is the basis of Birman-Schwinger principle.

Proposition 3.1 $\forall k \in \mathbb{C}^+ :\ \dim \mathrm{Ker}(H_{\alpha m} - k^2) = \dim \mathrm{Ker}\left(I + K^k_\alpha\right)$.

In particular, these results imply the crucial equivalence stated in the following corollary.

Corollary 3.1 $\forall k \in \mathbb{C}^+ :\ k^2 \in \sigma_{disc}(H_\alpha) \Leftrightarrow K^k_\alpha \psi = -\psi$

It tells us that to solve the partial differential equation is equivalent to find out when the integral operator K^k_α has eigenvalue -1. Notice that its kernel is analogous in dependency on the coupling function to the one in classical Birman-Schwinger principle (*cf.* (1) in [6]).

4 The Adiabatic Limit

These preliminaries we can use for investigation of bound states in the case of the weakly coupled Hamiltonian. In this situation α differs from α_0 slightly and it means that we can replace $\alpha-\alpha_0$ by $\lambda(\alpha-\alpha_0)$ and consider small λ.

Similarly as in [6], [1] and [3], Thm. 4.2., we find necessary and sufficient conditions for H_α to have a (unique) bound state for all sufficiently small positive λ. In addition we shall derive expansion of the bound-state energy at $\lambda = 0$.

Henceforth we assume that

(a1) $\alpha(x)-\alpha_0 \in L^{1+\varepsilon}(\mathbb{R}, dx)$ for some small $\varepsilon > 0$,

(a2) $\alpha(x)-\alpha_0 \in L^1(\mathbb{R}, |x|\,dx)$.

As a consequence of *(a1), (a2)*; the function $\alpha(x)-\alpha_0$ belongs to $L^1(\mathbb{R}, dx)$ too.

4.1 Preliminaries

The Green's function G_{α_0} in (6) is given by (4). In the following we consider k^2 as an eigenvalue of the weakly coupled Hamiltonian given by the replacement $\alpha - \alpha_0 \mapsto \lambda(\alpha - \alpha_0)$, and look for the function $\lambda \mapsto k^2$. Since we are interested in the discrete spectrum, we must take the Green's function for $k^2 < \nu_1(\alpha_0)$. Thus we have

$$K_\alpha(x, x'; k) = |\alpha(x) - \alpha_0|^{\frac{1}{2}} \sum_{n=1}^{\infty} \frac{|\chi_n(0;\alpha_0)|^2}{2\kappa_n} e^{-\kappa_n |x-x'|} (\alpha(x') - \alpha_0)^{\frac{1}{2}}, \qquad (8)$$

where $\kappa_n := \sqrt{\nu_n(\alpha_0) - k^2}$.

Next we note the elementary fact that $\lim_{\lambda\to 0} k^2 = \nu_1$ (*i.e.* $\kappa_1 \to 0$ as $\lambda \to 0$).

The key idea of the further process is that K_α^k has a well-behaved limit as $k^2 \to \nu_1$ except for a divergent rank-one piece. The singularity is contained in the first term of the expansion (8) and can be singled out by taking (as in [1], [3]) $K_\alpha^k = Q_\alpha + P_\alpha = Q_\alpha + A_\alpha + N_\alpha$ where

$$\begin{aligned} Q_\alpha(x, x') &= |\alpha(x) - \alpha_0|^{\frac{1}{2}} e^{-\kappa_1|x|} \frac{|\chi_1(0)|^2}{2\kappa_1} e^{-\kappa_1|x'|} (\alpha(x') - \alpha_0)^{\frac{1}{2}} \\ A_\alpha(x, x') &= |\alpha(x) - \alpha_0|^{\frac{1}{2}} \frac{|\chi_1(0)|^2}{\kappa_1} e^{-\kappa_1|x|_>} \sinh \kappa_1 |x|_< (\alpha(x') - \alpha_0)^{\frac{1}{2}} \qquad (9) \\ N_\alpha(x, x') &= |\alpha(x) - \alpha_0|^{\frac{1}{2}} \sum_{n=2}^{\infty} \frac{|\chi_n(0)|^2}{2\kappa_n} e^{-\kappa_n|x-x'|} (\alpha(x') - \alpha_0)^{\frac{1}{2}}. \end{aligned}$$

Here we have introduced $|x|_< := \max\{0, \min\{|x|, |x'|\} \operatorname{sgn}(xx')\}$ and $|x|_> := \max\{|x|, |x'|\}$.

Next we define

$$\begin{aligned} A_0(x, x') &:= |\alpha(x) - \alpha_0|^{\frac{1}{2}} |\chi_1(0)|^2 |x|_< (\alpha(x') - \alpha_0)^{\frac{1}{2}} \\ N_0^\beta(x, x') &:= |\alpha(x) - \alpha_0|^{\frac{1}{2}} \sum_{n=2}^{\infty} \frac{|\chi_n(0)|^2}{2} \frac{e^{-\beta\sqrt{\nu_n - \nu_1}|x-x'|}}{\beta\sqrt{\nu_n - \nu_1}} (\alpha(x') - \alpha_0)^{\frac{1}{2}} \end{aligned}$$

($\beta > 0$), then one has

Proposition 4.1 *Assume that the assumptions (a1), (a2) are valid, then*

$$\lim_{\kappa_1 \to 0} \|A_\alpha - A_0\|_{HS} = 0 \quad \text{and} \quad \lim_{\kappa_1 \to 0} \|N_\alpha - N_0^1\|_{HS} = 0. \qquad (10)$$

PROOF: **(a)** A_0 is Hilbert-Schmidt since

$$\begin{aligned} \|A_0\|_{HS}^2 &= \int_{\mathbb{R}^2} |A_0(x, x')|^2 \, dx \, dx' \\ &= |\chi_1(0)|^4 \int_{\mathbb{R}^2} |\alpha(x) - \alpha_0| \, |x|_<^2 \, |\alpha(x') - \alpha_0| \, dx \, dx' \\ &\le |\chi_1(0)|^4 \left(\int_{\mathbb{R}} |x| \, |\alpha(x) - \alpha_0| \, dx \right)^2 < \infty \end{aligned}$$

by assumption. Next we have $\lim_{\kappa_1 \to 0} A_\alpha = A_0$ and $|A_\alpha(x,x';\kappa_1)| \le |A_0(x,x')|$. This ensures that the assumptions of the dominated convergence theorem are valid, and thus we immediately get the first claim of the proposition.

(b) N_0^1 has a logarithmic singularity as $x' \to x$. Nevertheless, its Hilbert-Schmidt norm is still finite because by Hölder inequality we have

$$\begin{aligned} \|N_0^1\|_{HS}^2 = & \\ = & \sum_{m,n=2}^{\infty} \frac{|\chi_n(0)\,\chi_m(0)|^2}{4\sqrt{\nu_n-\nu_1}\sqrt{\nu_m-\nu_1}} \int_{\mathbb{R}^2} |\alpha(x)-\alpha_0|\; e^{-\frac{1}{2}(\sqrt{\nu_n-\nu_1}+\sqrt{\nu_m-\nu_1})|x-x'|}\,\cdot \\ & \qquad \cdot |\alpha(x')-\alpha_0|\; e^{-\frac{1}{2}(\sqrt{\nu_n-\nu_1}+\sqrt{\nu_m-\nu_1})|x-x'|}\, dx\, dx' \\ \le & \; C(\varepsilon)\, \|\alpha-\alpha_0\|_{1+\varepsilon}^2 \sum_{m,n=2}^{\infty} \frac{1}{\sqrt{\nu_n-\nu_1}\sqrt{\nu_m-\nu_1}\,(\sqrt{\nu_n-\nu_1}+\sqrt{\nu_m-\nu_1})^{2\varepsilon'}} \end{aligned}$$

ε' is the abbreviation for $\frac{\varepsilon}{1+\varepsilon} > 0$. The monotone convergence theorem justifies the interchange of summation and integration we have made. The last sum now converges because the sequence $\{\sqrt{\nu_n}\}$ grows linearly with n as $n \to \infty$. The assumptions of the dominated convergence theorem are fulfilled again because $\lim_{\kappa_1 \to 0} N_\alpha = N_0^1$ and $|N_\alpha(x,x';\kappa_1)| \le |N_0^1(x,x')|$. ∎

We will also need the boundedness of A_α, N_α for complex $z := \kappa_1$ but it is easy to prove

Lemma 4.1 *1.* $\forall z \in \mathbb{C}, \operatorname{Re} z \ge 0, |z| < \frac{\pi}{2}:\ |A_\alpha(z)| \le \sqrt{2}\,|A_0|$

2. $\exists C, \beta > 0\ \ \forall z \in \mathbb{C}, \operatorname{Re} z \ge 0, |z| < \sqrt{\nu_n - \nu_1}:\ |N_\alpha(z)| \le C|N_0^\beta|$.

4.2 The Main Theorem

In this section we want to prove the following technical result.

Theorem 4.1 *Assume that the hypotheses (a1), (a2) are valid. Then, for λ small, H_α has at most one simple eigenvalue $E(\lambda) < \nu_1$, and this happens if and only if*

$$\sqrt{\nu_1 - E} \equiv \kappa_1 = -\frac{\lambda}{2}\,|\chi_1(0)|^2 \left(e^{-\kappa_1|\cdot|}\,(\alpha-\alpha_0)^{\frac{1}{2}}, (I+\lambda P_\alpha)^{-1}\, e^{-\kappa_1|\cdot|}\,|\alpha-\alpha_0|^{\frac{1}{2}} \right) \tag{11}$$

has a solution $\kappa_1 > 0$.

PROOF: From the proof of Proposition 4.1 it is clear that $\|P_\alpha\| \le \|A_0\|_{HS} + \|N_0\|_{HS} < \infty$, and thus for λ sufficiently small $\|\lambda P_\alpha\| < 1$, so $I + \lambda P_\alpha$ is invertible and we may write

$$(I + \lambda K_\alpha^k)^{-1} = \left[I + (I+\lambda P_\alpha)^{-1} \lambda Q_\alpha \right]^{-1} (I + \lambda P_\alpha)^{-1}.$$

Hence λK_α has eigenvalue -1 if and only if the same is true for $(I+\lambda P_\alpha)^{-1}\lambda Q_\alpha$. Since Q_α is a rank-one operator (see (9)), $(I+\lambda P_\alpha)^{-1}\lambda Q_\alpha = (\psi, \cdot)\,\varphi$ with $\psi :=$

$\lambda \, \frac{|\chi_1(0)|^2}{2\kappa_1} \, e^{-\kappa_1|\cdot|} \, (\alpha(\cdot)-\alpha_0)^{\frac{1}{2}}$ and $\varphi := (I+\lambda P_\alpha)^{-1} \, e^{-\kappa_1|\cdot|} \, |\alpha(\cdot)-\alpha_0|^{\frac{1}{2}}$; it has just one eigenvalue (ψ,φ). Putting it equal to -1 we get the condition (11).

This proves the entire theorem except for the assertion that (11) has *at most* one positive solution for λ small and fixed. Since there is a one-to-one correspondence between eigenvalues of H_α and solutions of (11) and the number of eigenvalue can only go up with replacement $\alpha-\alpha_0 \mapsto -|\alpha-\alpha_0|$, we need only show that (11) has at most one solution when $\alpha-\alpha_0 \le 0$. In this case, (11) is equivalent to $z = G(z,\lambda)$ where

$$G(z,\lambda) := \frac{\lambda}{2} \, |\chi_1(0)|^2 \left(e^{-z|\cdot|} \, |\alpha-\alpha_0|^{\frac{1}{2}}, (I+\lambda P_\alpha)^{-1} \, e^{-z|\cdot|} \, |\alpha-\alpha_0|^{\frac{1}{2}} \right). \tag{12}$$

To complete the proof, we need the following lemmas.

4.2.1 Auxiliary Results

Lemma 4.2 *If $\alpha-\alpha_0 \not\equiv 0$, then $|z|^{-1} \le C_1\lambda^{-1}$ for λ small.*

PROOF: From (12), we see that any solution of $z = G(z,\lambda)$ for λ small must obey $z = \frac{\lambda}{2}|\chi_1(0)|^2 \int_{\mathbb{R}} |\alpha(x)-\alpha_0| \, dx + \mathcal{O}(\lambda^2)$, what gives us the assertion provided $\alpha-\alpha_0$ is not identically zero. ∎

Lemma 4.3 *For sufficiently small z, $\left\| \frac{\partial P_\alpha}{\partial z} \right\| < \frac{C_2}{|z|}$.*

PROOF: Let us choose a contour in complex plane, $\varphi : s = z + ze^{it}$; $t \in [0,2\pi]$. The operator-valued fuction $P_\alpha(z)$ is real-analytic in the region $\mathrm{Re}\, z > 0$ and has a bounded limit as $z \to 0+$. Hence Cauchy integral formula together with Lemma 4.1 gives

$$\left| \frac{\partial P_\alpha}{\partial z} \right| = \left| \frac{1}{2\pi i} \int_0^{2\pi} \frac{P_\alpha(s) \, i \, z \, e^{it} dt}{z^2 \, e^{2it}} \right| \le \frac{\sqrt{2}|A_0| + C|N_0^\beta|}{|z|},$$

but A_0, N_0^β have finite HS norms. ∎

Lemma 4.4 $\forall z_0 \in \mathbb{R} \quad \exists C_3 > 0 \quad \forall z \in [0,z_0]$:

$$\left| \left(e^{-z|\cdot|} \, |\alpha-\alpha_0|^{\frac{1}{2}}, \left[2\,|\cdot| + \frac{\partial P_\alpha}{\partial z} \right] e^{-z|\cdot|} \, |\alpha-\alpha_0|^{\frac{1}{2}} \right) \right| \le C_3$$

PROOF: An explicit calculation shows that the partial derivative is finite and remains bounded as $z \to 0+$. The whole scalar product is then finite because of the assumptions *(a1)*, *(a2)*. ∎

Lemma 4.5

$$\exists C_4 > 0 \ : \qquad \left| \left(e^{-z|\cdot|} \, |\alpha-\alpha_0|^{\frac{1}{2}}, \left[|\cdot| P_\alpha + P_\alpha |\cdot| \right] e^{-z|\cdot|} \, |\alpha-\alpha_0|^{\frac{1}{2}} \right) \right| \le \frac{C_4}{|z|}$$

PROOF: Since $\forall x > 0 : x\, e^{-x} \le e^{-1}$, using Schwarz inequality we infer that the expression is bounded by $\frac{e^{-1}}{|z|} \, \|\alpha-\alpha_0\|_1 \, \|P_\alpha\|_{HS}$. ∎

4.2.2 Completion of the Proof

By the elementary inequality $(I+\lambda P_\alpha)^{-1} \leq \frac{1-\lambda P_\alpha}{1-\lambda} \leq d\,(1-\lambda P_\alpha)$ held for λ small enough and by the preceding lemmas we have for $|z^{-1}| \leq C_1\lambda^{-1}$ and all sufficiently small λ:

$$\left|\frac{\partial G}{\partial z}\right| \leq C\lambda\,.$$

Let z_1, z_2 are two solutions of $z = G(z,\lambda)$. Because they have to fulfill

$$|z_1 - z_2| = \left|\int_{z_1}^{z_2} \frac{\partial G}{\partial z} dz\right| \leq \left|\int_{z_1}^{z_2} \left|\frac{\partial G}{\partial z}\right| dz\right| \leq C\lambda|z_1 - z_2|,$$

we have the uniqueness ensured for $\lambda < C^{-1}$. ∎

4.3 Condition for a Bound State

Now we introduce necessary and sufficient condition for there to be a bound state when λ is small.

Theorem 4.2 *Let α obeys hypotheses (a1), (a2) and $\alpha-\alpha_0 \not\equiv 0$. Then H_α has an eigenvalue $E(\lambda) < \nu_1$ for all small $\lambda > 0$ if and only if $\int_{\mathbb{R}}(\alpha(x)-\alpha_0)\,dx \leq 0$. If this condition holds, then the eigenvalue is unique and simple and obeys*

$$\begin{aligned}
\sqrt{\nu_1 - E(\lambda)} = &-\frac{\lambda}{2}\,|\chi_1(0)|^2 \int_{\mathbb{R}} (\alpha(x)-\alpha_0)\,dx \\
&-\frac{\lambda^2}{4}\Bigg\{ |\chi_1(0)|^4 \int_{\mathbb{R}^2} (\alpha(x)-\alpha_0)\,|x-x'|\,(\alpha(x')-\alpha_0)\,dx\,dx' \\
&-|\chi_1(0)|^2 \sum_{n=2}^{\infty} |\chi_n(0)|^2 \int_{\mathbb{R}^2} (\alpha(x)-\alpha_0)\,\frac{e^{-\sqrt{\nu_n-\nu_1}|x-x'|}}{\sqrt{\nu_n-\nu_1}}\,(\alpha(x')-\alpha_0)\,dx\,dx' \Bigg\} \\
&+\mathcal{O}(\lambda^3).
\end{aligned} \tag{13}$$

PROOF: By the implicit-function theorem we see that (11) has a unique solution for λ small and it is given by (13).

It remains to prove that such solution is strictly positive for λ small. It is clearly fulfilled, if $\int(\alpha(x)-\alpha_0)dx < 0$. If this integral is zero, the first term of order λ in the expansion (13) vanishes and we have to prove then that the term of order λ^2 is strictly positive. One can do it by the cunning trick as in the proof of Thm. 2.5 in [6] (see also the end of the proof of Thm. 4.2 in [3]). ∎

The work has been partially supported by the GA ASCR 1048801.

References

[1] R. Blackenbecler, M. L. Goldberger, and B. Simon, *The bound states of weakly coupled longe-range one-dimensional quantum hamiltonians*, Ann. Phys. **108** (1977), 69–78.

[2] J. F. Brasche, P. Exner, Yu. A. Kuperin, and P. Šeba, *Schrödinger operators with singular interactions*, J. Math. Anal. Appl. **184** (1994), 112–139.

[3] P. Duclos and P. Exner, *Curvature-induced bound states in quantum waveguides in two and three dimensions*, Rev. Math. Phys. **7** (1995), 73–102.

[4] P. Exner, P. Šeba, M. Tater, and D. Vaněk, *Bound states and scattering in quantum waveguides coupled laterally through a boundary window*, J. Math. Phys. **37** (1996), 4867–4887.

[5] J. Goldstone and R. L. Jaffe, *Bound states in twisting tubes*, Phys. Rev. **B 45** (1992), 14100–14107.

[6] B. Simon, *The bound state of weakly coupled Schrödinger operators in one and two dimensions*, Ann. Phys. **97** (1976), 279–288.

Operator Theory:
Advances and Applications, Vol. 108

On absolute continuity of spectra of periodic elliptic operators

Peter Kuchment
Mathematics and Statistics Department
Wichita State University
Wichita, KS 67260-0033, USA

Sergei Levendorskiĭ
Mathematics Department
Rostov State Academy of Economics
Rostov-on-Don, Russia

Abstract

The paper contains a brief description of a simplified version of A. Sobolev's proof of absolute continuity of spectra of periodic magnetic Schrödinger operators. This approach is in principle applicable to all periodic elliptic operators known to be of interest for math physics (including Maxwell), and in all these cases leads to the same model problem of complex analysis.

1 Introduction

Elliptic differential operators with periodic coefficients arise naturally in many areas of mathematical physics. One can mention quantum solid state theory, where the main operators of interest are the stationary Schrödinger operator

$$-\Delta + V(x) \tag{1}$$

and the magnetic Schrödinger operator

$$(i\partial - A(x))^2 + V(x). \tag{2}$$

Here the scalar electric potential $V(x)$ and the vector magnetic potential $A(x)$ are assumed to be periodic with respect to a lattice in $\mathbf{R^d}$ [1]. Another example, which gained importance due to recent advances in the photonic crystals theory [8]

---

[0] *1991 Mathematics Subject Classification.* Primary 35P99, Secondary 35J10
Key words and phrases. Periodic operator, elliptic operator, absolutely continuous spectrum, Schrödinger operator, magnetic and electric potential

is the periodic Maxwell operator $\nabla \times \varepsilon(x)^{-1}\nabla\times$ defined on zero-divergence vector fields in $\mathbf{R^3}$. Here the periodic function $\varepsilon(x)$ represents the electric permitivity of the medium. Scalar counterparts of this operator (arising also in acoustics) are the operators $-\nabla \cdot \varepsilon(x)^{-1}\nabla$ and $-\varepsilon(x)^{-1}\Delta$. Other examples are the anisotropic divergent type operators $\sum \partial_i a_{ij}(x)\partial_j$, where the magnetic and electric potential terms can also be included, and the periodic Dirac operator ([2], [6], [12]). In all what follows we do not attempt to find optimal conditions on the coefficients of operators. The reader can refer to the recent survey [5] for detailed discussion of this topic.

For all these operators, the structure of the spectrum is of major interest. Consider a periodic Schrödinger operator (1) in $\mathbf{R^d}$. Under mild conditions on the real potential the operator is self-adjoint [13]. It has been clear to physicists for a long time that the spectrum of this operator in $L_2(\mathbf{R^d})$ does not contain any eigenvalues. This conjecture had not been proven until the celebrated Thomas' paper [17] (see also its improved versions in [5], [13], and [15]). The paper [17] contained the proof of a more general statement: the spectrum of the selfadjoint periodic operator (1) is absolutely continuous. Absence of singular continuous spectrum for periodic elliptic operators holds in a very general setting and is a rather straightforward consequence of Floquet theory ([9], [13], [14]).

The question arises on whether absence of eigenvalues is shared by all periodic elliptic differential operators. It is known [9] that this is not true in general for elliptic operators of order four. Still, the common belief is that periodic elliptic operators of second order do not possess any eigenvalues. The talks presented at this conference by M. Birman and T. Suslina described several such results. There are non-standard situations in mesoscopic physics and photonic crystal theory when one can encounter point spectrum of periodic second order differential problems on graphs [10]. This, however, does not influence our belief in absolute continuity of spectra of such problems in $\mathbf{R^d}$. Let us briefly list some known results and refer the reader to [5] for details. The Thomas' theorem [17] was extended to a broader class of periodic Schrödinger operators in the M. Reed and B. Simon's book [13]. L. Danilov [6] proved absolute continuity of the spectrum for the case of the Dirac operator with a periodic scalar potential. The first result for the magnetic Schrödinger operator (2) was obtained by R. Hempel and I. Herbst in [7] in the case of small magnetic potentials. The full strength statement about the magnetic Schrödinger operator was proven in the remarkable papers by M. Birman and T. Suslina [2] and A. Sobolev [16]. The elegant algebraic approach of the paper [2] works only in dimension two. The paper [2] lead A. Sobolev to an ingenious proof in arbitrary dimension [16]. Recent results can be found in [3], [4], [5], [12], and [15].

Our goal was to get good grasp of the proof presented in [16]. We found that the initial proof can be significantly simplified, which in particular leads to some mild improvements in the result and, most importantly, to a possibility of broad generalizations. Our aim is mostly methodological: to provide a simplified and unified approach that treats in an uniform way all the operators mentioned above and, we hope, clarifies the situation. It also leads to somewhat weaker assumptions on the magnetic potential in comparison with [16], a stronger estimate from below, and validity in non-selfadjoint case. An interesting observation is that in all situations known to the authors the same model $\overline{\partial}$ problem arises, good understanding of which would lead to progress in all cases that are not treated yet (Maxwell, divergent, etc.). Most of the ideas of the proof were already contained in a more obscured form in papers [2] and [16]. This paper contains only a brief description of the results. The complete account will be given elsewhere.

2 The magnetic Schrödinger operator

Our main object of study is the magnetic Schrödinger operator in $\mathbf{R^d}$

$$H = (D + A(x))^2 + V(x)$$

with a scalar electric potential $V(x)$ and a vector magnetic potential $A(x)$. We will assume for simplicity of presentation that both potentials are periodic with respect to the integer lattice $\mathbf{Z^d}$; the case of general lattices does not require any significant changes in the proofs. According to Thomas [17] (see also [13], [9]), absence of eigenvalues will be proven if one is able to show existence of a *quasimomentum* $k \in \mathbf{C^d}$ such that the operator

$$H(k) = (D + k + A(x))^2 + V$$

has zero kernel on the torus $\mathbf{T^d} = \mathbf{R^d}/\mathbf{Z^d}$. Our key statement is the following theorem:

Theorem 1 *Let $A \in \left[H^s(\mathbf{T^d})\right]^d$ for some $s > 3d/2-1$. Then there exist constants $C > 0$, $\beta > 0$ and vector $e \in \mathbf{R^d}$, such that for sufficiently large $\rho \in \mathbf{R}$ and any $u \in H^2(\mathbf{T^d})$*

$$\left|\left|(D + k + A(x))^2 u\right|\right|_{L_2} \geq C(|\rho|\, ||u||_{L_2} + ||u||_{H^1}),$$

where $k = (\beta + i\rho)e \in \mathbf{C^d}$.

An immediate consequence of this estimate is the following statement.

Theorem 2 *Assume that A is like in the previous theorem and the electric potential V is such that*

$$||Vu||_{L_2(\mathbf{T^d})} \leq C_\varepsilon\, ||u||_{L_2(\mathbf{T^d})} + \varepsilon\, ||u||_{H^1(\mathbf{T^d})} \tag{3}$$

for arbitrary $\varepsilon > 0$ and $u \in H^1$ (for instance, $||Vu||_{L_2(\mathbf{T^d})} \leq C\, ||u||_{H^\alpha(\mathbf{T^d})}$ for some $\alpha < 1$). Then the periodic magnetic Schrödinger operator $H = (D + A(x))^2 + V(x)$ has no point spectrum in $L_2(\mathbf{R^d})$.

The condition (3) can be verified for different classes of potentials. It is not optimal and can be improved (see the survey [5] for details)

In the case of real potentials the operator is self-adjoint, so one obtains the following result.

Theorem 3 *The spectrum of the periodic magnetic Schrödinger operator* $H = (D + A(x))^2 + V(x)$ *in* $L_2(\mathbf{R^d})$ *is absolutely continuous.*

3 The scheme of the proof

Let $H(k) = (D + k + A(x))^2$ and Λ_ρ be the operator that multiplies the mth Fourier coefficient of a periodic function by $(\rho^2 + m^2)^{1/2}$. These are operators on the torus $\mathbf{T^d}$. Let

$$k = 2\pi(i\rho + \beta)e \in \mathbf{C^d},$$

where $\beta \in \mathbf{R}$ is fixed, $\rho \in \mathbf{R}$ is arbitrarily large, and $e \in \mathbf{R^d}$. We introduce the principal symbol of the operator $H(k)$ as

$$H_0(k, m) = (2\pi m + k)^2 = 4\pi^2 \left[(m + \beta e)^2 - \rho^2 + 2i\rho e \cdot (m + \beta e)\right].$$

Notice that we include some lower order differential terms with parameter k into the principal symbol, which is rather standard when working with pseudodifferential operators with parameters. The zero set of this symbol is

$$Z_\rho = \left\{m \middle| \, (m + \beta e)^2 = \rho^2, \, e \cdot (m + \beta e) = 0\right\}.$$

We choose a suitable finite multiplicity covering of the dual space $\mathbf{R^d_\xi}$ by sets $U_{\rho j}$ of diameter ρ^δ for an appropriately chosen $\delta \in (0, 1)$. The goal is to produce a set of local approximate inverse operators $R_{\rho,j}$ such that on functions whose Fourier series are supported in $U_{\rho j}$ the following properties are satisfied: $||R_{\rho,j}|| \leq C$, $R_{\rho,j}H(k)\Lambda_\rho^{-1} = I + T_{\rho,j}$, and $||T_{\rho,j}|| \to 0$ uniformly with respect to j when $\rho \to \infty$. Then a suitable partition of unity in the dual space finishes the job. Namely, an operator

$$R_\rho = \sum_j \phi_j(D) R_{\rho,j} \psi_j(D)$$

is constructed in such a way that

$$R_\rho H(k) \Lambda_\rho^{-1} = I + T_\rho,$$

where

$$\lim_{\rho \to \infty} ||T_\rho|| = 0. \tag{4}$$

Here $\phi_j(D)$ and $\psi_j(D)$ are operators that multiply Fourier coefficients of a function by cut-off functions ϕ_j and ψ_j. Existence of such an operator R_ρ proves the Theorem 1.

Now we see that the main task is to construct the local approximate inverse operators $R_{\rho j}$. The situation looks differently away from the set Z_ρ and closely to it. Namely, when the distance from $U_{\rho j}$ to Z_ρ is more than ρ^δ, the principal part will dominate the magnetic one, which analogously to the Thomas' case leads to invertibility. However, close to Z_ρ the magnetic potential part becomes of comparable strength with the principal part. In each of these open sets the operator $H(k)\Lambda_\rho^{-1}$ can be reduced to a model first order differential operator by linearizing its symbol at one point. This model operator happens to be of a generalized Cauchy-Riemann type

$$\frac{\partial}{\partial \bar{z}} + g(z) \tag{5}$$

on a complex plane, where the plane arises as a rational plane in $\mathbf{R^d}$ spanned by two integer vectors l and n, and the function $g(z)$ is periodic. The goal is to invert such an operator on the torus with controlled norm of the inverse. A. Sobolev invented a nice trick that does this [16]. Namely, one can gauge away most of the magnetic potential A, leaving only a small part of it, which allows to construct a controllable inverse operator. The construction of the gauge transform amounts to solving the Cauchy-Riemann equation $\overline{\partial} f = g(z) f$ for an invertible periodic function f on the plane. In general there are obstructions (some Fourier coefficients of g must vanish), but a clever choice of an $l-n$ plane guarantees that these obstructions are moved into a tail of the Fourier series of the potential. This allows one to gauge away most of the potential and to construct an approximate inverse with controllable norm.

4 Concluding remarks and acknowledgments

This scheme works uniformly for all classes of elliptic periodic operators of interest. It is also applicable to the cases where complete results are not yet known: the Maxwell operator, Dirac operator with a general matrix potential, and anisotropic divergent type operators. In all these three situations one discovers that the local model operator on Z_ρ (5) has a matrix "potential" g. One ends up with the following question: given a periodic matrix function $g(z)$ on the complex plane one needs to find an invertible periodic matrix function $f(z)$ such that $\overline{\partial} f = g(z) f$. So, we have to deal with a non-commutative version of the model problem that was discussed above. This problem has been studied in complex analysis [11]. The function g defines in some natural way a holomorphic vector bundle, and a required function f exists if and only if this bundle is trivial. What one needs to know is whether there is any "non-commutative" analog of Sobolev's trick mentioned above.

The authors express their gratitude to Professors M. Birman, A. Sobolev, T. Suslina, A. Tumanov, R. Novikov, A. Pankov, and V. Palamodov for information and comments. After this paper was submitted, we learned from Prof. Suslina that

Prof. Sobolev had also recently reduced conditions on smoothness of the magnetic potential to the ones of the Theorem 1.

This research was partly sponsored by the NRC through a COBASE Grant, NSF through the Grant DMS 9610444, and by the Department of Army, Army Research Office, through a DEPSCoR Grant. The authors thank the NRC, NSF, and the ARO for this support. The content of this paper does not necessarily reflect the position or the policy of the federal government, and no official endorsement should be inferred.

References

[1] N.W. Ashcroft and N.D. Mermin, *Solid State Physics,* Holt, Rinehart and Winston, New York-London 1976.

[2] M. Sh. Birman and T. A. Suslina, The two-dimensional periodic magnetic Hamiltonian is absolutely continuous. (Russian) Algebra i Analiz 9(1997), no. 1, 32–48; translation in St. Petersburg Math. J. 9(1998), no. 1, 21–32.

[3] M. Sh. Birman and T. A. Suslina, Absolute continuity of the two-dimensional periodic magnetic Hamiltonian with discontinuous vector-valued potential. (Russian) Algebra i Analiz 10(1998), no. 4, 1–36; translation in St. Petersburg Math. J. 10(1999), no. 4.

[4] M. Sh. Birman and T. A. Suslina, The periodic Dirac operator is absolutely continuous, Preprint ESI #603, 1998. http://www.esi.ac.at

[5] M. Sh. Birman and T. A. Suslina, Periodic magnetic Hamiltonian with a variable metric. The problem of absolute continuity, Preprint ESI #604, 1998. http://www.esi.ac.at

[6] L. Danilov, Spectrum of the Dirac operator in $\mathbf{R}^n$, Theor. Math. Phys. 85(1990), no.1, 41-53.

[7] R. Hempel and I. Herbst, Strong magnetic fields, Dirichlet boundaries, and spectral gaps, Comm. Math. Phys. 164(1995), 237-259.

[8] J.D. Joannopoulos, R.D. Meade, and J. N. Winn, *Photonic Crystals. Molding the Flow of Light,* Princeton Univ. Press, Princeton, NJ, 1995.

[9] P. Kuchment, *Floquet Theory for Partial Differential Equations,* Birkhäuser Verlag, Basel 1993.

[10] P. Kuchment and L. Kunyansky, Spectral Properties of High Contrast Band-Gap Materials and Operators on Graphs, to appear in *Experimental Mathematics*, Part II in preparation.

[11] B. Malgrange, *Lectures on the theory of functions of several complex variables*, Tata Institute of Fundamental Research, Bombay, Springer-Verlag, Berlin 1958.

[12] A. Morame, Absence of singular spectrum for a perturbation of a two-dimensional Laplace-Beltrami operator with periodic electro-magnetic potential, Preprint 98-327 in mp_arc.

[13] M. Reed, B. Simon, *Methods of Modern Mathematical Physics, Vol.IV: Analysis of Operators,* Academic Press, New York, 1978.

[14] J. Sjostrand, Microlocal analysis for the periodic magnetic Schrödinger equation and related questions, in : Microlocal Analysis and Applications, Lect. Notes in Phys., v.1495, 237-332, Springer-Verlag, Berlin, 1991.

[15] Z. Shen. On absolute continuity of the periodic Schrödinger operators, Preprint ESI 597, 1998. http://www.esi.ac.at

[16] A. Sobolev, Absolute continuity of the periodic magnetic Schrödinger operator, Preprint, 1997.Preprint ESI 495, 1997. http://www.esi.ac.at

[17] L. E. Thomas, Time dependent approach to scattering from impurities in a crystal, Comm. Math. Phys. 33(1973), 335-343.

Operator Theory:
Advances and Applications, Vol. 108

Hardy inequalities for magnetic Dirichlet forms

Ari Laptev[1] and Timo Weidl[1,2]

[1]Royal Institute of Technology, Department of Mathematics
S-10044 Stockholm, Sweden
[2]Universität Regensburg, Naturwissenschaftliche Fakultät I
D-93040 Regensburg, Germany

Abstract

It is known that the classical Hardy inequality fails in $\mathbb{R}^2$. We show that under certain non-degeneracy conditions on vector potentials, the Hardy inequality becomes possible for the corresponding magnetic Dirichlet form.
Keywords: magnetic field, Hardy's inequality, Schrödinger operators
MSC: 35R45, 35J10, 81Q10

0. Introduction. Let $\mathbf{a}$ be a magnetic vector potential and let ρ be a non-negative function on $\mathbb{R}^d$, $d \geq 2$. In this short note we consider the Hardy type estimates

$$\int_{\mathbb{R}^d} \rho |u|^2 dx \leq C h(\mathbf{a})[u] := C \int_{\mathbb{R}^d} |(i\nabla + \mathbf{a})u|^2 dx, \qquad u \in C_0^\infty(\mathbb{R}^d), \tag{1}$$

where the constant C might depend on $\mathbf{a}$, d and ρ but not on u. If $\mathbf{a} = 0$, $\rho = |x|^{-2}$ and $d \geq 3$, then (1) coincides with the classical Hardy inequality where $C = C(d) = 4(d-2)^{-2}$. It is also known from Kato's inequality [K] (see also [AHS]), that (1) for $\mathbf{a} = 0$ implies the same inequality (with the same constant C) for $\mathbf{a} \neq 0$.

If $d = 2$ then the classical Hardy inequality is no longer true. The standard form of this inequality can be given by (1) with $\rho(x) = |x|^{-2}(1 + \log^2 |x|)^{-1}$, $\mathbf{a} = 0$ and under some additional assumptions on u. For example we have

$$\int_{\mathbb{R}^2} \frac{|u|^2}{|x|^2(1+\log^2|x|)}\, dx \leq C \int_{\mathbb{R}^2} |\nabla u|^2\, dx, \qquad \int_{\{|x|=1\}} u(x)\, dx = 0. \tag{2}$$

It was observed in [S1] that the logarithmic factor in (2) is needed only for functions u depending on $|x|$ and can be removed for functions u satisfying $\int_{\{|x|=r\}} u(x)\, dx = 0$ for any $r > 0$.

The main result of this paper shows that by introducing a non-trivial magnetic field $\mathbf{a}$, we sometimes are able to remove the unpleasant logarithmic factor in (2) and prove the inequality

$$\int_{\mathbb{R}^2} \frac{|u|^2}{|x|^2}\, dx \le C \int_{\mathbb{R}^2} |(i\nabla + \mathbf{a})u|^2\, dx, \qquad u \in C_0^\infty(\mathbb{R}^2 \setminus \{0\}), \tag{3}$$

without any other additional assumptions on u. An important example where (3) can be used is the study of the negative spectrum of two-dimensional Schrödinger operators (see [BL], [S2] and [LN]). In [LN] the authors were forced to implant the Hardy term $|x|^{-2}$ into their class of Schrödinger operators in order to obtain a CLR-type inequality. The inequality (3) automatically gives this term and therefore leads to natural applications to the corresponding magnetic Schrödinger operator.

Another application of our results concerns the problem of the existence of resonance states. In particular, (3) implies that the class of vector potentials $\mathbf{a}$ considered in this paper, "takes off" the resonance state at 0 for the corresponding two-dimensional magnetic Schödinger operator. This is, however, a partial case of a more general result obtained in [W], where it was proved that any non-trivial vector potential $\mathbf{a}$ removes the resonance state at 0. This fact follows from the following Hardy-type inequality

$$\int_{|x|<1} |u|^2\, dx \le C \int_{\mathbb{R}^2} \omega |(i\nabla + \mathbf{a})u|^2\, dx\,, \qquad C = C(\omega, \mathbf{a})\,,$$

holding for any $\mathbf{a}$ which cannot be gauged away (see [W]) and any positive weight ω for which $\omega + \omega^{-1}$ is locally bounded.

Notice that the spectral properties of the operator K_r defined in (6) and the corresponding decompositions (9), (10) were used in [LS] when studying the spectral asymptotics of magnetic two-dimensional Schrödinger operators with respect to a small coupling constant at the corresponding vector potential.

Acknowledgments. Almost all the ideas of the proof of the main result of this paper were achieved during the two weeks conference at the International Erwin Schrödinger Institute in June 1998. The authors would like to express their gratitude to T. Hoffmann-Ostenhof for inviting them to participate in this conference. The first author is also grateful to I.Herbst for many stimulating discussions.

1. The main result. Most of our discussions will be described in polar coordinates $(r, \theta) \in [0, \infty) \times \mathbb{S}^1$. We denote by $e_r = x/r$ the radial unit vector and by e_θ the unit vector which completes e_r up to the oriented orthonormal basis at $x \neq 0$. We work in the transversal (or Poincaré) gauge (see [T, Section 8.4.2]). In this gauge the radial component of the vector potential $\mathbf{a}$ is equal to zero and therefore

$$(\mathbf{a}, e_r) = 0 \qquad \text{and} \qquad (\mathbf{a}, e_\theta) =: a.$$

Applying Stokes' formula we find

$$\Phi(r) := (2\pi)^{-1} \int_{\mathbb{S}} a(r,\theta)\, r\, d\theta = (2\pi)^{-1} \int_{|x|<r} b(x)\, dx, \qquad b = \operatorname{curl} \mathbf{a}, \tag{4}$$

where by $\Phi = \Phi(r)$ we denote the normalized magnetic flux through the disk $B(0,r) = \{x \in \mathbb{R}^2 : |x| < 1\}$. The quadratic form $h(\mathbf{a})$ introduced in (1) can be written in a more convenient form by using the polar coordinate

$$h(\mathbf{a})[u] = \int_0^\infty \int_0^{2\pi} \left(|u'_r|^2 + r^{-2} |i u'_\theta + r\, a\, u|^2 \right) r\, d\theta\, dr. \tag{5}$$

When studying the form (5) we use spectral properties of the selfadjoint differential operator K_r defined on $H^1(\mathbb{S})$ by

$$K_r \varphi = i \frac{\partial}{\partial \theta} \varphi + r a \varphi, \qquad 0 \le \theta < 2\pi. \tag{6}$$

The spectrum of this operator is discrete and its eigenvalues $\{\lambda_k\}_{k\in\mathbb{Z}}$ and the complete orthonormal system of eigenfunctions $\{\varphi_k\}_{k\in\mathbb{Z}}$ are given by

$$\lambda_k = \lambda_k(r) = k + (2\pi)^{-1} r \int_0^{2\pi} a(r,\theta)\, d\theta = k + \Phi(r) \tag{7}$$

and

$$\varphi_k = \varphi_k(r,\theta) = (2\pi)^{-1/2} e^{-i(\theta\, \lambda_k(r) - r \int_0^\theta a(r,\eta)\, d\eta)}. \tag{8}$$

For any function $u \in L^2(\mathbb{R}^2)$ we introduce the following decomposition

$$u(r,\theta) = \sum_{k\in\mathbb{Z}} u_k(r) \varphi_k(r,\theta). \tag{9}$$

Obviously if $u \in H^1(\mathbb{R}^2)$, then by Parseval's identity

$$h(\mathbf{a})[u] = \int_{\mathbb{R}^2} |u'_r|^2\, dx + \sum_{k\in\mathbb{Z}} \int_0^\infty |\lambda_k(r)|^2 |u_k(r)|^2\, r^{-1} dr. \tag{10}$$

Let $\varepsilon \in (0, 1/2)$. Denote

$$M(\varepsilon) = \{r > 0 : \min_{k\in\mathbb{Z}} |k - \Phi(r)| < \varepsilon\}. \tag{11}$$

If Φ is a continuous function (which normally follows from our assumptions), then the set $M(\varepsilon)$ consists of not more than a countable number of open intervals.

Theorem 1 *Let* $\mathbf{a}$ *be continuous on* $\mathbb{R}^2$ *and* curl $\mathbf{a} \in L^1_{\mathrm{loc}}(\mathbb{R}^2)$. *Assume that there exist* $A = A(\varepsilon) > 0$, $\varepsilon \in (0, 1/2)$ *and a finite or infinite number of open intervals* $I_j = (\alpha_j, \beta_j)$, $j = 1, \ldots, N$, $N \le \infty$, *possibly accumulating at infinity, such that*

1) $M(\varepsilon) \subset \cup_{j=1}^N I_j$,

2) $\beta_{j-1} < \alpha_j < \beta_j < \infty, \quad j = 1, \ldots, N$,

3) $|I_j| = \beta_j - \alpha_j \le A \min\{1 + \alpha_j, \alpha_j - \beta_{j-1}, \alpha_{j+1} - \beta_j\}, \quad j = 1, \ldots, N$,

where formally $\beta_0 := \beta_1 - \alpha_2$. *Then the following magnetic Hardy-type inequality holds*

$$\int_{\mathbb{R}^2} \frac{|u|^2}{1+|x|^2}\, dx \le C \int_{\mathbb{R}^2} |(i\nabla + \mathbf{a})\, u|^2\, dx, \qquad u \in C_0^\infty(\mathbb{R}^2), \tag{12}$$

where $C = C(\varepsilon, A)$.

Remark. The condition 3) forbids the flux Φ (see (4)) to stabilize at integers. For example, if the magnetic field b is compactly supported and its total flux is not an integer, then the number of intervals I_j is finite, all the conditions 1)-3) of Theorem 1 are satisfied and the inequality (12) holds true.

Before proving Theorem 1 we would like to illustrate the conditions 1)-3) by giving the following example: Assume that for each $j \in \mathbb{N}$ there exists a gauge such that $\mathbf{a}(x) = 0$ for $x \in \{\alpha_j < |x| < \beta_j\}$, where $\beta_j < \alpha_{j+1}$ and $\beta_j/\alpha_j \to \infty$ as $j \to \infty$. Introduce

$$\psi_j(x) = \min\{\ln_+(|x|\alpha_j^{-1}),\, 1,\, \ln_+(\beta_j|x|^{-1})\}, \quad j \in \mathbb{N}.$$

Then $h(\mathbf{a})[\psi_j] = h(0)[\psi_j] \le 4\pi$, while $\int |\psi_j|^2(1+|x|^2)^{-1}\, dx \to \infty$ as $j \to \infty$ and thus the inequality (12) fails. In this case the magnetic flux Φ satisfies $\Phi(r) \in \mathbb{Z}$ if $\alpha_j < r < \beta_j$, which is the course of the magnetic field being trivial on long intervals.

2. Proof of Theorem 1. Fix now a smooth function χ such that $0 \le \chi(t) \le 1$, $\chi(t) = 1$ for $0 \le t \le 1$ and $\chi(t) = 0$ for $t < -\frac{1}{2A}$ or $t > 1 + \frac{1}{2A}$. Denote

$$\chi_j(r) := \chi(|I_j|^{-1}(r - \alpha_j)) \quad \text{and} \quad \psi(r) := \sum_{j=1}^N \chi_j(r).$$

From the condition 3) it follows that the supports of the functions χ_j are disjoint. The function $1 - \psi$ "cuts off" the "bad" set where the eigenvalues λ_k introduced in (7) are less than ε. Thus $\psi(r) \ne 1$ implies $|\lambda_k(r)| > \varepsilon$ and we conclude

$$\begin{aligned}
\frac{1}{2}\int_{\mathbb{R}^2} \frac{|u|^2}{1+|x|^2}\, dx &\le \int_{\mathbb{R}^2} \frac{|\psi u|^2}{1+|x|^2} dx + \int_0^\infty \frac{(1-\psi(r))^2}{1+r^2} \sum_{k\in\mathbb{Z}} |u_k(r)|^2\, r\, dr \\
&\le \int_{\mathbb{R}^2} \frac{|\psi u|^2}{1+|x|^2}\, dx + \varepsilon^{-2} \int_0^\infty \sum_{k\in\mathbb{Z}} \lambda_k^2(r)|u_k(r)|^2 \frac{dr}{r} \\
&\le \int_{\mathbb{R}^2} \frac{|\psi u|^2}{1+|x|^2}\, dx + \epsilon^{-2} h(\mathbf{a})[u].
\end{aligned} \tag{13}$$

It remains to estimate the term $\int(1+|x|^2)^{-2}|\psi u|^2 dx$. Since the supports of the functions χ_j are disjoint, it is sufficient to consider their contributions separately. We now use that if $u \in H^1(\alpha,\beta)$, $u(\beta)=0$, $\beta>\alpha\geq 0$, then

$$\int_\alpha^\beta |u(r)|^2\, r\, dr \leq 2^{-1}(\beta-\alpha)^2 \int_\alpha^\beta |u'(r)|^2\, r\, dr$$

For any fixed $\theta \in \mathbb{S}$ we have $\operatorname{supp}\chi_j u(\cdot,\theta) \subset (\alpha_j - |I_j|/2A, \beta_j + |I_j|/2A)$ and thus the latter inequality implies

$$\begin{aligned} \int |\chi_j(r)u(r,\theta)|^2\, r\, dr &\leq 2^{-1}|I_j|^2(1+A^{-1})^2 \int \left|\frac{\partial}{\partial r}(\chi_j(r)u(r,\theta))\right|^2 r\, dr \\ &\leq 2^{-1}(1+A^{-1})^2\left(|I_j|^2 \int_{\chi_j\neq 0}\left|\frac{\partial u}{\partial r}\right|^2 r\, dr + (\max|\chi'|^2)\int_{\chi_j'\neq 0} |u|^2\, r\, dr\right). \end{aligned} \tag{14}$$

The condition 3) gives us, in particular, $|I_j| < 1+\alpha_j$. If we integrate the inequality (14) over $\mathbb{S}$ and estimate the values of $|x|$ according to the two side inequalities $\alpha_j - (2A)^{-1} < |x| < \beta_j + (2A)^{-1} = \alpha_j + |I_j| + (2A)^{-1}$ as $x \in \operatorname{supp}\chi_j$, then we obtain

$$\begin{aligned} \int_{\mathbb{R}^2} \frac{|\chi_j u|^2}{1+|x|^2} dx &= \int_0^{2\pi}\int_0^\infty \frac{|\chi_j u|^2}{1+|r|^2}\, r\, dr\, d\theta \\ &\leq C_1 \int_0^{2\pi}\int_{\chi_j\neq 0}\left|\frac{\partial u}{\partial r}\right|^2 r\, dr d\theta + C_2 \int_{\chi_j'\neq 0}\sum_{k\in\mathbb{Z}} |u_k(r)|^2 \frac{dr}{r}, \end{aligned}$$

where the constants $C_l = C_l(A)$, $l=1,2$. By using again $\chi_{j_1}\chi_{j_2}=0$ for $j_1\neq j_2$ and the inequality $|\lambda_k(r)| > \varepsilon$ as $\chi_k'(r)\neq 0$, we conclude

$$\begin{aligned} \sum_{j=1}^N \int_{\mathbb{R}^2} \frac{|\chi_j u|^2}{1+|x|^2}\, dx &\leq C_1 \int_{\mathbb{R}^2}\left|\frac{\partial u}{\partial r}\right| dx + C_2\varepsilon^{-2}\int_0^\infty \sum_{j=1}^N \lambda_j^2(r)|u_j(r)|^2\frac{dr}{r} \\ &\leq C_3 h(\mathbf{a})[u], \end{aligned}$$

where $C_3 = \max(C_1, C_2\varepsilon^{-2})$. This together with (13) completes the proof. □

3. A local Hardy inequality We can now easily obtain a version of Hardy's inequality for a set of functions with supports in a bounded set. By analogy with (11) we introduce

$$L(\varepsilon) = \{r>0 : \min_{k\in\mathbb{Z}} |k-\Phi(r)| < \varepsilon\}, \quad \varepsilon \in (0,1/2).$$

Theorem 2 *Let* $\mathbf{a}$ *be continuous on* $\mathbb{R}^2\setminus\{0\}$ *and* curl $\mathbf{a} \in L^1_{\text{loc}}(\mathbb{R}^2\setminus\{0\})$. *Assume that there exist* $A = A(\varepsilon) > 0$, $\varepsilon \in (0,1/2)$ *and a finite or infinite number of open intervals* $I_j = (\alpha_j,\beta_j)$, $j=1,\ldots,N$, $N\leq\infty$, *possibly accumulating at zero, such that*

1) $L(\varepsilon) \subset \cup_{j=1}^{N} I_j$,

2) $0 < \beta_{j+1} < \alpha_j < \beta_j < \infty, \quad j = 1, \dots, N$,

3) $|I_j| = \beta_j - \alpha_j \leq A \min\{\alpha_j, \alpha_j - \beta_{j+1}, \alpha_{j-1} - \beta_j\}, \quad j = 1, \dots, N$,

where $\alpha_0 := \alpha_1 + \beta_1 - \beta_2$. *Then the following magnetic Hardy-type inequality holds*

$$\int_{\mathbb{R}^2} \frac{|u|^2}{|x|^2}\, dx \leq C \int_{\mathbb{R}^2} |(i\nabla + \mathbf{a})\, u|^2\, dx, \qquad u \in C_0^\infty(B(0,R) \setminus \{0\}), \tag{15}$$

where $C = C(\varepsilon, A, R)$.

Proof. Let us consider the change of variables $x = 1/y$, $v(y) = u(1/y)$. Then the class of function $C_0^\infty(B(0,r_0) \setminus \{0\})$ maps onto $C_0^\infty(\mathbb{R}^2 \setminus \bar{B}(0, 1/r_0))$, all the conditions of Theorem 2 become equivalent to the corresponding conditions of Theorem 1 and the inequality (15) turns into (12). The theorem is proved. □

Remark. The conditions of Theorem 2 are satisfied only if the magnetic field $b = \operatorname{curl} \mathbf{a}$ has a singularity at $x = 0$. Otherwise for any $\varepsilon > 0$ and $r_0 > 0$ there exists $\delta > 0$ such that $[0, \delta) \subset L(\varepsilon, r_0)$. This contradicts the condition 2) which states, in particular, that all the intervals I_j are separated from 0.

Combining Theorems 1 and 2 we obtain a result concerning the inequality (3).

Corollary 1 *Under the conditions of Theorems 1 and 2 the two-dimensional magnetic Hardy inequality* (3) *holds.*

4. Aharonov-Bohm-type magnetic fields. Finally we would like to give here a simple example. Let $\mathbf{a}$ be an Aharonov-Bohm-type magnetic field, namely,

$$(\mathbf{a}, e_\theta) = a(r, \theta) = \frac{\Psi(\theta)}{r}, \qquad \Psi \in L^\infty(\mathbb{S}). \tag{16}$$

In this case the corresponding magnetic field b is equal to zero everywhere except $x = 0$. Denote by $\bar{\Psi}$ the mean value of the function Ψ over $\mathbb{S}$

$$\bar{\Psi} = (2\pi)^{-1} \int_0^{2\pi} \Psi(\theta)\, d\theta.$$

Theorem 3 *Let us assume that the vector potential* $\mathbf{a}$ *is given by* (16) *and* $\bar{\Psi} \neq k$, *for any* $k \in \mathbb{Z}$. *Then*

$$\int_{\mathbb{R}^2} \frac{|u|^2}{|x|^2}\, dx \leq A \int_{\mathbb{R}^2} |(i\nabla + \mathbf{a})u|^2\, dx, \qquad u \in C_0^\infty(\mathbb{R}^2 \setminus \{0\}),$$

where $A = (\min_{k \in \mathbb{Z}} |k - \bar{\Psi}|)^{-2}$. *The constant* A *is sharp.*

Proof. In this case the eigenvalues defined in (7) are independent of r and equal $\lambda_k = k + \bar{\Psi}$. Then using (9) and (10) we obtain

$$\int_{\mathbb{R}^2} \frac{|u|^2}{|x|^2}\, dx \le A \int_0^\infty \sum_{k\in\mathbb{Z}} \lambda_k^2 |u_k(r)|^2 \frac{dr}{r} \le A\, h(\mathbf{a})[u].$$

Let us assume that the constant A is achieved at k_0, $A = |k_0 + \bar{\Psi}|^{-2}$. Then it easy to see that the constant A is sharp on the class of functions $\phi(r)\exp(i\theta k_0)$, where $\phi \in C_0^\infty(0,\infty)$. The proof is complete. □

References.

[AHS] Y. Avron, I. Herbst, B. Simon: "Schrödinger Operators with Magnetic Fields I", *Duke Math. J.*, **45** (1978) 847-883.

[BL] M.Sh. Birman, A. Laptev: "The negative discrete spectrum of a two-dimensional Schrödinger operator" *Comm. Pure Appl. Math.*, **XLIX** (1996) 967–997.

[K] T. Kato: "Schrödinger Operators with singular Potentials", *Israel J. Math.*, **13** (1972) 135–148.

[LN] A. Laptev, Yu. Netrusov: "On the negative eigenvalues of a class of Schödinger operators", preprint KTH, Sweden, (1998).

[LS] A. Laptev, O. Safronov: "The negative discrete spectrum of a class of two-dimensional Schrödinger operators with magnetic fields", preprint KTH, Sweden, (1998).

[S1] M.Z. Solomyak: "A Remark on the Hardy Inequalities", *Integr. Equat. Oper. Th.*, **19** (1994) 120–124.

[S2] M.Z. Solomyak: "Piecewise-polynomial approximation of functions from $H^l((0,1)^d)$, $2l = d$, and applications to the spectral theory of Schrödinger operator", *Israel J. Math.*, **86** (1994) 253–276.

[T] B. Thaller: "The Dirac equation", *Texts and Monographs in Physics, Springer-Verlag* (1992).

[W] T. Weidl: "Remarks on virtual bound states for semi-bounded operators", to appear in *Comm. Partial Diff. Eq.*, 35 pp.

Operator Theory:
Advances and Applications, Vol. 108

Adiabatic curvature, chaos and the deformations of Riemann Surfaces

Péter Lévay
Institute of Theoretical Physics,
Technical University of Budapest H-1521 Budapest, Hungary

Abstract

Parametrized families of Landau Hamiltonians on leaky tori $\Sigma_{(h,c)}$ (i.e Riemann surfaces with genus h and number of cusps c, $2h+c \geq 3$) corresponding to classically chaotic families of geodesic motion in constant magnetic field are investigated. The parameters describe deformations of these surfaces. It is shown that the adiabatic curvature two-form of the lowest Landau level is the sum of two (three) terms in the compact $\Sigma_{(h,0)}$ (noncompact $\Sigma_{(h,c)}, c \neq 0$) case . The first term is proportional to the natural symplectic form on deformation space. For both the compact and non-compact cases we have a fluctuating term reflecting the chaotic nature of the geodesic motion. For $\Sigma_{1,0}$ (torus) we have no fluctuating term, the geodesic motion being integrable in this case. As an illustration, the $\Sigma_{1,1}$ case (Gutzwiller's leaky torus) is worked out in some detail.

1. Introduction

Parametrized families of quantum systems exhibit interesting geometric properties described by a quantity called adiabatic curvature. Describe these families by the Hamiltonians $H(x)$ where $x \in \mathcal{M}$ with $\mathcal{M}$ being the parameter space, having the structure of a smooth manifold. Denote by $P_n(x)$ the family of spectral projectors corresponding to the eigenvalue $E_n(x)$ (assumed to belong to the discrete part of the spectrum). We assume, moreover, that $H(x)$ is depending smoothly on the (local) coordinates x, and $E_n(x)$ is separated by an energy gap from the rest of the spectrum. Then adiabatic curvature is defined as

$$\omega(P_n) = Tr(P_n dP_n \wedge dP_n). \tag{1.1}$$

$\omega(P_n)$ is a two-form on $\mathcal{M}$ and $dP_n = \partial_j P_n dx^j, \quad j = 1, \dots dim\mathcal{M}$. For projectors of rank one the integral $\gamma_n(C) = \int_S \omega(P_n)$ where S is a surface bounded by a closed loop $C = \partial S$ in $\mathcal{M}$ gives the celebrated Berry's Phase [1] for systems cycled adiabatically around C.

A particularly interesting class of such systems arises when we consider those families $H(x)$ for which the corresponding classical dynamics is chaotic. Motivated by quantum chaology it would be nice to see how chaos manifests itself in the anholonomy properties of eigensubspaces over $\mathcal{M}$. In this paper we specialize this class further, by restricting our attention to the chaotic [2] geodesic motion in a constant magnetic field on two-dimensional surfaces of constant negative curvature. These surfaces are the finite area Riemann-surfaces $\Sigma_{h,c}$ of compact or non-compact type with a fixed number of holes (genus h) and points infinitely far away (horns or cusps c). By deforming these surfaces under the constraint of *preserving their area* we are given a family of classically chaotic systems. The parameter space $\mathcal{M} \equiv \mathcal{T}_{h,c}$ is the "space of shapes" for a particular $\Sigma_{h,c}$. The quantization of these classically chaotic families yields us for the need to study the properties of the spectral subbundles for Landau Hamiltonians over $\mathcal{T}_{h,c}$. A natural geometric quantity for this purpose is the adiabatic curvature $\omega(P_n)$.

Another important physical motivation comes from the observation of Avron et. al. [3] that these surfaces mimic the properties of mesoscopic systems with or without leads attached to them. By threading the holes and the leads (cusps) with Aharonov-Bohm fluxes we are given a family of quantum systems parametrized by the fluxes. It is natural to consider the adiabatic response of this quantum system to a small change in the fluxes. It turns out that the first-order correction is proportional to the change in flux in time, with the proportionality coefficient being just the adiabatic curvature. In this case $\omega(P)$ plays the role of a nondissipative conductivity for a quantum Hall fluid of a completely filled Landau level. In a further paper Avron et. al. [4] have shown that $\omega(P)$ calculated for the lowest Landau level as a two-form on the flux-torus is consisting of a term which is quantized and another one describing conductance fluctuations. In two dimensions when the surfaces are itself deformed $\omega(P)$ now regarded as a two-form on the space of shapes has the meaning of a *nondissipative viscosity* [5,6] for the aforementioned quantum Hall fluid. In this paper we calculate $\omega(P)$ living on $\mathcal{T}_{h,c}$ for the projector corresponding to the lowest Landau level. The material of this paper is an extension of our paper [7] covering only the case of compact ($\Sigma_{h,0}$) surfaces.

2. Landau Hamiltonians and their deformations

For the definition of our Riemann-surfaces $\Sigma_{h,c}$ ($2h + c \geq 3$) we make use of Riemann uniformisation. The simply connected universal cover of these surfaces is $\mathcal{H}$ the Poincaré upper-half-plane. Then $\Sigma_{h,c}$ is realized as the quotient of $\mathcal{H}$ by a discrete group Γ which is isomorphic to the fundamental group $\pi_1(\Sigma_{h,c})$. We take $\Gamma \subset PSL(2, \mathbf{R})$, acting on $\mathcal{H}$ as fractional linear transformations, to be a torsion free finitely generated Fuchsian group. Γ is generated by $2h$ hyperbolic and c parabolic transformations satisfying $U_1^{-1}V_1^{-1}U_1V_1 \dots U_h^{-1}V_h^{-1}U_hV_hS_1 \dots S_c =$

1. The fixed points of the parabolic elements in $\mathcal{H}$ are taken by the covering projection $\pi : \mathcal{H} \to \mathcal{H}/\Gamma$ to the cusps in $\Sigma_{h,c}$. The unique complex structure of $\mathcal{H}$ is determined by a metric of constant negative curvature. We take this metric to be the usual Poincaré metric with scalar curvature $R = -1$. The covering projection then induces a metric g on $\Sigma_{h,c}$ with $R_g = -1$.

Now we describe our parameter space. First we define $Metr_{const}(\Sigma_{h,c})$, as the space of metrics on $\Sigma_{h,c}$ with $R_g = -1$. We single out a metric g from this space, with line element to be of the form $\hat{g} = e^{2\sigma(x)}(dx_1^2 + dx_2^2)$ on $\Sigma_{h,c}$ by using suitable local coordinates [8]. Then we would like to describe the deformations of $\Sigma_{h,c}$ as the deformations of the metric g. It is clear that infinitesimal metric deformations span the tangent space of $Metr_{const}(\Sigma_{h,c})$ at g [9]. One can define a metric on the space of metrics as follows

$$\langle \delta g_1 | \delta g_2 \rangle_g = \int_{\Sigma_{h,c}} d^2x \sqrt{g} g^{ad} g^{bc} \delta g_{1ab} \delta g_{2cd}, \quad a, b, c, d = 1, 2. \tag{2.1}$$

Now let us consider the group $Diff_0(\Sigma_{h,c})$ i.e. the group of orientation preserving diffeomorphisms continuously connected to the identity. The action of this group on $Metr_{const}(\Sigma_{h,c})$ is described in [9], here we only remark that its Lie algebra is the space of smooth vector fields on $\Sigma_{h,c}$. When acting with this group, the first order (infinitesimal) change in the metric when going along the integral curve of some of such vector fields v is just the Lie derivative $L_v g$ of g along v. This group clearly preserves R hence via Gauss-Bonnet the area of $\Sigma_{h,c}$. However, these are obviously *not* the deformations we are interested in. Factoring out with the action of this group gives the space

$$\mathcal{T}_{h,c} = Metr_{const}(\Sigma_{h,c})/Diff_0(\Sigma_{h,c}) \tag{2.2}$$

which is our true parameter space. In order to introduce local coordinates at the point g for this space we proceed as follows. The area preserving deformations $g \mapsto g + \delta g$ have to preserve R i.e. we must have $\delta R = 0$. A calculation shows that $\delta R = -\frac{1}{2}(R + \nabla_c \nabla^c)(g^{ab}\delta g_{ab}) + \frac{1}{2}\nabla^a \nabla^b \delta g_{ab}$, where ∇_c is the covariant derivative corresponding to the Levi-Civita connection of the metric g. In order to preserve R we have to set up the constraints $\nabla^a \delta g_{ab} = g^{ab}\delta g_{ab} = 0$. Introducing complex coordinates via $\hat{g} = e^{2\sigma} dz d\bar{z} = 2g_{z\bar{z}} dz d\bar{z}$ these imply that $\delta\hat{g} = 2g^{z\bar{z}}\Re(f_{zz}dz^2)$ with $\partial_{\bar{z}} f = 0$. Moreover it can be shown that for this δg we have $\langle \delta g | \delta g_0 \rangle_{|g} = 0$ where $\delta g_0 \equiv L_v g$ is the change arising from the action of $Diff_0(\Sigma_{h,c})$. It then follows that the deformations describing (2.2) locally are related to holomorphic quadratic differentials $f dz^2$ on $\Sigma_{h,c}$. Using the Riemann-Roch theorem one can show that these differentials span a $3h - 3 + c$ dimensional *complex* vector space. Choosing a base f_j, $j = 1, \dots 3h - 3 + c$ in this space we can introduce the *local* complex coordinates τ^j via $f = \tau^j f_j$ corresponding to the deformation δg. It can be shown that in the overlapping neighbourhoods these coordinates transform complex analytically. (In this respect consult Ref.[10] for the compact, and the papers [11-12] and references therein for the noncompact case.) In fact this space is well-known to mathematicians and even to physicists. It is the Teichmüller space of the Riemann-surfaces $\Sigma_{h,c}$ extensively used in string theory [14]. This observation enables us to use the powerful techniques developed in this field to calculate $\omega(P)$.

One can project the (2.1) metric to a metric on $\mathcal{T}_{g,c}$ which is of the form (at the point g)

$$h_{j\bar{k}}(g) = \int_{\Sigma_{h,c}} f_j \bar{f}_{\bar{k}} g^{z\bar{z}} d^2x. \tag{2.3}$$

(see section 7.3.6 of Ref.[10] in this respect). This metric is Kähler and gives rise to the so called Weil-Petersson-Kähler form $\omega_{WP}(g) = ih_{j\bar{k}}(g)d\tau^j \wedge d\bar{\tau}^k$ on $\mathcal{T}_{h,c}$.

Having discussed the deformations we are interested in, now we turn to the description of Landau Hamiltonians (i.e. Hamiltonians describing the motion of a charged particle in a constant magnetic field) on $\Sigma_{h,c} = \mathcal{H}/\Gamma$, and their deformations. When one would like to study electrodynamics on topologically nontrivial configuration spaces $Q = \Sigma_{h,c}$ the field strength F and the vector potential A have to be regarded as the curvature and connection forms of a principal $U(1)$ bundle over Q. F is a two-form which is closed due to Maxwell's equation, however it cannot always be represented in the form $F = dA$ *globally* if $H^2(Q, \mathbf{R})$ is nontrivial. However, in spite of this we still have the chance to define electrodynamics in these cases provided $[\frac{1}{2\pi}F] \in H^2(Q, \mathbf{Z})$, i.e. if F suitably normalized defines an integral second cohomology class. More precisely one has [13] the following theorem: there exists a complex line bundle L over Q with connection ∇ with F being its curvature form iff $[\frac{1}{2\pi}F] \in H^2(Q, \mathbf{Z})$. Then in the local coordinates x_1, x_2 the connection reads as $\nabla_a = \partial_a - iA_a$ with A_a a local vector potential in a given gauge. Since we have also the metric g as defined above, we define our Landau Hamiltonian as $H_B = -\frac{1}{2}e^{-2\sigma}[(\partial_1 - iA_1)^2 + (\partial_2 - iA_2)^2]$, with $F = dA = Be^{2\sigma}dx_1 \wedge dx_2$, $B = const.$ Since $\int F = BV = 2\pi\mathbf{Z}$ where V is the area, the flux on $\Sigma_{h,c}$ has to be quantized. From the Gauss-Bonnet theorem the area is $2\pi(2h + c - 2)$ hence B must be rational. However, further arguments [3,15] show that a non-integer B can only be introduced consistently provided we also introduce fluxes through the cusps. Hence in the absence of fluxes we have $B \in \mathbf{Z}$. The next step is to use complex coordinates and to choose a gauge $A = A_z dz + A_{\bar{z}} d\bar{z} = -2iB(\partial_z\sigma)dz$, to write H_B as

$$H_B = -2e^{-2\sigma}[\partial_z\partial_{\bar{z}} - 2B(\partial_z\sigma)\partial_{\bar{z}}] + \frac{B}{2} = D_B^\dagger D_B + \frac{B}{2} = D_{B+1}D_{B+1}^\dagger - \frac{B}{2} \tag{2.4}$$

where $D_B = 2e^{-2\sigma}\partial_{\bar{z}}$ and $D_B^\dagger = -\partial_z + 2(B-1)(\partial_z\sigma)$ which are just the operators introduced in [14]. Using the Riemann-Roch theorem [3] we obtain the degeneracy of the lowest Landau level as $\frac{N}{2} = dim_{\mathbf{C}}KerD_B = (2B-1)(g-1) + (B-1)c$ for $B > 1$ the only case considered here. The next step is to deform the operators D_B and $D_B^\dagger$ by using the metric deformations described above. Here we meet a pleasant surprise: D_B ($D_B^\dagger$) varies holomorphically (antiholomorphically) with τ^j [12,14]. The same property is valid for $|\psi_\alpha\rangle$ ($\langle\psi_\alpha|$) $\alpha = 1, \dots N$ for the (unnormalized) eigenfunctions belonging to the kernel of the operator D_B. Note that these are local eigen-sections of a holomorphic vector bundle over $\mathcal{T}_{h,c}$.

3. Calculation of the adiabatic curvature

Now we see what the holomorphic structure found above can do for us. First one can prove using (1.1) that $\omega(P)$ can be written in the form

$$\omega(P) = i\partial\bar{\partial} \log det\langle\psi_\alpha|\psi_\beta\rangle, \tag{3.1}$$

where $\partial \equiv d\tau^j \frac{\partial}{\partial\tau^j}$, and $\bar{\partial} \equiv d\bar{\tau}^{\bar{j}} \frac{\partial}{\partial\bar{\tau}^{\bar{j}}}$.
Unfortunately the analytic form of eigenfunctions needed in (3.1) are not known so we have to proceed differently. The important observation is that the term (3.1) appears in formulas derived in string theory when calculating multiloop amplitudes. The obstruction of defining these amplitudes consistently is what string theorists call a *holomorphic anomaly*. From the mathematical point of view this obstruction is the nontrivial curvature of the determinant line bundle for our Landau Hamiltonian. A natural metric $||s||_Q$ on this bundle was given by Quillen [16](here s is a section of the bundle). The curvature of this line bundle for $B > 1$ is of the form [14]

$$\partial\bar{\partial} \log ||s||_Q^2 = \partial\bar{\partial} \log \frac{det' D_B^\dagger D_B}{det\langle\psi_\alpha|\psi_\beta\rangle}. \tag{3.2}$$

The expression (3.2) can be calculated provided an appropriate regularization is found for the calculation of $det' D_B^\dagger D_B$, where prime indicates the deletion of zero modes.

Theorem 1: For compact surfaces $\Sigma_{h,0}$ ($h \neq 0, 1$) we have

$$\omega(P) = -\frac{1}{12\pi}(6B^2 - 6B + 1)\omega_{WP} + i\partial\bar{\partial} \log det' D_B^\dagger D_B, \quad B > 1. \tag{3.3}$$

where ω_{WP} is the natural symplectic form on the space of shapes defined by (2.3).

Proof: The proof of this theorem is based on the definition of the determinant by zeta function regularization. Using the formula $\bar{\partial} \ln det' K = -\lim_{t\to 0} Tr((1 - P)\bar{\partial}K(K)^{-1}e^{-tK})$ [17], valid for parametrized families of operators K and short time Heat-kernel techniques the proof can be found in [14], adapted to Landau Hamiltonians in [7]. The (3.2) obstruction was first calculated by Belavin and Knizhnik [18].

Now we can use the result [19] $\partial\bar{\partial} \log det' D_B^\dagger D_B = \partial\bar{\partial} \log Z(B)$ where $Z(B)$ is Selberg's zeta function defined as

$$Z(B) = \prod_p \prod_{k=0}^{\infty} (1 - e^{-(B+k)l(p)}). \tag{3.4}$$

Here p are inconjugate primitive hyperbolic elements of Γ representing simple closed geodesics with hyperbolic length $l(p)$ with respect to the Poincaré metric on $\mathcal{H}$. This term (being an exact form) is describing viscosity fluctuations.

Our next task is to derive a similar formula for the non-compact case ($\Sigma_{h,c}$). However in this case the Landau Hamiltonians has c-fold continuous spectrum. Hence the question arises how to define a regularized determinant in this case. It is quite natural to adopt the *definition* [12] $det' D_B^{\dagger} D_B \equiv Z(B)$ for $B > 1$.

In order to formulate the theorem we proceed with some definitions. The Eisenstein series $E_r(z, s)$ [20] corresponding to the rth cusp ($r = 1, \ldots c$) of the group Γ uniformizing $\Sigma_{h,c}$, is defined as $E_r(z,s) = \sum_{\gamma \in \Gamma/\Gamma_r} \Im(\sigma_r^{-1}\gamma z)^s$. Here Γ_r are non conjugate parabolic subgroups in Γ generated by S_r, and the $\sigma_r \in PSL(2, R)$ take Γ_r via conjugation to Γ_∞, which is generated by the transformations $z \mapsto z + 1$. Then we have the following generalization of Theorem 1.

Theorem 2: For surfaces $\Sigma_{h,c}$ ($2h + c \geq 3$) we have

$$\omega(P) = -\frac{1}{12\pi}(6B^2 - 6B + 1)\omega_{WP} + \frac{\pi}{9}\omega_{CUSP} + i\partial\bar{\partial}\log Z(B), \quad B > 1. \tag{3.5}$$

where ω_{CUSP} is a new Kähler form modifying the symplectic structure on the space of shapes defined as

$$\omega_{CUSP}(g) \equiv i \left(\sum_{r=1}^{c} \int_{\Sigma_{h,c}} f_j \bar{f}_k g^{z\bar{z}} E_r(.,2) d^2x \right) d\tau^j \wedge d\bar{\tau}^k. \tag{3.6}$$

Proof: Theorem 2 is a trivial consequence of Theorem 1 in [12] proved by Takhtajan and Zograf. Their proof avoids the use of heat kernel techniques. Instead they derive variation formulas of the Selberg zeta function using Teichmüller theory.

In order to gain some insight to the question of how chaos reveals itself in the form of $\omega(P)$, we also include the result for the torus $\Sigma_{1,0}$ case. In this case $\mathcal{T}_{1,0}$ is just the complex upper half plane $\mathcal{H}$ [9].

Theorem 3: For the torus $\Sigma_{1,0}$ we have for the nth Landau level (each of them having $B \in \mathbf{Z}^+$-fold degeneracy ($n = 0, 1, \ldots$)

$$\omega(P_n) = -\frac{B}{4\pi}(n + \frac{1}{2})\frac{d\tau_1 \wedge d\tau_2}{{\tau_2}^2} \equiv -\frac{B}{2\pi}(n + \frac{1}{2})\omega_{WP} \tag{3.7}$$

Proof: The proof of this theorem can be found in [5] for $n = 0$ and in [7,21] for arbitrary n. (In this case the eigenfunctions are known.) As one can see for the torus we have no fluctuating term reflecting the well-known fact that the geodesic motion on $\Sigma_{1,0}$ is integrable.

In order to clarify the meaning of the terms appearing in the above theorems we introduce another set of coordinates on $\mathcal{T}_{h,c}$ which are more geometrical in origin. Unfortunately these coordinates have no apparent relationship to the complex structure of $\mathcal{T}_{h,c}$. It is a classical result in Teichmüller theory that $2d$ (d is the complex dimension of $\mathcal{T}_{h,c}$) closed curves α_j, $j = 1, 2 \ldots 2d$ from the free homotopy classes of $\Sigma_{h,c}$ can be chosen in such a way that the lengths $l_j \equiv l(\alpha_j)$ provide local real coordinates for $\mathcal{T}_{h,c}$. Then it is a result of Wolpert [11] that ω_{WP} has an expression in these coordinates

$$\omega_{WP} = \sum_{j<k} W_{jk}^{-1} dl_j \wedge dl_k, \quad W_{jk} = \sum_{q \in \alpha_j \# \alpha_k} \cos\theta_q \tag{3.8}$$

Here $\alpha_j \# \alpha_k$ denotes the intersection locus of the geodesics, and θ_q is the intersection angle at q measured from α_j to α_k. Moreover one can introduce vector fields t_j which are Hamiltonian i.e. $\omega_{WP}(t_j, .) = -dl_j$. These vector fields generate the Frenchel-Nielsen twists obtained by cutting the surface along l_j rotating one side of the cut with respect to the other, and then glue the sides together. From the geometric point of view these coordinates are easier to visualize as the ones describing the space of shapes. Moreover one can show that viscosity fluctuations are exponentially small for large magnetic field [7], provided we stay away from the boundary of Teichmüller space where one of the geodesic lengths is pinching down to zero. Hence for large magnetic field $\omega(P) \sim -\frac{B^2}{2\pi}\omega_{WP}$.

We close this paper by illustrating these result for Gutzwiller's leaky torus [22] ($\Sigma_{1,1}$). $\Sigma_{1,1}$ is described as a fundamental domain $\mathcal{H}/\Gamma$ for the Fuchsian group Γ freely generated by two hyperbolic transformations $U, V \in PSL(2, \mathbf{R})$ with $U^{-1}V^{-1}UV$ parabolic. This group can be represented by $SL(2, \mathbf{R})$ elements such that $a_1 \equiv trU > 2$, $a_2 \equiv trV > 2$, and $a_3 \equiv trUV > 2$. For a particular choice of these matrices see [22]. Using parabolicity one gets $trU^{-1}V^{-1}UV = -2$. By virtue of the identity $trUtrV = trUV + trU^{-1}V$ one has $a_1a_2a_3 = \sum_s a_s^2$. Since the traces of hyperbolic elements are related to the geodesic lengths as $a_s = 2\cosh l_s/2$ the numbers $a_s > 2$ with the constraint above will describe the *two dimensional* space of shapes. Introducing new coordinates $y_1 \equiv a_1/(a_2a_3)$, $y_2 \equiv a_2/(a_1a_3)$, $y_3 \equiv a_3/(a_1a_2)$ one can see that $\mathcal{T}_{1,1} = \{y_s > 0, s = 1, 2, 3 | y_1 + y_2 + y_3 = 1\}$ i.e. it is a simplex. Using (3.8) one can show that $\omega_{WP} = \frac{dy_1 \wedge dy_2}{y_1y_2y_3}$ [23]. As far as ω_{CUSP} is concerned a formula similar to (3.8) in terms of the lengths l_s to our knowledge is not available. It represents a modification of the symplectic structure as defined by ω_{WP}. The fluctuating term reflects the change in the geodesic length spectrum when changing the shape of $\Sigma_{1,1}$ by varying the independent coordinates y_1 and y_2.

Acknowledgements

Financial support from Országos Tudományos Kutatási Alap (OTKA), Grant Nos. T017179 and T021228 is gratefully acknowledged.

References

[1] A. Shapere and F. Wilczek (eds.), *Geometric Phases in Physics* (World Scientific, Singapore, 1989).

[2] N. L. Balázs and A. Vörös, Phys. Rep. **143**, 109 (1986).

[3] J. E. Avron, M. Klein, A. Pnueli and L. Sadun, Phys. Rev. Lett. **69**, 128 (1992).

[4] J. E. Avron, R. Seiler and P. G. Zograf, Phys. Rev. Lett. **73**, 325 (1994).

[5] J. E. Avron, R. Seiler and P. G. Zograf, Phys. Rev. Lett. **75**, 697 (1995).

[6] J. E. Avron, J. Stat. Phys. **92**, 543-557 (1998).

[7] P. Lévay, Phys. Rev. **E56**, 6173 (1997).

[8] M. Nakahara, *Geometry, Topology and Physics* (Adam Hilger IOP Publishing, Bristol, 1990) pp. 238.

[9] C. Nash, *Differential Topology and Quantum Field Theory* (Academic Press, London, 1991).

[10] Y. Imayoshi and M. Taniguchi, *An Introduction to Teichmüller Spaces* (Springer-Verlag, Tokyo, 1992).

[11] S. Wolpert, Annals of Mathematics **117**, 207 (1983).

[12] L. A. Takhtajan and P. G. Zograf, Commun. Math. Phys. **137**, 339 (1991).

[13] B. Kostant, *Quantization and Unitary Representations. Lecture in Modern Analysis III.* (Lecture Notes in Mathematics **170**) (Springer-Verlag, Berlin, 1970)

[14] E D'Hoker and D. H. Phong, Rev. Mod. Phys. **60**, 917 (1988).

[15] A. Comtet, B. Georgeot and S. Ouvry, Phys. Rev. Lett. **71**, 3786 (1993).

[16] D. Quillen, Funct. Anal. Appl. **19**, 31 (1985).

[17] N. Berline, E. Getzler and M. Vergne, *Heat Kernels and Dirac Operators* (Springer-Verlag Berlin 1996).

[18] A. A. Belavin and V. G. Knizhnik, Sov. Phys. JETP **64**, 214 (1986).

[19] E. D'Hoker and D. H. Phong, Commun. Math. Phys. **104**, 537 (1986).

[20] T. Kubota, *Elementary theory of Eisenstein series* (John Wiley N. Y. 1975).

[21] P. Lévay, J. Math. Phys. **36**, 2792 (1995).

[22] M. C. Gutzwiller, Physica **7D**, 341 (1983).

[23] S. Wolpert, Comment. Math. Helvetici **58**, 246 (1983).

Operator Theory:
Advances and Applications, Vol. 108

OPERATOR INTERPRETATION OF RESONANCES ARISING IN SPECTRAL PROBLEMS FOR 2 × 2 MATRIX HAMILTONIANS*

A. K. Motovilov[†‡], R. Mennicken[§]

We consider the analytic continuation of the transfer function for a 2×2 matrix Hamiltonian into the unphysical sheets of the energy Riemann surface. We construct non-selfadjoint operators representing operator roots of the transfer function which reproduce certain parts of its spectrum including resonances situated in the unphysical sheets neighboring the physical sheet. On this basis, completeness and basis properties for the root vectors of the transfer function (including those for the resonances) are proved.

1. Introduction. In this work we deal with Hamiltonians of the 2×2 matrix form

$$\mathbf{H} = \begin{pmatrix} A_0 & B_{01} \\ B_{10} & A_1 \end{pmatrix}$$

It is assumed that the operator $\mathbf{H}$ acts in an orthogonal sum $\mathcal{H} = \mathcal{H}_0 \oplus \mathcal{H}_1$ of the Hilbert spaces $\mathcal{H}_0$ and $\mathcal{H}_1$ while the entries $A_0 : \mathcal{H}_0 \to \mathcal{H}_0$, and $A_1 : \mathcal{H}_1 \to \mathcal{H}_1$, are selfadjoint operators. The couplings $B_{ij} : \mathcal{H}_j \to \mathcal{H}_i$, $i \neq j$, $B_{01} = B_{10}^*$, are assumed to be bounded operators. We are especially interested in the physically typical case where the spectrum of, say, A_1 is partly or totally embedded into the absolutely continuous spectrum of A_0 and the transfer function $M_1(z) = A_1 - z + V_1(z)$, where $V_1(z) = B_{10}(z - A_0)^{-1}B_{01}$, admits analytic continuation (as an operator-valued function) through the absolutely continuous spectrum of the entry A_0 into the unphysical sheet(s) of the energy z plane. Notice that the resolvent $(\mathbf{H} - z)^{-1}$ of the operator $\mathbf{H}$ can be expressed explicitly in terms of the inverse transfer function

*Financial support of this work by the DFG, INTAS and RFBR is kindly acknowledged
†Physikalishes Institut, Universität Bonn, Endenicher Allee 11-13, D-53115 Bonn, Germany
‡On leave of absence from the Joint Institute for Nuclear Research, 141980 Dubna, Russia
§Naturwiss. Fakultät I – Mathematik, Universität Regensburg, D-93040 Regensburg, Germany

$M_1^{-1}(z)$. Therefore, in studying the spectral properties of the transfer function M_1 one studies at the same time the spectral properties of the operator matrix $\mathbf{H}$.

We construct an operator-valued function $V(X)$ (see [9]) on the space of operators in $\mathcal{H}_1$ possessing the property: $V(X)\psi_1 = V(z)\psi_1$ for any eigenvector ψ_1 of X corresponding to an eigenvalue z and then study the equation

$$H_1 = A_1 + V(H_1). \tag{1}$$

In the case where the spectra of A_0 and A_1 have no intersection, Eq. (1) is reduced (cf. [9]) to an operator Riccati equation for a similarity transform which allows to find invariant subspaces of the matrix $\mathbf{H}$ admitting a graph representation [1, 2, 8] (also see [6]). Now we prove the solvability of Eq. (1) in the case where the spectra of A_1 and A_0 overlap, cf. also [8]. However in this case the solutions of (1) already represent non-selfadjoint operators.

The problem considered is closely related to the resonances generated by the matrix $\mathbf{H}$. Regarding a definition of the resonance and history of the subject see, e. g., the book [11]. A recent survey of the literature on resonances can be found in [10]. Throughout the paper we treat resonances as the discrete spectrum of the transfer function M_1 situated in the unphysical sheets of its Riemann surface since points of this spectrum automatically correspond to the poles of the analytic continuation of the resolvent $(\mathbf{H} - z)^{-1}$ (understood in the form sense). Using the fact that the root vectors of the solutions H_1 of the equation (1) are at the same time such vectors for M_1, we prove completeness and even basis properties for the root vectors including those for the resonances. The results obtained allow immediate applications in particular to the scattering problems for multichannel Schrödinger equation.

A detailed exposition of the material presented including proofs in the case of essentially more general spectral situation will be given in the extended paper [7].

2. Analytic continuation of the transfer function. For the sake of simplicity we assume in the report that all the spectrum $\sigma(A_0)$ of the entry A_0 is absolutely continuous consisting of an only interval $J^0 = (\mu_0^{(1)}, \mu_0^{(2)})$ with $-\infty \le \mu_0^{(1)} < \mu_0^{(2)} \le \infty$ while all the spectrum $\sigma(A_1)$ of the entry A_1 is totally embedded into the interval J^0. Therefore, we assume $\sigma(A_1) \subset J^0$.

Let E_0 be the spectral measure for the entry A_0, $A_0 = \int_{\sigma(A_0)} \lambda\, dE_0(\lambda)$. Then the function $V_1(z)$ can be written

$$V_1(z) = \int_{\sigma(A_0)} dK_B(\mu)(z-\mu)^{-1}$$

with

$$K_B(\mu) = B_{10}E^0(\mu)B_{01}$$

where $E^0(\mu)$ is the spectral function of A_0, $E^0(\mu) = E_0\Big((-\infty, \mu)\Big)$. We suppose that the function $K_B(\mu)$ is differentiable in $\mu \in J^0$ in the operator norm topology.

Further, we suppose that the derivative $K'_B(\mu)$ is continuous within the closed interval $\overline{J^0}$ and, moreover, that it admits analytic continuation from this interval to a simply connected domain situated, say, in $\mathbb{C}^-$. Let this domain be called D^-. We assume that the boundary of D^- includes the entire spectral interval J^0. Since $K'_B(\mu)$ represents a selfadjoint operator for $\mu \in J^0$ and $J^0 \subset \mathbb{R}$, the function $K'_B(\mu)$ also automatically admits analytic continuation from J^0 into the domain D^+, symmetric to D^- with respect to the real axis and $[K'_B(\mu)]^* = K'_B(\bar{\mu})$, $\mu \in D^\pm$.

Let Γ^l, $l = \pm 1$, be a rectifiable Jordan curve in D^l resulting from continuous deformation of the interval J^0, the (finite) end points of this interval being fixed. The quantity

$$\mathcal{V}_0(B, \Gamma^l) = \int_{\Gamma^l} |d\mu| \, \|K'_B(\mu)\|$$

where $|d\mu|$ stands for the Lebesgue measure on Γ^l, is called variation of the function $K_B(\mu)$ along the contour Γ^l. We suppose that there exists a contour (contours) Γ^l where the value $\mathcal{V}_0(B, \Gamma^l)$ is finite including also the case of the unbounded interval J^0. The contours Γ^l satisfying the condition $\mathcal{V}_0(B, \Gamma^l) < \infty$ are said to be K_B-bounded contours.

LEMMA 1 *The analytic continuation of the transfer function $M_1(z)$, $z \in \mathbb{C} \backslash \sigma(A_0)$, through the spectral interval J^0 into the subdomain $D(\Gamma^l) \subset D^l$, $l = \pm 1$, bounded by this interval and a K_B-bounded contour Γ^l is given by*

$$M_1(z, \Gamma^l) = A_1 - z + V_1(z, \Gamma^l) \quad \text{with} \quad V_1(z, \Gamma^l) = \int_{\Gamma^l} d\mu \, K'_B(\mu) \, (z - \mu)^{-1}. \quad (2)$$

For $z \in D^l \cap D(\Gamma^l)$ one has $M_1(z, \Gamma^l) = M_1(z) + 2\pi \mathrm{i} \, l K'_B(z)$.

Proof is reduced to the observation that the function $M_1(z, \Gamma^l)$ is holomorphic for $z \in \mathbb{C} \setminus \Gamma^l$ and coincides with $M_1(z)$ for $z \in \mathbb{C} \setminus \overline{D(\Gamma^l)}$. The last equation representing $M_1(z, \Gamma^l)$ via $M_1(z)$ is obtained from (2) using the Residue Theorem.

The latest statement of the lemma shows that the transfer function M_1 has at least two-sheeted Riemann surface. The sheet of the complex plane $\mathbb{C}$ where the transfer function $M_1(z)$ is considered together with the resolvent $(\mathbf{H} - z)^{-1}$ initially is said to be the physical sheet. The remaining sheets of the Riemann surface of M_1 are said to be unphysical sheets. In the present work we only deal with the unphysical sheets connected through the interval J^0 immediately to the physical sheet.

3. Example. To explain our reasons to introduce the function $K'_B(\mu)$ we will briefly consider an example closely related to the multichannel Schrödinger operator. Let $\mathcal{H}_0 = L_2(\mathbb{R}^n)$, $n \geq 1$, and A_0 be the Laplacian, $A_0 = -\Delta$, defined on the Sobolev space $W_2^2(\mathbb{R}^n)$. We impose no restrictions on the Hilbert space $\mathcal{H}_1$ and on the selfadjoint entry A_1 except for the assumption that $\sigma(A_1) \Subset \sigma(A_0) = [0, +\infty)$.

Let $\widetilde{b} \in L_2(\mathbb{R}^n, \mathcal{H}_1)$ be an $\mathcal{H}_1$-valued function of the variable $x \in \mathbb{R}^n$ whose Fourier transform

$$b(p) = \frac{1}{(2\pi)^{n/2}} \int_{\mathbb{R}^n} dx\, \widetilde{b}(x) \exp(\mathrm{i}\langle p, x\rangle), \quad p \in \mathbb{R}^n,$$

represents a continuous function of p with respect to the norm topology in $\mathcal{H}_1$. We define the coupling B_{10} as

$$B_{10}u_0 = \int_{\mathbb{R}^n} dx\, \widetilde{b}(x) u_0(x), \quad u_0 \in \mathcal{H}_0.$$

Obviously, the operator B_{10} is bounded and $\|B_{10}\| \leq \|\widetilde{b}\|_{L_2(\mathbb{R}^n, \mathcal{H}_1)}$. The adjoint operator B_{01} reads

$$(B_{01}u_1)(x) = \langle u_1, \widetilde{b}(x)\rangle_{\mathcal{H}_1}, \quad u_1 \in \mathcal{H}_1.$$

Denote by $\mathcal{B}_r$ the open ball in $\mathbb{R}^n$ centered at the origin and having radius r, $\mathcal{B}_r = \{p \in \mathbb{R}^n : |p| < r\}$. For $\mu > 0$ the value of the spectral function $E^0(\mu)$ of the Laplacian $-\Delta$ represents the integral operator in $L_2(\mathbb{R}^n)$ whose kernel reads as the Fourier transform of the characteristic function of the ball $\mathcal{B}_{\sqrt{\mu}}$ (see, e. g., Ref. [3], §4.2 of Ch. 8),

$$E^0(\mu; x, x') = \frac{1}{(2\pi)^n} \int_{\mathcal{B}_{\sqrt{\mu}}} dp \exp(\mathrm{i}\langle p, x - x'\rangle).$$

If $\mu \leq 0$ then $E^0(\mu; x, x') = 0$.

A simple computation in the example considered shows that, for $\mu > 0$,

$$K'_B(\mu) = \frac{d}{d\mu}\Big(B_{10}E^0(\mu)B_{01}\Big) = \frac{1}{2}\mu^{(n-2)/2} \int_{S^{n-1}} d\widehat{p}\ b(\mu^{1/2}\widehat{p})\ \langle \cdot, b(\mu^{1/2}\widehat{p})\rangle_{\mathcal{H}_1}$$

where S^{n-1} stands for the unit sphere in $\mathbb{R}^n$. Thus, for the analytic continuability of $K'_B(\mu)$ into a domain $D \subset \mathbb{C}$ surrounding the spectrum of the entry A_1 it suffices to require the analytic continuability of the function $b(p)$ into an appropriate domain of $\mathbb{C}^n$. In particular, if $b(p)$ admits an analytic continuation into a "strip" $|\operatorname{Im} p| < a$ for some $a > 0$, then the function $K'_B(\mu)$ is holomorphic in μ in the parabolic domain $\operatorname{Re}\mu > -a^2 + \frac{1}{4a^2}(\operatorname{Im}\mu)^2$ cut along the interval $(-a^2, 0]$. Notice that the holomorphy of $b(p)$ in a strip corresponds to the case of an exponentially decreasing $\widetilde{b}(x)$ as $x \to \infty$ which is often encountered in applications.

4. Factorization theorem. Let the spectrum of a linear operator $Y : \mathcal{H}_1 \to \mathcal{H}_1$ is separated from a K_B-bounded contour Γ. Then one can define the operator

$$V_1(Y, \Gamma) = \int_\Gamma d\mu\, K'_B(\mu)(Y - \mu)^{-1}.$$

Obviously, this operator is bounded,

$$\|V_1(Y,\Gamma)\| \leq \mathcal{V}_0(B,\Gamma) \sup_{\mu\in\Gamma} \|(Y-\mu)^{-1}\|.$$

In what follows we consider the equation (cf. [9])

$$X = V_1(A_1 + X, \Gamma). \tag{3}$$

This equation possesses the following important characteristic property: If X is a solution of (3) and u_1 is an eigenvector of $H_1 = A_1 + X$, $H_1 u_1 = z u_1$, then $z u_1 = A_1 u_1 + V_1(H_1,\Gamma) u_1 = A_1 u_1 + V_1(z,\Gamma) u_1$. This implies that any eigenvalue z of H_1 is automatically an eigenvalue for the continued transfer function $M_1(z,\Gamma^l)$ and u_1 is its eigenvector. Thus, having found the solution(s) of the equation (3) one obtains an effective means of studying the spectral properties of the transfer function $M_1(z,\Gamma)$ itself, referring to well known facts of Operator Theory [4, 5].

THEOREM 1 *Let a K_B-bounded contour Γ satisfy the condition*

$$\mathcal{V}_0(B,\Gamma) < \frac{1}{4}\, d_0^2(\Gamma) \tag{4}$$

where $d_0(\Gamma) = \operatorname{dist}\{\sigma(A_1),\Gamma\}$. Then Eq. (3) is uniquely solvable in any ball including operators $X : \mathcal{H}_1 \to \mathcal{H}_1$ the norms of which are bounded as $\|X\| \leq r$ with $r_{\min}(\Gamma) \leq r < r_{\max}(\Gamma)$ where

$$r_{\min}(\Gamma) = d_0(\Gamma)/2 - \sqrt{d_0^2(\Gamma)/4 - \mathcal{V}_0(B,\Gamma)}, \quad r_{\max}(\Gamma) = d_0(\Gamma) - \sqrt{\mathcal{V}_0(B,\Gamma)}. \tag{5}$$

In fact, the solution X belongs to the smallest ball $\|X\| \leq r_{\min}(\Gamma)$.

One can prove this statement making use of the Banach's Fixed Point Theorem (see [7]). One can even prove that if the index $l = \pm 1$ is fixed then, under the condition (4), the solution X does not depend on a concrete contour $\Gamma \subset D^l$. Moreover, this solution satisfies the inequality $\|X\| \leq r_0(B)$ with $r_0(B) = \inf_{\Gamma^l:\,\omega(B,\Gamma^l)>0} r_{\min}(\Gamma^l)$ where $\omega(B,\Gamma^l) = d_0^2(\Gamma^l) - 4\mathcal{V}_0(B,\Gamma^l)$. The value of $r_0(B)$ does not depend on l. But when l changes, the solution X can also change. For this reason we supply it in the following with the index l writing $X^{(l)}$. As a matter of fact, the operators $H_1^{(l)} = A_1 + X^{(l)}$, $l = \pm 1$, represent operator roots of the transfer function M_1.

THEOREM 2 *Let Γ^l be a contour satisfying the condition (4) and $H_1^{(l)} = A_1 + X^{(l)}$ where $X^{(l)}$ is the above solution of the basic equation (3). Then, for $z \in \mathbb{C} \setminus \Gamma^l$, the transfer function $M_1(z,\Gamma^l)$ admits the factorization*

$$M_1(z,\Gamma^l) = W_1(z,\Gamma^l)\,(H_1^{(l)} - z) \tag{6}$$

where $W_1(z,\Gamma^l)$ is a bounded operator in $\mathcal{H}_1$,

$$W_1(z,\Gamma^l) = I_1 - \int_{\Gamma^l} d\mu\, K'_B(\mu)(H_1^{(l)} - \mu)^{-1}(\mu - z)^{-1}.$$

Here, I_1 denotes the identity operator in $\mathcal{H}_1$. For $\operatorname{dist}\{z,\sigma(A_1)\} \le d_0(\Gamma^l)/2$ the operator $W_1(z,\Gamma^l)$ is boundedly invertible.

THEOREM 3 *The spectrum $\sigma(H_1^{(l)})$ of the operator $H_1^{(l)} = A_1 + X^{(l)}$ belongs to the closed $r_0(B)$-vicinity*

$$\mathcal{O}_{r_0}(A_1) = \{z \in \mathbb{C}:\ \operatorname{dist}\{z,\sigma(A_1)\} \le r_0(B)\}$$

of the spectrum of A_1. Moreover, the spectrum of $M_1(\,\cdot\,,\Gamma^l)$ in

$$\mathcal{O}_{d_0/2}(A_1) = \{z:\ z \in \mathbb{C},\ \operatorname{dist}\{z,\sigma(A_1)\} \le d_0(\Gamma^l)/2\}$$

is only represented by the spectrum of $H_1^{(l)}$ including the complex spectrum (in particular the resonances).

Let

$$\Omega^{(l)} = \int_{\Gamma^l} d\mu\, (H_1^{(-l)*} - \mu)^{-1} K'_B(\mu)\, (H_1^{(l)} - \mu)^{-1}$$

where Γ^l stands for a contour satisfying the condition (4).

THEOREM 4 *The operators $\Omega^{(l)}$, $l = \pm 1$, possess the following properties (cf. [8]):*

$$\|\Omega^{(l)}\| < 1, \qquad \Omega^{(-l)} = \Omega^{(l)*},$$

$$-\frac{1}{2\pi\mathrm{i}}\int_\gamma dz\, [M_1(z,\Gamma^l)]^{-1} = (I_1 + \Omega^{(l)})^{-1}, \tag{7}$$

$$-\frac{1}{2\pi\mathrm{i}}\int_\gamma dz\, z\, [M_1(z,\Gamma^l)]^{-1} = (I_1 + \Omega^{(l)})^{-1} H_1^{(-l)*} = H_1^{(l)}(I_1 + \Omega^{(l)})^{-1} \tag{8}$$

where γ stands for an arbitrary rectifiable closed contour going in the positive direction around the spectrum of $H_1^{(l)}$ inside the set $\mathcal{O}_{d_0(\Gamma)/2}(A_1)$. The integration over γ is understood in the strong sense.

The formulas (7) and (8) allow one, in principle, to find the operators $H_1^{(l)}$ and, thus, to resolve the equation (3) only using the contour integration of the inverse transfer function $[M_1(z,\Gamma^l)]^{-1}$. Also, Eq. (8) implies that the spectrum of $H_1^{(-)*}$ coincides with the spectrum of $H_1^{(+)}$.

5. Completeness and basis properties. In the following we restrict ourselves to the case where the resolvent $R_1(z) = (A_1 - z)^{-1}$ of the entry A_1 is a compact operator in $\mathcal{H}_1$ for $z \in \rho(A_1)$. Then, according to Theorem IV.3.17 of [5], the operators $H_1^{(l)}$ also have compact resolvents since $X^{(l)}$ are bounded operators.

Denote by $\mathcal{H}_{1,\lambda}^{(l)}$ the algebraic eigenspace of $H_1^{(l)}$ corresponding to an eigenvalue λ. Let m_λ be the algebraic multiplicity, $m_\lambda = \dim \mathcal{H}_{1,\lambda}^{(l)}$, $m_\lambda < \infty$, and $\psi_{\lambda,i}^{(l)}$, $i = 1, 2, \ldots, m_\lambda$, be the root vectors of $H_1^{(l)}$ forming a basis of the subspace $\mathcal{H}_{1,\lambda}^{(l)}$. Regarding these vectors we already have the following assertion representing a particular case of Theorem V.10.1 from [4].

THEOREM 5 *The closure of the linear span of the system* $\{\psi_{\lambda,i}^{(l)},\ \lambda \in \sigma(H_1^{(l)}),\ i = 1, 2, \ldots, m_\lambda\}$ *coincides with* $\mathcal{H}_1$.

We shall consider the case where the spectrum $\sigma(A_1)$ includes infinitely many points and the entry A_1 is semibounded from below. The eigenvalues $\lambda_i^{(A_1)}$, $i = 1, 2, \ldots$, of the operator A_1 are assumed to be enumerated in increasing order; $\lim\limits_{i\to\infty} \lambda_i^{(A_1)} = +\infty$. Since we assume $\sigma(A_1) \subset J^0$ this assumption means that the interval J^0 is infinite, $\mu_0^{(2)} = \infty$. Also, we suppose that there is a number i_0 such that for any $i \geq i_0$

$$\lambda_i^{(A_1)} - \lambda_{i-1}^{(A_1)} > 2r. \tag{9}$$

with some $r > r_0(B)$. Let γ_0 be a circle centered at $z = (\lambda_1^{(A_1)} + \lambda_{i_0-1}^{(A_1)})/2$ and having the radius $(\lambda_{i_0-1}^{(A_1)} - \lambda_1^{(A_1)})/2 + r$ while γ_i for $i \geq i_0$ be the circles centered at $\lambda_i^{(A_1)}$ and having the radius r. Let us introduce the projections

$$\mathrm{Q}_i^{(l)} = -\frac{1}{2\pi \mathrm{i}} \int_{\gamma_i} dz\, (H_1^{(l)} - z)^{-1}, \quad i = 0, i_0, i_0 + 1, \ldots .$$

The subspaces $\mathcal{N}_i^{(l)} = \mathrm{Q}_i^{(l)} \mathcal{H}_1$ are invariant under $H_1^{(l)}$; $\dim \mathcal{N}_i^{(l)}$ coincides with a sum of algebraic multiplicities of the eigenvalues $\lambda \in \sigma(H_1^{(l)})$ lying inside γ_i.

LEMMA 2 *Under the condition* (9) *the sequence* $\mathcal{N}_i^{(l)}$, $i = 0, i_0, i_0 + 1, \ldots,$ *is* ω*-linearly independent and complete in* $\mathcal{H}_1$.

The next theorem represents a slightly extended statement of Theorems V.4.15 and V.4.16 of [5] (the extension only concerns a possible degeneracy of the eigenvalues of A_1).

THEOREM 6 *Assume* $\lambda_{i+1}^{(A_1)} - \lambda_i^{(A_1)} \to \infty$ *as* $i \to \infty$. *Let* i_0 *be a number starting from which the inequality* $\lambda_i^{(A_1)} - \lambda_{i-1}^{(A_1)} > 4r_0$ *holds. Then the following limit exists*

$$s - \lim_{n\to\infty} \sum_{i=0, i\geq i_0}^{n} \mathrm{Q}_i^{(l)} = I_1. \tag{10}$$

Additionally, assume that $\sum_{i=1}^{\infty} (\lambda_{i+1}^{(A_1)} - \lambda_i^{(A_1)})^{-2} < \infty$. *Then* (10) *is true for any renumbering of* $\mathrm{Q}_i^{(l)}$ *and there exists a constant* C *such that* $\left\| \sum_{i\in\mathcal{I}} \mathrm{Q}_i^{(l)} \right\| \leq C$ *for any finite set* $\mathcal{I}$ *of integers* $i = 0$, $i \geq i_0$.

References

[1] ADAMJAN, V. M., and LANGER, H.: Spectral Properties of a Class of Operator-Valued Functions, J. Operator Theory **33** (1995), 259–277.

[2] ADAMYAN, V., LANGER, H., MENNICKEN, R., and SAURER, J.: Spectral Components of Selfadjoint Block Operator Matrices with Unbounded Entries, Math. Nachr. **178** (1996), 43–80.

[3] BIRMAN, M. S., and SOLOMJAK, M. Z.: Spektral'naya Teoria Samosopriazhennyh Operatorov v Gil'bertovom Prostranstve, Izdatel'stvo Leningr. Univ., Leningrad, 1980 [English translation: Spectral Theory of Selfadjoint Operators in Hilbert Space, Reidel, Dordrecht, 1987].

[4] GOHBERG, I. C., and KREIN, M. G.: Introduction to the Theory of Linear Non-selfadjoint Operators, American Mathematical Society, Providence, 1988.

[5] KATO, T. Perturbation Theory for Linear Operators, New York: Springer-Verlag, 1966.

[6] MALYSHEV, V. A., and MINLOS, R. A.: Invariant Subspaces of Clustering Operators. I., J. Stat. Phys. **21** (1979), 231–242; Invariant Subspaces of Clustering Operators. II., Comm. Math. Phys. **82** (1981), 211–226.

[7] MENNICKEN, R., and MOTOVILOV, A. K.: Operator interpretation of resonances arising in spectral problems for 2 × 2 operator matrices, LANL E-print `funct-an/9708001` (`http://xxx.lanl.gov/abs/funct-an/9708001`). Accepted for publication in Math. Nachr.

[8] MENNICKEN, R., and SHKALIKOV, A. A.: Spectral Decomposition of Symmetric Operator Matrices, Math. Nachr. **179** (1996), 259–273.

[9] MOTOVILOV, A. K.: Removal of the Resolvent-like Energy Dependence from Interactions and Invariant Subspaces of a Total Hamiltonian, J. Math. Phys. **36** (1995), 6647–6664.

[10] MOTOVILOV, A. K.: Representations for the Three–Body T–Matrix, Scattering Matrices and Resolvent on Unphysical Energy Sheets, Math. Nachr. **187** (1997), 147–210.

[11] REED, M., and SIMON, B.: Methods of Modern Mathematical Physics, IV: Analysis of Operators, Academic Press, N.Y., 1978.

AMS Classification Numbers: Primary 47A56, 47Nxx; Secondary 47N50, 47A40.

Operator Theory:
Advances and Applications, Vol. 108

On the operator-norm convergence of the Trotter-Kato product formula

H. Neidhardt[a], V.A. Zagrebnov[b,c]

[a] *Fachbereich Mathematik, Universität Potsdam, Postfach 601553, D-14415 Potsdam*
[b] *Centre de Physique Théorique, C.N.R.S., F-13288 Marseille Luminy*
[c] *Université de la Mediterranée (Aix-Marseille II), F-13288 Luminy*

1 Introduction

Let A and B be non-negative self-adjoint operators on the separable Hilbert space $\mathcal{H}$. Let

$$\mathcal{H}_0 := \overline{\mathrm{dom}(A^{1/2}) \cap \mathrm{dom}(B^{1/2})} \subseteq \mathcal{H}. \tag{1.1}$$

Then the form-sum $H = A \dot{+} B$ is a densely defined self-adjoint operator on the subspace $\mathcal{H}_0$. Kato has shown in [7] that under these conditions the Trotter product formula strongly converges to e^{-tH} uniformly in $t \in [0,T]$, $0 < T < +\infty$, i.e., one has

$$s - \lim_{n\to\infty} \left(e^{-tA/n} e^{-tB/n}\right)^n = e^{-tH} P_0, \qquad t \geq 0, \tag{1.2}$$

where P_0 is the orthogonal projection from $\mathcal{H}$ onto $\mathcal{H}_0$. Moreover, it was found by Kato in [6, 7] that the Trotter product formula is true not only for the exponential function e^{-x}, $x \geq 0$, but for a whole class of Borel measurable functions $f(\cdot)$ and $g(\cdot)$ which are defined on $\mathbf{R}^1_+ := [0,\infty)$ and characterized by

$$0 \leq f(x) \leq 1, \qquad f(0) = 1, \qquad f'(+0) = -1 \tag{1.3}$$

$$0 \leq g(x) \leq 1, \qquad g(0) = 1, \qquad g'(+0) = -1. \tag{1.4}$$

They are called Kato-functions in the following. In [7] it was proven that for Kato-functions $f(\cdot)$ and $g(\cdot)$ besides (1.2) the Trotter-Kato product formula

$$s - \lim_{n\to\infty} (f(tA/n)g(tB/n))^n = e^{-tH} P_0 \tag{1.5}$$

always converges uniformly in $t \in [0,T]$, $0 < T < +\infty$.

Naturally the problem arises whether the Trotter-Kato product formula converges in stronger topologies than the strong operator one, for example, in the operator-norm topology or in topologies of symmetrically-normed ideals of compact operators. By examples one knows that the strong operator topology cannot be replaced by the operator-norm convergence in general. One the other hand, in [8] we have found conditions on the couple A, B such that the Trotter-Kato product formula converges even in the trace norm topology. Hence it is seems naturally to ask for conditions which imply convergence in stronger topologies. Moreover, motivated by applications one is confronted with the problem how fast is the convergence, i.e., whether one can find error bounds for the convergence in a given topology. Rogava was the first who has indicated in [13] operator-norm error bounds for the exponential Trotter product formula when the operator B is relatively bounded with respect to A. This result was improved in [5] with a better error bound when the operator B is relatively bounded with respect to A^α for some α, $0 \leq \alpha < 1$. For the special case of the Schrödinger semigroup error bounds were found in [1], [3] and [4].

In the following we give a survey of our recent results [9]-[12] on convergence of the Trotter-Kato product formula in the operator-norm and in norms of symmetrically-normed ideals. In Section 2 we find necessary and sufficient conditions which guarantee that the Trotter-Kato product formula converges in the operator-norm. Using this criterion we derive more practicable conditions which imply operator-norm convergence of the Trotter-Kato product formula. These conditions can be characterized by the key words resolvent compactness, relative compactness, product resolvent compactness and smallness. Moreover, in the smallness case besides the convergence error bounds for the convergence in the operator-norm are presented. Finally, we extend the results to the convergence in symmetrically-normed ideals of compact operators.

2 Operator-norm convergence

We start by a key criterion [11] guaranteeing the *operator-norm* convergence. To this end we set

$$T(t) := f(tA)^{1/2} g(tB) f(tA)^{1/2}, \qquad t \geq 0, \tag{2.1}$$

$$R(t) := \frac{1}{t}(I - T(t)), \qquad t > 0, \tag{2.2}$$

$$F(t) := g(tB)^{1/2} f(tA) g(tB)^{1/2}, \qquad t \geq 0, \tag{2.3}$$

$$S(t) := \frac{1}{t}(I - F(t)), \qquad t > 0. \tag{2.4}$$

Further, we set $\mathcal{H}(t) := \overline{\operatorname{ran}(f(tA))} \subseteq \mathcal{H}$, $t \geq 0$. By $Q(t)$, $t \geq 0$, we denote the orthogonal projection from $\mathcal{H}$ onto $\mathcal{H}(t)$. By (1.3) one obviously has that

$$s - \lim_{t \to +0} Q(t) = I. \tag{2.5}$$

We note that for each $t \geq 0$ the projection $Q(t)$ commutes with A. We set $\tilde{f}(tA) := f(tA)|\mathcal{H}(t) : \mathcal{H}(t) \longrightarrow \mathcal{H}(t)$, $t \geq 0$. By definition we have $\ker(\tilde{f}(tA)) = \{0\}$ and

$$0 \leq \tilde{f}(tA) \leq I(t), \qquad t \geq 0, \tag{2.6}$$

where $I(t) := I|\mathcal{H}(t)$. Therefore, it makes sense to introduce the family of operators $\{M(t)\}_{t>0}$,

$$M(t) := \frac{1}{t}[\tilde{f}(tA)^{-1} - I(t)] + \frac{1}{t}Q(t)[I - g(tB)]Q(t) : \mathcal{H}(t) \longrightarrow \mathcal{H}(t), \tag{2.7}$$

with domain

$$\mathrm{dom}(M(t)) = \mathrm{ran}(f(tA)) \subseteq \mathcal{H}(t). \tag{2.8}$$

Finally, we put $I_0 := I|\mathcal{H}_0$, see (1.2). Note that in case $f(x) > 0$, $x \in \mathbf{R}^1_+$, one readily gets

$$\mathcal{H}(t) = \mathcal{H} \qquad \text{and} \qquad Q(t) = I, \qquad t \geq 0. \tag{2.9}$$

Theorem 2.1 *Let A and B be non-negative self-adjoint operators on the separable Hilbert space $\mathcal{H}$ and let $f(\cdot)$ and $g(\cdot)$ be Kato-functions. Then the following conditions are equivalent:*

(i) $\lim_{r\to+\infty} \sup_{t\in[a,b]} ||T(t/r)^r - P_0 e^{-tH} P_0|| = 0, \qquad [a,b] \subseteq (0,\infty),$

(ii) $\lim_{t\to+0} ||(\lambda I + R(t))^{-1} - P_0(\lambda I_0 + H)^{-1} P_0|| = 0, \qquad \lambda > 0,$

(iii) $\lim_{t\to+0} ||Q(t)(\lambda I(t) + M(t))^{-1} Q(t) - P_0(\lambda I_0 + H)^{-1} P_0|| = 0, \qquad \lambda > 0,$

(iv) $\lim_{t\to+0} ||(\lambda I + S(t))^{-1} - P_0(\lambda I_0 + H)^{-1} P_0|| = 0, \qquad \lambda > 0,$

(v) $\lim_{r\to+\infty} \sup_{t\in[a,b]} ||F(t/r)^r - P_0 e^{-tH} P_0|| = 0, \qquad [a,b] \subseteq (0,\infty),$

(vi) $\lim_{n\to\infty} \sup_{t\in[a,b]} ||(f(tA/n)g(tB/n))^n - P_0 e^{-tH} P_0|| = 0, \quad [a,b] \subseteq (0,\infty),$

(vii) $\lim_{n\to\infty} \sup_{t\in[a,b]} ||(g(tB/n)f(tA/n))^n - P_0 e^{-tH} P_0|| = 0, \quad [a,b] \subseteq (0,\infty).$

Below we say that the *Trotter-Kato product formula converges locally uniformly away from zero in the operator-norm for the family $\{F(t)\}_{t\geq 0}$* if the limit (*v*) exists. By Theorem 2.1 this is equivalent to the local uniform convergence in the operator-norm of the Trotter-Kato product formula for all families generated by $f(\cdot)$ and $g(\cdot)$. So we will simply say that the *Trotter-Kato product formula converges locally uniformly away from zero in the operator-norm* having in mind that it is true for all families generated by $f(\cdot)$ and $g(\cdot)$.

Using this theorem one proves more practicable theorems on operator-norm convergence of the Trotter-Kato product formula which rely on compactness assumptions. If $f(\cdot)$ is a Kato-function, then one can associate with $f(\cdot)$ another Kato-function $f_0(\cdot)$ setting

$$0 \leq \varphi_0(x) := \inf_{0<s\leq x} s^{-1}\left(\frac{1}{f(s)} - 1\right), \tag{2.10}$$

where $0^{-1} := +\infty$, and

$$f_0(x) := \begin{cases} 1 & x = 0 \\ (1 + x\varphi_0(x))^{-1} & x > 0, \end{cases} \tag{2.11}$$

In the following the set of compact operators on $\mathcal{H}$ is denoted by $\mathcal{L}_\infty(\mathcal{H})$.

Theorem 2.2 *Let A and B be non-negative self-adjoint operators on $\mathcal{H}$ and let $f(\cdot)$ and $g(\cdot)$ be Kato-functions. If condition $f_0(t_0 A) \in \mathcal{L}_\infty(\mathcal{H})$ is satisfied for some $t_0 > 0$, then the Trotter-Kato formula converges locally uniformly away from zero in the operator-norm. Moreover, one has*

$$(\lambda I_0 + H)^{-1} \in \mathcal{L}_\infty(\mathcal{H}_0) \tag{2.12}$$

for $\lambda > 0$.

This theorem possesses a converse. To this end one has to introduce the notion of a regular Kato-function. If $f(\cdot)$ is a Kato-function we set

$$b(x) := \sup_{0 \le s \le x} s f(s), \qquad x \in \mathbf{R}^1_+, \tag{2.13}$$

and

$$r(x) := \sup_{s \in [x,\infty)} f(s), \qquad x \in \mathbf{R}^1_+. \tag{2.14}$$

We note that $b(\cdot)$ is non-decreasing and $0 \le b(x) \le x$, $x \in \mathbf{R}^1_+$, while $r(x)$ is non-increasing such that $0 \le r(x) \le 1$, $x \in \mathbf{R}^1_+$. A Kato-function $f(\cdot)$ is called *regular* if the conditions

$$\lim_{x \to +\infty} \frac{b(x)}{x} = 0 \tag{2.15}$$

and

$$0 \le r(x) < 1, \qquad x > 0, \tag{2.16}$$

are satisfied. Regular Kato-functions behave very well: If the non-negative operator A is resolvent compact, i.e., one has $(I + A)^{-1} \in \mathcal{L}_\infty(\mathcal{H})$, then for each $t_0 > 0$ the condition $f_0(t_0 A) \in \mathcal{L}_\infty(\mathcal{H})$ is satisfied. This gives rise to the following theorem:

Theorem 2.3 *Let A be a non-negative self-adjoint operator. The Trotter-Kato product formula converges locally uniformly away from zero in the operator-norm for any regular Kato-function $f(\cdot)$, any Kato-function $g(\cdot)$ as well as for any non-negative self-adjoint operator B if and only if one has $(I + A)^{-1} \in \mathcal{L}_\infty(\mathcal{H})$.*

We learn from Theorem 2.3 that for the operator-norm convergence of the Trotter-Kato product formula no specific information about the non-negative self-adjoint B is necessary, if the non-negative operator A is resolvent compact. However, this gives an idea that having some information on B one can weaken the assumptions with respect to A. Indeed, this is true as the following theorem shows.

Theorem 2.4 *Let $A \ge I$ and $B \ge 0$ be two self-adjoint operators such that*

$$\mathrm{dom}(A^{1/2}) \subseteq \mathrm{dom}(B^{1/2}). \tag{2.17}$$

Further, let $f(\cdot)$ and $g(\cdot)$ be Kato-functions. If

$$C_0 := \sup_{x>0} \frac{x f(x)}{1 - f(x)} < +\infty \tag{2.18}$$

and

$$B^{1/2}A^{-1/2} \in \mathcal{L}_\infty(\mathcal{H}), \tag{2.19}$$

then the Trotter-Kato product formula converges locally uniformly away from zero in the operator-norm.

In other words, the theorem says that the *relative compactness* of the operators $A^{1/2}$ and $B^{1/2}$ implies norm-convergence of the Trotter-Kato product formula. Since the relative compactness of $A^{1/2}$ and $B^{1/2}$ does not assume that A is resolvent compact, the Theorem 2.4 is weaker than Theorem 2.3. However, the price is that A and B are interrelated. The same is true for the following theorem in which the *product* of the resolvents of the operators A and B is *compact.*

Theorem 2.5 *Let A and B be non-negative self-adjoint operators and let $f(\cdot)$ and $g(\cdot)$ be Kato-functions such that conditions (2.18) and*

$$S_0 := \sup_{x>0} \frac{xg(x)}{1-g(x)} < +\infty \tag{2.20}$$

are satisfied. If

$$(I+A)^{-1}(I+B)^{-1} \in \mathcal{L}_\infty(\mathcal{H}), \tag{2.21}$$

then the Trotter-Kato product formula converges locally uniformly away from zero in the operator-norm. Moreover, one has $(I_0+H)^{-1} \in \mathcal{L}_\infty(\mathcal{H}_0)$.

3 Smallness

Assumptions of the type that B^α is relatively bounded by A^α, $0<\alpha\le 1$, with an A^α-bound smaller than one, play a special rôle. This comes from the fact that besides the norm convergence of the Trotter-Kato product formula, in this case one can find operator-norm *error bounds.* More precisely, let us assume that

$$A \ge I \qquad \text{and} \qquad B \ge 0, \tag{3.1}$$

$$\mathrm{dom}(A^\alpha) \quad \subseteq \quad \mathrm{dom}(B^\alpha), \qquad \frac{1}{2} < \alpha \le 1, \tag{3.2}$$

and

$$\|B^\alpha f\| \le a\|A^\alpha f\|, \qquad f \in \mathrm{dom}(A^\alpha), \qquad 0<a<1. \tag{3.3}$$

Furthermore, we restrict the class of Kato-functions by the conditions

$$C_0 \quad := \quad \sup_{x>0} \frac{xf(x)^{1/2\alpha}}{1-f(x)} < +\infty, \tag{3.4}$$

$$C_2 \quad := \quad \sup_{x>0} \left| \left(f(x) - \frac{1}{1+x} \right) \frac{1}{x^2} \right| < +\infty, \tag{3.5}$$

$$S_2 \quad := \quad \sup_{x>0} \left| \left(g(x) - \frac{1}{1+x} \right) \frac{1}{x^2} \right| < +\infty. \tag{3.6}$$

Finally, we relate the Kato-functions $f(\cdot)$, $g(\cdot)$ and the relative A^α-bound of B^α by the condition

$$0 < a^{1/\alpha} C_0 S_1 < 1, \tag{3.7}$$

where the constant S_1 is given by

$$S_1 := \sup_{x>0} \frac{1-g(x)}{x} < +\infty. \tag{3.8}$$

First we consider the case $\alpha = 1$.

Theorem 3.1 *Let A and B be self-adjoint operators which satisfy (3.1) - (3.3) for $\alpha = 1$. If the Kato-functions $f(\cdot)$ and $g(\cdot)$ obey (3.4) - (3.6) and condition (3.7) is valid, then there are constants $C > 0$ and $C' > 0$ such that the estimates*

$$\left\|T(t/n)^n - e^{-tH}\right\| \le C\,\frac{\ln(n)}{n}, \qquad n = 2,3,\ldots, \tag{3.9}$$

and

$$\left\|(f(tA/n)g(tB/n))^n - e^{-tH}\right\| \le C'\,\frac{\ln(n)}{n}, \qquad n = 2,3,\ldots, \tag{3.10}$$

hold uniformly in $t \ge 0$.

Next we consider $1/2 < \alpha < 1$. In contrast to the case $\alpha = 1$, now the Theorem 3.1 splits up into several theorems.

Theorem 3.2 *Let A and B be two self-adjoint operators which obey (3.1) - (3.3) for some $\alpha \in (1/2, 1)$. If the Kato functions $f(\cdot)$ and $g(\cdot)$ satisfy (3.4) - (3.6) and (3.7), then there are constants $C > 0$ and $C' > 0$ such that*

$$\left\|T(t/n)^n - e^{-tH}\right\| \le C\frac{1}{n^{\alpha-1/2}}, \qquad n = 1,2,\ldots, \tag{3.11}$$

and

$$\left\|(f(tA/n)g(tB/n))^n - e^{-tH}\right\| \le C'\frac{1}{n^{\alpha-1/2}}, \qquad n = 1,2,\ldots, \tag{3.12}$$

uniformly in $t \ge 0$.

However, away from zero the error bound can be improved.

Theorem 3.3 *Let A and B be two self-adjoint operators which obey (3.1) - (3.3) for some $\alpha \in (1/2, 1)$. If the Kato functions $f(\cdot)$ and $g(\cdot)$ satisfy (3.4) - (3.6) and (3.7), then for each $\epsilon > 0$ there are constants $C_\epsilon > 0$ and $C'_\epsilon > 0$ such that*

$$\left\|T(t/n)^n - e^{-tH}\right\| \le C_\epsilon\frac{\ln(n)}{n^{2\alpha-1}}, \qquad n = 2,3,\ldots, \tag{3.13}$$

and

$$\left\|(f(tA/n)g(tB/n))^n - e^{-tH}\right\| \le C'_\epsilon\frac{\ln(n)}{n^{2\alpha-1}}, \qquad n = 2,3,\ldots, \tag{3.14}$$

uniformly in $t \ge \epsilon$.

One can improve the estimates if it is known that

$$\mathrm{dom}(H^\alpha) \subseteq \mathrm{dom}(A^\alpha). \tag{3.15}$$

In this case one has the following theorem:

Theorem 3.4 *Let A and B be self-adjoint operators which obey (3.1) - (3.3) for some $\alpha \in (1/2, 1)$. If the Kato functions $f(\cdot)$ and $g(\cdot)$ satisfy (3.4) - (3.6) and (3.7) and if in addition the condition (3.15) takes place, then there are constants $D > 0$ and $D' > 0$ such that*

$$\left\| T(t/n)^n - e^{-tH} \right\| \leq D \frac{1}{n^{2\alpha-1}}, \qquad n = 1, 2 \ldots, \tag{3.16}$$

and

$$\left\| (f(tA/n)g(tB/n))^n - e^{-tH} \right\| \leq D' \frac{1}{n^{2\alpha-1}}, \qquad n = 1, 2, \ldots, \tag{3.17}$$

uniformly in $t \geq 0$.

The condition (3.15) is satisfied if, for example, one has

$$\mathrm{dom}(A^\beta) \subseteq \mathrm{dom}(B^\alpha), \qquad 0 < \beta < \alpha < 1, \tag{3.18}$$

or, in particular,

$$\mathrm{dom}(A^{1/2}) \subseteq \mathrm{dom}(B^\alpha), \qquad \frac{1}{2} < \alpha < 1. \tag{3.19}$$

4 Symmetrically-normed ideals

In order to be self-contained we recall some of the basic notions of symmetrically-normed ideals [2]. Let c_0 be the space of all sequences $\xi = \{\xi_j\}_{j=1}^\infty$ of real numbers which tend to zero. We denote by c_f the subspace of c_0 consisting of all sequences with finite number of non-zero terms (finite sequences). A real-valued function $\phi(\xi) = \phi(\xi_1, \xi_2, \ldots)$ defined on c_f is called a *norming function* if it has the following properties:

$$\phi(\xi) > 0, \qquad \forall \xi \in c_f, \quad \xi \neq 0, \tag{4.1}$$

$$\phi(\alpha\xi) = |\alpha|\phi(\xi), \qquad \forall \xi \in c_f, \quad \forall \alpha \in bR, \tag{4.2}$$

$$\phi(\xi + \eta) \leq \phi(\xi) + \phi(\eta), \qquad \forall \xi, \eta \in c_f, \tag{4.3}$$

$$\phi(1, 0, \ldots) = 1. \tag{4.4}$$

A norming function ϕ is called *symmetric* if it has the additional property

$$\phi(\xi_1, \xi_2, \ldots, \xi_n, 0, 0, \ldots) = \phi(|\xi_{j_1}|, |\xi_{j_2}|, \ldots, |\xi_{j_n}|, 0, 0, \ldots) \tag{4.5}$$

for any $\xi \in c_f$ and any permutation $j_1, j_2, \ldots, j_n$ of integers $1, 2, \ldots, n$. Let $\xi = \{\xi_1, \xi_2, \ldots\} \in c_0$ and let

$$\xi^{(n)} := \{\xi_1, \xi_2, \ldots, \xi_n, 0, 0, \ldots\}. \tag{4.6}$$

If ϕ is a symmetric norming function we set

$$c_\phi := \{\xi \in c_0 : \sup_n \phi(\xi^{(n)}) < +\infty\}. \tag{4.7}$$

Therefore, one gets

$$c_f \subseteq c_\phi \subseteq c_0. \tag{4.8}$$

By Lemma III.3.1 of [2] it follows that the limit

$$\phi(\xi) := \lim_{n\to\infty} \phi(\xi^{(n)}), \qquad \xi \in c_\phi, \tag{4.9}$$

exists. The space c_ϕ is linear over real numbers.

If X is a compact operator, i.e. $X \in \mathcal{L}_\infty(\mathcal{H})$, then by

$$s(X) := \{s_1(X), s_2(X), \ldots\} \tag{4.10}$$

we denote the sequence of singular numbers of X counting multiplicities. We always assume that

$$s_1(X) \geq s_2(X) \geq \ldots \geq s_n(X) \geq \ldots . \tag{4.11}$$

Let ϕ be a symmetric norming function. We set

$$\mathcal{L}_\phi(\mathcal{H}) := \{X \in \mathcal{L}_\infty(\mathcal{H}) : s(X) \in c_\phi\}. \tag{4.12}$$

The set $\mathcal{L}_\phi(\mathcal{H})$ is a two-sided ideal in the algebra $\mathcal{L}(\mathcal{H})$ of all bounded operators on $\mathcal{H}$. Setting

$$\|X\|_\phi := \phi(s(X)), \qquad X \in \mathcal{L}_\phi(\mathcal{H}), \tag{4.13}$$

one obtains a norm on the two sided ideal $\mathcal{L}_\phi(\mathcal{H})$ such that $\mathcal{L}_\phi(\mathcal{H})$ becomes a Banach space. One always has that

$$\mathcal{L}_1(\mathcal{H}) \subseteq \mathcal{L}_\phi(\mathcal{H}) \subseteq \mathcal{L}_\infty(\mathcal{H}) \tag{4.14}$$

and

$$\|X\|_\phi \leq \|X\|_1, \qquad X \in \mathcal{L}_1(\mathcal{H}), \tag{4.15}$$

where $\mathcal{L}_1(\mathcal{H})$ denotes the set of trace-class operators on $\mathcal{H}$. We say that the *Trotter-Kato product formula converges locally uniformly away from $t_0 > 0$ in $\mathcal{L}_\phi(\mathcal{H})$ for the family* $\{F(t)\}_{t\geq 0}$ (similarly for the families $\{T(t)\}_{t\geq 0}$, $\{f(tA)g(tB)\}_{t\geq 0}$ and $\{g(tB)f(tA)\}_{t\geq 0}$) if for any bounded interval $[\tau, \tau_0] \subseteq (t_0, +\infty)$ there is a $r_0 \geq 1$ such that

$$e^{-tH} \in \mathcal{L}_\phi(\mathcal{H}_0) \qquad \text{and} \qquad F(t/r)^r \in \mathcal{L}_\phi(\mathcal{H}) \tag{4.16}$$

for $t \in [\tau_0, \tau]$ and $r \geq r_0$ as well as

$$\lim_{r\to\infty} \sup_{t\in[\tau_0,\tau]} \|F(t/r)^r - e^{-tH}P_0\|_\phi = 0. \tag{4.17}$$

Similarly to Theorem 2.1 we find that it is enough to consider only the sequence $\{F(t/r)\}_{r\geq 1}^{\infty}$. This results from the following theorem.

Theorem 4.1 *Let A and B be non-negative self-adjoint operators on the separable Hilbert space $\mathcal{H}$ and let $f(\cdot)$ and $g(\cdot)$ be Kato-functions. If the Trotter-Kato product formula converges locally uniformly away from $t_0 > 0$ in $\mathcal{L}_\phi(\mathcal{H})$ for one of the families $\{F(t)\}_{t\geq 0}$, $\{T(t)\}_{t\geq 0}$, $\{g(tB)f(tA)\}_{t\geq 0}$ or $\{f(tA)g(tB)\}_{t\geq 0}$, then the Trotter-Kato product formula converges locally uniformly away from t_0 in $\mathcal{L}_\phi(\mathcal{H})$ for all these families.*

Furthermore, it is useful to restrict the class of Kato-functions to the *dominated* Kato-functions: *Kato-functions $f(\cdot)$ and $g(\cdot)$ are called dominated by Borel measurable functions $f^D(\cdot) : \mathbf{R}^1_+ \longrightarrow \mathbf{R}^1_+$ and $g^D(\cdot) : \mathbf{R}^1_+ \longrightarrow \mathbf{R}^1_+$*, respectively, if

$$f(qx)^{1/q} \leq f^D(x) \qquad \text{and} \qquad g(qx)^{1/q} \leq g^D(x), \qquad 0 < q \leq 1, \tag{4.18}$$

for $x \geq 0$. Further, we use the notation

$$F^D(t_0) := g^D(t_0B)^{1/2} f^D(t_0A) g^D(t_0B)^{1/2}. \tag{4.19}$$

Now we can extend Theorem 2.3 to the symmetrically-normed ideals:

Theorem 4.2 *Let A and B be non-negative self-adjoint operators on the separable Hilbert space $\mathcal{H}$ and let $f^D(\cdot) : \mathbf{R}^1_+ \longrightarrow \mathbf{R}^1_+$ be a Borel measurable function such that $f^D(t_0A) \in \mathcal{L}_\phi(\mathcal{H})$ for some $t_0 > 0$. If the Kato-function $f(\cdot)$ is regular and dominated by $f^D(\cdot)$ and $g(\cdot)$ is any Kato-function, then the Trotter-Kato product formula converges locally uniformly away from t_0 in $\mathcal{L}_\phi(\mathcal{H})$ for all families generated by $f(\cdot)$ and $g(\cdot)$.*

Kato-functions $f(\cdot)$ and $g(\cdot)$ are called *self-dominated Kato-functions* if

$$f(qx)^{1/q} \leq f(x) \qquad \text{and} \qquad g(qx)^{1/q} \leq g(x), \qquad 0 < q \leq 1, \tag{4.20}$$

for $x \geq 0$. Using this notion Theorem 4.2 can be reformulated as follows:

Theorem 4.3 *Let A and B be non-negative self-adjoint operators on the separable Hilbert space $\mathcal{H}$ and let $f^D(\cdot)$ be a self-dominated Kato-function such that $f^D(t_0A) \in \mathcal{L}_\phi(\mathcal{H})$ for some $t_0 > 0$. Then for any Kato-function $f(\cdot)$ obeying $f(x) \leq f^D(x)$, $x \geq 0$, and any Kato-function $g(\cdot)$ the Trotter-Kato product formula converges locally uniformly away from t_0 in $\mathcal{L}_\phi(\mathcal{H})$ for all families generated by $f(\cdot)$ and $g(\cdot)$.*

Obviously, Theorem 2.4 has no extension to symmetrically-normed ideals since relative compactness does not imply that $\{F(t/r)^r\}_{r=1}^{\infty}$ belongs to some symmetrically-normed ideal for $t > 0$ and $r \geq 1$ large enough. However, Theorem 2.5 admits an extension to the symmetrically-normed ideals.

Theorem 4.4 *Let A and B be non-negative self-adjoint operators on the separable Hilbert space $\mathcal{H}$ and let $f^D(\cdot) : \mathbf{R}^1_+ \longrightarrow \mathbf{R}^1_+$ and $g^D(\cdot) : \mathbf{R}^1_+ \longrightarrow \mathbf{R}^1_+$ be Borel measurable functions such that $F^D(t_0) \in \mathcal{L}_\phi(\mathcal{H})$ for some $t_0 > 0$. If the Kato-functions $f(\cdot)$ and $g(\cdot)$ are dominated by $f^D(\cdot)$ and $g^D(\cdot)$, respectively, and satisfy conditions (2.18) and (2.20), then the Trotter-Kato product formula converges locally uniformly away from t_0 in $\mathcal{L}_\phi(\mathcal{H})$ for all families generated by $f(\cdot)$ and $g(\cdot)$.*

Similar to Theorem 4.3 the Theorem 4.4 admits a reformulation for *self-dominated* Kato-functions.

Corollary 4.5 *Let A and B be non-negative self-adjoint operators on the separable Hilbert space $\mathcal{H}$ and let $f^D(\cdot) : \mathbf{R}^1_+ \longrightarrow \mathbf{R}^1_+$ and $g^D(\cdot) : \mathbf{R}^1_+ \longrightarrow \mathbf{R}^1_+$ be self-dominated Kato-functions such that conditions (2.18) and (2.20) as well as $F^D(t_0) \in \mathcal{L}_\phi(\mathcal{H})$ for some $t_0 > 0$ are satisfied. If the Kato-functions $f(\cdot)$ and $g(\cdot)$ obey $f(x) \leq f^D(x)$ and $g(x) \leq g^D(x)$ for $x \geq 0$, then the Trotter-Kato product formula converges locally uniformly away from t_0 in $\mathcal{L}_\phi(\mathcal{H})$ for all families generated by $f(\cdot)$ and $g(\cdot)$.*

For the self-dominated Kato-functions $f^D(x) = g^D(x) = (1+x)^{-1}$ and $f^D(x) = g^D(x) = e^{-x}$ we find the following corollaries:

Corollary 4.6 *Let A and B be non-negative self-adjoint operators on the separable Hilbert space $\mathcal{H}$ such that*

$$(I+B)^{-1/2}(I+A)^{-1}(I+B)^{-1/2} \in \mathcal{L}_\phi(\mathcal{H}). \tag{4.21}$$

If $f(\cdot)$ and $g(\cdot)$ are Kato-functions such that $f(x) \leq \frac{1}{1+x}$ and $g(x) \leq \frac{1}{1+x}$ for $x \geq 0$, then the Trotter-Kato product formula converges locally uniformly away from zero (sic!) in $\mathcal{L}_\phi(\mathcal{H})$ for all families generated by $f(\cdot)$ and $g(\cdot)$.

and

Corollary 4.7 *Let A and B be non-negative self-adjoint operators on the separable Hilbert space $\mathcal{H}$ such that*

$$e^{-t_0 B/2} e^{-t_0 A} e^{-t_0 B/2} \in \mathcal{L}_\phi(\mathcal{H}). \tag{4.22}$$

If $f(\cdot)$ and $g(\cdot)$ are Kato-functions such that $f(x) \leq e^{-x}$ and $g(x) \leq e^{-x}$ for $x \geq 0$, then the Trotter-Kato product formula converges locally uniformly away from t_0 in $\mathcal{L}_\phi(\mathcal{H})$ for all families generated by $f(\cdot)$ and $g(\cdot)$.

Finally, it is possible to characterize the convergence in symmetrically-normed ideals in a *necessary and sufficient* manner:

Theorem 4.8 *Let A and B be non-negative self-adjoint operators on the separable Hilbert space $\mathcal{H}$ and let $f(\cdot)$ and $g(\cdot)$ be self-dominated Kato-function which satisfy conditions (2.18)and (2.20). Then there is a $t_0 > 0$ such that the Trotter-Kato product formula converges locally uniformly away from $t_0 > 0$ in $\mathcal{L}_\phi(\mathcal{H})$ for all families generated by $f(\cdot)$ and $g(\cdot)$ if and only if there is $s_0 > 0$ and an integer p such that $F(s_0)^p \in \mathcal{L}_\phi(\mathcal{H})$.*

Specified to the ideal of the trace-class operators the Theorem 4.8 takes the form:

Corollary 4.9 *Let A and B be non-negative self-adjoint operators on the separable Hilbert space $\mathcal{H}$ and let $f(\cdot)$ and $g(\cdot)$ be self-dominated Kato-function which satisfy conditions (2.18) and (2.20). Then there is a $t_0 > 0$ such that the Trotter-Kato product formula converges locally uniformly in $\mathcal{L}_1(\mathcal{H})$ away from $t_0 > 0$ for all families generated by $f(\cdot)$ and $g(\cdot)$ if and only if there is a $s_0 > 0$ and an integer p such that $F(s_0) \in \mathcal{L}_p(\mathcal{H})$.*

Notice that for the *self-dominated* Kato-functions the Theorem 2.5 can be reformulated as follows:

Theorem 4.10 *Let A and B be non-negative self-adjoint operators on the separable Hilbert space $\mathcal{H}$ and let $f(\cdot)$ and $g(\cdot)$ self-dominated Kato-function which satisfy conditions (2.18) and (2.20). If there is a $t_0 > 0$ such that $F(t_0) \in \mathcal{L}_\infty(\mathcal{H})$, then the Trotter-Kato product formula converges locally uniformly away from zero in the operator norm for all families generated by $f(\cdot)$ and $g(\cdot)$.*

Finally, Theorems 3.1 and 3.3 admit the following generalization to symmetrically-normed ideals:

Theorem 4.11 *Let A and B be self-adjoint operators which satisfy conditions (3.1)-(3.3). Further, let $f^D(\cdot) : \mathbf{R}^1_+ \longrightarrow \mathbf{R}^1_+$ and $g^D(\cdot) : \mathbf{R}^1_+ \longrightarrow \mathbf{R}^1_+$ be bounded Borel measurable functions such that $F^D(t_0) \in \mathcal{L}_\phi(\mathcal{H})$ for some $t_0 > 0$. Assume that the Kato-functions $f(\cdot)$ and $g(\cdot)$ are dominated by $f^D(\cdot)$ and $g^D(\cdot)$. If for some $\alpha \in (1/2, 1]$ the Kato-functions $f(\cdot)$ and $g(\cdot)$ satisfy conditions (3.4) - (3.7), then the Trotter-Kato product formula converges uniformly away from $2t_0$ in $\mathcal{L}_\phi(\mathcal{H})$ for $\{T(t)\}_{t\geq 0}$ and $\{f(tA)g(tB)\}_{t\geq 0}$. Moreover, there are constants $C_{\alpha,t_0} > 0$ and $C'_{\alpha,t_0} > 0$ (depending on α and t_0) such that for any bounded interval $[\tau_0, \tau] \subseteq (2t_0, \infty)$ and sufficiently large n one has the estimates*

$$\sup_{t\in[\tau_0,\tau]} \|T(t/n)^n - e^{-tH}\|_\phi \leq C_{\alpha,t_0} \frac{\ln(n)}{n^{2\alpha-1}} \tag{4.23}$$

and

$$\sup_{t\in[\tau_0,\tau]} \|(f(tA/n)g(tB/n))^n - e^{-tH}\|_\phi \leq C'_{\alpha,t_0} \frac{\ln(n)}{n^{2\alpha-1}}. \tag{4.24}$$

If $\alpha \in (1/2, 1)$ and in addition the condition (3.15) is satisfied, then for any bounded interval $[\tau_0, \tau] \subseteq (2t_0, \infty)$ and sufficiently large n the stronger estimates

$$\sup_{t\in[\tau_0,\tau]} \|T(t/n)^n - e^{-tH}\|_\phi \leq C_{\alpha,t_0} \frac{1}{n^{2\alpha-1}} \tag{4.25}$$

and

$$\sup_{t\in[\tau_0,\tau]} \|(f(tA/n)g(tB/n))^n - e^{-tH}\|_\phi \leq C'_{\alpha,t_0}\frac{1}{n^{2\alpha-1}} \tag{4.26}$$

are valid.

References

[1] Doumeki, A.; Ichinose, T.; Tamura, H.: Error bounds on exponential product formulas for Schrödinger operators. *J. Math. Soc. Japan.* **50** (1998), 359-377.

[2] Gohberg, I.C.; Krein, M.G.: *Introduction to the theory of linear nonselfadjoint operators.* AMS, Providence, Rhode Island 02904, 1969.

[3] Helffer, B.: Around the transfer operator and the Trotter-Kato formula. In *Operator Theory: Advances and Appl.* **78** (1995), 161-175.

[4] Ichinose, T.; Takanobu, S.: Estimates of the difference between the Kac operator and the Schrödinger semigroup. *Commun. Math. Phys.* **27** (1997), 167-197.

[5] Ichinose, T.; Tamura, H.: Error estimate in operator norm for Trotter-Kato product formula. *Integr. Equ. Oper. Theory* **27** (1997), 195-207.

[6] Kato, T: On the Trotter-Lie product formula. *Proc. Japn. Acad.* **50** (1974), 694-698.

[7] Kato, T.: Trotter's product formula for an arbitrary pair of self-adjoint contraction semigroups. *Topics in Funct. Anal., Ad. Math. Suppl. Studies* Vol. **3**, 185-195 (I.Gohberg and M.Kac eds.). Acad. Press, New York 1978.

[8] Neidhardt, H.; Zagrebnov, V.A.: The Trotter-Kato product formula for Gibbs semigroups. *Commun. Math. Phys.* **131** (1990), 333-346.

[9] Neidhardt, H.; Zagrebnov, V.A.: On error estimates for the Trotter-Kato product formula. *Lett. Math. Phys.* **44** (1998), 169-186.

[10] Neidhardt, H.; Zagrebnov, V.A.: Fractional powers of self-adjoint operators and Trotter-Kato product formula. Submitted to *Integr. Equ. Oper. Theory.*

[11] Neidhardt, H.; Zagrebnov, V.A.: Trotter-Kato product formula and operator norm convergence. Preprint CPT-98/P.3670, Marseille, June 1998.

[12] Neidhardt, H.; Zagrebnov, V.A.: Trotter-Kato product formula and symmetrically-normed ideals. Preprint CPT-98/P.3671, Marseille, June 1998.

[13] Rogava, D.L.: Error bounds for Trotter-type formulas for self-adjoint operators. *Funct. Anal. Application* **27**, No. 3 (1993), 217-219.

Operator Theory:
Advances and Applications, Vol. 108

A Particular Case of The Inverse Problem for The Sturm-Liouville Equation with Parameter-Dependent Potential

Vyacheslav Pivovarchik
Odessa State Academy of Civil Engineering and Architecture
Didrihson str. 3, 270045 Odessa, Ukraine

Abstract

An inverse problem for the Sturm-Liouville equation with a potential depending on the spectral parameter is considered. It is proved that the spectra of the corresponding Dirichlet problems on $[0, b]$ and $[b, a]$ and that of the Dirichlet-Neumann problem on $[0, a]$ $(a > b)$ uniquely determine the potential almost everywhere. These problems arise when one considers small vibrations of a smooth inhomogeneous partially damped string clamped at an inner point.

1 Introduction

Consider the problem

$$\frac{\partial}{\partial s}\left(A(s)\frac{\partial v}{\partial s}\right) - p(s)\frac{\partial v}{\partial t} - \frac{\partial^2 v}{\partial t^2} = 0 \tag{1.1}$$

$$v(0,t) = \left.\frac{\partial v(s,t)}{\partial s}\right|_{s=l} = 0 \tag{1.2}$$

describing small transverse vibrations of a string of stiffness $A(s) \geq \varepsilon > 0$ and damping coefficient $p(s) \geq 0$. Here $v(s,t)$ is the transverse displacement and $l > 0$ is the length of the string. The left end of the string is fixed while the right end is free. Substituting $v(s,t) = u(\lambda, s)e^{i\lambda t}$ into (1.1)-(1.2) we obtain

$$\frac{\partial}{\partial s}\left(A(s)\frac{\partial u}{\partial s}\right) - i\lambda p(s)u + \lambda^2 u = 0 \tag{1.3}$$

$$u(\lambda, 0) = \left.\frac{\partial u(\lambda, s)}{\partial s}\right|_{s=l} = 0. \tag{1.4}$$

Let the stiffness be smooth: $A(s) \in W_2^2(0,l)$. Then by means of Liouville transform

$$x(s) = \int_0^s A^{-\frac{1}{2}}(s')ds', \tag{1.5}$$

$$y(\lambda, x) = A^{\frac{1}{4}}(s(x))\, u(\lambda, s(x)), \tag{1.6}$$

we obtain

$$y'' + (\lambda^2 - i\lambda p(x) - q(x))\, y = 0, \tag{1.7}$$

$$y(0) = y'(a) + \beta y(a) = 0, \tag{1.8}$$

where

$$q(x) = A^{-\frac{1}{4}}(x)\frac{d^2}{dx^2}A^{\frac{1}{4}}(x),$$

$$\beta = -\frac{1}{4}A^{-1}(x)\left.\frac{dA(x)}{dx}\right|_{x=a},$$

$$a = \int_0^l A^{-\frac{1}{2}}(s)ds,$$

and the operator F_β defined by the formulae $F_\beta = -y'' + q(x)y$; $D(F_\beta) = \{y; y \in W_2^2(0,a), y(0) = y'(a) + \beta y(a) = 0\}$ is strictly positive due to the condition $A(x) \geq \varepsilon > 0$.

Inverse problems for equation (1.7) on a semiaxis and on axis were considered in [1]-[3] where the Marchenko method [4] was generalized for the case of a potential linear in spectral parameter. Sufficient conditions for uniqueness of solution of the mentioned inverse problem on an axis were obtained in [5]. For the case of a purely imaginary $p(x)$ the inverse problem on a finite interval by two spectra was solved in [6].

In the present paper we consider a simplified version of the problem (1.7), (1.8), namely

$$y'' + (\lambda^2 - i\lambda p(x) - q(x))\, y = 0, \tag{1.9}$$

$$y(\lambda, 0) = y'(\lambda, a) = 0, \tag{1.10}$$

where

1)

$$p(x) = \begin{cases} p = const > 0, & if \quad x \in [0,b] \\ 0, & if \quad x \in (b,a], \end{cases}$$

and

2) a real-valued $q(x) \in L_2(0,a)$ and the corresponding operator F_0 is strictly positive.

From the physical point of view $\beta = 0$ means that $\left.\frac{dA(s)}{ds}\right|_{s=l} = 0$. The condition 2) means that the damping is piecewise constant and it is absent on $[b,a]$. Here $b = \int_0^{l_1} A^{-1/2}(s)ds$ and $l_1 \in (0,l)$. The spectrum of problem (1.9), (1.10) consists of normal eigenvalues which we denote $\{\lambda_k\}_{-\infty}^{\infty}{}_{k\neq 0}$ ($\lambda_{-k} = -\bar{\lambda}_k$ for λ_k being not purely imaginary, $Re\ \lambda_{k+1} \geq Re\ \lambda_k$). The spectrum does not determine uniquely the set $\{a, b, p, q(x)\}$. Hence we consider two additional problems

$$y'' + \left(\lambda^2 - i\lambda p - q(x)\right) y = 0, \tag{1.11}$$

$$y(\lambda, 0) = y(\lambda, b) = 0, \tag{1.12}$$

and

$$y'' + \left(\lambda^2 - q(x)\right) y = 0, \tag{1.13}$$

$$y(\lambda, b) = y(\lambda, a) = 0. \tag{1.14}$$

The two problems arise if the same string is clamped at the point l_1 separating the damped part of the string from the free one and if the right end of the string is fixed.

The spectra of problems (1.11), (1.12) and (1.13), (1.14) consist of normal eigenvalues accumulating to infinity. Denote $\{\nu_k\}_{-\infty}^{\infty}{}_{k\neq 0}$ $\left(\left\{\nu_k^{(1)}\right\}_{\substack{k\neq 0\\-\infty}}^{\infty}\right)$ the spectrum of problem (1.11), (1.12) (that of problem (1.13), (1.14), respectively).

Inverse problems by three spectra were considered in [7, 8] (with $p = 0$). In what follows we consider the direct problem, i.e. the description of the triplet of spectra $\{\lambda_k\}_{-\infty}^{\infty}{}_{k\neq 0}$, $\{\nu_k\}_{-\infty}^{\infty}{}_{k\neq 0}$, $\left\{\nu_k^{(1)}\right\}_{\substack{k\neq 0\\-\infty}}^{\infty}$ and show that the triplet determines uniquely the constants a, b, p and the potential $q(x)$ almost everywhere. We describe also the procedure constructing the set $\{a, b, p, q(x)\}$.

2 Direct problem

Lemma 2.1.
For all k, $Im\ \nu_k^{(1)} = 0$, $Im\ \nu_k > 0$, $Im\ \lambda_k > 0$ holds.
Proof.
The first equality is a consequence of the inequality $F_0 >> 0$. The inequality $Im\ \nu_k > 0$ can be easily proved reducing the problem (1.11), (1.12) to the self-adjoint one by means of the transformation $\tau^2 = \lambda^2 - ip\lambda$. Using the results of [9] it can be easily proved that $Im\ \lambda_k \geq 0$. To prove the inequality $Im\ \lambda_k > 0$ suppose λ_k to be a real eigenvalue of problem (1.9), (1.10), and y_k to be the corresponding eigenvector. Then

$$\int_0^a y_k'' \bar{y}_k dx + \int_0^a \left(\lambda_k^2 - q(x)\right) |y(k)|^2 dx + i\lambda_k p \int_0^b |y_k|^2 dx = 0,$$

and, consequently, $\int_0^b |y_k|^2 dx = 0$, and $y_k(b) = y_k'(b) = 0$ what is impossible. Lemma 2.1 is proved.

Consider the following boundary problem

$$y'' + \left(\lambda^2 - i\lambda p(x) - q(x)\right) y = 0, \tag{2.1}$$

$$y(\lambda, 0) = y(\lambda, a) = 0, \tag{2.2}$$

and

$$y'' + \left(\lambda^2 - q(x)\right) y = 0, \tag{2.3}$$

$$y(\lambda, b) = y'(\lambda, a) = 0, \tag{2.4}$$

Denote $\{\xi_k\}_{\substack{k\neq 0\\ -\infty}}^{\infty}$ $\left(\left\{\mu_k^{(1)}\right\}_{\substack{k\neq 0\\ -\infty}}^{\infty}\right)$ the spectrum of problem (2.1), (2.2) (that of problem (2.3), (2.4), respectively).

Lemma 2.2.
For all k, $Im\ \mu_k^{(1)} = 0,\ Im\ \xi_k > 0$ holds.
The proof of Lemma 2.2 is analogous to that of Lemma 2.1.

Lemma 2.3.
$\left\{\nu_k^{(1)}\right\}_{\substack{k\neq 0\\ -\infty}}^{\infty} \bigcap \{\lambda_k\}_{\substack{k\neq 0\\ -\infty}}^{\infty} = \emptyset$, $\left\{\nu_k^{(1)}\right\}_{\substack{k\neq 0\\ -\infty}}^{\infty} \bigcap \{\nu_k\}_{\substack{k\neq 0\\ -\infty}}^{\infty} = \emptyset$, $\{\nu_k\}_{\substack{k\neq 0\\ -\infty}}^{\infty} \bigcap \{\lambda_k\}_{\substack{k\neq 0\\ -\infty}}^{\infty} = \emptyset$.

Proof.
The first and the second statements of Lemma 2.2 follow from Lemma 2.1. Denote $s(\lambda, x)$ $(s_1(\lambda, x))$ the solution of equation (1.9) satisfying the conditions $s(\lambda, 0) = s'(\lambda, 0) - 1 = 0$ $(s_1(\lambda, b) = s_1'(\lambda, b) - 1 = 0)$. Denote $c_1(\lambda, x)$ the solution of (1.9) satisfying the conditions $c_1(\lambda, b) - 1 = c_1'(\lambda, b) = 0$. It is easy to check the identities

$$s(\lambda, a) = s'(\lambda, b)s_1(\lambda, a) + s(\lambda, b)c_1(\lambda, a), \tag{2.5}$$

$$s'(\lambda, a) = s'(\lambda, b)s_1'(\lambda, a) + s(\lambda, b)c_1'(\lambda, a). \tag{2.6}$$

Multiplying (2.5) by $s_1'(\lambda, a)$ and using (2.6) and the identity

$$c_1(\lambda, a)s_1'(\lambda, a) = 1 + s_1(\lambda, a)c_1'(\lambda, a), \tag{2.7}$$

we obtain

$$s(\lambda, a)s_1'(\lambda, a) = s_1(\lambda, a)s'(\lambda, a) + s(\lambda, b). \tag{2.8}$$

Suppose $\nu_k = \lambda_p$. Then $s'(\lambda_p, a) = s(\lambda_p, b) = s(\nu_k, b) = 0$, and, consequently, $s(\lambda_p, a)s_1'(\lambda_p, a) = 0$. But $s(\lambda_p, a) \neq 0$ because $s'(\lambda_p, a) = 0$, and $s_1'(\lambda_p, a) \neq 0$ because all the zeroes of $s_1'(\lambda, a)$ are real and $Im\ \lambda_p > 0$. This is a contradiction. Lemma 2.3 is proved.

3 Inverse problem

Here we describe how to construct the set $\{a, b, p, q(x)\}$ by the triplet of spectra and prove the uniqueness of the set $\{a, b, p, q(x)\}$. Denote $T(a, b, p, q(x))$ the triplet of spectra $\{\lambda_k\}_{\substack{k\neq 0\\-\infty}}^{\infty}$, $\{\nu_k\}_{\substack{k\neq 0\\-\infty}}^{\infty}$, $\left\{\nu_k^{(1)}\right\}_{\substack{k\neq 0\\-\infty}}^{\infty}$ of problems (1.9), (1.10); (1.11), (1.12); (1.13), (1.14) respectively, with the parameters $a > 0$, $b \in (0, a)$, $p > 0$ and the potential $q(x)$ satisfying the condition 2).

Theorem.
If $T(a_1, b_1, p_1, q_1(x)) = T(a_2, b_2, p_2, q_2(x))$, then $a_1 = a_2$, $b_1 = b_2$, $p_1 = p_2$ and $q_1(x) \overset{a.e.}{=} q_2(x)$.
Proof and construction of $\{a, b, p, q(x)\}$ from $T(a, b, p, q(x))$.
Let
$T(a_1, b_1, p_1, q_1(x)) = T(a_2, b_2, p_2, q_2(x)) = \left\{\{\lambda_k\}_{\substack{k\neq 0\\-\infty}}^{\infty}, \{\nu_k\}_{\substack{k\neq 0\\-\infty}}^{\infty}, \left\{\nu_k^{(1)}\right\}_{\substack{k\neq 0\\-\infty}}^{\infty}\right\}$.
Then using known asymptotic formulae ([4], p.233) we obtain

$$b_1 = b_2 = \lim_{k\to\infty} \frac{\pi k}{\nu_k}, \quad a_1 = a_2 = b_1 + \lim_{k\to\infty} \frac{\pi k}{\nu_k^{(1)}}, \quad p_1 = p_2 = -2i \lim_{k\to\infty} \left(\nu_k - \frac{\pi k}{b_1}\right).$$

The proof of the next equality can be found in [4] (p.225)

$$s_1(\lambda, a) = (b - a) \prod_{k=1}^{\infty} \left(\frac{a - b}{\pi k}\right)^2 \left(\nu_k^{(1)2} - \lambda^2\right). \tag{3.1}$$

Making the parameter transformation $\tau^2 = \lambda^2 - ip\lambda$ and using the same result of [4] (p.225) we obtain

$$s(\lambda, b) = b \prod_{k=1}^{\infty} \left(\frac{b}{\pi k}\right)^2 \left(\nu_k^2 - ip\nu_k - \lambda^2 + ip\lambda\right). \tag{3.2}$$

In the same way (2.1) implies that $s'(\lambda, a)$ is sinus-type function (see for definition of such functions [10]) and consequently

$$s'(\lambda, a) = K \lim_{n\to\infty} \prod_{\substack{k\neq 0\\k=-n}}^{n} \left(1 - \frac{\lambda}{\lambda_k}\right), \tag{3.3}$$

where the constant K may be derived from the condition

$$\lim_{\tau\to\infty} \frac{c(i\tau, a)}{\cosh \tau a} = 1.$$

It is clear that under the conditions of the Theorem the obtained functions $s(\lambda, b)$, $s_1(\lambda, a)$ and $s'(\lambda, a)$ are the same for the both sets $\{a_i, b_i, p_i, q_i(x)\}$ $(i =$

1, 2). Substituting (3.1) and (3.2) into (2.7) we find the product $s(\lambda,a)s_1'(\lambda,a)$ (the same for both (i=1, 2) sets $\{a_i, b_i, p_i, q_i(x)\}$). Due to Lemma 2.2 we can identify the set of real zeroes of the product $s(\lambda,a)s_1'(\lambda,a)$ as $\left\{\mu_k^{(1)}\right\}_{\substack{k\neq 0\\ -\infty}}^{\infty}$ and the set of complex zeroes of $s(\lambda,a)s_1'(\lambda,a)$ as $\left\{\xi_k^{(1)}\right\}_{\substack{k\neq 0\\ -\infty}}^{\infty}$. We can construct the projection of $q(x)\in L_2(b,a)$ onto (b,a) via a procedure described in [4, 11] using the two spectra $\left\{\nu_k^{(1)}\right\}_{\substack{k\neq 0\\ -\infty}}^{\infty}$ and $\left\{\mu_k^{(1)}\right\}_{\substack{k\neq 0\\ -\infty}}^{\infty}$. Hence $q_1(x) \overset{a.e.}{=} q_2(x)$ for $x\in(b,a)$.

Construct the function

$$s(\lambda,a) = M \lim_{n\to\infty} \prod_{\substack{k\neq 0\\ k=-n}}^{n} \left(1-\frac{\lambda}{\xi_k}\right), \tag{3.4}$$

where the constant M can be derived from the condition

$$\lim_{\tau\to\infty} \frac{i\tau s(i\tau,a)}{\sinh \tau a} = 1,$$

which follows from (2.5) and the known asymptotics of the solutions ([4], p.233). To find the projection of $q(x)$ onto $[0,b]$ we derive $s'(\lambda,a)$ from the system of equations (2.5), (2.6) using (2.7):

$$s'(\lambda,b) = s'(\lambda,a)c_1(\lambda,a) - s(\lambda,a)c_1'(\lambda,a), \tag{3.5}$$

where $c_1(\lambda,a)$ and $c_1'(\lambda,a)$ can be found by solving the direct problem. Namely, solving equation (2.3) with the boundary conditions $c_1(\lambda,b)-1 = c_1'(\lambda,b) = 0$ we find $c_1(\lambda,x)$ and, consequently, $c_1(\lambda,a)$ and $c_1'(\lambda,a)$. Substituting the found functions $c_1(\lambda,a)$, $c_1'(\lambda,a)$ and the functions $s'(\lambda,a)$ and $s(\lambda,a)$ determined by (3.3) and (3.4) into (3.5) we can find $s'(\lambda,b)$, whose set of zeroes, $\{\mu_k\}_{\substack{k\neq 0\\ -\infty}}^{\infty}$, coincides with the set of eigenvalues of the problem

$$y'' + \left(\lambda^2 - i\lambda p - q(x)\right) y = 0,$$

$$y(\lambda,0) = y'(\lambda,b) = 0.$$

Denote $\tau_k^2 = \mu_k^2 - i\mu_k p$ and $\theta_k^2 = \nu_k^2 - i\nu_k p$. Then $\tau_k^2 > 0$ and $\theta_k^2 > 0$ because of the condition $F_0 \gg 0$. Hence we can construct the projection of $q(x)$ onto $[0,b]$ via the mentioned procedure [4, 11] using the two spectra $\left\{\tau_k^2\right\}_1^{\infty}$, $\left\{\theta_k^2\right\}_1^{\infty}$. At all the steps of the reconstruction we obtained unique results. Thus the obtained projection $q(x)$ belongs to $L_2(0,b)$. The Theorem is proved.

Remark.
Unfortunately, we are not able to identify the points of the set $\left\{\mu_k^{(1)}\right\}_{\substack{k\neq 0\\ -\infty}}^{\infty}$ among the zeroes of the product $s(\lambda,a)s_1'(\lambda,a)$ in the case $p=0$ and the solution of the inverse problem may be not unique.

The research described in this publication was made possible in part by Award No. UM1-298 of the Governments of Ukraine and U.S. Civilian Research and Development Foundation for the Independent States of the Former Soviet Union (CRDF).

References

[1] M.Jaulent, C.Jean, The Inverse S-Wave Scattering Problems for a Class of Potential Depending on Energy, Commun. Math. Phys., vol.28, (1972), 177-220.

[2] M.Jaulent, Inverse Scattering Problems in Absorbing Media, J. Math. Phys., vol.17, No 7, (1976), 1351-1360.

[3] M.Jaulent, C.Jean, The Inverse Problem for the One-Dimensional Schrödinger Equation with an Energy-Dependent Potential I, II, Ann. Inst. Henri Poincaré, sec.A, vol.25, No 2, (1976), 105-137.

[4] V.A.Marchenko, Sturm-Liouville Operators and Applications, Birkhauser, OT, 22, (1986), 367 pp.

[5] T.Aktosun, M.Klaus, C. van der Mee, Wave Scattering in One Dimension with Absorption, J.Math.Phys., vol.39, No 4, (1998), 1957-1992.

[6] M.G.Gasymov, G.Sh.Guseinov, Determination of the Operator of Diffusion from Spectral Data, (in Russian), Doklady AN Azerbaijan SSR, vol.37, No 2, (1981), 19-23.

[7] V.N.Pivovarchik, An Inverse Sturm-Liouville problem by Three Spectra, to appear in Integral Equations and Operator Theory.

[8] F.Gesztesy, B.Simon, On the Determination of a Potential from Three Spectra, to be published in Birman Birthday Volume in Advances in Mathematical Sciences, V.Buslaev and M.Solomyak (eds.), Amer. Math. Soc., Providence, RI.

[9] M.G.Krein, H.Langer, On Some Mathematical Principles in the Linear Theory of Damped Oscillations of Continua II, Integral Equations and Operator Theory, vol.1/4, (1978), 87-110.

[10] B.Ja.Levin, On bases of exponential functions in L_2 (in Russian), Zapiski fiz. mat. facult. Kharkov Univ. and Kharkov mat. obsch., Vol.27, ser. A (1961), 39-48.

[11] B.M.Levitan, Inverse Sturm-Liouville problems (in Russian), Moscow, Nauka, GRFML (1984), 240 pp.

Operator Theory:
Advances and Applications, Vol. 108

One-dimensional Schrödinger operators with decaying potentials

Christian Remling

Universität Osnabrück, Fachbereich Mathematik/Informatik,
49069 Osnabrück, GERMANY

E-mail: cremling@mathematik.uni-osnabrueck.de

Abstract

We consider one-dimensional Schrödinger operators $H = -d^2/dx^2 + V(x)$ with power-decaying potentials $V(x) = O(x^{-\alpha})$, and we discuss results and open problems concerning the structure of the essential spectrum of H.

1 Results

In this note, we will discuss one-dimensional Schrödinger equations,

$$-y''(x) + V(x)y(x) = Ey(x). \tag{1}$$

We will always assume that V can be bounded by

$$|V(x)| \leq \frac{C}{(1+x)^\alpha} \tag{2}$$

with α positive. There will be no additional assumption on V. (Some of the results I am going to present actually hold under hypotheses more general than (2), but things are particularly easy to describe in this setting.) We are interested in the spectral properties of the corresponding operators $H_\beta = -d^2/dx^2 + V(x)$ on $L_2(0,\infty)$, say, with boundary conditions $y(0)\cos\beta + y'(0)\sin\beta = 0$.

Since V in particular tends to zero as x goes to infinity, we have that $\sigma_{ess} = [0,\infty)$. That leaves the question of the type of the spectrum on $(0,\infty)$. I will try to give an overview on the known results (including recent contributions by myself), and, in the last section, I will also point out some problems that remain to be solved. The proofs of my own results will be discussed in Section 2; because of lack of space, it will be impossible to include all details, but then there will be references.

It will turn out that there are two critical values of the exponent α from eq. (2), namely $\alpha = 1$ and $\alpha = 1/2$. The first result is

Theorem 1.1 **([13])** *Suppose* $V(x) = o(x^{-1})$. *Then the spectrum is purely absolutely continuous on* $(0, \infty)$.

This may be thought of as a strengthening (on the power scale) of the well known statement that the spectrum is purely absolutely continuous if the potential is integrable. On the other hand, such results cannot hold for $\alpha \le 1$. Indeed, the classical counterexample of von Neumann and Wigner [20] shows that potentials $V(x) = O(x^{-1})$ can have embedded eigenvalues below a certain threshold energy. In fact, the borderline case $\alpha = 1$ is rather subtle; there still is the following positive result.

Theorem 1.2 **([8])** *If* $\alpha = 1$, *then* (1) *has an* L_2 *solution for at most countably many positive energies* E_n. *Moreover,* $\sum E_n < \infty$.

However, one can have *dense* point spectrum with only slightly weaker bounds on the potential:

Theorem 1.3 **([10, 17])** *Suppose that* $C(x) \to \infty, C(x) \ge 1$. *Then there is a potential* V *satisfying* $|V(x)| \le C(x)/(1+x)$ *so that* $\sigma_{pp} = [0, \infty)$.

Despite all this, it is still true that the absolutely continuous part is stable as long as $\alpha > 1/2$. In fact, more detailed statements are possible. Recall that the quasiclassical (or WKB) approximation (which is, in general, not applicable if we assume only (2)!) yields solutions of (1) of the asymptotic form

$$\begin{pmatrix} y(x,E) \\ y'(x,E) \end{pmatrix} = \begin{pmatrix} 1 \\ i\sqrt{E} \end{pmatrix} e^{i\omega_0(x,E)} + o(1) \qquad (x \to \infty), \tag{3}$$

where $\omega_0(x, E) = \int_0^x \sqrt{E - V(t)}\, dt$. Now define the exceptional set S by

$$S = \{E > 0 : (1) \text{ does not have a solution } y \text{ satisfying } (3)\}.$$

Note that if $E \notin S$ and y is of the form (3), then $\overline{y}$ is a linearly independent solution, so we have complete control over the solution space in this case. The following Theorem is a sharp version of earlier results of Kiselev [6, 7].

Theorem 1.4 **([1, 13])** *If* $\alpha > 1/2$, *then* $|S| = 0$.

(Here, $|\cdot|$ denotes Lebesgue measure.) By this result and the remark preceding it, all solutions of (1) are bounded if $E \in (0, \infty) \setminus S$. By [18], Theorem 1.4 therefore has the following consequence:

Corollary 1.5 **([1, 13])** *If* $\alpha > 1/2$, *then the absolutely continuous part of the spectral measure is equivalent to* $\chi_{(0,\infty)}(E)\, dE$. *In particular,* $\sigma_{ac} = [0, \infty)$.

Recently, Deift and Killip proved the following spectacular result:

Theorem 1.6 ([3]) *If $V \in L_2 + L_1$, then the absolutely continuous part of the spectral measure is equivalent to $\chi_{(0,\infty)}(E)\, dE$.*

Of course, Corollary 1.5 also follows from this. On top of that, the proof of this theorem given in [3] is much more transparent than the analysis of [1, 13]. However, there is no statement on the asymptotics of the solutions of (1).

Actually, in the situation where V satisfies (2), one can say even more on the exceptional set S, namely, there is the following bound on its Hausdorff dimension:

Theorem 1.7 ([14]) $\dim S \leq 2(1-\alpha)$.

Note that this implies Theorem 1.4 because a set with Hausdorff dimension strictly less than 1 is of Lebesgue measure zero. Since, again by [18], possible singular spectral measure on $(0, \infty)$ is supported by S, Theorem 1.7 gives restrictions on the structure of such spectra.

There are examples of potentials (with randomness, though) [4, 16] so that (2) holds with $\alpha = 1/2$ but the corresponding Schrödinger operator has purely singular spectrum. So the bound in Theorem 1.7 must take the value 1 for $\alpha = 1/2$. By the same token, the hypothesis $\alpha > 1/2$ of Theorem 1.4 cannot be weakened.

2 Discussion of the proofs

In this section, I will explain the ideas of the proofs of Theorems 1.1, 1.4, and 1.7. These proofs are very typical of the spectral theory of *one-dimensional* operators in that one analyzes the asymptotic behavior of the solutions of (1). One main ingredient is the use of the (modified) Prüfer transform (see also [8, 9] for further information). So, let $y(x,k)$ be a solution of (1) with $E = k^2$ with prescribed initial values at $x = 0$. Introduce the Prüfer variables R, ψ by writing $y = R \sin \psi/2$, $y' = kR \cos \psi/2$. Integration of the differential equations that R, ψ obey yields

$$2k(\ln R(y,k) - \ln R(x,k)) = \int_x^y V(t) \sin \psi(t,k)\, dt, \tag{4}$$

$$k(\theta(y,k) - \theta(x,k)) = \int_x^y V(t) \cos \psi(t,k)\, dt, \tag{5}$$

where $\theta = \psi - \omega$, and ω is essentially (twice) the phase factor ω_0 of the WKB approximation (compare Section 1):

$$\omega(x,k) = 2kx - \frac{1}{k} \int_0^x V(t)\, dt.$$

This break-up of ψ is useful because one has control over oscillatory integrals with phase ω (see Theorem 2.1 below) and, most importantly, it is possible to set up an iteration scheme that efficiently exploits this property of ω. More specifically, proceed as follows. Decompose $(0,\infty)$ into subintervals (x_{n-1},x_n). Then, according to (4), (5), we will have to study

$$\sum_{n=1}^{\infty} \int_{x_{n-1}}^{x_n} V(x) e^{i\psi(x,k)}\, dx = \sum_{n=1}^{\infty} \int_{x_{n-1}}^{x_n} V(x) e^{i\omega(x,k)} e^{i\theta(x,k)}\, dx. \tag{6}$$

In fact, it is easy to see that whenever this sum converges, (1) has a solution of the form (3), at least for $x \in \{x_n\}$. Now we need

Theorem 2.1 *Suppose μ is a finite Borel measure, supported on a compact subset of $(0,\infty)$. If, for some $D \in [0,1]$, we have $\mu(I) \le C_0 |I|^D$ for all intervals I, then*

$$\int d\mu(k) \left| \int_c^d dx\, f(x) e^{i\omega(x,k)} \right|^2 \le C(d-c)^{1-D} \int_c^d dx\, |f(x)|^2$$

for all $f \in L_2(c,d)$, $0 \le c < d < \infty$.

Note that if $V \equiv 0$, then this reduces to results from [19] on the Fourier transform. Theorem 2.1 was first formulated and proved in the special case $D = 1$ in [7]; the extension to singular D-dimensional measures μ (which is crucial to proving Theorem 1.7) is from [14].

Returning to (6), we integrate by parts, integrating $Ve^{i\omega}$ and differentiating $e^{i\theta}$. We get

$$\sum_{n=1}^{\infty} \Bigg(e^{i\theta(x_n,k)} \int_{x_{n-1}}^{x_n} V(x) e^{i\omega(x,k)}\, dx \quad - \\ \frac{i}{k} \int_{x_{n-1}}^{x_n} dx\, V(x) \cos(\omega(x,k)+\theta(x,k)) e^{i\theta(x,k)} \int_{x_{n-1}}^{x} dt\, V(t) e^{i\omega(t,k)} \Bigg).$$

Now Theorem 2.1 together with the Cauchy-Schwarz inequality (in $L_2(d\mu)$) show that the μ-integral of the first term in the sum can be estimated by

$$\int d\mu(k) \left| \int_{x_{n-1}}^{x_n} V(x) e^{i\omega(x,k)}\, dx \right| \le C(x_n - x_{n-1})^{(1-D)/2} \|V\chi_{(x_{n-1},x_n)}\|_2.$$

In particular, if the x_n's are chosen in such a way that this bound is summable, then also, as desired,

$$\sum_{n=1}^{\infty} \left| \int_{x_{n-1}}^{x_n} V(x) e^{i\omega(x,k)}\, dx \right| < \infty$$

for μ-almost every k by monotone convergence.

The remaining terms are dealt with by an iteration scheme also based on these two devices: integration by parts and the control over integrals with respect to $d\mu$ provided by Theorem 2.1. More specifically, one starts with the intervals (x_{n-1}, x_n), and at each step, every interval is subdivided into smaller subintervals. The point is that the double integrals are over smaller and smaller regions (modulo contributions that are summable μ-almost everywhere). Finally, an elementary estimate shows that these terms are also summable almost everywhere (after a sufficiently large number of steps). It turns out that this procedure works if μ is as in Theorem 2.1 with $D > 2(1-\alpha)$. As the final step of this part of the proof, one extends the asymptotic formula (3) from $x \in \{x_n\}$ to x tending to infinity arbitrarily. For the technical details of these rather lengthy arguments, consult [13, 14].

Now a measure theoretic argument (see [14]) allows one to convert convergence μ-almost everywhere to convergence off a set of Hausdorff dimension $\leq 2(1-\alpha)$. This proves Theorem 1.7 (which contains Theorem 1.4). Christ and Kiselev give a completely different proof of Theorem 1.4 in [1] (see also [2]).

To prove Theorem 1.1, notice that Theorem 1.7 with $\alpha = 1$ ensures that possible singular spectral measure is supported by a set of zero Hausdorff dimension. (In fact, this special case of Theorem 1.7 can be obtained with considerably less work from a corresponding result on Fourier integrals, cf. [13].) But (4) shows that for any non-trivial solution of (1), $\ln R(x) = o(\ln x)$, and it is a consequence of the refined subordinacy theory [5] that this asymptotic behavior is incompatible with zero-dimensional spectral measures. So there cannot be any singular spectrum on $(0, \infty)$.

3 Further remarks and open questions

Let us summarize the picture provided by the results discussed in Section 1. We have $\sigma_{ac} = [0, \infty)$ as long as $\alpha > 1/2$. If $\alpha > 1$ (or if only $V(x) = o(x^{-1})$), the spectrum is *purely* absolutely continuous on $(0, \infty)$. If $1/2 < \alpha \leq 1$, there may also be embedded singular spectrum, but, according to Theorem 1.7, there is a gap $(2(1-\alpha), 1)$ of forbidden dimensions for the support of the corresponding part of the spectral measure.

The first set of open questions is connected with the fact that we do not know any example with $\alpha > 1/2$ and embedded singular *continuous* spectrum. For instance, as discussed above, Theorem 1.7 implies that there always is a support of the singular continuous part of the spectral measure of dimension $\leq 2(1-\alpha)$ (namely, S), but it is not clear if this bound is optimal. This problem seems to be rather difficult. Somewhat less ambitiously, one can ask if $2(1-\alpha)$ is optimal as a bound on $\dim S$. Here, some progress is possible. Using ideas from [12], I could construct examples which show that Theorem 1.7 is indeed sharp at least if $\alpha > 4/5$ (see [15]).

The case $\alpha = 1$ has some extra subtleties, as already pointed out in the first section. Again, we don't know much about possible singular continuous spectrum. As for the point spectrum, I was able to show that Theorem 1.2 is optimal in a certain sense [15]. Namely, based on the construction of [17], one can give examples of potentials $V(x) = O(x^{-1})$ so that (1) has an L_2 solution for $E \in \{E_n\}$ and $\sum E_n^p = \infty$ for all $p < 1$.

Finally, the statement of Corollary 1.5 is equivalent to saying that the absolutely continuous parts of $H = -d^2/dx^2 + V(x)$ and $H_0 = -d^2/dx^2$ are unitarily equivalent, and this is exactly the conclusion one would also get from scattering theory if the (modified) wave operators existed and were complete (see, e.g., [11]). (These latter properties are, of course, stronger than what Corollary 1.5 states.) I do not know of any result or counterexample addressing these issues.

References

[1] M. Christ and A. Kiselev, *Absolutely continuous spectrum for one-dimensional Schrödinger operators with slowly decaying potentials: some optimal results*, to appear in J. Amer. Math. Soc.

[2] M. Christ, A. Kiselev, and C. Remling, *The absolutely continuous spectrum of one-dimensional Schrödinger operators with decaying potentials*, Math. Research Letters **4** (1997), 719–723.

[3] P. Deift and R. Killip, in preparation.

[4] F. Delyon, B. Simon, and B. Souillard, *From power pure point to continuous spectrum in disordered systems*, Ann. Inst. H. Poincaré **42** (1985), 283–309.

[5] S. Jitomirskaya and Y. Last, *Dimensional Hausdorff properties of singular continuous spectra*, Phys. Rev. Letters **76** (1996), 1765–1769.

[6] A. Kiselev, *Absolutely continuous spectrum of one-dimensional Schrödinger operators and Jacobi matrices with slowly decreasing potentials*, Commun. Math. Phys. **179** (1996), 377–400.

[7] A. Kiselev, *Preservation of the absolutely continuous spectrum of Schrödinger equations under perturbations by slowly decreasing potentials and a.e. convergence of integral operators*, to appear in Duke Math. J.

[8] A. Kiselev, Y. Last, and B. Simon, *Modified Prüfer and EFGP transforms and the spectral analysis of one-dimensional Schrödinger operators*, Commun. Math. Phys. **194** (1998), 1–45.

[9] A. Kiselev, C. Remling, and B. Simon, *Effective perturbation methods for one-dimensional Schrödinger operators*, to appear in J. Diff. Eq.

[10] S.N. Naboko, *Dense point spectra of Schrödinger and Dirac operators*, Theor. and Math. Phys. **68** (1986), 646–653.

[11] M. Reed and B. Simon, *Methods of Modern Mathematical Physics, III. Scattering Theory*, Academic Press, London-San Diego, 1979.

[12] C. Remling, *Some Schrödinger operators with power-decaying potentials and pure point spectrum*, Commun. Math. Phys. **186** (1997), 481–493.

[13] C. Remling, *The absolutely continuous spectrum of one-dimensional Schrödinger operators with decaying potentials*, Commun. Math. Phys. **193** (1998), 151–170.

[14] C. Remling, *Bounds on embedded singular spectrum for one-dimensional Schrödinger operators*, to appear in Proc. Amer. Math. Soc.

[15] C. Remling, *Schrödinger operators with decaying potentials: some examples*, in preparation.

[16] B. Simon, *Some Jacobi matrices with decaying potentials and dense point spectrum*, Commun. Math. Phys. **87** (1982), 253–258.

[17] B. Simon, *Some Schrödinger operators with dense point spectrum*, Proc. Amer. Math. Soc. **125** (1997), 203–208.

[18] G. Stolz, *Bounded solutions and absolute continuity of Sturm-Liouville operators*, J. Math. Anal. Appl. **169** (1992), 210–228.

[19] R. Strichartz, *Fourier asymptotics of fractal measures*, J. Funct. Anal. **89** (1990), 154–187.

[20] J. von Neumann and E. Wigner, *Über merkwürdige diskrete Eigenwerte*, Z. Phys. **30** (1929), 465–467.

Operator Theory:
Advances and Applications, Vol. 108

Stability of Limiting Absorption Principle under Singular Perturbations

Walter Renger
Institut für Mathematik, TU-Clausthal,
38678 Clausthal-Zellerfeld, Germany
mawr@tu-clausthal.de

Abstract

Given an operator H_1 for which a limiting absorption principle holds, we study operators H_2 which are produced by perturbing H_1 in the sense that the difference between some powers of the resolvents is compact. We show that (except for possibly a discrete set of eigenvalues) a limiting absorption principle holds for H_2.

Suppose H is a self-adjoint operator in a Hilbert space $\mathcal{H}$, $R(z) = (H - z)^{-1}$ its resolvent, and λ some value in its spectrum $\sigma(H)$. In order to study spectral and scattering properties of H it is useful to have some control over the behavior of $R(z)$ as $z = \lambda \pm i\varepsilon$ tends to λ. We say that a limiting absorption principle (LAP) holds for H if there is an open set $\Delta \subset \mathbb{R}$ and a dense subspace X of $\mathcal{H}$ ($X \subset \mathcal{H} \subset X^*$) such that $\lim_{\varepsilon\downarrow 0} R(\lambda \pm i\varepsilon)$ exists in the norm topology of $\mathcal{B}(X, X^*)$, the space of bounded operators from X to X^*.

We study the stability of such a limiting absorption principle under singular perturbations. That is, we suppose a LAP holds for some operator H_1 and try to establish that it holds for a second operator H_2 which is constructed by perturbing H_1 in some way. In particular we are interested in the case where a differential operator is perturbed by introducing Dirichlet boundary conditions on some set. Related results can for instance be found in Iwashita [7] and Leis and Roach [9].

We present an abstract theory that includes these as well as other types of regular and singular perturbations. Our approach is similar to the one presented by Ben-Artzi and Devinatz [2], but more general: The abstract theory in [2] is geared towards potential perturbations; while it does cover other situations as well it is not applicable in our case.

Our key condition on the perturbation is that some difference of powers of the resolvents is compact from X^* to X: $R_2(z)^m - R_1(z)^m \in \mathcal{B}_\infty(X^*, X)$ for some $m \in \mathbb{N}$. (Actually we will have to modify this slightly, since we will work in a two-space framework.) We work with powers of the resolvents rather than the resolvents themselves because in some cases (cf. [12]) it is easier to show compactness for the difference of the powers.

While we try to keep this presentation as self-contained as possible we have to make concessions to brevity. Most proofs are little more than sketches and some had to be omitted entirely. Our results will be published elsewhere [12] in full detail. There we will also discuss how this abstract theory can be applied to perturbations of Feller operators (generators of Markov processes with certain additional properties, cf. [4] [5] [12]).

Notation. Let $(\mathcal{H}_1, \langle\cdot,\cdot\rangle_1)$ and $(\mathcal{H}_2, \langle\cdot,\cdot\rangle_2)$ be Hilbert spaces, J_2 a bounded map from $\mathcal{H}_1$ to $\mathcal{H}_2$ with $J_2 J_2^*$ equal to the identity on $\mathcal{H}_2$, $J_1 = J_1^* = 1$ the identity in $\mathcal{H}_1$. Let the Hilbert space $(X, \langle\cdot,\cdot\rangle_X)$ be densely and continuously embedded in $\mathcal{H}_1$, $(X^*, \langle\cdot,\cdot\rangle_{X^*})$ its dual, so $X \subset \mathcal{H}_1 \subset X^*$ with dense and continuous embeddings.

The main example to keep in mind here is $\mathcal{H}_1 = L^2(\mathbb{R}^d)$, $\mathcal{H}_2 = L^2(\Sigma)$ with $\Sigma \subset \mathbb{R}^d$ and J_2 the restriction from $\mathbb{R}^d$ to Σ. X will then usually be a weighted L^2-space $X = L^2_\varphi := \{u \in L^2 : \|\varphi u\| < \infty\}$ with weight function $\varphi : \mathbb{R}^d \to [1,\infty)$ and $X^* = L^2_{\varphi^{-1}}$.

Norms and scalar products will be written with the corresponding space as a subscript, for $\mathcal{H}_1$ and $\mathcal{H}_2$ we simply use the subscripts 1 and 2. $\mathcal{B}$ denotes the (space of) bounded operators, $\mathcal{B}_\infty$ compact operators.

Our basic assumptions on the operators H_1 and H_2 are

Hypothesis 1. (Basic Properties) *Suppose that for $k = 1,2$ the operators H_k act as self-adjoint semibounded operators in $\mathcal{H}_k$ with domains $D(H_k)$. Moreover, suppose there exists a constant $c_R \in \mathbb{R}$ such that for all $z \in \mathbb{C}$ with* $\operatorname{Re} z < c_R$

$$J_k^* R_k(z) J_k \in \mathcal{B}(X, X) \tag{1}$$

the resolvents being continuous functions of z in $\mathcal{B}(X,X)$.

Hypothesis 2. (LAP for H_1) *Let Δ be a given open subset of $\mathbb{R}$: Assume that for all $\lambda \in \Delta$*

$$R_1^\pm(\lambda) := \lim_{\varepsilon \downarrow 0} R_1(\lambda \pm i\varepsilon) \tag{2}$$

exists and is continuous with respect to λ in $\mathcal{B}(X, X^*)$.

Hypothesis 3. (Compact perturbation) *Suppose there exists a constant $c_E \in \mathbb{R}$ and an $m \in \mathbb{N}$ such that for all $a < c_E$*

$$E(a) := J_2^* R_2(a)^m J_2 - R_1(a)^m \in \mathcal{B}_\infty(X^*, X). \tag{3}$$

We will need one more hypothesis, but first we have to introduce new operators $F_k(a,z)$ which will be key objects in our analysis. For $a \in \mathbb{R}$ small enough and $z = \lambda \pm i\varepsilon$ close to $\lambda \in \Delta$ one can define

$$F_k(a,z) := \left[R_k(a)^m - (z-a)^{-m}\right]^{-1} \tag{4}$$

in $\mathcal{B}(\mathcal{H}_k, \mathcal{H}_k)$. This operator can be factored as

$$F_k(a,z) = (z-a)^m \prod_{j=0}^{m-1} \Big\{ (-w_j)\big[1 + w_j(z-a)R_k\big(a + w_j(z-a)\big)\big] \Big\}, \qquad (5)$$

where $w_j = e^{2\pi i(j/m)}$. There are some technicalities associated with how small precisely one has to choose a, we refer to [12] for details.

Moreover, one verifies that $J_k^* F_k^{\pm}(a,\lambda) J_k := \lim_{\varepsilon \downarrow 0} J_k^* F_k(a, \lambda \pm i\varepsilon) J_k$ exists and is continuous in $\mathcal{B}(X, X^*)$ if and only if $J_k^* R_k^{\pm}(\lambda) J_k$ exists and is continuous.

With this we can formulate

Hypothesis 4. *Suppose that for each $\beta > 0$ there exists an $a \in \mathbb{R}$ small enough such that for each compact subset K of $\Delta_\beta = \Delta \cap (-\beta, \beta)$, there is a constant $c_K > 0$ with the following property: If $\lambda \in K$ and $u \in X$ satisfies $\big(1 + E(a) F_1^{\pm}(a, \lambda)\big)u = 0$ then $R_1^{\pm}(\lambda)u$ lies in $\mathcal{H}_1$ with*

$$\|R_1^{\pm}(\lambda)u\|_1 \le c_K \|u\|_X \,. \qquad (6)$$

Hypothesis 4, as it is presented here, is rather awkward, because it intertwines conditions on H_1 and conditions on H_2 via the operator $E(a)$. We will give sufficient, much more convenient, conditions that guarantee Hypothesis 4 in the case of weighted L^2-spaces (Hypotheses 8 and 9).

Now fix $\beta > 0$, $a \in \mathbb{R}$ sufficiently small, set $\Delta_\beta = \Delta \cap (-\beta, \beta)$ and abbreviate $E := E(a) = J_2^* R_2(a)^m J_2 - R_1(a)^m$, $F_k(z) := F_k(a, z)$.

The basic idea of our approach is to derive properties of $F_2(z)$ from those of $F_1(z)$. One can calculate

$$\begin{aligned} J_2^* J_2 F_1(z) &= J_2^* F_2(z) J_2 + J_2^* F_2(z) J_2 \big[J_2^* R_2(a)^m J_2 - R_1(a)^m\big] F_1(z) \\ &= J_2^* F_2(z) J_2 \big[1 + E F_1(z)\big]. \end{aligned} \qquad (7)$$

This operator identity holds in $\mathcal{B}(\mathcal{H}_1, \mathcal{H}_1)$ and by extension in $\mathcal{B}(X, X^*)$. Since $EF_1^{\pm}(\lambda) \in \mathcal{B}_\infty(X, X)$ we may use the Fredholm alternative: For fixed λ either $(1 + EF_1^{\pm}(\lambda))$ is invertible in $\mathcal{B}(X, X)$ or there exists an $u \in X$, $u \neq 0$ with $(1 + EF_1^{\pm}(\lambda))u = 0$. (One can show that if for some $\lambda \in \Delta_\beta$ one of $1 + EF_1^{+}(\lambda)$ and $1 + EF_1^{-}(\lambda)$ is not invertible neither of them is.)

One verifies that if $1 + EF_1^{\pm}(\lambda_0)$ is invertible in $\mathcal{B}(X, X)$ for some $\lambda_0 \in \Delta_\beta$ then there exists a neighborhood $U \subset \Delta_\beta$ of λ_0 such that for all $\lambda \in U$ the limit $J_2^* F_2^{\pm}(\lambda) J_2$ exists and is continuous with respect to λ in $\mathcal{B}(X, X^*)$, hence the same holds for $R_2^{\pm}(\lambda)$.

If $J_2^* R_2^{\pm}(\lambda) J_2$ exists in $\mathcal{B}(X, X^*)$ in some neighborhood $U \subset \Delta_\beta$ of λ_0 we know (cf. [11], Thm. XIII.19 and XIII.20) that the spectrum $\sigma(H_2)$ of H_2 is purely absolutely continuous in U. Hence we need to analyze the set where $1 + EF_1^{\pm}(\lambda)$ is not invertible in more detail. We begin with a simple observation.

Lemma 5. *If $\lambda \in \Delta_\beta$ is an eigenvalue of H_2 with eigenvector ψ then $u := -EJ_2^*\psi \in X$ satisfies $\big(1 + EF_1^\pm(\lambda)\big)u = 0$, $u \neq 0$.*

Proof. $u \in X$ by definition, factoring E one establishes $u \neq 0$, because otherwise $\zeta = a + \frac{1}{w_j}(\lambda - a)$ would have to be an eigenvalue of H_1 for some $w_j = e^{2\pi i(j/m)}$. To verify $\big(1 + EF_1^\pm(\lambda)\big)u = 0$ one calculates

$$EF_1(\lambda \pm i\varepsilon)u = EJ_2^*\psi + EF_1(\lambda \pm i\varepsilon)\big[(\lambda \pm i\varepsilon - a)^{-m} - (\lambda - a)^{-m}\big]J_2^*\psi\,. \tag{8}$$

The claim follows once we establish

$$EF_1(\lambda \pm i\varepsilon)\big[(\lambda \pm i\varepsilon - a)^{-m} - (\lambda - a)^{-m}\big]J_2^*\psi \underset{\varepsilon\downarrow 0}{\longrightarrow} 0.$$

Here $F_1(\lambda \pm i\varepsilon)[(\lambda \pm i\varepsilon - a)^{-m} - (\lambda - a)^{-m}]$ converges to 0 in $\mathcal{B}(X, X^*)$ and can be shown to be bounded in $\mathcal{B}(\mathcal{H}_1, \mathcal{H}_1)$ as ε tends to 0, hence it is weakly convergent in $\mathcal{B}(\mathcal{H}_1, \mathcal{H}_1)$ which is sufficient to establish the claim. □

The converse uses Hypothesis 4.

Lemma 6. *Suppose $\big(1 + EF_1^\pm(\lambda)\big)$ is* not *invertible in $\mathcal{B}(X, X)$ for some $\lambda \in \Delta_\beta$; let $u \in X$, $u \neq 0$ satisfy $u = -EF_1^\pm(\lambda)u$. Then λ is an eigenvalue of H_2 with eigenvector $\psi = R_2(a)J_2F_1^\pm(\lambda)u \in D(H_2)$, $\psi \neq 0$.*

Proof. Rewrite $u = -EF_1^\pm(\lambda)u$ in terms of $v = F_1^\pm(\lambda)u \in \mathcal{H}_1$, then multiply by the projection $1 - J_2^*J_2$. This yields $J_2^*J_2v \neq 0$ and $J_2^*\big[R_2(a)^m - (\lambda - a)^{-m}\big]J_2v = 0$. From that we find $(\lambda - H_2)R_2(a)J_2v = 0$. □

Lemmas 5 and 6 lead to our main theorem.

Theorem 7. *Suppose H_1 and H_2 satisfy Hypotheses 1, 2, 3, and 4. Then $\sigma(H_2) \cap \Delta$ is, except for possibly a discrete set $\sigma_p(H_2) \cap \Delta$ of eigenvalues of finite multiplicity, purely absolutely continuous. For all $\lambda \in \Delta \setminus \sigma_p(H_2)$*

$$J_2^*R_2^\pm(\lambda)J_2 := \lim J_2^*R_2(\lambda \pm i\varepsilon)J_2 \tag{9}$$

exists and is continuous in $\mathcal{B}(X, X^*)$.

Proof. It is sufficient to establish the claim on $\Delta_\beta = \Delta \cap (-\beta, \beta)$ for all $\beta > 0$. We choose a small enough.

Suppose $\sigma_p(H_2)$ is not discrete in Δ_β. Then there exists a sequence $(\lambda_k)_{k\in\mathbb{N}} \subset \big(\Delta_\beta \cap \sigma_p(H_2)\big)$ with $\lambda_k \underset{k\to\infty}{\longrightarrow} \lambda \in \Delta_\beta$. ($\lambda$ may or may not be an eigenvalue.) By Lemma 5 for each λ_k there exists a vector $u_k \in X$ with $(1 + EF_1^\pm(\lambda_k))u_k = 0$, $\|u_k\|_X = 1$. By boundedness a subsequence, without loss the sequence $\{u_k\}_{k\in\mathbb{N}}$ itself, is weakly Cauchy in X. Since $u_k = -EF_1^\pm(\lambda_k)u_k$ with E compact it is in fact strongly Cauchy. Then $u = \lim_{k\to\infty} u_k$ satisfies $\|u\|_X = 1$ and $\big(1 + EF_1^\pm(\lambda)\big)u = 0$. By Lemma 6 there exist vectors $\psi_k, \psi \in D(H_2)$ with $(H_2 - \lambda_k)\psi_k = 0$, $(H_2 -$

$\lambda)\psi = 0$. One verifies $\|\psi_k\|_2 \le c$ so by orthogonality $\psi_k \xrightarrow[k\to\infty]{w} 0$ in $\mathcal{H}_2$ and a fortiori in X^*. On the other hand $u_k \to u$ and the continuity of $F_1^{\pm}(\cdot)$ imply $J_2^*\psi_k = J_2^* R_2(a) J_2 F_1^{\pm}(\lambda_k) u_k \xrightarrow[k\to\infty]{} J_2^*\psi \neq 0$ in X^*, which is a contradiction.

A similar argument may be used to show that each eigenvalue can have at most finite multiplicity.

If $\lambda_0 \in \Delta_\beta$ is not an eigenvalue of H_2 then by Lemma 6 $\left(1 + EF_1^{\pm}(\lambda_0)\right)$ is invertible in $\mathcal{B}(X, X)$ and hence the limit $J_2^* R_2^{\pm}(\lambda) J_2$ exists for all λ in some neighborhood $U \subset \Delta_\beta$ of λ_0. By the general theory (cf., e.g., [11], Thm. XIII.19, XIII.20) this establishes the claim. □

In most applications of the limiting absorption principle some weighted L^2-space is used for X: Let M be a measure space with Borel field $\mathcal{M}$, $\Sigma \in \mathcal{M}$ a measurable set. Define $\mathcal{H}_1 = L^2 := L^2(M)$, $\mathcal{H}_2 = L^2(\Sigma)$, $X = L^2_\varphi := \{u \in L^2(M) : \|\varphi u\| < \infty\}$ with $\varphi : M \to [1, \infty)$ measurable. (Norm and scalar product without a subscript denote the usual L^2-norm and scalar product.) $L^2_{\varphi^s}$, $s \ge 0$ is defined similarly, $L^2_{\varphi^{-s}}$ as its dual. Then we can replace Hypotheses 2, 3, and 4 by

Hypothesis 8. *Suppose there is an open set $\Delta \subset \mathbb{R}$ such that for all $\lambda \in \Delta$,*

$$R_1^{\pm}(\lambda) := \lim_{\varepsilon \downarrow 0} R_1(\lambda \pm i\varepsilon) \tag{10}$$

exists and is continuous in $\mathcal{B}(L^2_\varphi, L^2_{\varphi^{-1}})$. Assume that for each compact subset K of Δ there exists a constant $c_K > 0$ such that for all $\lambda \in K$ and all $u \in L^2_{\varphi^2}$ with $R_1^+(\lambda)u = R_1^-(\lambda)u$

$$\|R_1^{\pm}(\lambda)u\| \le c_K \|u\|_{\varphi^2}\,. \tag{11}$$

Hypothesis 9. *Furthermore, suppose there exists a constant $c_E \in \mathbb{R}$ and an $m \in \mathbb{N}$ such that for all $a < c_E$*

$$E := J_2^* R_2(a)^m J_2 - R_1(a)^m \in \mathcal{B}_\infty(L^2, L^2_{\varphi^{2+\gamma}}) \tag{12}$$

for some $\gamma > 0$.

Using interpolation theory as presented by Lions and Magenes [10] together with a bootstrap argument one can prove (cf. [12])

Proposition 10. *Hypotheses* 8 *and* 9 *imply Hypotheses* 2, 3, *and* 4.

Let us illustrate how this theory may be applied with the simplest example, the Laplacian.

Example 11. Domain perturbations of $-\Delta$ on $\mathbb{R}^d$. Suppose $\mathcal{H}_1 = L^2(\mathbb{R}^d)$, $H_1 = -\Delta$ the Laplacian, $\mathcal{H}_2 = L^2(\Sigma)$ with $\Sigma \subset \mathbb{R}^d$ open, $\Gamma = \mathbb{R}^d \setminus \Sigma$, and $H_2 = -\Delta_\Sigma$ the Dirichlet Laplacian on Σ.

Using Fourier transformation one verifies that Hypothesis 8 holds with $\Delta = (0, \infty)$ and $\varphi = (1 + |x|)^s$, $s > \frac{1}{2}$, see [2]. Similarly one can verify that Hypothesis 1 is satisfied. So the remaining question is, what sets Γ can one allow if one wants to make sure that Hypothesis 9 remains valid. Using the connection between $-\Delta$ and Brownian motion one can verify (see [12]) that a sufficient condition is

$$\varphi^{4+\varepsilon} v_\Gamma \in L^1(M). \tag{13}$$

Here $\Gamma = M \setminus \Sigma$, $\varepsilon > 0$, and $v_\Gamma = E_x[e^{-S},\ S < \infty]$ is the (modified) equilibrium potential with $S = \inf\{t > 0 : \int_0^t 1_\gamma(X_s)ds > 0\}$ the penetration time of Γ.

This means that Γ may be an arbitrary bounded set. (No regularity of the boundary is needed.) One can also allow certain unbounded sets with finite measure; condition (13) is slightly stronger than requiring that Γ has finite capacity.

Let us compare this with known results. Existence and completeness of the wave operators for this scattering system has been established in [3] for arbitrary obstacles of finite capacity, but of course that says nothing about the singular continuous spectrum. There are results on absence of the singular continuous spectrum here as well; to our knowledge the most general ones can be found in [1] and [6]. While [1] requires some regularity of the boundary of Γ, [6] allows arbitrary *bounded* sets Γ.

So our approach yields a small gain in generality insofar as it allows small unbounded obstacles Γ. More important the results in [1] and [6] are specifically tailored to Schrödinger operators, ours are for more general operators.

References

[1] A. Boutet de Monvel-Berthier, V. Georgescu, and A. Soffer, *N-body hamiltonians with hard-core interactions*, Rev. Math. Phys. **6** (1994), 515–596.

[2] M. Ben-Artzi and A. Devinatz, *The limiting absorption principle for partial differential operators*, Memoirs Amer. Math. Soc. **66** (1987), #364.

[3] J. van Casteren and M. Demuth, *Completeness of Scattering systems with obstacles of finite capacity*, in Oper. Theory Adv. Appl., Vol. 102, pp. 39–50, Birkhäuser, Basel, 1998.

[4] M. Demuth and J. van Casteren, *Stochastic Spectral Theory for Self-Adjoint Feller Operators*, Birkhäuser, to appear.

[5] M. Demuth and J. van Casteren, *Framework and results of stochastic spectral analysis*, in Oper. Theory Adv. Appl., Vol. 70, pp. 123–132, Birkhäuser, Basel, 1994.

[6] M. Griesemer, *N-Body Quantum Systems with Hard-Core Interactions*, dissertation, ETH Zürich, 1996 (Diss. ETH No. 11644.

[7] H. Iwashita, *Spectral theory for symmetric systems in an exterior domain, II*, J. Funct. Anal. **82** (1989), 91–112.

[8] T. Kato, *Perturbation Theory for Linear Operators*, Springer, Berlin, 1980.

[9] R. Leis and G. F. Roach, *An initial boundary-value problem for the Schrödinger equation with long-range potential*, Proc. R. Soc. Lond. **A 417** (1988), 353–362.

[10] J. L. Lions and E. Magenes, *Non-Homogeneous Boundary Value Problems and Applications I*, Springer, Berlin, 1972.

[11] M. Reed and B. Simon, *Methods of Modern Mathematical Physics, Volume IV, Analysis of Operators*, Academic Press, New York, 1978.

[12] W. Renger, *Limiting absorption principle for singularly perturbed Feller operators*, preprint, 1997.

Operator Theory:
Advances and Applications, Vol. 108

Existence of Averaging Integrals for Self-Adjoint Operators

Wolfgang Scherer*
Institute for Theoretical Physics A, TU Clausthal
Leibizstraße 10, 38678 Clausthal-Zellerfeld
Germany

Abstract

The existence of averaging integrals of self-adjoint operators (which appear in normal form and KAM-type perturbation theories) is proven for certain classes of operators with purely discrete spectrum.

1 Introduction and motivation

Transformations to normal form or KAM-type iterations thereof are used in classical mechanics to construct perturbation expansions of systems close to integrable ones. Perturbation techniques using similar ideas have also been constructed for certain quantum systems [1, 2, 3, 4]. A major obstacle in the classical as well as in the quantum case is to find well-defined solutions of certain equations (the so-called homological equations) needed to construct the algorithms. The difficulty to establish these solutions arises from the appearance of small divisor problems [2]. To motivate the quantum construction, let us briefly recall the classical situation. The unperturbed Hamiltonian h_0 is a function on phase space M which is equipped with a Poisson bracket structure $\{\cdot,\cdot\}$. h_0 is assumed to be sufficiently smooth, integrable and non-degenerate in the sense of Liouville–Arnold. Let $h(\epsilon) := \sum_{p=0}^{\infty} \frac{\epsilon^p}{p!} h_p$ be the perturbed Hamiltonian where the perturbations h_p, $p \geq 1$ are assumed to be sufficiently smooth functions on M. The idea of perturbation theory is to look for an ϵ–dependent generating function $w(\epsilon) := \sum_{p=0}^{\infty} \frac{\epsilon^p}{p!} w_{p+1}$ (with ϵ–independent smooth functions w_l) such that $-w(\epsilon)$ generates a canonical flow $\xi(\epsilon)$ with "time" ϵ. Then $\varphi(\epsilon) := \xi(\epsilon)^{-1}$ acts with via pull-back on the perturbed Hamiltonian $h(\epsilon)$ and gives a new Hamiltonian $k(\epsilon) := \varphi(\epsilon)^* h(\epsilon)$ which is assumed to be analytic in ϵ: $k(\epsilon) = \sum_{p=0}^{\infty} \frac{\epsilon^p}{p!} k_p$ and for which one finds $k_0 = h_0$ and

$$k_p = f_p - ad_{h_0}(w_p), \qquad p \geq 1$$

*E-mail:ptws@pt.tu-clausthal.de

with $f_p = f_p(h_0, \ldots, h_p; w_1, \ldots, w_{p-1})$ and $ad_f(g) := \{f, g\}$. In choosing $w(\epsilon)$ one aims to make the dynamics generated by $k(\epsilon)$ as simple as possible, because it is related to the motion generated by $h(\epsilon)$ by conjugation. In general it is not possible to find $w(\epsilon)$ such that the motion generated by $k(\epsilon)$ may be found explicitly. However, we may choose the w_l successively in such a manner that each k_p at least commutes with h_0, i.e. such that $ad\, h_0(k_p) = 0\, \forall p$. Thus we have to find solutions of the so-called (classical) homological equation, i.e. given functions h and g find a function w such that

$$ad_h(\, g - ad_h(w)\,) = 0. \tag{1}$$

The solutions of this problem (which are not unique) can be written as averaging integrals along the flow of h (i.e. in its application to perturbation theory: along the flow of the unperturbed system, which is assumed known). Let $h, g \in C^\infty(M)$ and let $\phi_h(t) : M \to M$ be the flow generated by h. Suppose h and g are such that

$$\overline{g}^{(h)} := \lim_{T\to\infty} \frac{1}{T} \int_0^T ds\ \overset{-1}{\phi_h}(s)^* g \quad \in C^\infty(M), \tag{2}$$

$$s^{(h)}(g) := \lim_{T\to\infty} \frac{1}{T} \int_0^T dt \int_0^t ds\ \overset{-1}{\phi_h}(s)^*(g - \overline{g}^{(h)}) \quad \in C^\infty(M) \tag{3}$$

exist. Then it follows that

$$\begin{aligned} g - ad_h\left(s^{(h)}(g)\right) &= \overline{g}^{(h)} \\ ad_h(\overline{g}^{(h)}) &= 0. \end{aligned}$$

providing $w = s^{(h)}(g)$ as the solution of the classical homological equation (1).

Recently the author has used the fact that the Schrödinger equation can be formulated as an infinite-dimensional Hamiltonian system and has formally extended the averaging integrals (2, 3) to self-adjoint operators and constructed formal analogs of the classical Poincaré normal form as well as a KAM algorithm for both stationary and time-dependent systems [5, 6, 7]. Here we shall restrict ourselves to the time-independent case in which the equivalent of $\varphi(\epsilon)$ above becomes a unitary ϵ-dependent transformation generated by a self-adjoint operator $W(\epsilon)$. Following the procedure outlined for the classical case, the (quantum) homological equation then takes the following form. Given self-adjoint operators H, G find $S^{(H)}(G)$ such that

$$[H, G + i[S^{(H)}(G), H]] = 0 \tag{4}$$

where $[\cdot, \cdot]$ denotes the commutator. In the stationary case this formal construction of the solution of (4) can be written as

$$S^{(H)}(G) := \lim_{T\to\infty} \frac{1}{T} \int_0^T dt \int_0^t ds\, U(s)\left(G - \overline{G}^{(H)}\right)U(s)^{-1} \tag{5}$$

where $U(t) := \exp(-itH)$ and

$$\overline{G}^{(H)} := \lim_{T\to\infty} \frac{1}{T} \int_0^T dt\, U(t) G U(t)^{-1}. \tag{6}$$

The purpose of this note is to rigorously establish the existence of $\overline{G}^{(H)}$ and $S^{(H)}(G)$ as strong limits in the case of a bounded G and an H with purely discrete spectrum satisfying some kind of gap condition on the eigenvalue distance and allowing for non-uniformly bounded multiplicity, but with some restrictions on the growth-rate of their multiplicity.

2 Convergence results for quantum averaging integrals

For the proof of the main result on the convergence of the averaging integrals we will use the following preparatory Lemma, the proof of which is elementary.

Lemma 1 *Let H be self-adjoint operator on a separable Hilbert space $\mathbb{H}$ with a purely discrete spectrum whose eigenvalues are denoted by $\{E_j\}_{j\in\mathbb{N}}$, let $U(t) := \exp(-itH)$ and let P_j denote the orthogonal projection on* $\mathrm{Eig}(E_j, H)$.

Then

$$s\text{-}\lim_{m\to\infty} \sum_j^m P_j = \mathbb{I} \tag{7}$$

$$U(t) = s\text{-}\lim_{m\to\infty} \sum_j^m e^{-iE_j t} P_j \tag{8}$$

$$U(t) P_j = P_j U(t) = e^{-iE_j t} P_j \tag{9}$$

$$U(t) G U(t)^{-1} = s\text{-}\lim_{m,n\to\infty} \sum_j^m \sum_k^n e^{i(E_k - E_j)t} P_j G P_k \tag{10}$$

and the last limit is uniform *in t.*

The main result of this note establishing the existence of the averaging integrals (5) and (6) as strong limits can now be formulated as the following theorem, the proof of which uses the so-called Schur-Holmgren condition (see e.g. [8, p. 328]) and ideas laid out by Turner [9] and Howland [10].

Theorem 1 *Let $H, \mathbb{H}, \{E_j\}_{j\in\mathbb{N}}, U(t)$ and P_j be as above and suppose that*

$$\dim \mathrm{Eig}(E_k, H) \le a k^\delta \tag{11}$$

for some $a, \delta \ge 0$ and

$$|E_k - E_j| \ge b |k^\tau - j^\tau| \tag{12}$$

for some $b > 0$ *and*

$$\tau \geq 1 + \frac{\delta}{2}\,. \tag{13}$$

Then for any bounded *self-adjoint operator* G *on* $\mathbb{H}$ *the operators*

$$\overline{G}^{(H)} \;:=\; s\text{-}\lim_{T\to\infty} \frac{1}{T}\int_0^T dt\, U(t) G U(t)^{-1} \tag{14}$$

$$S^{(H)}(G) \;:=\; s\text{-}\lim_{T\to\infty} \frac{1}{T}\int_0^T dt \int_0^t ds\, U(s)\Big(G - \overline{G}^{(H)}\Big)U(s)^{-1} \tag{15}$$

exist as bounded self-adjoint operators, are given by

$$\overline{G}^{(H)} \;=\; s\text{-}\lim_{n\to\infty} \sum_j^n P_j G P_j \tag{16}$$

$$S^{(H)}(G) \;=\; s\text{-}\lim_{n,m\to\infty} \sum_j^n \sum_{k\neq j}^m \frac{P_j G P_k}{i(E_j - E_k)} \tag{17}$$

and satisfy

$$||\overline{G}^{(H)}|| \;\leq\; ||G|| \tag{18}$$

$$U(t)\,\overline{G}^{(H)}\,U^{-1}(t) \;=\; \overline{G}^{(H)} \tag{19}$$

$$||S^{(H)}(G)|| \;\leq\; C_H ||G|| \tag{20}$$

with some constant C_H *depending on* H.

Proof: We will first show that $\sum_j \sum_{k\neq j} \frac{P_j G P_k}{i(E_k - E_j)}$ is a bounded operator with a bound given by (20) (self-adjointness is then a trivial consequence). This result will then be used in evaluating the right hand side of (14). For $j \in \mathbb{N}$ and $\alpha \in D_j := \{1, \ldots, \dim \mathrm{Eig}(E_j, H)\}$ let $\psi_{j\alpha} \in \mathbb{H}$ be a basis of eigenvectors. In this basis $\sum_j \sum_{k\neq j} \frac{P_j G P_k}{i(E_k - E_j)}$ has the matrix elements $\frac{G_{j\alpha k\beta}}{i(E_k - E_j)}$ if $j \neq k$ and 0 if $j = k$ According to a theorem by Schur [8, p. 328]

$$M := \sup\Big\{ \sum_{k;k\neq j} \sum_{\beta\in D_k} \left|\frac{G_{j\alpha k\beta}}{i(E_k - E_j)}\right| \mid j \in \mathbb{N}, \alpha \in D_j \Big\} < \infty \tag{21}$$

then implies

$$||\sum_j \sum_{k\neq j} \frac{P_j G P_k}{i(E_k - E_j)}|| \leq M\,.$$

To show that indeed $M < \infty$ we note that for a given $j \in \mathbb{N}, \alpha \in D_j$

$$\sum_{k;k\neq j} \sum_{\beta\in D_k} \left|\frac{G_{j\alpha k\beta}}{i(E_k - E_j)}\right| \;\leq\; \Big(\sum_{k;k\neq j} \sum_{\beta\in D_k} |G_{j\alpha k\beta}|^2\Big)^{\frac{1}{2}} \Big(\sum_{k;k\neq j} \sum_{\beta\in D_k} \frac{1}{|E_k - E_j|^2}\Big)^{\frac{1}{2}}$$

$$\leq \quad ||G||\sqrt{a}\Big(\sum_{k;k\neq j} \frac{k^\delta}{|E_k - E_j|^2}\Big)^{\frac{1}{2}}$$
$$\leq \quad \frac{\sqrt{a}}{b}||G||\Big(\sum_{k;k\neq j} \frac{k^\delta}{|k^\tau - j^\tau|^2}\Big)^{\frac{1}{2}}$$

where we have used the assumptions (11,12). The remaining sum on the right hand side can be estimated by the integral

$$\sum_{k;k\neq j} \frac{k^\delta}{|k^\tau - j^\tau|^2} \leq c \int_{x\geq 1,|x-j|\geq 1} dx \frac{x^\delta}{(x^\tau - j^\tau)^2}$$

and for τ satisfying assumption (13) this integral can be shown to have finite bounds uniform in j. Hence M in (21) is of the form $M = C_H||G||$ which is then also an upper bound for the norm of $\sum_j \sum_{k\neq j} \frac{P_j G P_k}{i(E_k - E_j)}$ and the bound in (20) is established.

Now turn to the proof of (16). For this we note that the uniform convergence in (10) implies for any $\psi \in \mathbb{H}$

$$\begin{aligned}
\frac{1}{T}\int_0^T dt U(t)\, G\, U(t)^{-1}\psi &= \frac{1}{T}\sum_j \sum_{k\neq j} \frac{e^{i(E_k - E_j)t} - 1}{i(E_k - E_j)} P_j G P_k \psi + \sum_j P_j G P_j \psi \\
&= \frac{1}{T}\sum_j \sum_{k\neq j} \frac{U(T)P_j G P_k U(T)^{-1} - P_j G P_k}{i(E_k - E_j)}\psi \\
&\quad + \sum_j P_j G P_j \psi
\end{aligned}$$

and the boundedness of $\sum_j \sum_{k\neq j} \frac{P_j G P_k}{i(E_k - E_j)}$ shown above then gives

$$\overline{G}^{(H)}\psi = \lim_{T\to\infty} \frac{1}{T}\int_0^T dt\, U(t) G U(t)^{-1}\psi = \sum_j P_j G P_j \psi\,.$$

The estimate $||\overline{G}^{(H)}|| \leq ||G||$ given in (18) is a straightforward consequence of the orthogonality of the P_j. Self-adjointness of $\overline{G}^{(H)}$ is now an immediate consequence and (19) is then easily established with the help of (9).

It remains to show (17). Due to the boundedness of $G, \overline{G}^{(H)}$ and the uniform convergence in (10) we have for any $\psi \in \mathbb{H}$

$$\begin{aligned}
\int_0^t ds U(s)(G - \overline{G}^{(H)})U(s)^{-1}\psi &= \sum_j \sum_{k\neq j} \frac{e^{i(E_k - E_j)t} - 1}{i(E_k - E_j)} P_j G P_k \psi \\
&= U(t)\sum_j \sum_{k\neq j} \frac{P_j G P_k}{i(E_k - E_j)} U(t)^{-1}\psi
\end{aligned}$$

$$-\sum_j \sum_{k\neq j} \frac{P_j G P_k}{i(E_k - E_j)}\psi$$

and using once more (10) thus

$$\frac{1}{T}\int_0^T dt \int_0^t ds U(s)(G - \overline{G}^{(H)})U(s)^{-1}\psi = \frac{1}{T}\sum_j \sum_{k\neq j} \frac{P_j G P_k}{(E_k - E_j)^2}\psi$$

$$-\frac{1}{T}U(T)\Big(\sum_j \sum_{k\neq j} \frac{P_j G P_k}{(E_k - E_j)^2}\Big)U(T)^{-1}\psi$$

$$+\sum_j \sum_{k\neq j} \frac{P_j G P_k}{i(E_j - E_k)}\psi .$$

Here one can show that $\sum_j \sum_{k\neq j} \frac{P_j G P_k}{(E_k - E_j)^2}$ is a bounded operator by exactly the same procedure as the one used to show the boundedness of $\sum_j \sum_{k\neq j} \frac{P_j G P_k}{i(E_k - E_j)}$. Hence

$$\lim_{T\to\infty} \frac{1}{T}\int_0^T dt \int_0^t ds U(s)(G - \overline{G}^{(H)})U(s)^{-1}\psi = \sum_j \sum_{k\neq j} \frac{P_j G P_k}{i(E_j - E_k)}\psi$$

which proves (17). ■

As a final remark we note that the case of non-degenerate equidistant spectrum, i.e. $E_{k+1} - E_k = const$ which corresponds to $\delta = 0$ and $\tau = 1$ is included in the conditions of Theorem 1.

References

[1] M. KUMMER, *Nuovo Cimento* **1 B**, 123 (1971).

[2] J. BELLISARD, in: *Trends and Developments in the Eighties* (Eds. S. Albeverio and P. Blanchard), 1–106, Singapore (1985), World Scientific.

[3] M. COMBESCURE, *Ann. Inst. H. Poincaré* **47** (1), 63 (1987).

[4] P. DUCLOS and P. ŠŤOVÍČEK, *Comm. Math. Phys.* **177**, 327 (1996).

[5] W. SCHERER, *J. Phys. A: Math. Gen.* **27**, 8231 (1994).

[6] W. SCHERER, *J. Phys. A: Math. Gen.* **30**, 2825 (1997).

[7] W. SCHERER, *J. Math. Phys.* **39** (5), 2597 (1998).

[8] A. E. TAYLOR, *Introduction to Functional Analysis*, John Wiley & Sons, New York (1958).

[9] R. E. L. TURNER, *Comm. Pure Appl. Math.* **18**, 519 (1965).

[10] J. S. HOWLAND, *Ann. Inst. H. Poincaré* **49**, 309 and 325 (1989).

Operator Theory:
Advances and Applications, Vol. 108

Monotonicity versus Non-Monotonicity in Random Operators

Günter Stolz
University of Alabama at Birmingham,
Department of Mathematics,
CH 452, Birmingham, AL 35294–1170, USA
e-mail: stolz@math.uab.edu

1 Introduction

The past two decades have brought widespread interest and a large number of publications in the mathematical theory of random operators. While this has lead to many deep results and considerable mathematical insights, there are still important open problems which are far from being solved. This becomes particularly clear when looking at what is known for models other than the relatively well understood Anderson model and its generalizations. Of course, there are big challenges left in the Anderson model as well (delocalization!), but for other physically significant models (e.g. Poisson model, random displacement model) even the seemingly better understood phenomenon of localization is mathematically far from being settled.

The purpose of this note is twofold: First, we want to shed light on the mathematical reason for this discrepancy in knowledge between various models by discussing monotonicity versus non-monotonicity in random Schrödinger operators. Second, we will present some new methods which allow to handle non-monotonous operators in dimension $d = 1$ and, in particular, lead to proofs of exponential localization at all energies for the one-dimensional Poisson and random displacement models. For brevity, we will concentrate here on sketching the main ideas in the case of the Poisson model, which is technically somewhat easier to handle. More details and also a complete treatment of the displacement model will be provided in joint work with D. Buschmann [1].

2 Monotonicity versus Non-Monotonicity

The discrete Anderson model is described by the random operator

$$h_\omega^A = h_0 + \sum_{n\in\mathbf{Z}^d} \lambda_n(\omega)\delta_n \tag{1}$$

in $\ell^2(\mathbf{Z}^d)$, where h_0 is the discrete Laplacian, ω a random parameter and the λ_n are independent, identically distributed (i.i.d.) random variables. A continuum analog is given by

$$H_\omega^A = -\Delta + \sum_{n\in\mathbf{Z}^d} \lambda_n(\omega) f(x-n) \tag{2}$$

in $L^2(\mathbf{R}^d)$ with λ_n i.i.d. and the single site potential f is assumed to be non-negative and (for simplicity) bounded and compactly supported. A suitable restriction on the distribution of λ_n makes H_ω^A almost surely self-adjoint.

Both, h_ω^A and H_ω^A, depend monotonously on the random parameters λ_n, i.e. if ω and ω' are such that $\lambda_n(\omega) \geq \lambda_n(\omega')$ for all n, then $h_\omega^A - h_{\omega'}^A \geq 0$ and $H_\omega^A - H_{\omega'}^A \geq 0$ in the sense of quadratic forms. A monotonicity property of this type enters most of the known proofs of localization for Anderson-type models since it is commonly used in establishing so-called *Wegner estimates* respectively the phenomenon of *spectral averaging*. For a notable exception see [4] where a Wegner estimate for the Anderson model without sign restriction on f (i.e. no monotonicity) is established.

Wegner estimates and/or spectral averaging are used in almost all known proofs of localization and, in fact, in proving other phenomena of recent interest like absence of diffusion or dynamical localization.

As an example, let us state the result on spectral averaging which was used by Kotani and Simon in [5] to establish localization for the continuum Anderson model H_ω^A in $d=1$. Independence of the λ_n in (2) and Fubini's theorem allow to study H_ω^A by fixing all λ_n with the exception of λ_0, which leads to a one-parameter operator of the form

$$H^A(\lambda) = -\frac{d^2}{dx^2} + V(x) + \lambda f(x). \tag{3}$$

Assume that V is continuous and $|V(x)| \leq C(1+x^2)$ and that $f \neq 0$ is continuous, non-negative and compactly supported, and let ρ_λ^A be the Weyl-Titchmarsh spectral measure for H_λ (i.e. the trace of the 2×2-matrix-valued spectral measure for H_λ, compare [2]). For arbitrary $\lambda_1 < \lambda_2$ define the averaged spectral measure ρ^A by

$$\rho^A(B) := \int_{\lambda_1}^{\lambda_2} \rho_\lambda^A(B)\, d\lambda \tag{4}$$

for Borel sets $B \subset \mathbf{R}$. It is proven in [5] that ρ^A is absolutely continuous (one can get by with weaker assumptions on V and f, see [6]). The proof uses monotonicity of certain Prüfer angles in λ which in turn is a consequence of the monotonicity of $H^A(\lambda)$ in the parameter λ (note $f \geq 0$).

While spectral averaging, i.e. absolute continuity of averaged spectral measures, carries over to the multi-dimensional Anderson model [5, 3], it does not hold for the Poisson model, at least not in its most elementary form as will be discussed next.

The one-dimensional Poisson model is given by

$$H_\omega^P = -\frac{d^2}{dx^2} + \sum_i f(x - x_i(\omega)), \tag{5}$$

where the $x_i(\omega)$ are the points of a Poisson process on $\mathbf{R}$ with constant density $\alpha > 0$. They may be labeled by $i \in \mathbf{Z} \setminus \{0\}$ such that $\ldots < x_{-1}(\omega) < 0 < x_1(\omega) < x_2(\omega) < \ldots$. Setting also $x_0 := 0$ and introducing the random distances $p_n := x_{n+1} - x_n$, $n \in \mathbf{Z}$, one has that the p_n are independent and identically distributed with density $\alpha e^{-\alpha t}$, $t \geq 0$.

The p_n are the natural analogs to the coupling constants λ_n in the Anderson model. Fixing all but one of the p_n leads us (up to a linear parameter change) to an operator family of the form

$$H^P(a) = -\frac{d^2}{dx^2} + V_a(x) \tag{6}$$

for $a > 0$, where

$$V_a(x) = \begin{cases} V(x) & \text{for } x < 0, \\ 0 & \text{for } 0 \leq x < a, \\ V(x-a) & \text{for } x \geq a. \end{cases} \tag{7}$$

Here $V \in L^1_{loc}(\mathbf{R})$ is a given real-valued potential such that $-\frac{d^2}{dx^2} + V$ is limit point at $\pm\infty$, guaranteeing self-adjointness of $H^P(a)$ for all a. Proceeding as with the Anderson model define

$$\rho^P(B) := \int_{a_1}^{a_2} \rho_a^P(B)\, da, \tag{8}$$

where $a_2 > a_1 > 0$ are arbitrary and ρ_a^P is the Weyl-Titchmarsh spectral measure for $H^P(a)$.

Is ρ^P absolutely continuous?

This should not necessarily be expected since the family of potentials V_a is not monotonous in a. Nevertheless, it was proven in [7] that ρ^P is at least absolutely continuous at energies $E > 0$, which lead to a proof of localization for the one-dimensional Poisson model at positive energies. However, localization at negative energies remained an open problem.

Indeed, ρ^P is **not** necessarily absolutely continuous at negative energies as can be seen from the following counterexample: Fix a number $E < 0$. Then pick a potential V on the negative half line $(-\infty, 0)$ such that $-u'' + Vu = Eu$ has a solution $u_- \in L^2(-\infty, 0)$ which also satisfies the boundary condition $u_-'(0)/u_-(0) =$

$-\sqrt{|E|}$ (some finite negative well will do for V). Then fix V on $(0,\infty)$ such that $-u'' + Vu = Eu$ has a solution $u_+ \in L^2(0,\infty)$ with $u_+'(0)/u_+(0) = -\sqrt{|E|}$ (even $V = 0$ on $(0,\infty)$ can be used here, but less trivial choices of V are possible if desired). u_- and u_+ can now be suitably matched by

$$u(x) = \begin{cases} u_-(x), & x < 0 \\ c_1 \exp(-\sqrt{|E|}x), & 0 \le x < a, \\ c_2 u_+(x-a), & x \ge a \end{cases} \tag{9}$$

to yield a globally square-integrable solution of $-u'' + V_a u = Eu$ for **every** $a > 0$. Thus E is an eigenvalue of $H^P(a)$ for every a, i.e. $\rho_a^P(\{E\}) > 0$ and therefore $\rho^P(\{E\}) = \int \rho_a^P(\{E\})\, da > 0$. ρ^P is not absolutely continuous!

3 Localization for the Poisson model

It violates physical arguments that the breakdown of spectral averaging for $H^P(a)$ at $E < 0$ should correspond with the breakdown of localization for the 1-d Poisson model at $E < 0$, taking into account that localization is already proven at $E > 0$: If there is a mobility edge at all, then delocalized states should appear at high rather than low energy. Thus our failure must have technical reasons. In fact, we will outline now how these problems can be overcome and the following theorem is proven.

Theorem 1 *In the Poisson model H_ω^P defined by (5) assume that $f \in L^2(\mathbf{R})$ is real-valued, compactly supported and $f \not\equiv 0$. Then H_ω^P has almost surely pure point spectrum at all energies with exponentially decaying eigenfunctions.*

The first key insight in getting this result is to realize that fixing all but one of the p_n to arrive at the family $H^P(a)$ is too restrictive when dealing with negative energies in the Poisson model. We have all the independent random parameters p_n to play with. Considering the above counterexample, we argue that it should be less likely for an eigenvalue E to be stationary in, say, two independent parameters rather than just one. We thus attempt to prove *two-parameter spectral averaging* for the Poisson model:

Fixing all but two of the p_n, say p_0 and p_1, leads (again up to linear parameter changes) to the two-parameter model

$$H^P(a,b) = -\frac{d^2}{dx^2} + V_{a,b}, \tag{10}$$

where for $a > 0$ and $b > 0$

$$V_{a,b}(x) = \begin{cases} V(x), & x < 0 \\ 0, & 0 \le x < a \\ f(x-a), & a \le x < a+D \\ 0, & a+D \le x < a+D+b \\ V(x-(a+D+b)), & a+D+b \le x. \end{cases} \tag{11}$$

Here it is assumed w.l.o.g. that supp $f \subset [0, D]$. The parameters a and b are positive if p_0 and p_1 are sufficiently large, which happens with positive probability. This will eventually lead to a proof of localization for H_ω^P with positive probability and thus with probability one since spectral types are almost surely deterministic.

We would like to prove absolute continuity of the measure $\tilde{\rho}^P$ defined by

$$\tilde{\rho}^P(B) = \int_{b_1}^{b_2} \int_{a_1}^{a_2} \rho_{a,b}^P(B)\, da\, db, \tag{12}$$

where $a_2 > a_1 > 0$ and $b_2 > b_1 > 0$ are arbitrary and $\rho_{a,b}^P$ is the Weyl-Titchmarsh spectral measure for $H^P(a,b)$.

Unfortunately, the mechanism underlying the above counterexample causes trouble again: Take $E < 0$ and V as constructed in the counterexample. In addition, pick f such that there exists a solution of $-u'' + fu = Eu$ satisfying

$$u'(0)/u(0) = -\sqrt{|E|} \quad \text{and} \quad u'(D)/u(D) = -\sqrt{|E|}. \tag{13}$$

It is readily seen that this solution can be "implanted" in the solution constructed in the counterexample in a way such that a global L^2-solution of $-u'' + V_{a,b}u = Eu$ results independent of a and b. The eigenvalue E is stationary under both parameters!

It is at this point where the second key insight enters our proof: Given $E < 0$ it is easy to construct non-trivial f such that $-u'' + fu = Eu$ has a solution satisfying (13). But in Theorem 1 an $f \not\equiv 0$ is given, and the more relevant task is to determine the set of those negative E's where solutions satisfying (13) exist. The following result says that this type of "spectral parameter dependent boundary value problem" has discrete "spectrum" (also taking into account that our counterexample works in the same way if $-\sqrt{|E|}$ is replaced by $+\sqrt{|E|}$).

Proposition 2: *Let $f \in L^1(0, D)$ be real-valued and $f \not\equiv 0$. Then the set M_0 of all those $E < 0$ such that there exists a solution u of $-u'' + fu = Eu$ on $[0, D]$ which satisfies*

$$\frac{u'(0)}{u(0)} \in \left\{-\sqrt{|E|}, +\sqrt{|E|}\right\} \quad \textit{and} \quad \frac{u'(D)}{u(D)} \in \left\{-\sqrt{|E|}, +\sqrt{|E|}\right\} \tag{14}$$

is discrete in $(-\infty, 0)$.

This has to be considered as a result from inverse spectral theory. In one possible proof f is extended periodically to a potential F on $\mathbf{R}$. Suppose that M_0 is not discrete. Then an analyticity argument shows that $M_0 = (-\infty, 0)$ and Floquet theory can be used to show that $\sigma(-d^2/dx^2 + F) = [0, \infty)$ (for details, see [1]). By a classical result of Borg this would imply the contradiction $F \equiv 0$.

We have understood that spectral averaging (even over two parameters) can not hold on the set M_0. It turns out that this and $\{0\}$ are the only sets which need to be excluded:

Proposition 3 *Let the family $H^P(a,b)$ be defined by (10) and (11) with $f \in L^1(0,D)$, $f \not\equiv 0$, and $V \in L^1_{loc}(\mathbf{R})$ such that $-d^2/dx^2 + V$ is limit point at $\pm\infty$. Let $M_0 \subset (-\infty, 0)$ be the discrete set from Proposition 2. Then the measure $\tilde{\rho}^P$ defined by (12) is absolutely continuous on $\mathbf{R} \setminus (M_0 \cup \{0\})$.*

Equipped with Proposition 3, the proof of Theorem 1 can be completed along well known lines of reasoning as used for example in [5] and [7]. Since f is compactly supported it follows that H^P_ω is non-deterministic in the sense of Kotani theory. Again by Kotani theory this implies positivity of the Lyapunov exponent $\gamma(E)$ for almost every E. Osceledec's theorem yields the existence of exponentially decaying solutions, separately at $+\infty$ and $-\infty$. Proposition 3 is the key ingredient into "matching" these solutions to get almost surely pure point spectrum with exponentially decaying eigenfunctions for energies in $\mathbf{R}\setminus(M_0 \cup \{0\})$. The remaining set $M_0 \cup \{0\}$ is countable and closed by Proposition 2 and thus can not support any continuous spectrum either. Detailed arguments using the Gilbert-Pearson theory of subordinate solutions in the "matching process" (replacing the use of Shnol-type results on polynomial boundedness of generalized eigenfunctions in earlier treatments) are provided in [1].

We also refer to [1] for the proof of Proposition 3. Is is based on explicit formulas for Weyl-Titchmarsh spectral measures which were first found to be useful in studying 1-d random operators by Carmona in the early 80s and rely heavily on Prüfer transform methods. At negative energies outside the set M_0 it is seen that the Prüfer phases appearing in this formalism are either non-stationary in a or non-stationary in b. This is used to prove spectral averaging.

References

[1] D. Buschmann and G. Stolz: Two-Parameter Spectral Averaging and Localization for Non-Monotoneous Random Schrödinger Operators. Preprint

[2] E.A. Coddington and N. Levinson: Theory of ordinary differential equations. McGraw-Hill, New York 1955

[3] J.-M. Combes and P.D. Hislop: Localization for continuous random Hamiltonians in d-dimensions. *J. Funct. Anal.* **124**, 149 – 180 (1994)

[4] F. Klopp: Localization for some continuous random Schrödinger operators. *Commun. Math. Phys.* **167**, 553 – 569 (1995)

[5] S. Kotani and B. Simon: Localization in General One-Dimensional Random Systems. *Commun. Math. Phys.* **112**, 103 – 119 (1987)

[6] G. Stolz: Spectral theory of Schrödinger operators with potentials of infinite barriers type. Habilitationsschrift, Frankfurt 1994

[7] G. Stolz: Localization for random Schrödinger operators with Poisson potential. *Ann. Inst. Henri Poincaré* **63**, 297 – 314 (1995)

Operator Theory:
Advances and Applications, Vol. 108

A model in perturbation theory

D. Yafaev
Université de Rennes

Abstract

We consider a model in the perturbation theory where the Hamiltonian $\mathbf{H}$ is a sum of a multiplication operator $\mathbf{H}_0$ by $|x|^{2l}$ in the space $L_2(\mathbb{R}^d)$ and of an integral operator $\mathbf{V}$ with oscillating kernel. We find conditions for the coincidence of the essential spectra of the operators $\mathbf{H}_0$ and $\mathbf{H}$ and for the existence and completeness of the wave operators relating $\mathbf{H}_0$ and $\mathbf{H}$.

1. Our aim here is to introduce a new model in the perturbation theory. Let $\mathbf{H}_0$ be multiplication by $|x|^{2l}$, $l > 0$, in the space $L_2(\mathbb{R}^d)$ and let $\mathbf{V}$ be an integral operator defined by

$$(\mathbf{V}u)(x) = \int_{\mathbb{R}^d} v(\langle x, y\rangle)u(y)dy, \tag{1}$$

where v is a real periodic function

$$v(t) = \sum_{n=-\infty}^{\infty} v_n e^{int}, \quad v_{-n} = \bar{v}_n, \quad v_0 = 0. \tag{2}$$

The operator $\mathbf{V}$ can also be viewed as a Fourier integral operator. For example, if

$$v(t) = \gamma \cos t \quad \text{or} \quad v(t) = \gamma \sin t, \tag{3}$$

then (1) is a linear combination of the classical Fourier operator and of its adjoint. In this paper we study spectral properties of the Hamiltonian $\mathbf{H} = \mathbf{H}_0 + \mathbf{V}$ in the space $L_2(\mathbb{R}^d)$. The case of a constant function $v(t)$ corresponds [1] to the point interaction and is excluded by the condition $v_0 = 0$. Passing to the spherical coordinates, considering the space $L_2(\mathbb{R}^d)$ as $L_2(\mathbb{R}_+; L_2(\mathbb{S}^{d-1}))$ and making the change of variables $\lambda = |x|^{2l}$, we can fit the operator $\mathbf{H}$ into the framework of the Friedrichs model. However, the results of the paper [5] are not applicable to perturbations (1) (even in the case $d = 1$) since the kernels obtained do not satisfy its assumptions either for $\lambda = 0$ or for $\lambda \to \infty$ (or for both these points).

For the case (3), the model we consider first appeared in [4]. It was checked in [4] that $\sigma_{ess}(\mathbf{H}) = \sigma_{ess}(\mathbf{H}_0)$ for arbitrary $l > 0$ and that the wave operators $W_\pm(\mathbf{H}, \mathbf{H}_0)$ exist and are complete if $2l > d$. The discrete spectrum of $\mathbf{H}$ for such functions $v(t)$ was studied in [8].

We show that under the assumption

$$\sum_{n=1}^{\infty} |v_n| n^{-d/2} < \infty \tag{4}$$

the operator $\mathbf{V}$ is bounded in $L_2(\mathbb{R}^d)$ and $\sigma_{ess}(\mathbf{H}) = \sigma_{ess}(\mathbf{H}_0)$ for arbitrary $l > 0$. Moreover, if

$$\sum_{n=1}^{\infty} |v_n| n^{-d/2+1+\varepsilon} < \infty \tag{5}$$

for some $\varepsilon > 0$, then the wave operators $W_\pm(\mathbf{H}, \mathbf{H}_0)$ exist and are complete provided $2l > 1$. Note that both conditions (4) and (5) are getting less restrictive as d increases.

We found also necessary and sufficient conditions for the negative spectrum of the operator $\mathbf{H}$ to be finite and calculated explicitly the total number of its negative eigenvalues. These results will be published elsewhere.

2. Let us first of all rewrite (1) in operator notation. Let $\mathbf{\Phi}$,

$$(\Phi u)(x) = (2\pi)^{-d/2} \int_{\mathbb{R}^d} e^{-i\langle x,y\rangle} u(y)dy,$$

be the Fourier operator and let $\mathbf{T}_n$,

$$(\mathbf{T}_n u)(x) = n^{d/2} u(nx),$$

be the unitary operator of dilation s in $L_2(\mathbb{R}^d)$. Then equality (1) can be rewritten as

$$\mathbf{V} = (2\pi)^{d/2} \sum_{n=1}^{\infty} n^{-d/2} \mathbf{T}_n \,(v_n \Phi^* + \bar{v}_n \Phi). \tag{6}$$

To be precise, we take (6) for the definition of $\mathbf{V}$. Under assumption (4), series (6) is convergent in norm so that the operator $\mathbf{V}$ is bounded. This implies of course that the operator $\mathbf{H}$ is self-adjoint on the domain $\mathcal{D}(\mathbf{H}_0)$ of the operator $\mathbf{H}_0$.

As a side remark, we note that the operator

$$\tilde{\mathbf{V}} = \Phi \mathbf{V} \Phi^* = (2\pi)^{d/2} \sum_{n=1}^{\infty} n^{-d/2} \mathbf{T}_{1/n} \,(v_n \Phi^* + \bar{v}_n \Phi)$$

has the structure similar to that of $\mathbf{V}$. In particular, if (2) contains only a finite number of terms, then $\tilde{\mathbf{V}}$ can be written as an integral operator (1) with kernel $\tilde{v}(\langle x,y\rangle)$ where $\tilde{v}(t)$ is again a periodic function of t (but of course its period is not necessarily 2π). Thus, all our results can be reformulated in terms of the Hamiltonian $\tilde{\mathbf{H}} = \tilde{\mathbf{H}}_0 + \tilde{\mathbf{V}}$ where $\tilde{\mathbf{H}}_0 = (-\Delta)^l$.

Let Q be multiplication in the space $L_2(\mathbb{R}^d)$ by some bounded function $q(x)$ which tends to 0 as $|x| \to \infty$. Then the operator $Q\Phi Q$ is compact. It follows that operators $Q\mathbf{T}_n\Phi Q$ and $Q\mathbf{T}_n\Phi^*Q$ are compact for all n. According to (6), under assumption (4), this implies compactness of operators $Q\mathbf{V}Q$ and, in particular, of $(\mathbf{H}_0 + cI)^{-1}\mathbf{V}(\mathbf{H}_0 + cI)^{-1}$. Let us now use the resolvent identity

$$(\mathbf{H}+cI)^{-1} - (\mathbf{H}_0+cI)^{-1} = -\Big((\mathbf{H}+cI)^{-1}(\mathbf{H}_0+cI)\Big)\Big((\mathbf{H}_0+cI)^{-1}\mathbf{V}(\mathbf{H}_0+cI)^{-1}\Big), \tag{7}$$

where $c > 0$ is sufficiently large. The first factor in the right-hand side is bounded because $\mathbf{V}$ is bounded and the second is compact. Hence operator (7) is also compact which, by the Weyl theorem, ensures that the essential spectra of the operators $\mathbf{H}_0$ and $\mathbf{H}$ coincide. This yields the first result.

Theorem 1 *Let condition* (4) *hold. Then the operator* $\mathbf{V}$ *is bounded and the operator* $\mathbf{H}$ *is self-adjoint on the domain of the operator* $\mathbf{H}_0$. *Moreover,* $\sigma_{ess}(\mathbf{H}) = [0, \infty)$.

3. The operators $\mathbf{H}_0$ and $\mathbf{V}$ are rotationally invariant. This allows us to separate variables in the spherical coordinates which is necessary for a more detailed study of the operator $\mathbf{H}$. Let $\mathfrak{h}_m$ be the subspace of spherical functions $Y_m(\omega)$, $\omega \in \mathbb{S}^{d-1}$, of order m. Let $\mathcal{K}$ be the L_2-space with weight r^{d-1} of functions defined on $\mathbb{R}_+$ and let $\mathfrak{H}_m = \mathcal{K} \otimes \mathfrak{h}_m$. To put it differently, $\mathfrak{H}_m \subset \mathbb{R}^d$ is the subspace spanned by functions

$$u_m(x) = |x|^{-\delta} f(|x|) Y_m(\hat{x}), \quad \hat{x} = x|x|^{-1}, \quad \delta = (d-1)/2, \tag{8}$$

where $f \in L_2(\mathbb{R}_+)$ and $Y_m \in \mathfrak{h}_m$. Then

$$L_2(\mathbb{R}^d) = \bigoplus_{m=0}^{\infty} \mathfrak{H}_m, \quad \mathfrak{H}_m = \mathcal{K} \otimes \mathfrak{h}_m, \tag{9}$$

and every subspace $\mathfrak{H}_m$ is invariant with respect to the Fourier operator Φ, which reduces to the Fourier-Bessel transform on $\mathfrak{H}_m$. More precisely, set

$$(\Phi_m f)(r) = \int_0^\infty (rs)^{1/2} \mathcal{I}_{m+(d-2)/2}(rs) f(s) ds, \tag{10}$$

where $\mathcal{I}_{m+(d-2)/2}$ is the Bessel function. Then, for function (8),

$$(\Phi u_m)(x) = i^{-m} |x|^{-\delta} (\Phi_m f)(|x|) Y_m(\hat{x}). \tag{11}$$

The operator Φ_m is of course unitary on $L_2(\mathbb{R}_+)$ and $\Phi_m = \Phi_m^*$. Clearly,

$$(\mathbf{T}_n u_m)(x) = |x|^{-\delta} (T_n f)(|x|) Y_m(\hat{x}), \tag{12}$$

where T_n,

$$(T_n f)(r) = n^{1/2} f(nr),$$

is the unitary operator of dilations in $L_2(\mathbb{R}_+)$.

Set

$$v_m(t) = (2\pi)^{d/2} \sum_{n=1}^{\infty} n^{-\delta}(v_n i^m + \bar{v}_n i^{-m})(nt)^{1/2} \mathcal{I}_{m+(d-2)/2}(nt)$$

and

$$(V_m f)(r) = \int_0^{\infty} v_m(rs) f(s) ds.$$

According to (10)

$$V_m = (2\pi)^{d/2} \sum_{n=1}^{\infty} n^{-d/2}(v_n i^m + \bar{v}_n i^{-m}) T_n \Phi_m. \tag{13}$$

This equality should be considered as the precise definition of the operator V_m.

Obviously, $\mathfrak{H}_m$ is invariant with respect to the operator $\mathbf{H}_0$. It follows from (8), (11) and (12) that every subspace $\mathfrak{H}_m$ is also invariant with respect to operator (6). It reduces on $\mathfrak{H}_m$ to the operator $\Omega V_m \Omega^* \otimes I_m$ where $\Omega : L_2(\mathbb{R}_+) \to \mathcal{K}$ is a unitary operator defined by $(\Omega g)(r) = r^{-\delta} g(r)$ and I_m is the identity operator in the space $\mathfrak{h}_m$. Let H_0 be multiplication by x^{2l} and $H^{(m)} = H_0 + V_m$ in the space $L_2(\mathbb{R}_+)$. Then

$$\mathbf{H} = \bigoplus_{m=0}^{\infty} \Omega H^{(m)} \Omega^* \otimes I_m, \tag{14}$$

and hence the spectral analysis of the operator $\mathbf{H}$ reduces to that for the family of operators $H^{(m)}$.

4. Now we can check that the wave operators

$$W_{\pm}(\mathbf{H}, \mathbf{H}_0) = s - \lim_{t \to \pm\infty} \exp(i\mathbf{H}t) \exp(-i\mathbf{H}_0 t)$$

exist and are complete, that is their ranges coincide with the absolutely continuous subspace of the operator $\mathbf{H}$. According to (9), (14), to that end it suffices to verify the same facts for all pairs $H_0, H^{(m)}$, $m = 0, 1, \ldots$. By the Birman-Krein theorem [2] (see also [7]), we need only to show that operator

$$(H^{(m)} + cI)^{-1} - (H_0 + cI)^{-1} =$$

$$-\Big((H^{(m)} + cI)^{-1}(H_0 + cI)\Big)\Big((H_0 + cI)^{-1} V_m (H_0 + cI)^{-1}\Big),$$

belongs to the trace class $\mathfrak{S}_1$ or that

$$(H_0 + I)^{-1} V_m (H_0 + I)^{-1} \in \mathfrak{S}_1. \tag{15}$$

According to (13) it suffices to check that the operators $(H_0+I)^{-1}T_n\Phi_m(H_0+I)^{-1}$ belong to $\mathfrak{S}_1$ and the corresponding trace norms satisfy

$$\sum_{n=1}^{\infty}|v_n|n^{-d/2}||(H_0+I)^{-1}T_n\Phi_m(H_0+I)^{-1}||_1<\infty. \tag{16}$$

Since

$$(H_0+I)^{-1}T_n=n^{2l}T_n(H_0+n^{2l}I)^{-1},$$

condition (16) is equivalent to

$$\sum_{n=1}^{\infty}|v_n|n^{-d/2+2l}||(H_0+n^{2l}I)^{-1}\Phi_m(H_0+I)^{-1}||_1<\infty. \tag{17}$$

Next we use the following

Lemma 2 *Let the operator Φ_m be defined by formula* (10) *and let P and Q be multiplications by functions p and q. Then the operator $P\Phi_m Q$ acting in the space $L_2(\mathbb{R}_+)$ belongs to the trace class and*

$$||P\Phi_m Q||_1^2\le C\int_0^\infty|p(r)|^2(1+r^2)^\sigma dr\int_0^\infty|q(s)|^2(1+s^2)^\sigma ds,\quad C=C(m,\sigma), \tag{18}$$

provided the right-hand side is finite for some $\sigma>1/2$.

A proof can be obtained similarly to the case when the role of Φ_m is played by the usual Fourier transform (see [3] or [6]), so it will be given in a short form. Actually, we follow closely the proof of T. Kato published in [6]. Let G be multiplication by $(1+r^2)^{1/2}$. We have that $P\Phi_m Q=K_1LK_0$ where

$$K_1=PG^\sigma\Phi_m G^{-\sigma},\quad L=G^\sigma\Phi_m G^{-\sigma}\Phi_m G^{-\sigma}\Phi_m G^\sigma,\quad K_0=G^{-\sigma}\Phi_m G^\sigma Q.$$

Clearly, the Hilbert-Schmidt norm

$$||K_0||_2^2=\int_0^\infty\int_0^\infty(1+r^2)^{-\sigma}rs\mathcal{I}^2_{m+(d-2)/2}(rs)(1+s^2)^\sigma|q(s)|^2drds$$

$$\le C\int_0^\infty|q(s)|^2(1+s^2)^\sigma ds,$$

since the function $t\mathcal{I}^2_{m+(d-2)/2}(t)$ is bounded. Quite similarly,

$$||K_1||_2^2\le C\int_0^\infty|p(r)|^2(1+r^2)^\sigma dr.$$

To conclude the proof of (18), it remains to check that the operator $L=L_\sigma$ is bounded. Let $\mathbf{G}$ be multiplication by $(1+|x|^2)^{1/2}$ in the space $L_2(\mathbb{R}^d)$ and

$$\mathbf{L}_\sigma=\mathbf{G}^\sigma\Phi\mathbf{G}^{-\sigma}\Phi^*\mathbf{G}^{-\sigma}\Phi\mathbf{G}^\sigma.$$

By a direct calculation, one easily verifies that the operator $\mathbf{L}_2$ is bounded in $L_2(\mathbb{R}^d)$. Then, by the complex interpolation, this results extends to all $\sigma \in (0,2)$. The restriction of the operator $\mathbf{L}_\sigma$ on the subspace $\mathfrak{H}_m$ equals $i^{-m}\Omega L_\sigma \Omega^* \otimes I_m$. Hence the operator L_σ is bounded in $L_2(\mathbb{R}_+)$.

It follows from Lemma 2 that, in the case $2l > 1$, all operators

$$(H_0 + n^{2l} I)^{-1} \Phi_m (H_0 + I)^{-1}$$

belong to the trace class and

$$||(H_0 + n^{2l} I)^{-1} \Phi_m (H_0 + I)^{-1}||_1 \leq C n^{-2l+1+\varepsilon}, \quad C = C(m, \varepsilon),$$

for any $\varepsilon > 0$. Therefore, under assumption (5), the series (17) converges which implies (15). Let us formulate the result obtained.

Theorem 3 *Let $l > 1/2$ and let condition* (5) *hold. Then the wave operators $W_\pm(\mathbf{H}, \mathbf{H}_0)$ exist and are complete. In particular, the absolutely continuous part of $\mathbf{H}$ is unitarily equivalent to the operator $\mathbf{H}_0$.*

Compared to [4] we allow sufficiently more general functions $v(t)$ and relax the condition $2l > d$ imposed in [4] because separation of variables was not taken into account there.

REFERENCES

[1] F. A. Berezin, L. D. Faddeev, A remark on Schrödinger's equation with a singular potential, Soviet Math. Dokl. **2** (1961), 372-375.

[2] M. Sh. Birman and M. G. Krein, On the theory of wave operators and scattering operators, Soviet Math. Dokl. **3** (1962), 740-744.

[3] M. Sh. Birman and M. Z. Solomyak, Estimates for the singular numbers of integral operators, Russian Math. Surveys, **32**, no. 1 (1977), 15-89.

[4] M. Sh. Birman and D. R. Yafaev, On the trace-class method in potential scattering theory, J. Soviet Math. **56**, no. 2 (1991), 2285-2299.

[5] L. D. Faddeev, On the Friedrichs model in the theory of perturbations of the continuous spectrum, Amer. Math. Soc. Transl. Ser.2 **62** (1967).

[6] M. Reed and B. Simon, Methods of modern mathematical physics, v.3, Academic Press, 1979.

[7] D. R. Yafaev, Mathematical scattering theory, AMS, 1992.

[8] D. R. Yafaev, The discrete spectrum in the singular Friedrichs model, Prépublication 98-06, Université de Rennes 1, 1998.

Operator Theory:
Advances and Applications, Vol. 108

BAND GAP OF THE SPECTRUM IN PERIODICALLY CURVED QUANTUM WAVEGUIDES

KAZUSHI YOSHITOMI

Graduate School of Mathematics, Kyushu University
6-10-1 Hakozaki Higashiku Fukuoka, 812-8581, Japan
yositomi@math.kyushu-u.ac.jp

In this paper we study the spectral gap of the Dirichlet Laplacian on a periodically curved strip. We begin with the definition of the planar strip and the Dirichlet Laplacian. For $\gamma \in C(\mathbf{R};\mathbf{R})$, let

$$\kappa_\gamma \; : \mathbf{R} \ni s \mapsto (a_\gamma(s), b_\gamma(s)) \in \mathbf{R}^2$$

be a curve parameterized by its arc length whose curvature at s is $\gamma(s)$. For $d > 0$, we define

$$\Omega_{d,\gamma} \equiv \left\{ \left(a_\gamma(s) - u\frac{d}{ds}b_\gamma(s),\; b_\gamma(s) + u\frac{d}{ds}a_\gamma(s) \right) \in \mathbf{R}^2 \;;\; s \in \mathbf{R},\; -d < u < d \right\}.$$

$\Omega_{d,\gamma}$ is the region obtained by sliding the segment of length $2d$ which intersects κ_γ perpendicularly at its midpoint. Let $-\Delta^D_{\Omega_{d,\gamma}}$ be the Dirichlet Laplacian on $\Omega_{d,\gamma}$, i.e., the Friedrichs extension of the operator

$$-\Delta \quad \text{in} \quad L^2(\Omega_{d,\gamma}) \quad \text{with domain} \quad C_0^\infty(\Omega_{d,\gamma}).$$

This operator is the Hamiltonian for an electron confined in a quantum wire. Exner and Šeba proved the existence of bound states below the essential spectrum under suitable decay conditions on $\gamma(s)$ (see [3]). We also note that there are many works about elliptic differential equations on thin manifolds (see [2] and their references).

We consider the case that γ is periodic. We impose the following assumptions on γ.

$(A.1)$ $\gamma \in C^\infty(\mathbf{R})$.
$(A.2)$ $\gamma(s+\pi) = -\gamma(s)$ *for any* $s \in \mathbf{R}$.
$(A.3)$ *There exists* $d_0 > 0$ *such that*
(i) $\max_{s\in[0,\pi]} |\gamma(s)| < 1/d_0$,

(ii) $\Omega_{d_0,\gamma}$ *is not self-intersecting.*

Besides we assume $d \in (0, d_0]$. Under these conditions, the spectrum of $-\Delta^D_{\Omega_{d,\gamma}}$ has a band structure. For $\theta \in [0,2]$, we define

$$H_{\theta,d} \equiv -\frac{\partial}{\partial s}(1+u\gamma(s))^{-2}\frac{\partial}{\partial s} - \frac{\partial^2}{\partial u^2} + V(s,u) \quad \text{in} \quad L^2((0,\pi)\times(-d,d))$$

with domain

$$\begin{aligned} D_{\theta,d} \equiv \Big\{ & v(s,u) \in H^2((0,\pi)\times(-d,d)); \\ & v(\cdot,-d) = v(\cdot,d) = 0 \text{ in } L^2((0,\pi)), \\ & v(\pi,u) = e^{i\pi\theta}v(0,-u) \text{ in } L^2((-d,d)), \\ & \frac{\partial}{\partial s}v(\pi,u) = e^{i\pi\theta}\frac{\partial}{\partial s}v(0,-u) \text{ in } L^2((-d,d)) \Big\}, \end{aligned}$$

where

$$V(s,u) \equiv \frac{1}{2}(1+u\gamma(s))^{-3}u\gamma''(s) - \frac{5}{4}(1+u\gamma(s))^{-4}u^2\gamma'(s)^2 - \frac{1}{4}(1+u\gamma(s))^{-2}\gamma(s)^2.$$

We denote by $\mathcal{E}_j(\theta;d)$ the j-th eigenvalue of $H_{\theta,d}$ counted with multiplicity. Using the Floquet-Bloch reduction scheme, we have

$$\sigma(-\Delta^D_{\Omega_{d,\gamma}}) = \bigcup_{j=1}^{\infty}\bigcup_{\theta\in[0,2]}\{\mathcal{E}_j(\theta;d)\}.$$

We consider the asymptotic behavior of $\mathcal{E}_j(\theta;d)$ as d tends to 0. For $\theta \in [0,2]$, let

$$K_\theta \equiv -\frac{d^2}{ds^2} - \frac{1}{4}\gamma(s)^2 \quad \text{in} \quad L^2((0,\pi))$$

with domain

$$F_\theta \equiv \{v \in H^2((0,\pi))\ ; v(\pi) = e^{i\pi\theta}v(0),\ v'(\pi) = e^{i\pi\theta}v'(0)\}.$$

We call K_θ the reference operator for $H_{\theta,d}$. We denote by $k_j(\theta)$ the j-th eigenvalue of K_θ counted with multiplicity. Then we have the following.

Theorem. *For $\theta \in [0,2]$ and $j \in \mathbf{N}$, we have*

$$\mathcal{E}_j(\theta;d) = \left(\frac{\pi}{2d}\right)^2 + k_j(\theta) + O(d) \ \ (as\ d \to 0),$$

where the error term is uniform with respect to $\theta \in [0,2]$.

Proof (Sketch). In this proof, we mainly use the min-max principle. We first introduce approximate operators similar to those of [3]. We set

$$\Sigma_d \equiv (0,\pi) \times (-d,d).$$

For $\theta \in [0,2]$, we define

$$H^{\pm}_{\theta,d} \equiv -(1 \mp d\gamma_+)^{-2}\frac{\partial^2}{\partial s^2} - \frac{\partial^2}{\partial u^2} + V_{\pm}(s) \quad \text{in} \quad L^2(\Sigma_d) \quad \text{with domain} \quad D_{\theta,d},$$

where

$$V_+(s) \equiv \frac{1}{2}(1-d\gamma_+)^{-3}d\gamma_+'' - \frac{1}{4}(1+d\gamma_+)^{-2}\gamma(s)^2,$$

$$V_-(s) \equiv -\frac{1}{2}(1-d\gamma_+)^{-3}d\gamma_+'' - \frac{5}{4}(1-d\gamma_+)^{-4}d^2(\gamma_+')^2 - \frac{1}{4}(1-d\gamma_+)^{-2}\gamma(s)^2,$$

$$\gamma_+ \equiv \max_{s\in[0,\pi]}|\gamma(s)|, \quad \gamma_+' \equiv \max_{s\in[0,\pi]}|\gamma'(s)|, \quad \text{and} \quad \gamma_+'' \equiv \max_{s\in[0,\pi]}|\gamma''(s)|.$$

Then we have

$$H^-_{\theta,d} \le H_{\theta,d} \le H^+_{\theta,d}. \tag{1}$$

We estimate the eigenvalues of $H^+_{\theta,d}$ and $H^-_{\theta,d}$. For this purpose we introduce the following operators. For $\theta \in [0,2]$, we define

$$T^{\pm}_{\theta,d} \equiv -(1 \mp d\gamma_+)^{-2}\frac{d^2}{ds^2} + V_{\pm}(s) \quad \text{in} \quad L^2((0,\pi)) \quad \text{with domain} \quad F_\theta.$$

For $j \in \mathbf{N}$, we denote by $\mathcal{E}_j^{\pm}(\theta;d)$ the j-th eigenvalue of $T^{\pm}_{\theta,d}$ counted with multiplicity respectively. We note that the k-th eigenvalue of the operator

$$-\frac{d^2}{du^2} \quad \text{in} \quad L^2((-d,d)) \quad \text{with domain} \quad \{v \in H^2((-d,d))\ ;\ v(-d) = v(d) = 0\}$$

is $(\frac{\pi k}{2d})^2$. We set

$$\mu^{\pm}(j;k;\theta;d) \equiv \begin{cases} (\frac{\pi k}{2d})^2 + \mathcal{E}_j^{\pm}(\theta;d) & \text{(for odd } k\text{)}, \\ (\frac{\pi k}{2d})^2 + \mathcal{E}_j^{\pm}(\theta+1;d) & \text{(for even } k \text{ and } \theta \in [0,1]\text{)}, \\ (\frac{\pi k}{2d})^2 + \mathcal{E}_j^{\pm}(\theta-1;d) & \text{(for even } k \text{ and } \theta \in [1,2]\text{)}. \end{cases}$$

Then $\{\mu^{\pm}(j;k;\theta;d)\}_{j,k\in\mathbf{N}}$ is the set of all eigenvalues of $H^{\pm}_{\theta,d}$ counted with multiplicity. So it suffices to estimate $\mathcal{E}_j^{\pm}(\theta;d)$. Let us show the following.

For any $j \in \mathbf{N}$, we have

$$\mathcal{E}_j^{\pm}(\theta; d) = k_j(\theta) + O_j(d) \quad (d \to 0), \tag{2}$$

where each error term is uniform with respect to $\theta \in [0,2]$.

We rewrite $T_{\theta,d}^{+}$ and $T_{\theta,d}^{-}$ as follows.

$$\begin{aligned} T_{\theta,d}^{+} &= (1-d\gamma_+)^{-2}K_\theta + \frac{d\gamma_+}{(1+d\gamma_+)^2(1-d\gamma_+)^2}\gamma(s)^2 + \frac{1}{2}(1-d\gamma_+)^{-3}d\gamma_+'' \\ &= (1-d\gamma_+)^{-2}K_\theta + O(d) \quad \text{in} \quad L^2((0,\pi)). \end{aligned}$$

$$\begin{aligned} T_{\theta,d}^{-} =&(1+d\gamma_+)^{-2}K_\theta - \frac{d\gamma_+}{(1+d\gamma_+)^2(1-d\gamma_+)^2}\gamma(s)^2 - \frac{1}{2}(1-d\gamma_+)^{-3}d\gamma_+'' \\ &- \frac{5}{4}(1-d\gamma_+)^{-4}(\gamma_+')^2 \\ =&(1+d\gamma_+)^{-2}K_\theta + O(d) \quad \text{in} \quad L^2((0,\pi)). \end{aligned}$$

Applying the min-max principle, we get

$$\mathcal{E}_j^{\pm}(\theta; d) = (1 \mp d\gamma_+)^{-2}k_j(\theta) + O(d) \quad (d \to 0),$$

where each error term is uniform with respect to $\theta \in [0,2]$. Because $k_j(\cdot)$ is continuous on $[0,2]$, we get (2).

Next we show the following.

For any $j_0 \in \mathbf{N}$, there exists $\tilde{d} = \tilde{d}(j_0)$ such that for any $d \in (0, \tilde{d}]$, $k \geq 2$, $j \geq 1$, and $\theta \in [0,2]$,

$$\mu^{\pm}(j; k; \theta; d) \geq \mu^{\pm}(j_0; 1; \theta; d) + 1. \tag{3}$$

We fix any $j_0 \in \mathbf{N}$. We have for any $k \geq 2$, $j \geq 1$, and $\theta \in [0,2]$,

$$\begin{aligned} &\mu^{\pm}(j; k; \theta; d) - \mu^{\pm}(j_0; 1; \theta; d) \\ \geq&\frac{4\pi^2}{d^2} + \min_{\theta\in[0,2]} \mathcal{E}_1^{\pm}(\theta; d) - \frac{\pi^2}{d^2} - \max_{\theta\in[0,2]} \mathcal{E}_{j_0}^{\pm}(\theta; d) \\ \geq&\frac{3\pi^2}{d^2} + \min_{\theta\in[0,2]} k_1(\theta) + O(d) - \max_{\theta\in[0,2]} k_{j_0}(\theta) + O_{j_0}(d), \end{aligned} \tag{4}$$

where we used (2) in the third line. Therefore, we have (3).

(3) implies that for $d \in (0, \tilde{d}(j_0)]$ and $\theta \in [0,2]$, the j_0-th eigenvalue of $H^{\pm}_{\theta,d}$ is $\mu^{\pm}(j_0; 1; \theta; d)$. Using (1) and the min-max principle, we have for $d \in (0, \tilde{d}(j_0)]$ and $\theta \in [0,2]$,

$$\mu^{-}(j_0; 1; \theta; d) \leq \mathcal{E}_{j_0}(\theta; d) \leq \mu^{+}(j_0; 1; \theta; d).$$

This together with (2) implies the assertion of the theorem. □

Combining this theorem with the classical Borg's theorem (see [1]) about the inverse problem for Hill's equation, we have the following.

Corollary. *If γ is not identically 0, there exists some $j_0 \in \mathbf{N}$ and $C_{j_0} > 0$ such that*

$$\min_{\theta \in [0,2]} \mathcal{E}_{j_0+1}(\theta, d) - \max_{\theta \in [0,2]} \mathcal{E}_{j_0}(\theta, d) = C_{j_0} + O(d) \quad (d \to 0). \tag{5}$$

This corollary says that if γ is not identically 0, at least one band gap appears in the spectrum for sufficiently small d.

Remark. Using the scaling $\gamma \mapsto \epsilon\gamma$, where ϵ is a small parameter, we can specify the value of $j_0 \in \mathbf{N}$ such that (5) holds. For details see [5].

Example. We consider a quantum wire made of GaAs. We use the notations in the proof of the theorem. We assume that the width is 2 nm and the antiperiodicity length of the reference curve is 200π nm. For convenience, we take a unit length as $1 \equiv 200$ nm. Then we have $d = 5 \times 10^{-3}$. In this scale, we also assume that the curvature of the reference curve is given by

$$\gamma(s) = \cos s.$$

Using the method in the proof of [5, Lemma 3.6], we can see that γ satisfies $(A.1) - (A.3)$ with $d_0 > 5 \times 10^{-3}$.

We compute a lower bound for the width of the 1st band gap for this quantum wire. Now the reference operator takes the following form:

$$K_\theta = -\frac{d^2}{ds^2} - \frac{1}{8}\cos 2s - \frac{1}{8} \quad \text{in} \quad L^2((0,\pi)) \quad \text{with domain} \quad F_\theta.$$

Because the potential term is even, we can estimate the width of the first gap of this Mathieu operator by utilizing the analytic perturbation theorem for a simple eigenvalue (see [4]). Let $G_1(\beta)$ be the width of the first gap of the Mathieu operator

$$-\frac{d^2}{ds^2} + \beta\cos 2s \quad \text{in} \quad L^2(\mathbf{R}) \quad \text{with domain} \quad H^2(\mathbf{R}).$$

Then we have

$$G_1(\beta) \geq |\beta| \cdot \frac{256 - 64|\beta| - 64|\beta|^2 - |\beta|^3}{256 - 64|\beta| - 16|\beta|^2} =: m(\beta) \quad \text{for} \quad |\beta| < \beta_0,$$

where β_0 ($\fallingdotseq 1.547$) is the only one positive root of $t^3 + 64t^2 + 64t - 256 = 0$.

Applying this to K_θ, we have

$$k_2(1) - k_1(1) \geq m(1/8) \ (\fallingdotseq 0.1246).$$

Using the method in the proof of the theorem, we have

$$\max_{\theta\in[0,2]} \mathcal{E}_1^+(\theta;d) \leq (1-d)^{-2}k_1(1) + r_1, \quad \min_{\theta\in[0,2]} \mathcal{E}_2^-(\theta;d) \geq (1+d)^{-2}k_2(1) - r_2, \quad (6)$$

where

$$r_1 \equiv \frac{1}{2}(1-d)^{-3}d + \frac{d}{(1+d)^2(1-d)^2},$$

$$r_2 \equiv \frac{1}{2}(1-d)^{-3}d + \frac{d}{(1+d)^2(1-d)^2} + \frac{5}{4}(1-d)^{-4}d^2.$$

Because $k_1(1) < k_2(1) \leq 1$, we have

$$0 \leq (1-d)^{-2}k_1(1) - k_1(1) \leq (1-d)^{-2} - 1 =: r_3,$$

$$0 \leq k_2(1) - (1+d)^{-2}k_2(1) \leq 1 - (1+d)^{-2} =: r_4.$$

Using (4) and similar estimates to (6), we have for any $k \geq 2$, $j \geq 1$, and $\theta \in [0,2]$,

$$\mu^\pm(j;k;\theta;d) > \mu^\pm(2;1;\theta;d).$$

Combining these estimates, we get

$$\min_{\theta\in[0,2]} \mathcal{E}_2(\theta;d) - \max_{\theta\in[0,2]} \mathcal{E}_1(\theta;d) \geq m(1/8) - \sum_{j=1}^{4} r_j \ (\fallingdotseq 0.08951).$$

We convert this lower bound by using the following values.

$m^* \fallingdotseq 6.10 \times 10^{-32}[Kg]$ (the effective mass of a conduction electron in GaAs), $\hbar \fallingdotseq 1.05 \times 10^{-34}[\mathrm{J\cdot s}]$ (Planck's constant divided by 2π), $1[eV] \fallingdotseq 1.60 \times 10^{-19}[\mathrm{J}]$.

Then, the width of the first gap for this quantum wire is larger than

$$\frac{\hbar^2}{2m^*} \times (200 \times 10^{-9})^{-2} \times \left(m(1/8) - \sum_{j=1}^{4} r_j \right) \fallingdotseq 2.02 \times 10^{-25}\,[\mathrm{J}] \fallingdotseq 1.26\,[\mu eV].$$

References

1. G. Borg, *Eine Umkehrung der Sturm-Liouvilleschen Eigenwertaufgabe. Bestimmung der Differentialgleichung durch die Eigenwerte*, Acta Math. **78** (1946), 1–96.
2. P. Ciarlet, *"Mathematical elasticity, vol. I-III,"* North-Holland, Amsterdam, 1997.
3. P. Exner and P. Šeba, *Bound states in curved quantum waveguides*, J. Math. Phys. **30** (1989), 2574–2580.
4. T. Kato, *"Perturbation theory for linear operators,"* Springer-Verlag, Berlin, 1966.
5. K. Yoshitomi, *Band gap of the spectrum in periodically curved quantum waveguides*, J. Differential Equations **142** (1998), 123–166.

A list of other talks

Here we collect titles of the talks presented at the conference but not included in this volume.

Plenary Talks

J. Bellissard (Toulouse): *A kinetic theory for electronic transport: anomalous transport*
G. Casati (Milano): *Quantum nonergodic behaviour*
Y. Colin de Verdière (Grenoble): *Singular Bohr-Sommerfeld quantization condition*
R. Hempel (Braunschweig): *Spectral properties of strongly inhomogeneous periodic media*
E.H. Lieb (Princeton): *The ground state of low density Bose gas*
R. Seiler (Berlin): *Integrals of the discrete quantum pendulum*
B. Simon (Pasadena): *A new approach to inverse spectral theory*

Parallel Sessions Talks

T. Alferova (Gomel): *Relativistic two-particle one-dimensional scattering problem for superposition of delta potentials*
F. Brambila Paz (Mexico City): *Lax-Phillips scattering theory for short and long range perturbations*
J.F. Cariñena (Zaragoza): *The Riccati equation and exactly solvable Hamiltonians*
G. Chadzitaskos (Prague): *Coherent states over* C^M
M. Combescure (Orsay): *A rigorous proof of Gutzwiller's trace formula*
D. Damanik (Frankfurt a.M.): *Singular continuous spectrum for substitution Hamiltonians*
M. Damnjanović (Beograd): *Generalized Heisenberg-Dirac Hamiltonian*
S. De Bièvre (Lille): *On the semi-classical analysis of quantized "chaotic" maps on the torus*

O. Deryuzkova (Gomel): *Solving of the main equation of Poincaré invariant quantum mechanics for model potential*

J. Dittrich (Řež): *Massive scalar field in an oscillating bounded region*

Y. Fyodorov (Essen): *S-matrix statistics in quantum chaotic scattering: poles, phaseshifts and time delays*

S. Gnutzmann (Essen): *Quantum chaos and SU(3) dynamics*

G. Goldin (Rutgers): *Quantum kinematics of vortex filaments*

B. Grebert (Nantes): *KAM theorem for nonlinear Schrödinger operator*

M. Griesemer (Regensburg): *Instability for relativistic matter with self-generated magnetic fields*

M. Gruber (Berlin): *Spectral nature of the Schrödinger operator with periodic magnetic field (rational flux)*

J.-C. Guillot (Paris): *Inverse scattering at fixed energy for layered media*

Z. Haba (Wroclaw): *The classical limit of quantum dissipative systems*

A. Hassell (Canberra): *Scattering calculus and the N-body problem*

D. Herrmann (Nijmegen): *On spectral properties of Harper-like models*

H. Hirata (Tokyo): *An example of a second-kind phase transition for some discrete Schrödinger operators with magnetic potential*

P. Hislop (Lexington): *Correlated Wegner inequalities and localization for long-range and correlated potentials*

M. Horváth (Budapest): *Convergence and localization properties of the spectral expansion of Schrödinger and Dirac operators*

L.P. Horwitz (Tel Aviv): *Representation of quantum mechanical resonances in the framework of Lax-Phillips scattering theory*

Th. Hupfer (Erlangen): *On upper bounds on the density of states for continuum Schrödinger operators with Gaussian random potentials and magnetic fields*

V.I. Inozemtsev (Dubna): *Quantum Heisenberg chain with elliptic exchange*

V. Ivrii (Toronto): *Asymptotics of the ground state energy and ionization energy for atoms and molecules in strong magnetic field*

A. Kargol (Famagusta): *A semiclassical method for Coulomb scattering*

W. Karwowski (Wrocław): *Schrödinger operators perturbed by operators related to null sets*

M. Kiessling (Rutgers): *On eigenvalue density of random matrices*

W. Kirsch (Bochum): *On the spectral theory of random Schrödinger operators*

B. Kónya (Debrecen): *Charged particle in combined Coulomb plus homogeneous magnetic field*

E. Korotyaev (St. Petersburg): *Inverse problem for the Hill operator and estimates*

V. Koshmanenko (Kiev): *Construction and spectral properties of singularly perturbed operators*

I. Krasovsky (Dresden): *Explicit solution to some second-order differential and q-difference eigenvalue equations related to* sl_2 *and* $U_q(sl_2)$

J. Krause (Santiago do Chile): *Solvable models in non-Abelian quantum kinematics and dynamics*

U. Kuhl (Marburg): *A microwave realization of the Hofstadter butterfly*

P. Kurasov (Stockholm): *Few-body Krein's formula*

P. Leach (Durban): *A classicist's opinion on quantum chaos*

G. Lévai (Debrecen): *Exactly solvable quantum mechanical potential problem*

B. Meller (Satiago de Chile): *Resonances in a box*

V. Mikhailets (Białystok): *Schrödinger operators with local point interactions in dimension one*

I. Milošević (Beograd): *Symmetry classification of carbon nanotubes*

A. Moroz (Amsterdam): *"Gapology" for photonic crystals*

P. Müller (Göttingen): *Aspects of a localization proof for continuum Schrödinger operators with Gaussian random potentials*

O. Mustafa (Famagusta): *The shifted-l expansion technique to get eigenvalues of Schrödinger, Dirac, and Klein-Gordon wave equation*

M. Novitskii (Kharkov): *Nonexponential estimates of the angle between stable and unstable separatrices for Taylor-Chirikov-Green mapping*

P. O'Hara (Chicago): *Bell's inequality and the Pauli exclusion principle*

A. Onipko (Kiev): *Quantum conductance of molecular wires: Green function description of real systems*

Z. Papp (Debrecen): *The three-body Coulomb problem in three-potential formalism*

S. Pascazio (Bari): *Temporal evolutions in quantum mechanics and quantum Zeno effect*

I. Popov (St. Petersburg): *Solvable models for serially connected Aharonov-Bohm rings and localization effects*

G. Raikov (Sofia): *Asymptotic properties of the "magnetic" density of states*

L. Remezo (Liège): *Symmetries of a completely integrable Hamiltonian system*

D. Robert (Nantes): *Spectral shift function for the Dirac operator*

G. Rozenblioum (Göteborg): *Eigenvalue estimates for Schrödinger-like operators with magnetic field*

M.B. Ruskai (Lowell): *One-dimensional models for many-electron atoms in strong magnetic fields*

L. Sadun (Austin): *Generic behavior of topological quantum numbers*

O. Safronov (Stockholm): *The discrete spectrum in the gaps of the continuous one for sign indefinite perturbations with a large coupling constant*

Ch. Schulte (Clausthal): *Quantum-mechanical symmetries and self-adjoint extensions on the pointed plane*

P. Šeba (Řež): *Resonance trapping in a weakly open quantum dot*

I. Shereshevsky (Nizhni Novgorod): *Vortices in Ginzburg-Landau equation: numerical results and analytical problems*

G. Sobczyk (Puebla, Mexico): *Mathematics of quantum computing*

S.B. Sontz (Mexico City): *A reverse log-Sobolev inequality in the Segal-Bargmann space*

W. Spitzer (Princeton): *Hydrodynamics of quasi-free quantum systems*

A. Suzko (Dubna): *Bargmann-Darboux transformations for nonstationary Schrödinger equations*

S. Tcheremchantsev (Orléans): *Transport properties of Markovian Anderson model*

A. Tip (Amsterdam): *Quantization of conservative and lossy dielectrics*

N. Topor (Schenectady): *Perturbation theory for boundary S-matrix in 2D quantum field theory*

O. Vakhnenko (Kiev): *Bend-imitating approach in quantum wire-like nanostructures*

C. Villegas-Blas (Cuernavaca Morelos): *The Segal-Bargmann transform and canonical transformations*

J. Voigt (Dresden): *The non-autonomous Kato class*

K. Watanabe (Tokyo): *Some applications of the* H_{-2}*-construction*

J. Yanez (Santiago de Chile): *Variational principle for the chemical potential in the Thomas-Fermi theory*

N. Zettili (Dhahran): *Construction of an exactly solvable fermion model*

G.M. Zhislin (Nizhni Novgorod): *The present state of the study of the discrete spectrum of many-particle Hamiltonians with homogeneous magnetic field for particles with finite masses*

E. Zhizhina (Moscow): *A spectral analysis of stochastic disordered Ising model*

M. Znojil (Řež): *Asymptotically decreasing potentials with discrete spectra*

A list of participants

R. Adami, *Roma*	adami@marte.mat.uniroma1.it
T. Alferova, *Gomel*	alferova@gsu.unibel.by
J. E. Avron, *Technion*	avron@physics.technion.ac.il
J. Bellissard, *Toulouse*	jeanbel@irsamc2.ups-tlse.fr
F. Bentosela, *Marseille*	bento@cptsu5.univ-mrs.fr
M.S. Birman, *Sankt Petersburg*	tanya@petrov.stoic.spb.su
S. Böcker, *Bochum*	boecksbk@ruhr-uni-bochum.de
F. Brambila Paz, *Mexico*	fbp@hp.fciencias.unam.mx
J. Brasche, *Bonn*	brasche@riemann.iam.uni-bonn.de
Č. Burdík, *Prague*	burdik@dec1.fjfi.cvut.cz
J.F. Cariñena, *Zaragoza*	jfc@wigner.unizar.es
G. Casati, *Milano*	casati@fis.unico.it
G. Chadzitaskos, *Prague*	chadzita@br.fjfi.cvut.cz
T. Cheon de Caracorum, *Kochi*	cheon@mech.kochi-tech.ac.jp
Y. Colin De Verdiere, *Grenoble*	Yves.Colin-de-Verdiere@ujf-grenoble.fr
J.-M. Combes, *Marseille and Toulon*	combes@cptsu5.univ-mrs.fr
M. Combescure, *Orsay*	monique@qcd.th.u-psud.fr
D. Damanik, *Frankfurt/Main*	damanik@math.uni-frankfurt.de
M. Damnjanovic, *Beograd*	yqoq@afrodita.rcub.bg.ac.yu
S. De Bièvre, *Lille*	debievre@gat.univ-lille1.fr
M. Demuth, *Clausthal*	demuth@math.tu-clausthal.de
O. Deryuzhkova, *Gomel*	dom@gsu.unibel.by
J. Dittrich, *Řež near Prague*	dittrich@ujf.cas.cz
E. Elizalde, *Barcelona*	elizalde@ieec.fcr.es
L. Erdös, *Atlanta*	erdos@cims.nyu.edu
P. Exner, *Řež near Prague*	exner@ujf.cas.cz
P. Facchi, *Bari*	Paolo.Facchi@ba.infn.it
Y. Fyodorov, *Essen*	yan@next26.theo-phys.uni-essen.de
F. Germinet, *Paris*	germinet@math.jussieu.fr

F. Gesztesy, *Columbia*	fritz@math.missouri.edu
V.A. Geyler, *Saransk*	root@mathan.mordovia.su
E. Giere, *Bochum*	eckhard@mathphys.ruhr-uni-bochum.de
S. Gnutzmann, *Essen*	sven@next29.theo-phys.uni-essen.de
G.A. Goldin, *Rutgers*	gagoldin@joyce.rutgers.edu
G.M. Graf, *Zürich*	gmgraf@itp.phys.ethz.ch
B. Grebert, *Nantes*	grebert@math.univ-nantes.fr
M. Griesemer, *Regensburg*	marcel.griesemer@rphs1.physik.uni-regensbu
M. Gruber, *Berlin*	Gruber@mathematik.hu-berlin.de
R. Guardiola, *Valencia*	R.Guardiola@kate.fisato.uv.es
Ch.-A. Guerin, *Marseille*	caguerin@math.chalmers.se
J.C. Guillot, *Paris*	guillot@math.univ-paris13.fr
Z. Haba, *Wrocław*	zhab@proton.ift.uni.wroc.pl
E. Harrell, *Atlanta*	harrell@math.gatech.edu
A. Hassell, *Canberra*	hassell@math.mit.edu
M. Havlíček, *Prague*	havlicek@km1.fjfi.cvut.cz
R. Hempel, *Braunschweig*	hempel@badbit.math2.nat.tu-bs.de
D. Herrmann, *Nijmegen*	daniel@sci.kun.nl
H. Hirata, *Tokyo*	h-hirata@mm.sophia.ac.jp
P. Hislop, *Lexington*	hislop@ms.uky.edu
M. Horváth, *Budapest*	horvath@euromath.math.bme.hu
L.P. Horwitz, *Tel Aviv*	larry@post.tau.ac.il
Th. Hupfer, *Erlangen*	Hupfer@theorie1.physik.uni-erlangen.de
V.I. Inozemtsev, *Dubna*	inozv@thsun1.jinr.ru
V. Ivrii, *Toronto*	ivrii@math.toronto.edu
A. Jensen, *Aalborg*	matarne@math.auc.dk
A. Kargol, *Famagusta*	kargol@mozart.emu.edu.tr
W. Karwowski, *Wrocław*	wkar@proton.ift.uni.wroc.pl
M. Kiessling, *Rutgers*	miki@math.rutgers.edu
W. Kirsch, *Bochum*	werner@bigWerner.mathphys.ruhr-uni-boch
F. Kleespies, *Frankfurt/Main*	kleespie@math.uni-frankfurt.de
F. Klopp, *Paris*	klopp@math.univ-paris13.fr
B. Kónya, *Debrecen*	konyab@indigo.atomki.hu
E. Korotyaev, *Sankt Petersburg*	Evgeni.Korotyaev@pobox.spbu.ru
V. Koshmanenko, *Kiev*	kosh@imath.kiev.ua
I.V. Krasovsky, *Dresden*	ivk@idefix.mpipks-dresden.mpg.de
J. Krause, *Santiago de Chile*	jkrause@chopin.fis.puc.cl
D. Krejčiřík, *Řež near Prague*	david@ujf.cas.cz

P. Kuchment, *Wichita* kuchment@twsuvm.uc.twsu.edu
U. Kuhl, *Marburg/Lahn* kuhl@physik.uni-marburg.de
P. Kurasov, *Stockholm* pak@matematik.su.se
Kuroda Shigetoshi , *Tokyo* kuroda@math.gakushuin.ac.jp
P. Leach, *Durban* leach@scifs1.und.ac.za
D. Lenz, *Frankfurt/Main* dlenz@math.berkeley.edu
H. Leschke, *Erlangen* leschke@faupt100.physik.uni-erlangen.de
G. Lévai, *Debrecen* levai@cseles.atomki.hu
P. Lévay, *Budapest* levay@phy.bme.hu
E.H. Lieb, *Princeton* lieb@math.Princeton.edu
Ph.A. Martin, *Lausanne* phmartin@sdpdec9.epfl.ch
B. Meller, *Santiago de Chile* bmeller@bohr.fis.puc.cl
V. Mikhailets, *Kiev and Białystok* mikhai@math.uw.bialystok.pl
I. Milošević, *Beograd* ivag@afrodita.rcub.bg.ac.yu
A. Moroz, *Amsterodam* moroz@amoth2.amolf.nl
A.K. Motovilov, *Dubna* motovilv@thsun1.jinr.ru
E. Mourre, *Marseille* Eric.Mourre@cptsu5.univ-mrs.fr
P. Müller, *Göttingen* mueller@theorie.physik.uni-goettingen.de
O. Mustafa, *Famagusta* omustafa@mozart.emu.edu.tr
H. Najar, *Paris* najar@math.univ-paris13.fr
F. Nakano, *Sendai* nakano@math.tohoku.ac.jp
H. Neidhardt, *Potsdam* neidhard@rz.uni-potsdam.de
M.V. Novitskii, *Kharkov* novitskii@ilt.kharkov.ua
P. O'Hara, *Chicago* P-Ohara1@neiu.edu
A. Onipko, *Kiev* alex@ifm.liu.se
Z. Papp, *Debrecen* pz@indigo.atomki.hu
S. Pascazio, *Bari* Saverio.Pascazio@ba.infn.it
L.A. Pastur, *Paris* pastur@math.jussieu.fr
S. Perez-Oyarzun, *Santiago de Chile* sperez@chopin.fis.puc.cl
V.N. Pivovarchik, *Odessa* vnp@dtp.odessa.ua
M. Polakovič, *Bratislava* polakovic@kmat.elf.stuba.sk
I.Yu. Popov, *Sankt Petersbug* popov@cts.ifmo.ru
G. Raikov, *Sofia* gdraikov@omega.bg
L. Remezo, *Liège* ladislas.remezo@student.ulg.ac.be
Ch. Remling, *Osnabrück* cremling@mathematik.uni-osnabrueck.de
W. Renger, *Clausthal* mawr@math.tu-clausthal.de
Ch. Riebling, *Bochum* chris@mathphys.ruhr-uni-bochum.de
R. del Rio Castillo, *Mexico* delrio@servidor.unam.mx

D. Robert, *Nantes*	robert@lebesgue.math.sciences.univ-nantes.f
J. Ros, *Valencia*	Jose.Ros@uv.es
G. Rozenblioum, *Göteborg*	grigori@math.chalmers.se
M.B. Ruskai, *Lowell*	bruskai@esi.ac.at
L. Sadun, *Austin*	sadun@fireant.ma.utexas.edu
O. Safronov, *Stockholm*	safronov@math.kth.se
W. Scherer, *Clausthal*	ptws@pt.tu-clausthal.de
Ch. Schulte, *Clausthal*	ptcs@pt.tu-clausthal.de
P. Šeba, *Řež and Slemeno*	seba@kostelec.czcom.cz
R. Seiler, *Berlin*	seiler@math.tu-berlin.de
I. Shereshevsky, *Nizhni Novgorod*	ilya@ipm.sci-nnov.ru
B. Simon, *Pasadena*	bsimon@bigfoot.com
R. Sims, *Birmingham*	sims@vorteb.math.uab.edu
G. Sobczyk, *Puebla, Mexico*	sobczyk@mail.udlap.mx
S.B. Sontz, *Mexico*	sontz@xanum.uam.mx
W. Spitzer, *Princeton*	spitzer@phoenix.Princeton.edu
G. Stolz, *Birmingham*	stolz@vorteb.math.uab.edu
P. Šťovíček, *Prague*	stovicek@km1.fjfi.cvut.cz
T. Suslina, *Sankt Petersburg*	tanya@petrov.stoic.spb.su
A. Suzko, *Dubna*	suzko@thsun1.jinr.ru
M. Tater, *Řež near Prague*	tater@ujf.cas.cz
S. Tcheremchantsev, *Orléans et Bourges*	Serguei.Tcherem@labomath.univ-orleans.fr
A. Teta, *Roma*	teta@mercurio.mat.uniroma1.it
A. Tip, *Amsterdam*	tip@amolf.nl
J. Tolar, *Prague*	tolar@br.fjfi.cvut.cz
N. Topor, *Schenectady*	toporn@union.edu
O. Vakhnenko, *Kiev*	vakhnenko@gluk.apc.org
I. Veselić, *Bochum*	veselibp@mailhost.rz.ruhr-uni-bochum.de
C. Villegas-Blas, *Cuernavaca Morelos*	villegas@matcuer.unam.mx
J. Voigt, *Dresden*	voigt@math.tu-dresden.de
S. Warzel, *Erlangen*	warzel@theorie1.physik.uni-erlangen.de
K. Watanabe, *Tokyo*	watanabe@math.gakushuin.ac.jp
T. Weidl, *Stockholm*	weidl@math.kth.se
D.R. Yafaev, *Rennes*	Dimitri.Yafaev@univ-rennes1.fr
J. Yanez, *Santiago de Chile*	jyanez@chopin.fis.puc.cl
J. Yngvason, *Vienna*	yngvason@thor.thp.univie.ac.at
K. Yoshitomi, *Fukuoka*	yositomi@math.kyushu-u.ac.jp
V.A. Zagrebnov, *Marseille*	Valentin.Zagrebnov@cptsu5.univ-mrs.fr

N. Zettili, *Dhahran* nzettili@kfupm.edu.sa
G.M. Zhislin, *Nizhni Novgorod* greg@nirfi.sci-nnov.ru
E. Zhizhina, *Moscow* ejj@ippi.ras.ru
M. Znojil, *Řež near Prague* znojil@ujf.cas.cz
J.L. Zuleta, *Genève* zuleta@kalymnos.unige.ch

92 A. Ya. Shklyar
Complete Second Order Linear Differential Equations in Hilbert Spaces
1997. ISBN 3-7643-5377-5

93 Y. Egorov / B.-W. Schulze
Pseudo-Differential Operators, Singularities, Applications
1997. ISBN 3-7643-5484-4

94 M. I. Kadets / V. M. Kadets
Series in Banach Spaces. Conditional and Unconditional Convergence.
1997. ISBN 3-7643-5401-1

95 H. Dym / V. Katsnelson / B. Fritzsche / B. Kirstein (Ed.)
Topics in Interpolation Theory
1997. ISBN 3-7643-5723-1

96 D. Alpay / A. Dijksma / J. Rovnyak / H. de Snoo
Schur Functions, Operator Colligations, and Reproducing Kernel Pontryagin Spaces
1997. ISBN 3-7643-5763-0

97 M. L. Gorbachuk / V. I. Gorbachuk
M.G. Krein's Lectures on Entire Operators
1997. ISBN 3-7643-5704-5

98 I. Gohberg / Yu. Lyubich (Ed.)
New Results in Operator Theory and its Applications
The Israel M. Glazman Memorial Volume
1997. ISBN 3-7643-5775-4

99 T. Ayerbe Toledano / T. Dominguez Benavides / G. López Acedo
Measures of Noncompactness in Metric Fixed Point Theory
1997. ISBN 3-7643-5794-0

100 C. Foias / A. E. Frazho / I. Gohberg / M.A. Kaashoek
Metric Constrained Interpolation, Commutant Lifting and System
1998. ISBN 3-7643-5889-0

101 S. D. Eidelman / N. V. Zhitarashu
Parabolic Boundary Value Problems
1998. ISBN 3-7643-2972-6

102 I. Gohberg / R. Mennicken / C. Tretter (Ed.)
Differential and Integral Operators. International Workshop on Operator Theory and Applications, IWOTA 95, in Regensburg, July 31 – August 4, 1995.
1998. ISBN 3-7643-5890-4

103 I. Gohberg / R. Mennicken / C. Tretter (Ed.)
Recent Progress in Operator Theory. International Workshop on Operator Theory and Applications, IWOTA 95, in Regensburg, July 31 – August 4, 1995.
1998. ISBN 3-7643-5891-2

104 H. Bercovici / C. Foias (Ed.)
Nonselfadjoint Operator Algebras, Operator Theory, and Related Topics. The Carl M. Pearcy Anniversary Volume.
1998. ISBN 3-7643-5954-4

105 S. Bagdasarov
Chebyshev Splines and Kolmogorov Inequalities
1998. ISBN 3-7643-5984-6

106 A. Dijksma / I. Gohberg / M. A. Kaashoek / R. Mennicken (Ed.)
Contributions to Operator Theory in Spaces with an Indefinite Metric
The Heinz Langer Anniversary Volume
1998. ISBN 3-7643-6003-8

MIX
Papier aus verantwortungsvollen Quellen
Paper from responsible sources
FSC® C105338

If you have any concerns about our products,
you can contact us on
ProductSafety@springernature.com

In case Publisher is established outside the EU,
the EU authorized representative is:
Springer Nature Customer Service Center GmbH
Europaplatz 3, 69115 Heidelberg, Germany

Printed by Libri Plureos GmbH
in Hamburg, Germany